7th International LISA Symposium 2008

Journal of Physics: Conference Series Volume 154

Barcelona, Spain
16-20 June 2008

Editors:

Alberto Lobo

Carlos F Sopuerta

ISBN: 978-1-61738-379-3

Printed from e-media with permission by:

Curran Associates, Inc.
57 Morehouse Lane
Red Hook, NY 12571
www.proceedings.com

Some format issues inherent in the e-media version may also appear in this print version.

Printed by Curran Associates, Inc. (2010)

Institute of Physics
Dirac House, Temple Back
Bristol BS1 6BE UK

Phone: 44 1 17 929 7481
Fax: 44 1 17 920 0979

techtracking@iop.org

TABLE OF CONTENTS

PREFACE

In June 2006 the LISA International Science Team (LIST) accepted the bid presented by the Institut d'Estudis Espacials de Catalunya (IEEC) to host the 7th International LISA Symposium. This was during its 11th meeting at the University of Maryland, just before the 6th edition of the Symposium started in NASA's Goddard Space Flight Center.

The 7th International LISA Symposium took place at the city of Barcelona, Spain, from 16–20 June 2008, in the premises of CosmoCaixa, a modern Science Museum located in the hills near Tibidabo. Almost 240 delegates registered for the event, a record breaking figure compared to previous editions of the Symposium. Many of the most renowned world experts in LISA, Gravitational Wave Science, and Astronomy, as well as Engineers, attended LISA 7 and produced state-of-the-art presentations, while everybody benefited from the opportunity to have live discussions during the week in a friendly environment.

The programme included 31 invited plenary lectures in the mornings, and 8 parallel sessions in the afternoons. These were classified into 7 major areas of research: LISA Technology, LISA PathFinder, LISA PathFinder Data Analysis, LISA Data Analysis, Gravitational Wave sources, Cosmology and Fundamental Physics with LISA and Other Gravitational Wave Detectors. 138 abstracts for communications were received, of which a selection was made by the session convenors which would fit time constraints. Up to 63 posters completed the scientific programme. More details on the programme, including some of the talks, can be found at the Symposium website: http://www.ice.cat/research/LISA_Symposium. There was however a remarkable add-on: Professor Clifford Will delivered a startling presentation to the general public, who completely filled the *Auditori*—the main Conference Room, 320 seats—and were invited to ask questions to the speaker who had boldly guided them through the daunting world of *Black Holes, Waves of Gravity, and other Warped Ideas of Dr Einstein.*

The Proceedings of the 7th International LISA Symposium are jointly published by *Classical and Quantum Gravity* (CQG) and *Journal of Physics: Conference Series* (JPCS). This formula has a precedent in the last *Amaldi Conference* (Sydney 2007), and was motivated by the impossibility to fit all communications into a single CQG volume. Plenary speakers were invited to submit their contributions to CQG, and so were a number of parallel session authors chosen by the session convenors and the Science Organising Committee (SOC). Authors of the other parallel session presentations and posters were invited to submit to JPCS. All papers have been peer reviwed prior to being accepted for publication in either journal, and the whole set is well representative of the talks we heard during the Symposium.

Thanks are accordingly due to all authors for their collaborative attitude and, more generally, to all delegates who came to Barcelona and made of the Symposium a first class scientific event. The LISA community has been steadily growing since the Symposium took off in Chilton, near Oxford (UK) back in 1996. The support of such community strongly endorses a complex mission Project, whose short term future requires such support for a much longer term new era of Gravitational Wave Astronomy and Fundamental Physics. In this sense, the number of attendees and their active interest in the LISA mission sparks optimism.

The 7th International LISA Symposium sponsors are also sincerely acknowldged. They are: the Albert Einstein Institute (Hannover), the Spanish Ministry of Science and Innovation, the Generalitat de Catalunya (AGAUR), the Barcelona Institute of High Energy Physics (IFAE), the University of Barcelona (UB), the Polytechnique University of Catalunya (UPC), the Spanish Society of General Relativity and Gravitation (SEGRE), CosmoCaixa, NASA and the European Space Agency (ESA). The latter provided the LISA PathFinder model, a 1:4 scale model whose primer display we enjoyed during the Symposium.

Finally, the Local Organising Committee and the IEEC staff have given their enthusiastic support to the organisation in every detail, and have efficiently worked for months to make the Symposium happen. Many thanks to all of them, and congratulations.

Alberto Lobo and Carlos F Sopuerta
Institut de Ciències de l'Espai (CSIC-IEEC)
Guest Editors

This is a co-publication with *Classical and Quantum Gravity*. The bulk of the papers, after peer review, are published in *Journal of Physics: Conference Series*. However a selection of papers, are published separately in a special issue of *Classical and Quantum Gravity*.

7th International LISA Symposium IOP Publishing
Journal of Physics: Conference Series **154** (2009) 012001 doi:10.1088/1742-6596/154/1/012001

Magnetic polarisation effects of temperature sensors and heaters in *LISA* Pathfinder

J Sanjuán[1,2], A Lobo[1,2], M Nofrarias[3], J Ramos-Castro[4], N Mateos[1,2] and X Xirgu[1,2]

[1] Institut de Ciències de l'Espai, *CSIC*, Facultat de Ciències, Torre C5 parell, 08193 Bellaterra, Spain
[2] Institut d'Estudis Espacials de Catalunya (*IEEC*), Edifici Nexus, Gran Capità 2-4, 08034 Barcelona, Spain
[3]Max-Planck-Institut für Gravitationsphysik (Albert-Einstein-Institut), Callinstrasse 38, D-30167 Hannover, Germany
[4]Departament d'Enginyeria Electrònica, *UPC*, Campus Nord, Edifici C4, Jordi Girona 1-3, 08034 Barcelona, Spain

E-mail: sanjuan@ieec.fcr.es

Abstract. Temperature sensors and heaters belong in the diagnostics subsystem of the *LISA* Technology Package (*LTP*) on board *LISA* Pathfinder, the technology demonstrator for *LISA*. A number of these diagnostics items are placed at short distances from the *LTP* proof masses, and are negative temperature coefficient (*NTC*) thermistors. By design, these devices have tiny amounts of ferromagnetic materials which therefore constitute a potential source of disturbance to the performance of the *LTP*. We present a detailed magnetic characterisation of the *NTC*'s, and use the data to evaluate their impact on the acceleration noise budget of the *LTP*. The effect is seen to be small, and can be further reduced if the *NTC*'s are submitted to a de-magnetisation process before they are attached. Re-magnetisation is unlikely, as rather strong fields (mili-Tesla) are required to re-magnetise the *NTC*'s

1. Introduction

LISA Pathfinder [1] is a technological mission intended to demonstrate that two proof masses can be put into free-fall to a certain level of accuracy. This idea is reflected in the differential acceleration noise requirement for the *LTP*

$$S_{\delta a}^{1/2}(\omega) \leq 3\times 10^{-14}\left[1+\left(\frac{\omega/2\pi}{3\,\mathrm{mHz}}\right)^2\right]\,\mathrm{m\,s^{-2}\,Hz^{-1/2}} \tag{1}$$

in the frequency band between 1 mHz to 30 mHz.

This requirement implies stringent limitations on *internal* environment fluctuations. Specifically, thermal and magnetic fluctuations must be very low not to exert any noise force on the test masses (*TM*'s) [2].

The diagnostics subsystem [3] responsible of measuring such low temperature fluctuations includes 8 thermistors[1] (4 as temperature sensors and 4 as heaters) surrounding each of the

[1] BetaTherm thermistors.

TM's. Thermistors are manufactured mixing and synthesising oxides doped with ferromagnetic materials [4] which make them potentially dangerous as they are placed close[2] to the *TM*'s —see figure 1— since magnetic field and magnetic field gradient in the *TM* translate into a force due to its non-zero susceptibility and remanent magnetic moment —see section 2.

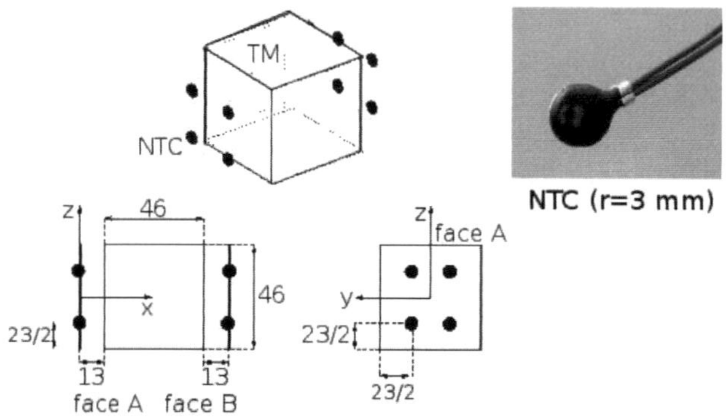

Figure 1. Situation of the thermistors surrounding the *TM*. Units in mm.

The *LTP* noise acceleration budget assigned to magnetic effects in the *TM* is [5]

$$S^{1/2}_{\delta a,\,\mathrm{magn}}(\omega) \leq 12\,\mathrm{fm\,s^{-2}\,Hz^{-1/2}} \tag{2}$$

in the measurement bandwidth (*MBW*). For this reason magnetic field, magnetic field gradient and their fluctuations must be well kept under a certain limits. The presence of the *NTC*'s might add noticeable noise to the differential acceleration. This paper describes the steps performed in order to quantify this effect. It is organised as follows: in section 2 the expression of the acceleration noise as a function of the magnetic field and magnetic field gradient is given together with the nominal properties of the *TM* and background magnetic conditions. Section 3 shows the the measured hysteresis curve of the thermistors. Section 4 describes the numerical calculations performed to obtain the magnetic field and magnetic field gradient in the *TM* considering the 8 *NTC*'s surrounding it. Section 5 shows the measurement of the first magnetisation curve (*FMC*) of the thermistors. Section 6 details the excess acceleration noise caused by the presence of the *NTC*'s surrounding the *TM*. Finally, in section 6 the conclusions are given.

2. Force fluctuations in the *TM* due to magnetic issues

The force fluctuation due to magnetic effects is [6]:

$$\begin{aligned} S_{\delta F_x}(\omega) &= V^2|\langle\mathbf{M}\rangle|^2 S_{\nabla B_x}(\omega) + \left(\frac{\chi V}{\mu_0}\right)^2 |\langle\nabla B_x\rangle|^2 S_{\mathbf{B}}(\omega) + \\ &+ \left(\frac{\chi V}{\mu_0}\right)^2 |\langle\mathbf{B}\rangle|^2 S_{\nabla B_x}(\omega) \end{aligned} \tag{3}$$

where V is the *TM* volume, $\mathbf{M}$ is the remanent magnetisation of the *TM*, χ is the magnetic susceptibility of the *TM*, $\langle\mathbf{B}\rangle$ is the magnetic field average over the *TM* volume and $\langle\nabla B_x\rangle$ the same for the magnetic field gradient. $S_{\mathbf{B}}$ and $S_{\nabla B_x}$ are the magnetic field and magnetic field gradient fluctuations over the *TM*[3]. The nominal values for these parameters are [1]:

[2] The distance from the sensors to the closer face of the *TM* is 13 mm.
[3] Assuming the fluctuations are homogeneous in all the *TM* volume.

- *TM* magnetic properties: $|\chi| = 10^{-5}$ and $\mathbf{M}=10^{-4}\,\mathrm{A\,m^{-1}}$,
- Background *dc* values: $|\mathbf{B}_{\rm bg}| \leq 10\ \mu\mathrm{T}$ and $|\boldsymbol{\nabla} B_{{\rm bg},x}| \leq 5\sqrt{3}\,\mu\mathrm{T\,m^{-1}}$,
- Magnetic fluctuation values: $S^{1/2}_{\mathbf{B}}(\omega) \leq 650$ nT $\mathrm{Hz}^{-1/2}$ and $S^{1/2}_{\nabla B_x}(\omega) \leq 250\sqrt{3}$ nT $\mathrm{m^{-1}\,Hz^{-1/2}}$ in the frequency band from 1 mHz to 30 mHz[4]

By using these values and equation (3) the noise acceleration due to the magnetic effects in the absence of thermistors is calculated and shown in table 1.

Table 1. Foreseen differential acceleration noise in the *LTP* due to magnetic effects in the *TM* (in the absence of thermistors).

Term	$S_{\Delta a_x}$ [$\mathrm{fm\,s^{-2}\,Hz^{-1/2}}$]
$V\lvert\langle\mathbf{M}\rangle\rvert S^{1/2}_{\boldsymbol{\nabla} B_x}(\omega)$	2.15
$(\chi V/\mu_0)\lvert\langle\boldsymbol{\nabla} B_x\rangle\rvert S^{1/2}_{\mathbf{B}}(\omega)$	2.22
$(\chi V/\mu_0)\lvert\langle\mathbf{B}\rangle\rvert S^{1/2}_{\nabla B_x}(\omega)$	1.70
$S^{1/2}_{\rm total\ mag}(\omega)$	4.46

The presence of the thermistors implies that

$$\mathbf{B} = \mathbf{B}_{\rm background} + \mathbf{B}_{NTC} \tag{4}$$

and, thus, the effect of $\mathbf{B}_{NTC}$ (and its associated gradient) in the *TM* acceleration needs to be quantified.

3. *NTC*'s magnetic characterisation: the hysteresis curve

The hysteresis curve for a set of thermistors is shown in figure 2. The measurement was performed by using a SQUID (Quantum Design MPMS XL Squid) at the Serveis Científico-Tècnics of the Universitat de Barcelona [7, 8].

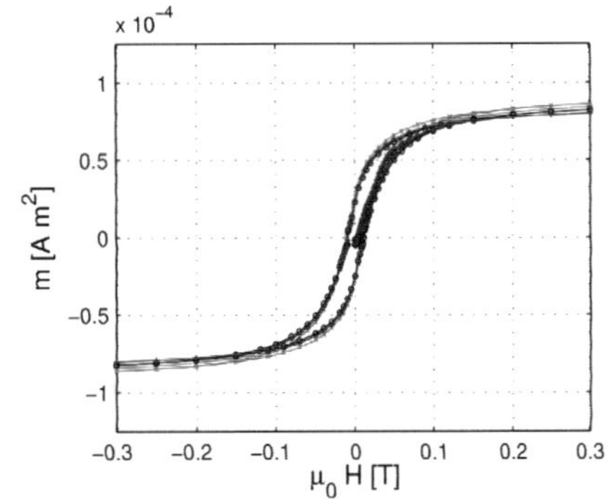

Figure 2. Hysteresis curve at 300 K for the set (four samples) of 10 kΩ *NTC* thermistors of BetaTherm.

The relevant information that can be extracted from figure 2 is shown in table 2.

[4] These values are superseded.

Table 2. Magnetic properties of the BetaTherm *NTC* thermistors. (i) $|\mathbf{m}_{\rm r}|$ is the remanent magnetic moment after saturation and at zero magnetic field, (ii) $|\mathbf{m}_{\rm sat}|$ is the magnetic moment at the saturating magnetic field and (iii) $\mu_0|\mathbf{H}_{\rm coer}|$ is the coercive field.

$\lvert\mathbf{m}_{\rm r}\rvert$	$\lvert\mathbf{m}_{\rm sat}\rvert$	$\mu_0\lvert\mathbf{H}_{\rm coer}\rvert$
$24\pm2\,\mu$A m^2	$83\pm2.5\,\mu$A m^2	10 mT

4. Numerical calculations

Numerical calculations to obtain the average magnetic field, $\langle\mathbf{B}\rangle$, and the average magnetic field gradient, $\langle\boldsymbol{\nabla} B_x\rangle$, considering the 8 *NTC*'s surrounding the *TM* have been performed assuming: (i) *NTC*'s behave like magnetic dipoles and (ii) the remanent magnetic moment is the one measured after *NTC* magnetic saturation[5].

Calculations considering different orientations of the 8 magnetic moments of the *NTC*'s allow us to know the worst possible combinations. A simple scheme of the magnetic moment orientations (arrows in the figure) surrounding the *TM* and the numerical values obtained are given in figure 3 and table 3.

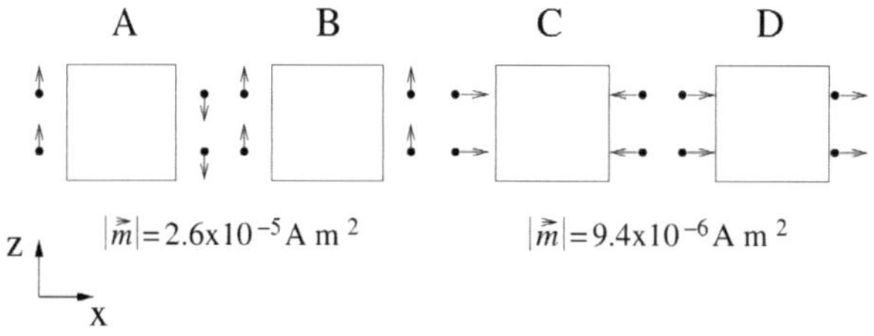

Figure 3. Configuration analysed for the magnetic field and magnetic field gradient calculations. This four configurations cover the worst cases involved in the problem. Intermediate configurations lead to smaller values of the magnetic field and magnetic field gradient in the *TM*. Arrows stand for the *NTC*'s magnetic moment orientation [6].

Table 3. Average value modulus of the magnetic field and magnetic field gradient caused by the eight *NTC*'s in the *TM* using a remanent magnetic moment of 24 μA m^{-2}.

Conf.	$\lvert\langle\mathbf{B}\rangle\rvert$ [μT]	$\lvert\langle\boldsymbol{\nabla} B_x\rangle\rvert$ [μT m^{-1}]
A	0	15.5
B	0.25	0
C	0	11.3
D	0.18	0

Graphical representations of the magnetic field and magnetic field gradient in the *TM* volume are shown in figure 4.

[5] This is a very worst case scenario since exposing the thermistors to their saturating magnetic field ($\approx$0.3 T) is unlikely.

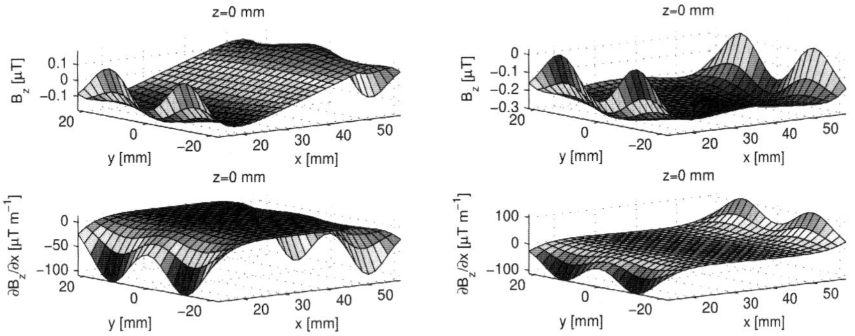

Figure 4. Representation of the z-component of the magnetic field and magnetic field gradient in the equatorial plane of *TM*. Left: Configuration A (see figure 3). Right: Configuration B (see figure 3).

5. *NTC*'s first magnetisation curve (*FMC*)

The numbers obtained in section 4 are very pessimistic since the magnetic field and magnetic field gradient have been calculated using the remanent magnetic moment, $|\mathbf{m}_\mathrm{r}|$, after saturating the *NTC*'s. To obtain more realistic estimations, the *FMC* was measured. The *FMC* consists in measuring the magnetic moment vs. applied field relationship after the de-magnetisation of the item [9]. The *FMC* is shown in figure 5 [7, 8].

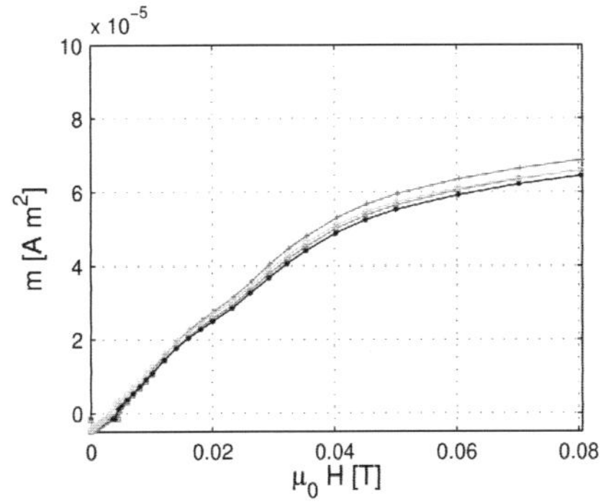

Figure 5. Measured *FMC* for a set of BetaTherm thermistors.

The remanent magnetic moment measured after the de-magnetisation is:

$$|\mathbf{m}_\mathrm{demag}| = 1.4 \pm 0.2 \; [\mu\mathrm{A\,m}^2] \tag{5}$$

which is about one order of magnitude less than the one measured after saturating the thermistor.
The *FMC* for the *NTC*'s can be linearised near the de-magnetisation zone as —see figure 5:

$$|\mathbf{m}_\mathrm{FMC,\,BetaTherm}| \simeq 1.45 \times 10^{-3}\,\mu_0\,|\mathbf{H}_\mathrm{FMC}| \tag{6}$$

This relationship will be useful for calculations of the maximum magnetic field the *NTC*'s can be exposed in order not to increase the acceleration noise —see section 6.

6. Excess *TM* noise calculations

The acceleration noise for three different scenarios will now be calculated and compared. The analysed scenarios are:

- absence of thermistors
- presence of thermistors with the maximum possible remanent magnetic moment, i.e., after saturation, $|\mathbf{m}_\mathrm{r}| = 24\,\mu\mathrm{A\,m^{-2}}$,
- presence of thermistors after de-magnetisation, $|\mathbf{m}_\mathrm{demag}| = 1.4\,\mu\mathrm{A\,m^{-2}}$.

The numbers obtained are summarised in table 4.

Table 4. Acceleration noise for three different scenarios. Units: $\mathrm{fm\,s^{-2}\,Hz^{-1/2}}$. Δ stands for the increase of the noise with respect to the "NO NTC's" noise numbers. The frequency band is between 1 mHz and 30 mHz.

Term	No NTC's	$\lvert\mathbf{m}_\mathrm{r}\rvert=$ 26 $\mu\mathrm{A\,m^2}$	$\lvert\mathbf{m}_\mathrm{demag}\rvert=$ 1.4 $\mu\mathrm{A\,m^2}$
$V\lvert\langle\mathbf{M}\rangle\rvert S^{1/2}_{\boldsymbol{\nabla}B_x}(\omega)$	2.10	2.10	2.10
$(\chi V/\mu_0)\lvert\langle\boldsymbol{\nabla}B_x\rangle\rvert S^{1/2}_{\mathbf{B}}(\omega)$	2.22	6.12	2.44
$(\chi V/\mu_0)\lvert\langle\mathbf{B}\rangle\rvert S^{1/2}_{\nabla B_x}(\omega)$	1.70	1.70	1.70
$S^{1/2}_{\text{total mag}}(\omega)$	4.45	7.29	4.54
Δ	—	64%	2%

If a desired Δ is chosen then the maximum magnetic field that the thermistors can tolerate can be calculated using equation 6. For instance, if $\Delta < 10\%$ is required, the thermistors should be de-magnetised and not exposed to magnetic fields higher than 5 mT.

7. Conclusions

The investigations carried out with respect to the magnetic characteristics of the *NTC*'s and their impact on the performance of the *LTP* experiment are the following:

- *NTC*'s show ferromagnetic behaviour.
- The magnetic properties of the *NTC*'s can degrade the performance of the *LTP*, increasing the magnetic noise by $\sim 65\,\%$ relative to the background in the very worst possible situation. However, even in such extreme conditions the budgeted magnetic noise ($12\,\mathrm{fm\,s^{-2}}/\sqrt{\mathrm{Hz}}$) is not reached.
- de-magnetisation of the *NTC*'s produces very good results: the magnetic noise can be reduced by about an order of magnitude, which makes it mostly negligible.

Acknowledgments

Support for this work came from Project ESP2004-01647 of Plan Nacional del Espacio of the Spanish Ministry of Education and Science (MEC). JS acknowledges a grant from MEC.

References

[1] Vitale S 2005 Science Requirements and Top-level Architecture Definition for the LISA Technology Package (*LTP*) on Board *LISA* Pathfinder (SMART-2), report no. LTPA-UTN-ScRD-Iss003-Rev1
[2] Araújo H *et al* 2007 *Journal of Physics: Conference Series* **66** 012003

[3] Sanjuán J, Lobo A, Nofrarias M, Ramos-Castro J and Riu P 2007 *Rev Sci Instr* **78** 104904
[4] Pallàs R and Webster JG 2000 *Sensors and signal conditioning* (New York)
[5] Vitale S *et al* 2005 Science requirement and top-level architecture definition for the *LISA* Technology Package (*LTP*) on board *LISA* Pathfinder LTPA-UTN-ScRD Issue 3.1
[6] Sanjuán J *et al* 2008 *Rev Sci Instr* **79** 084503
[7] Nofrarias M and Sanjuán J 2007 Magnetic characterisation of thermal diagnostic items S2-IEC-TN-3036
[8] Sanjuán J and Nofrarias M 2008 Characterisation and impact of NTC magnetic properties on the LTP noise budget S2-IEC-TN-3038
[9] Sears FW and Zemansky MW 1991 *College physics* (MA)

Gravitational Science with LISA Pathfinder

C Trenkel and S Kemble

Astrium Ltd, Gunnels Wood Road, Stevenage SG1 2AS, UK

E-mail: christian.trenkel@astrium.eads.net

Abstract. We investigate the potential of conducting interesting gravitational science experiments with LISA Pathfinder, by executing well defined de-orbiting manoeuvres following the nominal mission. Preliminary work suggests that the residual control authority of the micropropulsion system is sufficient to follow trajectories that cross the region surrounding the Sun-Earth saddle point, and also include one or multiple Earth flybys. Crossing the saddle point region may allow tests of Modified Newtonian Dynamics (MOND), while the flybys may potentially shed some light on the so-called flyby anomaly. We present some sample trajectories and discuss the limitations of the current model. Finally, we discuss the work required to take these ideas from the proof of principle presented here, to a concrete proposal for an extended mission.

1. Introduction

The LISA Pathfinder (LPF) spacecraft, currently being built as a technology demonstrator for the future gravitational wave observatory LISA, has the potential to be in itself a very valuable tool to conduct gravitational experiments in the Solar System.

Some of the main properties of relevance for the discussions that follow are:

- DC free-fall quality of $\approx 10^{-9}\text{ms}^{-2}$ (limited by spacecraft self-gravity)
- Free-fall quality of $\approx 10^{-13}\text{ms}^{-2}/\sqrt{\text{Hz}}$ around 1mHz
- Gravity gradiometer sensitivity $\approx 10^{-14}\text{s}^{-2}/\sqrt{\text{Hz}}$ around 1mHz

A comprehensive discussion of what LPF has to offer as a laboratory for gravitational experiments is provided in [1]. In that reference, several possible gravitational experiments, including tests of MOdified Newtonian Dynamics (MOND), are mentioned. It has been pointed out recently that MOND could be tested around gravitational saddle points within the Solar System [2]. Nominally, LPF will be placed in a Lissajous orbit around L1, and possible effects due to a theory such as MOND at that location are estimated to be very small [2].

Here, we look at the potential to execute de-orbiting manoeuvres from the nominal LPF orbit around L1, such that the spacecraft is sent back closer to Earth. The possible trajectories that the spacecraft would then follow include some that cross the region around the Sun-Earth saddle point, and all of them include a single or multiple Earth flybys.

Trajectories crossing the region around the saddle point could be used to test MOND, while the Earth flybys could potentially be used to shed some light on the so-called flyby anomaly [3].

2. Targets for LISA Pathfinder Trajectories

2.1. MOdified Newtonian Dynamics

For the Sun-Earth system, and ignoring all other gravitational perturbations, the saddle point is located around 2.59×10^8m from Earth, along the Earth-Sun line. According to[2], anomalous gravitational gradients in excess of $10^{-13}s^{-2}$ will be experienced within a MOND "bubble" of semi-major axis 766km (perpendicular to the Earth-Sun line), and semi-minor axis 383km (along the Earth-Sun line). Such gradients, if applied at appropriate frequencies, are certainly within LPFs sensitivity.

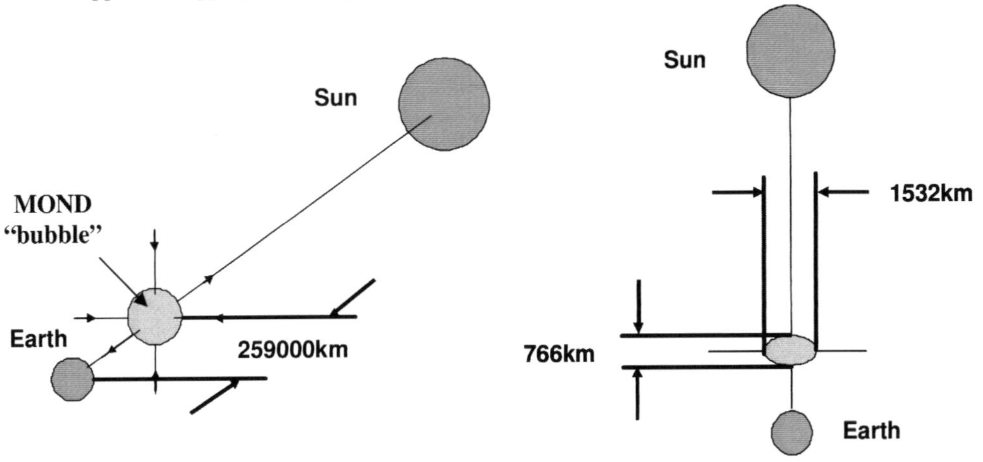

Figure 1 Schematic location and size of LPF target region (MOND "bubble") for MOND tests. Relative dimensions are not to scale.

In practice, other Solar System bodies modify both the location and the size of this unperturbed MOND bubble. The dominant effect, due to the Moon, can be estimated as follows: The background terrestrial gradients are of the order of $4 \times 10^{-11}s^{-2}$. During a solar eclipse, the Moon would get to within approximately 1.2×10^8m of the unperturbed saddle point. The gravitational acceleration due to the Moon at that point would then be of the order of $3.4 \times 10^{-4}ms^{-2}$, and the true gravitational saddle point (now of the Earth – Moon – Sun system) would shift by approximately $3.4 \times 10^{-4} / 4 \times 10^{-11} \approx 8.5 \times 10^6$m, or 8500km, towards the Earth. This is clearly a significant effect.

The effect on the bubble size can be estimated from the lunar gravity gradients, which for the example above are of the order of $6 \times 10^{-12}s^{-2}$. This amounts to about 15% of the Earth gradient, and this is also the magnitude of the bubble dimension reduction that can be expected.

For Jupiter, the equivalent saddle point position shift is of the order of 10km, and the impact on the bubble size is completely negligible.

For the proof of principle sought in this very preliminary work, the shift of the target region as a function of Moon position has not been considered.

2.2. Flyby Anomaly

The so-called "Flyby Anomaly" has now been observed in six Earth flybys, and manifests itself in an apparent change in orbital energy of the satellites before and after close Earth flybys [3]. These satellites were on unbound orbits, with v_∞ ranging between 4km/s and 16km/s. The altitudes at closest approach varied from 300km to 2500km.

The anomalous energy change, expressed as a change in v_∞ before and after the encounter, is present in both Doppler tracking and ranging data, and is a large effect with signal-to-noise-ratios of up to 10^3. An empirical expression has been derived for this change as a function of incoming and outgoing geocentric latitudes. To date, there is no known explanation for this effect.

Any de-orbiting manoeuvres as will be described below will inevitably result in LPF executing a single or multiple Earth flybys. Careful tracking of the spacecraft might help to shed some light on the flyby anomaly, or at least could provide some complementary data.

LPF Earth flybys would be characterised by the following properties:

- Flybys would not have $v_\infty > 0$, but $v_\infty <= 0$ (weakly bound orbits)
- For low altitude, "fast" flybys (1000s – 10000s), free-fall quality around single Test Mass could be significantly better than 10^{-9}ms^{-2} during that time
- Low (≈1000km) and high (>10000km) altitude flybys could be compared
- Geocentric latitudes could be chosen to maximise the effect

In this preliminary work, the main emphasis has been placed on crossing the MOND bubble, and Earth flybys are effectively a "spin-off".

3. Assumptions regarding LISA Pathfinder

3.1. Modifications to nominal LPF Hardware

None. Although modifications to the micropropulsion system might in principle be of advantage, LPF is deep into its implementation phase and it is much too late to consider any changes now.

3.2. Allowed interference with nominal LPF Mission

None. Again, the primary goal of LPF is to demonstrate the technology for LISA, and it will be assumed that no interference with the nominal mission is allowed. Any extended scientific operations proposed as a result of this work will have to be carried out in an extended phase, once the nominal mission is completed.

3.3. LPF Micropropulsion dV budgets

Simple estimates of the remaining total dV authority at the end of the nominal LPF mission have been carried out. For the FEEP system, total dVs between 6m/s and around 15m/s have been derived. However, it is noted that the FEEP micropropulsion system development is not yet completed.

For the purposes of this preliminary work, a conservative assumption of a total dV budget of 1m/s has been made. The DRS thrusters should also be capable of providing some additional dV, but no assumptions have been made about this.

It is also important to note that trajectory correction manoeuvres for LPF may be limited by factors other than total control authority, and that therefore this assumption is not as restrictive as it may, at first, seem.

4. Targeting the Earth-Sun saddle point from the LISA Pathfinder Lissajous orbit

4.1. Principle

LISA-Pathfinder flies in a large amplitude Lissajous orbit (in fact close to Halo orbit dimensions). Like all such orbits, it is unstable if adversely perturbed from its ideal state. This instability can be exploited to enable the spacecraft to leave the orbit with only a small manoeuvre. A DeltaV applied in a critical direction of 28 deg from the Earth-Sun direction ensures that the spacecraft state reaches an 'unstable manifold'. This means that over a period comparable with that of the Lissajous orbit, it will strongly diverge from its nominal motion. Possible motions include low energy escape trajectories and also trajectories returning to Earth with a low perigee. The problem to be solved is to find an appropriate manoeuvre after which it is possible to target the Earth-Sun gravitational equilibrium zone.

Following the discussions above, the size of the initial manoeuvre was limited to DeltaV <=1m/s. Furthermore, it was decided to propagate the trajectories for a maximum of two years from the initial manoeuvre. Within these constraints, some interesting solutions have been found.

4.2. Orbit Propagation Method

A standard multi-body gravity field numerical integrator was used for the orbit propagation. The software offers the opportunity to include a range of environmental models. Solar radiation pressure was not included in this preliminary work but can also be incorporated later.

It is important to stress that the primary goal of the work here was to show that a family of solutions exists to bringing LPF back to the saddle point following the nominal mission, given the constraints from the micropropulsion system. The existence of this family of solutions will not be affected by any future refinement of the orbit propagation, but the details of the optimum solution certainly will.

4.3. One Parameter Search

The first investigation concentrated on a varying a single parameter, namely the time of departure, while maintaining a constant dV magnitude (1m/s) and direction (28 deg from Earth-Sun direction). Some of the resulting trajectories are shown in **Figure 2**:

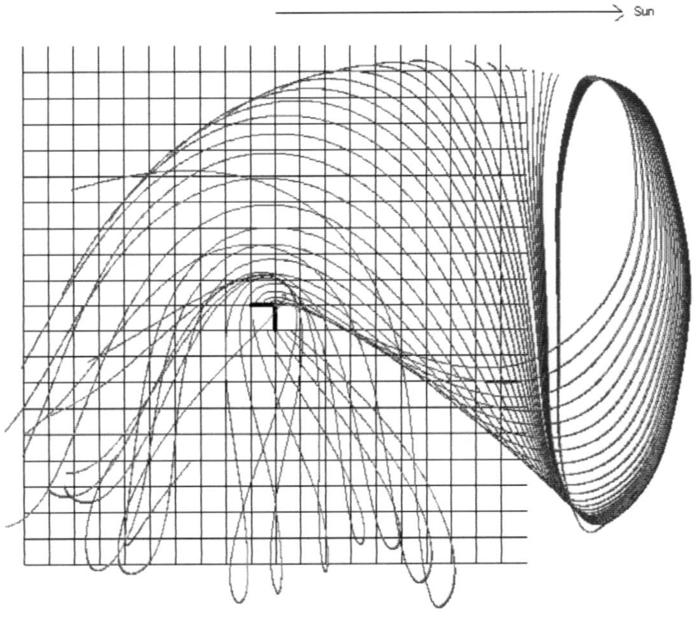

Figure 2 A range of orbits leaving the nominal Lissajous orbit, with a DeltaV of 1 m/s. The time application for each trajectory is incremented in 5 day steps. The grid is in the ecliptic using an Earth-Sun rotating reference. Grid squares are 100000km

This initial search showed that the large out of ecliptic motion of the nominal orbit makes it difficult to target the saddle point on the first 'leg' of the trajectory, where the spacecraft starts its return towards Earth. However the subsequent motion after Earth perigee is weakly bound and offers numerous further possibilities to reach the required region.

4.4. Two Parameter Search
A two parameter search was also conducted. The control parameters employed are the time at which the DeltaV is applied and the size of the DeltaV.

The resulting miss distances, as a function of time of application relative to an arbitrary reference, are shown in **Figure 3**:

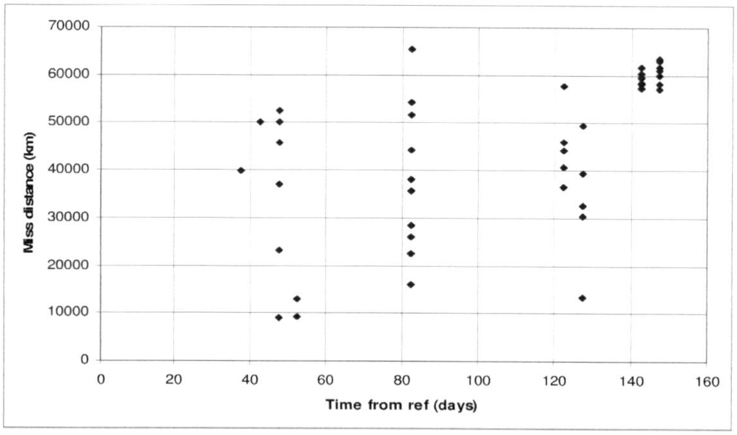

Figure 3 Target miss distance using a coarse 2 parameter set.

The coarse search indicates some clustering of low miss distances with time at which the manoeuvre is made, relative to an arbitrary reference.

As can be seen, some solutions have target miss distances smaller than 10000km, and with fine tuning miss distances of around 5000km can be achieved for selected cases. In view of the shift of the target region as a function of Moon position, as well as the uncertainty in the orbit propagation itself, no attempt was made at reducing this distance further.

4.5. Example Trajectories
As an illustration of what is possible, some selected trajectories from the two parameter search will be shown here. The solution with the smallest miss distance, around 5000km, is shown in **Figure 4**:

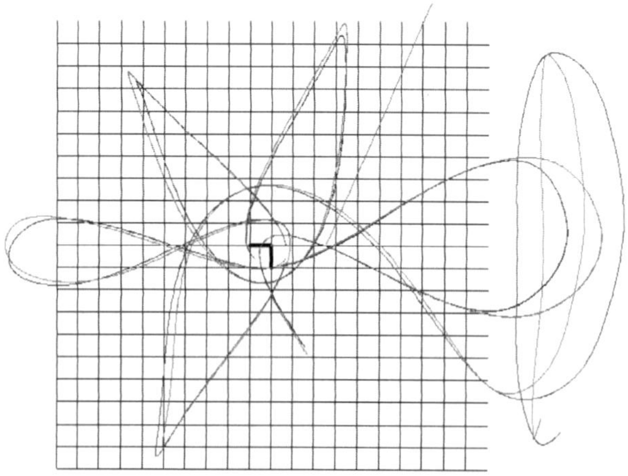

Figure 4 Two parameter solution with smallest miss distance

The time taken to reach the equilibrium region is approximately 480 days after the manoeuvre in the Lissajous orbit, and the initially applied dV is 0.87m/s.

An example of a "fast" transfer is shown in **Figure 5**:

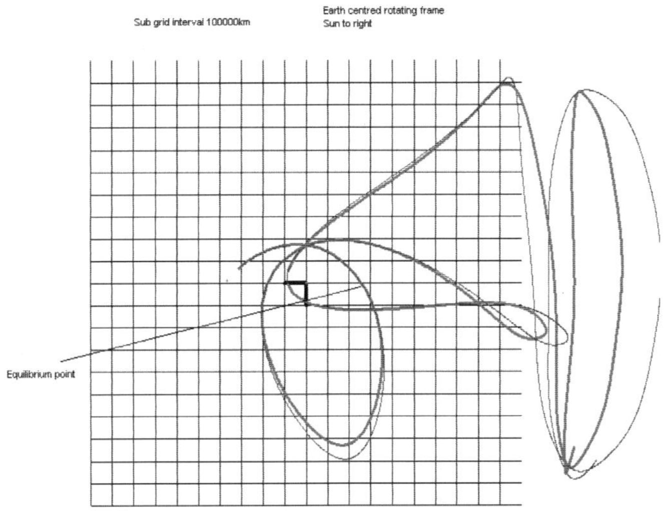

Figure 5 "Fast" transfer from Lissajous orbit to saddle point

The applied dV was 1m/s, and the transfer time was approximately 340 days. The miss distance was 30000km, but could in principle be improved with additional manoeuvres.

As a final example of what is possible, **Figure 6** shows a solution that includes a lunar flyby:

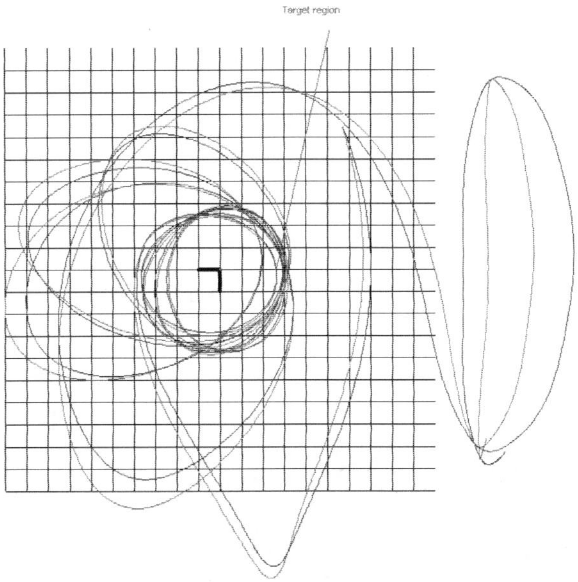

Figure 6 Solution including lunar flyby

A close lunar approach occurs (20000km) which results in the near circular orbit seen in the trajectory. One pass of this trajectory approaches the equilibrium zone with a miss distance of approximately 7000km.

4.6. Frequency Band for anomalous Gravity Gradient Measurements
In general, the LPF spacecraft will cross the region around the saddle point at a speed which is of the order of a few km/s. With a target region size of the order of 1000km, this means that anomalous gravity gradients will be applied for the duration of a few hundred seconds or so. This frequency band constitutes almost an ideal match to the LPF sensitivity.

4.7. Improvements
From a mission analysis point of view, numerous options exist to achieve closer approaches. The principle involves the application of further control parameters, which in general means a second manoeuvre. This could be applied at some time after the first manoeuvre, whilst still close to the initial Lissajous orbit, or at one of the subsequent apogees.

In order to find an optimum solution, it is certainly necessary to better define the target location, and also to refine the environmental models (gravitational and solar radiation pressure) used in the orbit propagation.

5. Conclusions and Future Work
We have shown that in principle, the residual control authority on LISA Pathfinder, following its nominal mission, is sufficient to bring LISA Pathfinder back to cross the region around the Sun-Earth saddle point, following one or more Earth flybys. These trajectories could be used for tests of MOdified Newtonian Dynamics (MOND), and perhaps to investigate the so-called flyby anomaly. The speed of the LPF spacecraft is also such that any anomalous effects would be applied in the appropriate frequency band.

Clearly, a significant amount of work is still required, before any concrete proposal for an extended scientific phase can be put forward. The many issues that still need to be resolved include:

- consolidation of dV budgets
- actual MOND bubble position and size as a function of Moon position
- possible constraints from LPF navigation accuracy
- possible constraints from LPF attitude requirements
- possible constraints from test mass re-caging or instrument commissioning
- detailed MOND gravitational gradient signature within target region
- …

In any event, we believe that the prospect of LISA Pathfinder being used to test gravitational theories is extremely exciting, and certainly deserves to be investigated in some detail.

References
[1] ESA-SCI (2007) 1: LISA Pathfinder: Einstein's Geodesic Explorer – The Science Case for LISA Pathfinder
[2] J.Bekenstein and J.Magueijo, Modified Newtonian Dynamics habitats within the solar system, Phys Rev D 73, 103513 (2006)
[3] J.D. Anderson et al, Anomalous Orbital-Energy Changes Observed during Spacecraft Flybys of Earth, Phys Rev Lett 100, 091102 (2008)

Optical metrology alignment and impact on the measurement performance of the LISA Technology Package

M Hirth[1], W Fichter[1], N Brandt[1,2], A Schleicher[2], D Gerardi[1,2] and G Wanner[3]

[1] iFR, Universitaet Stuttgart, Pfaffenwaldring 7a, 70569 Stuttgart, Germany
[2] Astrium GmbH, 88039 Friedrichshafen, Germany
[3] Albert Einstein Institut, Callinstrasse 38, 30167 Hannover, Germany

E-mail: `marc.hirth@ifr.uni-stuttgart.de`

Abstract. Aside from LISA Pathfinder's top-level acceleration requirement, there is a stringent independent requirement for the accuracy of the optical metrology system. In case of a perfectly aligned metrology system (optical bench and test masses) it should rather be independent of residual displacement jitter due to control. However, this ideal case will not be achieved as mechanical tolerances and uncertainties lead to a direct impact of test mass and spacecraft displacement jitter on the optical measurement accuracy.
In this paper, we present a strategy how to cover these effects for a systematic requirement breakdown. We use a simplified nonlinear geometrical model for the differential distance measurement of the test masses which is linearized and linked to the equations of motion for both the spacecraft and the two test masses. This leads from test mass relative displacement to a formulation in terms of applied force/torque and thus allows to distinguish the absolute motion of each of the three bodies. It further shows how motions in each degree of freedom couple linearly into the optical measurement via DC misalignments of the laser beam and the test masses. This finally allows for deriving requirements on the alignment accuracy of components and on permittable closed-loop acceleration noise.
In the last part a budget for the expected measurement performance is compiled from simulations as no measurement data is available yet.

1. Measurement model

The configuration of the two test masses (TMs) and the optical bench (OB) for LISA Pathfinder is shown in figure 1. Whenever the TM x-faces and the optical metrology system (OMS) are not nominally aligned, the differential distance measurement Δx along the sensitive axis may be degraded due to coupling of residual control jitter in other degrees of freedom into the sensitive axis. That means that DC misalignments in position and attitude of the TMs and the laser beams couple linearly via linear and angular displacement noise of either TMs or the spacecraft (SC).
To gain insight into this issue a measurement model for the differential distance readout of the two TMs was derived. This model does not aim for an exact description of the distance measurement of the laser beam as the whole optical bench setup would have to be taken into account. It rather describes the distance between the laser beam reflection points on the x-faces

of the TMs. The measurement direction of this model is the x-axis of the OB frame, i.e. the connecting line between the nominal center of mass (CoM) positions of the two TMs.
Besides the OB frame, additional coordinate frames were introduced. The XMRFi frame serves as reference frame on the reflecting surface of TM i (i=1,2). Its origin is defined as the intersection of the reflecting surface and its normal passing through the CoM of the respective TM, the x-axis is aligned with this normal. The RPFi origins are located at the laser beam reflection points on each TM with all axes aligned to the so-called interferometer axis frame (IAFi). The origin of the latter lies in the as-built reflection point on the dummy TM with the x-axis along the angle bisector of the as-built incident and reflected laser beams when differential wave front sensing (DWS) senses zero angles. For all these frames positive axes directions are defined 'along' OB frame axes. Figure 1 depicts the frames and notations in use exemplary for the xy-plane.

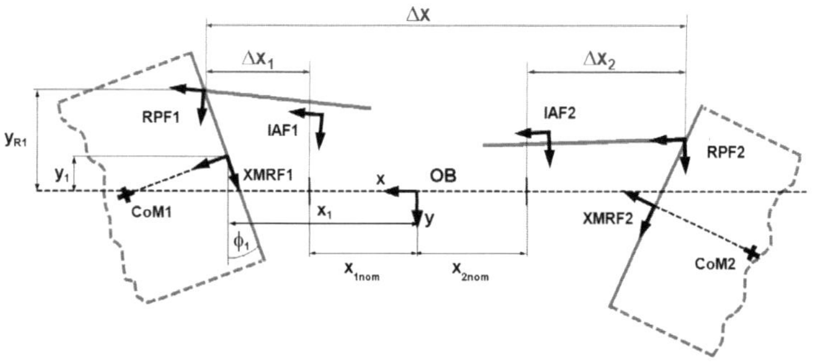

Figure 1. Coordinate frames and notations used in the measurement model, x_{1nom} and x_{2nom} are the nominal distances between the XMRF frames and the OB frame

For ideally aligned x-faces (i.e. $\phi_i = 0$) the measurement would obviously be the desired measurement $\Delta x = x_1 - x_2$ which is simply the distance between the known XMRFi frames on the TM faces. However, angular misalignments of the TMs additionally contribute to Δx if the reflection points do not correspond with the XMRFi frames. The final measurement equation can be written as:

$$\begin{aligned}\Delta x &= b_{x1}x_1 + b_{x2}x_2 - (y_{R1} - y_1)\phi_1 + (y_{R2} - y_2)\phi_2 + (z_{R1} - z_1)\eta_1 - (z_{R2} - z_2)\eta_2 \\ &\quad + b_{\phi_1^2}\phi_1^2 + b_{\phi_2^2}\phi_2^2 + b_{\eta_1^2}\eta_1^2 + b_{\eta_2^2}\eta_2^2 \end{aligned} \quad (1)$$

In order to account for quadratic terms in angles and for the non-ideal sensitivity of the interferometer readout (which is not resolvable with the simple geometrical model) an additional set of model parameters b_j was introduced. The relevant parameters can be extracted from complex alignment simulations that take into account the complete setup of all optical elements and properties of the laser beam.
Generally each XMRF coordinate in the measurement equation may comprise a DC and a noise component, e.g. $y_1 = y_1^{DC} + \overline{y}_1$. Thus, it is possible to rewrite equation (1) with all coordinates separated into terms linear in DC and displacement noise and quadratic terms in DC or displacement noise. All quadratic terms can be neglected as on the one hand pure DC terms do not introduce additional errors in the measurement bandwidth and on the other hand - as the DC components are much larger than the noise contributions - quadratic terms in noise have no driving impact. The remaining expression is linear in noise and can be written in vector form:

$$\Delta x = \mathbf{v}_1^T \mathbf{r}_1 + \mathbf{w}_1^T \boldsymbol{\varphi}_1 + \mathbf{v}_2^T \mathbf{r}_2 + \mathbf{w}_2^T \boldsymbol{\varphi}_2 \quad (2)$$

where $\mathbf{r}_i$ and $\boldsymbol{\varphi}_i$ are the position (x,y,z) and attitude (θ,η,ϕ) noise components relative to OB frame while the DC position and attitude alignments are arranged in the vectors $\mathbf{v}_i$ and $\mathbf{w}_i$. This equation shows how displacement noise relative to the OB frame couples via DC alignments of the TM x-faces into the differential distance measurement.

2. Relative vs. absolute coordinates

From equation (2) it is obvious that if the measurement error should be kept below an appropriate overall requirement, both DC misalignments and displacement noise have to be bounded. While the first directly leads to design requirements for OB, TMs and interfaces, the latter is related to requirements on control. In case of LISA Pathfinder, control is subdivided in attitude, drag-free and suspension loops [1]. Attitude and drag-free control is performed using colloid thrusters or FEEPs while suspension control acts directly on the TMs using electrostatic forces and torques. In addition, different "external" noise sources (i.e. not due to actuation) act on the SC and each TM and different readout noise is present in the signals of the sensors in use for each loop. With a description in terms of relative coordinates as in equation (2) it is impossible to distinguish whether displacement noise is induced by force noise on the TM or on the SC and thus how control has to be improved to mitigate the impact on the measurement accuracy. As a consequence it is necessary to separate the relative motion into absolute motion of SC/OB and the two TMs.

This separation is done by replacing the relative TM coordinates with the relations for the relative TM motion that come from the linearized equations of motion [3]. These are modified such that SC and TM CoMs are replaced by the OB and XMRF origins respectively. Similar to the relative coordinates in equation (1) acceleration terms (i.e. applied forces/torques per unit mass) in the equations of motion contain DC and noise components. Again only noise components in the measurement bandwidth are of interest and DC contributions are neglected. Substitution into the measurement model finally gives another equation which is linear in noise:

$$\Delta\ddot{x} = \mathbf{C}_{DC}^T \begin{bmatrix} \mathbf{a}_{OB} \\ \boldsymbol{\alpha}_{OB} \\ \mathbf{a}_{TM1} \\ \boldsymbol{\alpha}_{TM1} \\ \mathbf{a}_{TM2} \\ \boldsymbol{\alpha}_{TM2} \end{bmatrix} \qquad (3)$$

where the vector $\mathbf{C}_{DC}$ summarizes all misalignment contributions that couple with linear and angular closed-loop acceleration noise $\mathbf{a}$ or $\boldsymbol{\alpha}$ in the 18 OB (i.e. SC) and TM degrees of freedom. A detailed description of the involved terms can be found in [2].

3. Requirement breakdown strategy

Given the alignment vector $\mathbf{C}_{DC}$ from equation (3), requirements can be allocated to the magnitude of each row, i.e. the sum of all misalignment terms that couple with noise in one degree of freedom. Subsequently, this total is broken down to the single contributions of each entry. The resulting requirements can generally be assigned to three groups: requirements on the inertial sensor design (e.g. TM surface non-orthogonality), on the OMS design (IAF-alignments, readout sensitivities and DWS sensing bias) and requirements on the final integrated OMS-TM system (XMRF-IAF relations). The first two groups represent fixed systems (pre-flight) whereas the latter remains adjustable to some extent due to control during operation. Now as requirements on the alignment accuracy are available (in the range of some tens of μm or μrad), requirements on the acceleration noise can be derived from the top-level requirement on the measurement error which is given as $\sqrt{S_{\Delta x,req}} = 6\ \text{pm}/\sqrt{\text{Hz}}$ at 30 mHz (the frequency shape is shown in figure 2). As it is given in terms of a spectral density, the differential distance readout error equation (3) has to be converted using:

$$S_{\Delta\ddot{x}} = \mathbf{C}_{DC}^T \mathbf{S}_a \mathbf{C}_{DC} = (2\pi f)^4 S_{\Delta x} \qquad (4)$$

In this general form, $\mathbf{S_a}$ denotes the spectral density matrix of the acceleration noise in the considered 18 degrees of freedom. For the breakdown of requirements on acceleration noise it is assumed that - even if the different DoF will be coupled to some extent due to control - all noise contributions are uncorrelated. Then above equation collapses to

$$S_{\Delta x} = \frac{1}{(2\pi f)^4} \sum_{j=1}^{18} C_{DC,j}^2 S_{a_j} = \frac{1}{(2\pi f)^4} \sum_{j=1}^{18} S_{\Delta \ddot{x}_j} = \sum_{j=1}^{18} S_{\Delta x_j} \quad (5)$$

where S_{a_j} denote the auto-spectra of the acceleration noise in each degree of freedom, i.e. the main diagonal elements of $\mathbf{S}_a$. This assumption allows a direct apportioning of the total $S_{\Delta x,req}$ to each $S_{\Delta x_j}$ that comes from a coupling with a (non-zero) misalignment component $C_{DC,j}$. Then, according to equation (5) requirements on single acceleration noise terms can be derived using the magnitude of the respective alignment requirement on $C_{DC,j}$.

Preliminary investigations using the requirement values for the misalignments and acceleration noise data from a LTP closed-loop frequency analysis revealed that $S_{\Delta x_j}$ contributions that come from SC acceleration noise have a driving impact and the individual requirements could not be met over the whole measurement bandwidth. This mainly thruster-induced noise led to a redesign of the drag-free loops (that use the FEEPs as actuators) including the additional requirements from the alignment coupling. As DFACS uses specific control coordinates (i.e. the SC is controlled w.r.t. selected TM DoF [1]), above decomposition into OB and TM acceleration noise is not directly applicable to derive control requirements. The strategy and results for the successful loop redesign was shown in [4].

4. Performance results

As no measurements are available all relevant information for a performance budget is extracted from models. As mentioned above, acceleration noise data comes from a closed-loop frequency domain model of LISA Pathfinder implemented in *Matlab*. From this model we get:

$$\mathbf{S}_a(f) = \mathbf{H}^{*T}(f)\mathbf{S}_n(f)\mathbf{H}(f). \quad (6)$$

Here $\mathbf{H}$ is the matrix of the frequency response functions of the closed loop system ($\mathbf{H}^{*T}$ denotes its complex conjugate transpose) and $\mathbf{S}_n$ is the spectral density matrix of the noise sources under consideration. These are thruster (FEEP) noise, star tracker noise, capacitive actuation/readout noise and internal readout noise of the optical metrology system. $\mathbf{S}_n$ is modeled as uncorrelated white noise input with magnitude one, noise scaling and noise shape is already included in the frequency response functions.

Including measurement data into the budget will not be possible for the entire set of misalignment terms as they partially depend on the real in-flight orientation and position of the TMs. For orthogonality of the TMs and OB flight model properties, no data exists yet but will be available before launch. As an intermediate stage, alignments are taken that come from a simulation of the OB bonding procedure and the in-flight alignment of the TMs carried out the Albert Einstein Institute [5].

They use a $C++$ program to set up position and attitude of the single optical elements on the OB. This setup may represent the nominal bench or a misaligned layout. The latter is obtained by placing the optical bench elements and dummy test masses one after another (according to the OB alignment plan) randomly within their expected tolerance range. Clipped normal distributions are employed to generate this random positioning. In a second step the Fortran program *OptoCad* is used to trace the laser beams through the optical setup. It allows for computing beam parameters (at desired positions), reflection points and incident angles on optical elements taking into account their refractive indices. Finally a third C program derives the photodiode signals of each interferometer using similar routines as implemented in the data

management unit.
As mentioned above the sequence for the setup of the misaligned bench is chosen in a way that it represents the strategy of how the real elements on the bench will be bonded. Afterwards the dummy TMs are replaced by the 'real' TMs for which the in-flight alignment procedure is simulated, i.e. the TMs are rotated until 'DWS zero' condition is reached (IAFs/RPFs coincide with XMRFs). The whole procedure is carried out as Monte Carlo simulation (>4000 runs). Combining the data of frequency analysis and alignment simulation, it is possible to

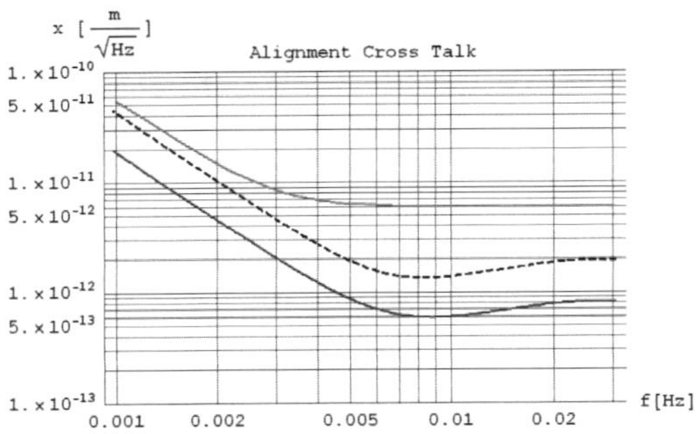

Figure 2. Performance results gained from Monte Carlo simulation data for misalignments and noise from closed-loop frequency domain analysis. Lines in plot are:
—— Requirement
—— RMS
- - - Mean+3σ

compute the misalignment-induced measurement error $\sqrt{S_{\Delta x}}$ according to equation (4). This was done for all Monte Carlo runs and the result is shown in Fig. 2. Concerning the RMS of all performance results the top level requirement can safely be met with large margin. If the 3σ-curve is considered as representative worst case, still a factor of about 3 is available at the upper end of the measurement bandwidth of 30mHz. Towards lower frequencies the margin decreases but still remains at about 25%.
It should be noted here that the performance results do not account for possible improvements due to any kind of calibration procedure using known misalignment values.

5. Correlation of acceleration noise

For the requirement breakdown, it was assumed that the acceleration noise contributions are uncorrelated, i.e. the non-diagonal elements of $\mathbf{S_a}$ are zero to allow the application of equation (5). However, the spectral density matrix $\mathbf{S_a}$ computed for the closed loop system is not purely diagonal. Thus, for an exact breakdown of $\sqrt{S_{\Delta x}}$ also the cross spectra of $\mathbf{S_a}$ would have to be considered. To estimate the error due to this neglection, the relative difference between both results is investigated:

$$\Delta = \left(\sqrt{S_{\Delta x,uc}} - \sqrt{S_{\Delta x,real}}\right) / \sqrt{S_{\Delta x,uc}} \tag{7}$$

Here $\sqrt{S_{\Delta x,uc}}$ denotes the simplified result of equation (5) while $\sqrt{S_{\Delta x,real}}$ gives the real contribution from equation (4). For $\Delta < 0$ the real performance is worse than predicted from the uncorrelated model. As the results are highly dependent on the magnitude and signs of the underlying alignment combinations, we use all data of the AEI Monte Carlo simulation for the comparison. The black set of lines in figure 3 show the Δ of each run, the red line represents the respective mean difference of all runs. The upper bound of the relative difference is approximately +30% at very low frequency and +25% at about 10 mHz. Thus, in best case the measurement error is overestimated up to these levels. This would even allow a relaxation of the single axes acceleration noise requirements. The lower bound varies down to about -20% around 10 mHz and results to an underestimation of the measurement error. In these cases the cross-spectra contributions would lead to more stringent requirements on acceleration noise. However, the mean difference over all runs remains below 5% over the whole measurement bandwidth.

Journal of Physics: Conference Series **154** (2009) 012003 doi:10.1088/1742-6596/154/1/012003

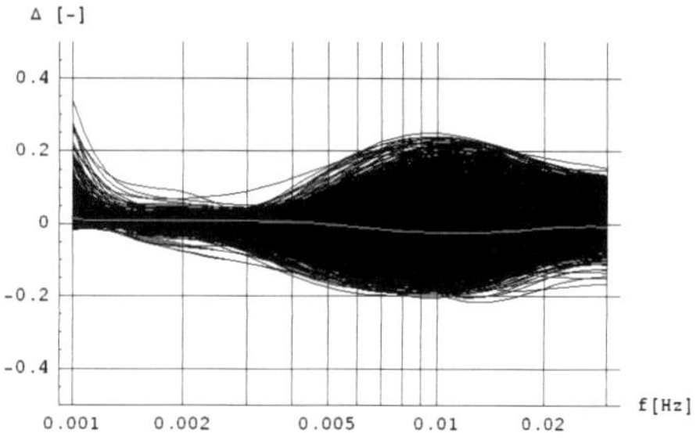

Figure 3. Relative difference (noise in OB coordinates), red line represents mean

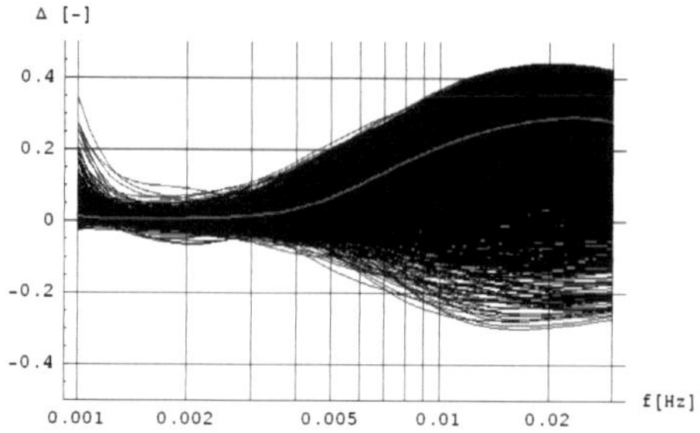

Figure 4. Relative difference (noise in DF coordinates), red line represents mean

This makes the applied breakdown strategy and lower level requirements on $\sqrt{S_{a_j}}$ safely applicable. Even more if it is kept in mind that the expected overall performance contains enough margin to the top-level requirement, see figure 2.
As additional controller requirements were derived for drag-free coordinates, the same investigation was carried out for drag-free rather than OB coordinates, see figure 4. At low frequencies the situation is similar. However for higher frequencies there is a larger spread in Δ, ranging from about -30% to 45%. In particular, the mean is no longer around zero, but raises up to about 30% at high frequencies. Though this is still not critical for the noise budget as the impact on the real performance is overestimated compared and the derived acceleration noise requirements on DF-axes tend to be conservative. The wider spread in Δ using the drag-free axes description was expectable: Angular motion of the OB gets directly mapped into lateral motion of two DF-axes (e.g. ϕ-motion into y_1 and y_2), thus the same noise source is obviously present on multiple axes and increases correlation. This also arises from the following:
For both DF and OB coordinate description the main contributors to the relative difference are FEEP noise (at frequencies larger 3 mHz) and star tracker noise (at very low frequencies). If these noise sources (both related to SC motion) are intentionally removed from the system, figures 3 and 4 become nearly identical (with nearly zero mean and $|\Delta|$ well below 20% for all frequencies).

6. Conclusion

In this paper it was presented how misalignments between the OMS and the inertial sensor of LISA Pathfinder that degrade the differential distance measurement of LISA Pathfinder can be covered using a simple nonlinear measurement model. Moreover it was depicted how the requirement on the measurement error could be broken down into different alignment and noise requirements. A strategy was shown how to separate the measurement errors due to noise inducing TM displacements from noise inducing SC displacement. The latter was identified as driver of misalignment effects and led to a redesign of the drag-free loops.
Subsequently a budget was presented that includes the noise performance of the updated controller design and Monte Carlo simulation data of the expected misalignments from a complex optical model. The results showed that measurement error is likely to be well below the requirement over the whole measurement bandwidth. Finally, it was shown that the assumption of uncorrelated acceleration noise in all degrees of freedom that was used for the requirement breakdown is no significant misjudgement and derived requirements can properly be used.

References

[1] Fichter W, Schleicher A and Vitale S 2007 *Lasers, Clocks and Drag-Free*, ed H Dittus, C Laemmerzahl et al (Berlin: Springer Verlag) pp 365–80

[2] Brandt N, Hirth M, Fichter W, Schubert R, Warren C, Wealthy R 2008, *Experiment Performance Budget*, S2-ASD-TN-3036, Issue 2.2, LISA Pathfinder Project Documentation

[3] Fichter W, Schleicher A, Schlotterer M, Saage R 2008, *DFACS General Design*, S2-ASD-TN-3001, Issue 3, LISA Pathfinder Project Documentation

[4] Schleicher A, Brandt N, Hirth M, Saage R and Fichter W 2008, *Drag-free control design for misaligned cubic test masses, Proc. of the 7th ESA Conf. on Guidance, Navigation and Control Systems*, unpublished

[5] Wanner G 2008, *LTP-OBI Alignment Simulation*, S2-AEI-TN-3051, Issue 2, LISA Pathfinder Project Documentation

Thermal coupling within LTP dynamics control loop

M Nofrarias[1], AF García Marín[1], G Heinzel[1], M Hewitson[1], A Lobo[2,3], J Sanjuán[2,3], J Ramos-Castro[4], K Danzmann[1]

[1] Max-Planck-Institut für Gravitationsphysik, Albert Einstein Institut (AEI), Callinstrasse 38, 30167 Hannover, Germany
[2] Institut de Ciències de l'Espai (ICE-CSIC), Facultat de Ciències , Torre C5, 08193 Bellaterra, Spain
[3] Institut d'Estudis Espacials de Catalunya (IEEC), Edifici Nexus, Gran Capità 2-4, 08034 Barcelona, Spain
[4] Departament d'Enginyeria Electrònica, UPC, Campus Nord, Edifici C4, Jordi Girona 1-3, 08034 Barcelona, Spain

E-mail: `miquel.nofrarias@aei.mpg.de`

Abstract. The Diagnostic Subsytem in the LISA Technology Package (LTP) on board the LISA Pathfinder mission (LPF) will characterise those external disturbances with a potential impact on the performance of the experiment coming from either thermal, magnetic or charged particles perturbations. A correct design of the experiments to measure these effects in flight requires a closed loop analysis that takes into account the dynamics of the test masses, the force applied by the controllers and those noisy terms (coming from sensing or force noise) that enters into the loop. We describe this analysis in the thermal case and we give a first numerical example of the instrument response to controlled thermal inputs.

1. Introduction

The LISA Technology Package Experiment (LTP) [1] on board LISA Pathfinder (LPF) mission will test key technologies for the future space borne gravitational wave detector LISA [2], a collaborative project between ESA and NASA. The LTP main concept consists in two test masses in nominally geodesic motion inside two separate high vacuum chambers; an optical metrology subsystem (OMS) [3] senses the relative distance to a given level of resolution that fixes the instrumental sensitivity. The test mass position relative to the spacecraft is controlled by the surrounding electrodes which act both as sensing and actuator elements; jointly with the micropopulsion thrusters they conform the Gravitational Reference Sensor (GRS) [4]. The high gain drag-free control loop forces the spacecraft to follow one test mass whereas the low frequency suspension loop allows to keep the differential acceleration noise between both test masses in the required top level performance sensitivity. For the LPF this value is expressed as

$$S_{\Delta a}^{1/2}(\omega) \leq 3 \times 10^{-14} \left[1 + \left(\frac{\omega/2\pi}{3\,\mathrm{mHz}}\right)^2\right] \mathrm{m\,s^2}/\sqrt{\mathrm{Hz}} \tag{1}$$

in the bandwidth $1\,\mathrm{mHz} \leq \omega/2\pi \leq 30\,\mathrm{mHz}$. There are various noise sources contributing to this noise budget and, among these, there are some that will be monitored by the Diagnostic Subsystem of the LTP [5], which is designed to measure temperature and magnetic fluctuations in those places where these magnitudes have a particular strong coupling with the main geodesic measurement, and that also will characterise the charged particle shower hitting the test mass. Diagnostics will play an active role as well in the noise model investigation process during the mission lifetime. The temperature and magnetic subsystem will count with heaters and coils which will allow, not only an environmental characterisation, but a full noise projection campaign. In particular, we will address here the thermal diagnostic case, consisting on 24 NTC (Negative Temperature Coefficient) sensors, 14 heaters and the front-end electronics [6] already tested to achieve the required performance [7]. Although on ground characterisation of these environmental coupling have been already performed, the experiment design on board the satellite needs to take into account the in-loop operation of the mission. In the following we will briefly describe how thermal experiments enter into the LTP control loop and give some numerical examples of the kind of signals planned to be injected. Before that, we first summarise the experimental results of some thermal coupling factors obtained so far.

2. Open loop characterisation of thermal effects

Thermal experiments on board the LTP can be split in three different groups: (1) experiments on the Electrode Housing (EH), (2) experiments on the Optical Window (OW) and (3) experiments on the CFRP (Carbon Fiber Reinforced Plastic) struts holding the LTP Core Assembly inside the thermal shield. Among these, experimental results exist for the two first ones and those will be the ones addressed here. The third one is more of a structural kind and has been studied by means of FEM simulations [8].

Heaters and temperature sensors in the electrode housing will test the coupling of the test mass motion against temperature dependent effects as radiation pressure, radiometer effect and outgassing. Experiments on ground set the coupling factor related to this effects to be of the order $\simeq$100 pN/K in a torsion pendulum experiment [9].

The optical window is an interface which seals the vacuum enclosure containing the electrode housing and, at the same time, lets the laser beam go through for interferometric metrology with the test mass. The optical window is the only non-bonded optical element in the optical metrology subsystem and, since it is clamped in a titanium flange, it will be sensitive to thermal and stress fluctuations. Experiments on ground have allowed modelling the response of the interferometer phase when a heat pulse is applied to the window. The analysis showed that the response can be described by an ARMA(2,1) transfer function relating temperature and phase [10]:

$$G(\alpha_0, \alpha_1, \beta_1) = \frac{\alpha_0 + \alpha_1 z^{-1}}{1 + \beta_1 z^{-1}} \tag{2}$$

The model parameters were found to be related with the mechanical (Young modulus) and thermal properties (thermal expansion) of the materials forming the window.

3. Closed loop analysis of thermal disturbances

The LTP dynamics control loop is shown in figure 1. The scheme shows two control loops for the two measurement channels: the distance from the first test mass to the spacecraft (x_1) and the distance between the two test masses (x_{12}). Following the notation in [11], we can describe the dynamics of the two measurement channels in terms of the following set of equations

$$\mathbf{D} \cdot \vec{q} = \vec{g} \tag{3a}$$

$$\vec{g} = -\mathbf{C} \cdot \vec{o} - \vec{g_n} \tag{3b}$$

$$\vec{o} = \mathbf{S} \cdot \vec{q} + \vec{o_n} \tag{3c}$$

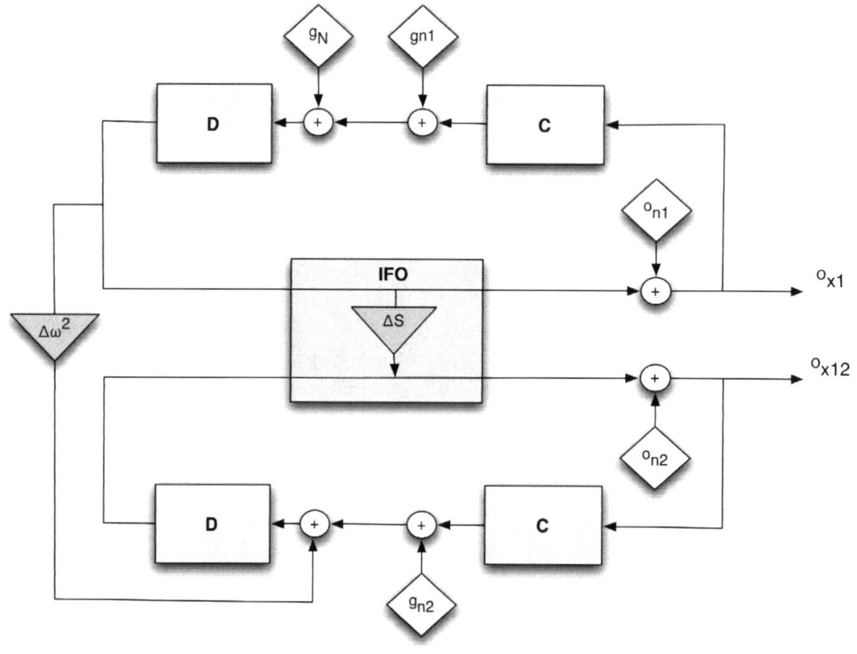

Figure 1. Simplified scheme for the dynamics control loop in the LTP. Boxes describe the interferometer (IFO), controllers (C) and dynamics (D). Rhombus represent noise contribution and triangles stand for cross-coupling between first (o_{x1}) and second channel (o_{x12}).

where the **D** is the dynamical matrix, **C** the controller, and **S** stands for the inteferometer, i.e. the matrix translating the position of the TM, $\vec{q}$, into interferometer read out $\vec{o}$. Subindex n stands for noise quantities, either sensing noise ($\mathrm{o_n}$) or force noise ($\mathrm{g_n}$). These matrices read as

$$\mathbf{D} = \begin{pmatrix} \mathrm{s}^2 + \omega_1^2 & 0 \\ \omega_2^2 - \omega_1^2 & \mathrm{s}^2 + \omega_2^2 \end{pmatrix} \quad \mathbf{C} = \begin{pmatrix} \mathrm{H_{df}} & 0 \\ 0 & \mathrm{H_{sus}} \end{pmatrix} \quad \mathbf{S} = \begin{pmatrix} 1 & 0 \\ \delta\mathrm{S} & 1 \end{pmatrix} \tag{4}$$

where the diagonal terms describes each channel dynamics (e.g. $s^2 + \omega_1^2$ is Newton's law in Laplace domain for the first test mass, being ω_1 the test mass stiffness) and control law (e.g. $\mathrm{H_{df}}$ stands for the drag free transfer function controller on the first test mass). The non-diagonal terms are cross-coupling between the two channels appearing as triangles in figure 1. From the previous expressions we can work out the response of the interferometer once all the dynamical and noise parameters are given,

$$\vec{o} = (\mathbf{D} \cdot \mathbf{S}^{-1} + \mathbf{C})^{-1}(\vec{g}_n + \mathbf{D} \cdot \mathbf{S}^{-1}\vec{o}_n) \tag{5}$$

This equation defines a *nominal output* for the interferometer and this will be the variable that we will use to the effect of thermal forces in the experiment in next section.

3.1. Closed loop thermal transfer functions

In order to evaluate the interferometric response of the experiment to an input signal we need to translate its effect into a measured variable in our previous nomenclature. This effect will differ depending on the heating location under consideration. When heating the electrode housing

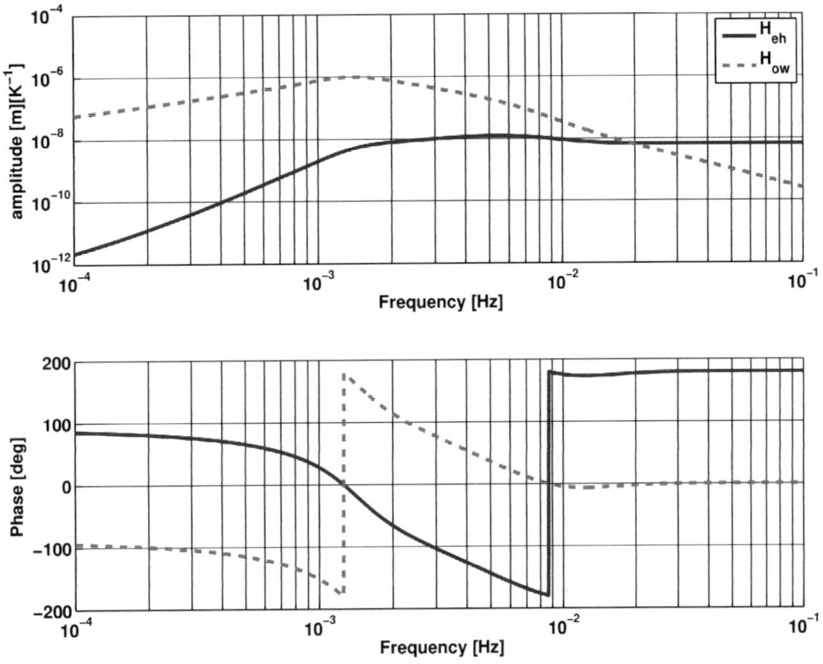

Figure 2. Closed loop thermal transfer functions for the electrode housing (H_{eh}) and the optical window (H_{ow}).

a net force will be applied to the test mass coming from radiation pressure, the radiometer or outgassing effects. Hence we need to introduce a term

$$\vec{g}_{eh} = \mathbf{G}_{eh} \cdot \vec{T}_{eh} \tag{6}$$

in equation (3b), where $\mathbf{G}_{eh}$ is a matrix with the thermal coupling factors and T_{eh} refers to the temperatures measured in the electrode housing. Following the same scheme, a thermal gradient in the optical window will produce a fake test mass displacement, i.e the interferometer will measure a displacement of the test mass though it will not be caused by any displacement but from mechanical stress and thermal distortions on the optical window glass [10]. The coupling in this case is described as a term

$$\vec{q}_{ow} = \mathbf{G}_{ow} \cdot \vec{T}_{ow} \tag{7}$$

in equation (3c). Introducing thermal contributions into the dynamics equation and working out the expressions we end with one equation gathering all the relevant information,

$$\vec{o} = (\mathbf{D} \cdot \mathbf{S}^{-1} + \mathbf{C})^{-1}(\vec{g}_n + \vec{g}_{eh} + \mathbf{D} \cdot \vec{q}_{ow} + \mathbf{D} \cdot \mathbf{S}^{-1} \cdot \vec{o}_n) \tag{8}$$

from where we derive two closed loop transfer functions for heat inputs coming from the electrode housing or the optical window to the output interferometer read-out,

$$\mathbf{H}_{eh} = (\mathbf{D} \cdot \mathbf{S}^{-1} + \mathbf{C})^{-1} \cdot \mathbf{G}_{eh} \tag{9}$$

$$\mathbf{H}_{ow} = (\mathbf{D} \cdot \mathbf{S}^{-1} + \mathbf{C})^{-1} \cdot \mathbf{D} \cdot \mathbf{G}_{ow} \tag{10}$$

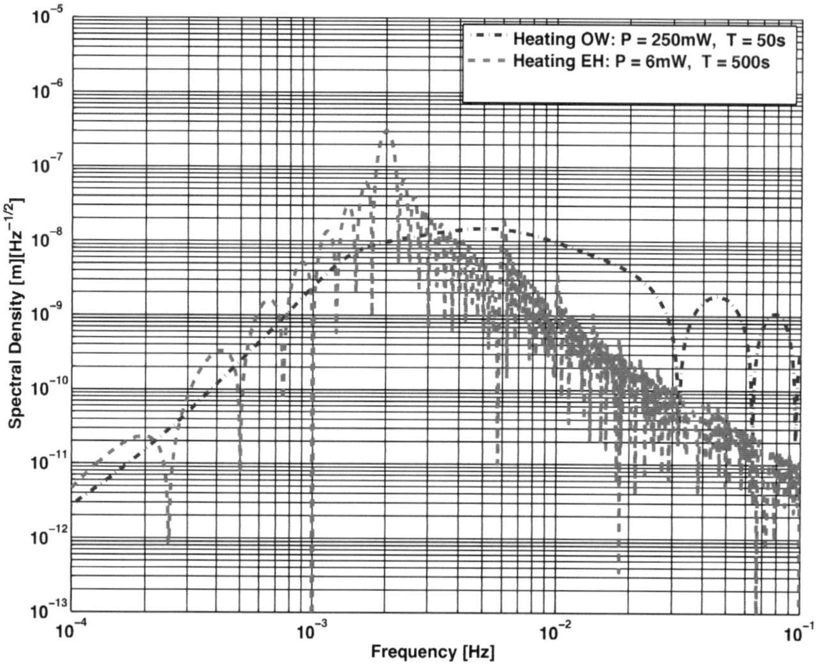

Figure 3. Typical thermal signal spectrum. Dot-dashed line represents a 50 s pulse at 250 mW applied to the optical window whereas dashed line is obtained by applying a series of 500 s pulses at 6 mW alternatively to both electrode housing walls to produce a 2 mHz signal.

These expression, evaluated with typical dynamical parameters in figure 2, will help to correctly design thermal experiments in flight. We next show how, according to the derived expressions, the heaters that form part of the thermal diagnostic subsystem can produce a signal clear enough to evaluate the thermal model parameters.

3.2. Thermal experiments

The evaluation of the interferometer output during thermal experiments requires the definition of an input thermal signal according to the available hardware in each location. In the electrode housing case, since heaters must be inside the GRS vacuum enclosure and strong vacuum restrictions applies, NTC sensors will be used as heaters and consequently the available power will be only of 90 mW. The heating strategy in this location is to alternatively switch on and off heaters on each side of the test mass in order to apply a periodic force in the test mass so that it peaks in the measuring bandwidth. As seen figure 3 and assuming that the expected in-loop noise floor in the second interferometer channel is $S_{o_{12}}^{1/2}(\omega) \simeq 10^{-10}\,\mathrm{m}/\sqrt{\mathrm{Hz}}$ at 1 mHz, the high sensitivity of the test mass to electrode housing thermal gradients makes it possible to measure this effect with a signal-to-noise ratio high enough to measure the model parameters with high accuracy.

In the optical window case, the available power for the heaters is of 1 W. As before, when compared with the expected instrument noise level, a typical input in this case produces an output strong enough to allow a good parameter estimation.

4. Conclusions

The thermal diagnostic subsytem on board the LTP will play a key role in understanding the noise model and to optimise the performace of the experiment. Here we have described the thermal diagnostic subsystem in closed loop, taking into account different noise contributions and the dynamics of the test masses. We have applied this formalism to the electrode housing and the optical window case, and obtained the transfer functions relating a thermal signal input to the measured interferometer signal. These transfer functions will allow a precise definition of the input signal that must optimise the measured output.

According to the current design, we have shown that typical thermal signals applied by the heaters on board the LTP will allow the determination of the parameters describing thermal effects with a good accuracy. The detailed analysis of these signals and the related parameters estimation is ongoing work and will be reported elsewhere.

Acknowledgments

We gratefully acknowledge support by Deutsches Zentrum für Luft- und Raumfahrt (DLR) (references 50 OQ 0501 and 50 OQ 0601) and financial support from the Spanish Ministry of Education, contracts ESP2004-01647 and ESP2007-61712. JS acknowledges a grant from MEC, and MN a Beatriu de Pinós grant from Generalitat de Catalunya.

References

[1] Anza S *et al* 2005 The LTP Experiment on the LISA Pathfinder Mission *Class. Quantum Grav.* **22** S125-38

[2] Bender P *et al* 2000, Laser Interferometer Space Antenna: A Cornerstone Mission for the observation of Gravitational Waves, ESA Report no. ESA-SCI(2000)11

[3] Heinzel G *et al* 2004 The LTP interferometer and phasemeter *Class. Quantum Grav.* **21** S581-88

[4] Dolesi R *et al* 2003 Gravitational sensor for LISA and its technology demonstration mission *Class. Quantum Grav.* **20** S99-108

[5] Lobo A *et al* 2003 In-flight Diagnostics in LISA Pathfinder *AIP Conf. Proc.* **873** 522-28

[6] Sanjuan J *et al* 2007 Thermal diagnostics front-end electronics for LISA Pathfinder *Rev. Sci. Instrum.* **78** 104904

[7] Lobo S *et al* 2006 On-ground tests of the LISA Pathfinder thermal diagnostic system *Class. Quantum Grav.* **23** 5177-193

[8] Brandt N *et al* 2005 LISA Pathfinder E2E performance simulation: optical and self-gravity stability analysis *Class. Quantum Grav.* **22** S493-99

[9] Carbone L *et al* 2007 Thermal gradient-induced forces on geodesic reference masses for LISA *Phys. Rev. D* **76** 102003

[10] Nofrarias M *et al* 2007 Thermal diagnostics of the Optical Window on board LISA Pathfinder *Class. Quantum Grav.* **24** 5103-121

[11] Vitale S *et al* 2007 Measurement of LTP dynamical coefficients by system identification, S2-UTN-TN-3045 v1.1, LTP Technical note.

Towards an improved magnetic diagnostic system for *LISA*

I Mateos[1], A Lobo[1], J Ramos-Castro[2], J Sanjuán[1] and M Nofrarias[3]

[1] Institut de Ciències de l'Espai (CSIC-IEEC), Edifici Nexus, Gran Capità 2-4, 08034 Barcelona, Spain

[2] Departament d'Enginyeria Electrònica, Universitat Politècnica de Catalunya (UPC), Campus Nord, Edifici C4, Jordi Girona 1-3, 08034 Barcelona, Spain

[3] Max-Planck-Institut für Gravitationsphysik (Albert-Einstein-Institute), Callinstrasse 38, D-30167 Hannover, Germany

E-mail: mateos@ice.csic.es

Abstract. The current design, and material implementation of the magnetic field sensing in the *LISA* Technology Package (*LTP*) on board *LISA* Pathfinder (*LPF*), is based on a set of 4 high-precision 3-axis fluxgate magnetometers. In order to avoid magnetic disturbances on the *LTP* proof masses (*TM*'s), originated by the sensors themselves, these are placed somewhat far from the *TM*'s, which results in partial field information losses. We are currently investigating alternative magnetic sensing techniques, based on *AMR* (Anisotropic Magnetoresistive) devices. These are much smaller in size than fluxgates, therefore a more numerous array can be thought of for flight. In addition, there is a chance that they may be attached closer to the *TM*'s, thereby enhancing magnetic field sensing spacial resolution. Several issues need to be addressed, such as real sensitivity (including electronics noise) and set/reset trigger procedures. A brief overview about the stability of the magnetic fields and gradients generated in the *LTP* by means of the coil will also be given. This paper show the latest results of our research.

1. Introduction

LISA Pathfinder [1] is a technologically sophisticated Mission to demonstrate that two proof masses can be put into free-fall to a certain level of accuracy. This idea is reflected in the differential acceleration noise requirements for the *LTP*

$$S^{1/2}_{\delta a,\mathrm{LPF}}(\omega) \leq 3\times 10^{-14}\left[1+\left(\frac{\omega/2\pi}{3\ \mathrm{mHz}}\right)^2\right]\ \mathrm{ms}^{-2}\mathrm{Hz}^{-1/2} \tag{1}$$

in the frequency band between 1 mHz and 30 mHz. *LISA* has a increased acceleration noise budget ($3\,\mathrm{fms}^{-2}\mathrm{Hz}^{-1/2}$) in the frequency band between 0.1 mHz and 0.1 Hz. This noise is the result of various disturbances which limit the performance of the instrumentation on-board. Magnetic disturbances are foreseen to contribute a significant fraction of the total acceleration budget. The effect of magnetic disturbances takes places as a consequence of the magnetic properties of the test masses, more precisely their remmant magnetic moment ($\mathbf{m}_0$) and their magnetic susceptibility (χ). Therefore, magnetometers have been required to be embarked with the purpose to produce an as accurate as possible map of magnetic field and gradient in the region of the *TM*s and also a pair of coaxial magnetic coils will generate controlled magnetic fields, which add linearly to the environmental one and gives rise to a total force per TM given by

$$F_x = \left\langle \left[\left(\mathbf{M} + \frac{\chi}{\mu_0} \mathbf{B} \right) \cdot \nabla \right] B_x \right\rangle V \tag{2}$$

These coils are intended to perform in flight measurements of $\mathbf{m}_0$ and χ, which will assess eventual deviations of the values of these parameters (measured in ground) generated during launch.

1.1. Magnetic Field Measurements

The magnetometers used in *LISA* Pathfinder are fluxgate type (Billingsley TFM100G4-VQS) —see figure 1. These magnetometers are able to perform very low noise measurements in the bandwidth between 1 mHz and 30 mHz, however, they are bulky and present other disadvantages such as:

- The spatial resolution of the magnetic field measurement is limited due to the large head sensor.
- They need to be placed far from the *TM*'s due to the quantity of permalloy (nickel-iron) contained in the fluxgate's core that can provoke magnetic disturbances in the *TM*'s.

In order to overcome the above limitations, we have done some preliminary research to assess the performance of another type of magnetometer, Anisotropic Magnetoresistive (*AMR*) sensors —see figure 1, in the *LTP* and *LISA* measurement bandwidth (*MBW*).

Figure 1. Left: Billingsley fluxgate magnetometer. Right: Honeywell *AMR* magnetometer.

1.2. Magnetic Fields and Magnetic Field Gradients Generation

Stability of the magnetic field and gradients created by the coils must comply with requirements [1] depending on the *TM*'s distance. This stability is required in order to determine accurately the magnetic properties of the *TM*'s (($\mathbf{m}_0$ and(χ)). The stability of the created magnetic field is equivalent to the stability of the driving current, hence the requirement can be expressed in terms of the current fluctuation of the current source —see table 1.

Table 1. Requirements of the magnetic field produced by the coils in the *LTP MBW*.

Magnitude		$S_I^{1/2}$
$S_{B_z}^{1/2}$	$5\ \mathrm{nT}/\sqrt{\mathrm{Hz}}$	$1.12\ \mu\mathrm{A}/\sqrt{\mathrm{Hz}}$
$S_{\partial B_z/\partial x}^{1/2}$	$12\ \mathrm{nT/m}/\sqrt{\mathrm{Hz}}$	$110\ \mathrm{nA}/\sqrt{\mathrm{Hz}}$

2. Magnetic field measurement

2.1. Proposal Front-End Electronics for the AMR

The analog signal conditioning circuit for the magnetic field sensing is shown in figure 2. A Wheatstone bridge is powered with an ac square signal ($f_B \approx 5$ Hz). This switching reduces the $1/f$ noise of the Instrumentation Amplifier (*IA*) by modulating the low frequency signals of interest to a higher band. The output signal from the bridge is amplified, low-pass filtered and sampled [2]

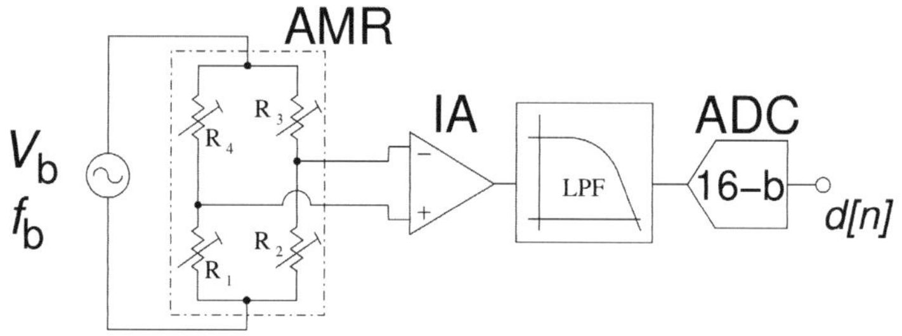

Figure 2. Analog signal processing scheme.

The resistance-magnetic field dependence of the *AMR* can be modeled as [3]:

$$R(H_y) = R_0 + \Delta R \frac{H_y}{H_0}\sqrt{1 - \left(\frac{H_y}{H_0}\right)^2} \tag{3}$$

where H_0 is the total anisotropic field, H_y is the measured field (perpendicular to the easy-axis), $R_0 \simeq 850\,\Omega$ is the nominal resistance of the AMR. The output voltage of the four sensor Wheatstone bridge can be described by:

$$V_o = V_B \frac{\Delta R}{R_0}\frac{H_y}{H_0}\sqrt{1 - \left(\frac{H_y}{H_0}\right)^2} \tag{4}$$

where $V_B = 5$ V is the bridge voltage excitation. Therefore the sensitivity (dV_o/dH_y) of the bridge is $s_B = \frac{\Delta R}{R_0}\frac{V_B}{H_0} = 160\mu\mathrm{V}/\mu\mathrm{T}$, in the temperature range of -25 to 100 °C. The gain of the *IA* has been set to 200, therefore the *global* sensitivity is $s_{\mathrm{total}} = s_B \cdot G_{IA} = 32\,\mathrm{mV}/\mu\mathrm{T}$.

Once the signal has been sampled by the analog-to-digital converter (*ADC*), the digital demodulating process begins. This process is shown in figure 3 and it consists, basically, on a decimator and a subsequent difference averaging.

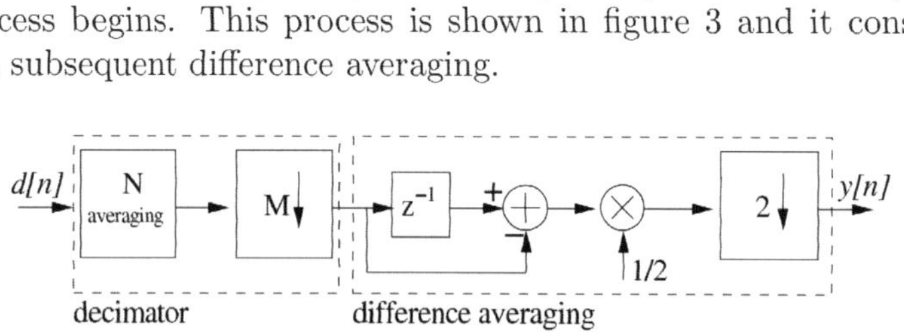

Figure 3. Digital de-modulation process.

2.2. Noise density in the AMR sensor

The total noise of the measurement chain, $S_H^{1/2}$, in terms of magnetic field spectral density can be written as:

$$S_H^{1/2}(H,\omega) = [S_{H,\mathrm{B}}(H,\omega) + S_{H,IA}(H,\omega) + S_{H,ADC}(H,\omega)]^{1/2} \tag{5}$$

where (all the terms are in $\mathrm{T/Hz}^{1/2}$ units):

- $S_{H,\,\mathrm{B}}(H,\omega)$ is the bridge noise density
- $S_{H,IA}(H,\omega)$ is the *IA* noise density refered to the input
- $S_{H,ADC}(H,\omega)$ is the *ADC* noise density refered to the input

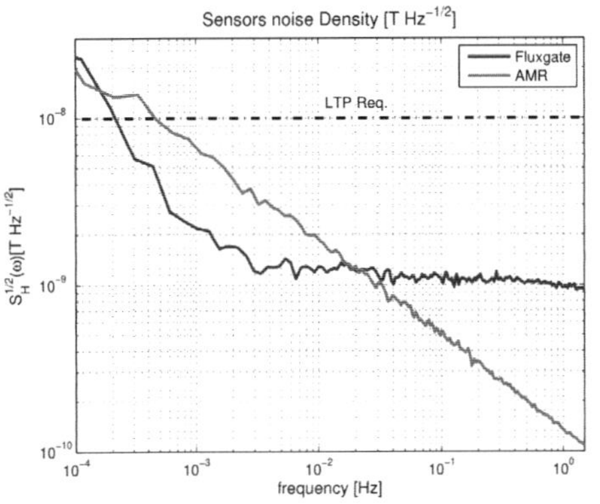

Figure 4. Magnetic field noise density for Fluxgate (blue) and *AMR* (red) sensors.

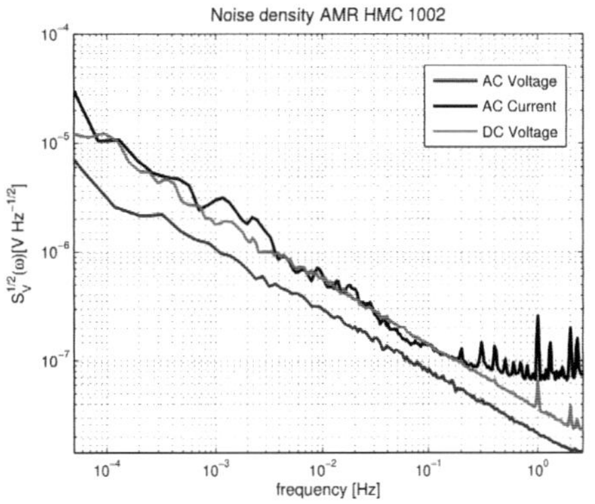

Figure 5. Voltage Noise density of the *AMR* with different Wheatstone Bridge excitation schemes. AC Voltage= 5 V, AC Current= 10 mA, DC Voltage=5 V.

Measurements using flux-gate and AMR magnetometers are shown in Figure 4 where it can be noticed that the magnetic field noise for both sensors is similar at the frequencies of interest for *LISA* (0.1 mHz). Between 1 mHz and 30 mHz both are below the *LTP* requirements. Above 20 mHz the differences are due to the measurement resolution, both have a 16 bits *ADC*, however, in the digital processing of the *AMR* the samples are averaged ($N = 3072$) and, thus, an equivalent resolution of 22 bits ($16 + \frac{1}{2}\log_2 N$) is obtained.

2.3. Magnetic characterisation of the AMR

AMR sensors are made of a nickel-iron thin film deposited on a silicon wafer and is patterned as a resistive strip. These materials can show ferromagnetic behaviours (except Silicon that shows diamagnetism) and could be potentially dangerous if they are to be attached near the test mass. Moreover the *AMR* film properties are well behaved only when the film's magnetic domains are aligned in the same direction (material saturated).

Most low field magnetic sensors will be affected by a large magnetic disturbing field (>300 μT) that may lead to output signal degradation. This disturbing field actually breaks down the magnetisation alignment in the Permalloy film. In order to reduce this effect, a large current Set/Reset pulse can be applied during $\simeq 2\mu$s to eliminate the effect of past magnetic history.

The state of these magnetic domains can retain for years as long as there is no magnetic disturbing field present ($\simeq 6 \cdot H_{\rm Earth}$).

Due to the strong magnetic requirement [1] of the mission —see table 2, the magnetic moment of the *AMR* HMC1001 was measured in a SQUID (Quantum Design MPMS XL Squid). We are interested in the saturation magnetic moment because it will be the normal operation state of the *AMR*.

Table 2. Magnetic requirements in the *LTP*

Magnitude	dc Req.	PSD Req.
B_z	10 μT	650 nT/$\sqrt{\rm Hz}$
$\partial B_z/\partial x$	5 μT/m	250 nT/m/$\sqrt{\rm Hz}$

The measured hysteresis curve (see Figure 6) shows a saturation magnetic moment of $m_{\rm sat} \simeq 0.44\ \mu{\rm Am}^2$, more than an order of magnitude below the remanent magnetic moment of the *NTC* ($m_{\rm r} \simeq 26\,\mu{\rm Am}^2$) placed in the electrode housing of the *LTP* and at a distance of $\simeq 13$ mm of the *TM* [4].

In order to estimate the magnetic field and magnetic field gradient caused by the *AMR* we shall make the assumptions that the sensors behave like magnetic dipoles. Figure 7 shows magnetic field and magnetic field gradient created by a *AMR* are within the requirements.

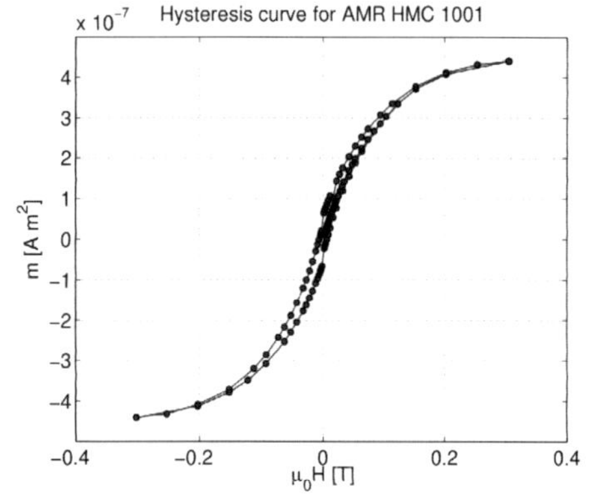

Figure 6. Hysteresis curve at 300 K for the *AMR*.

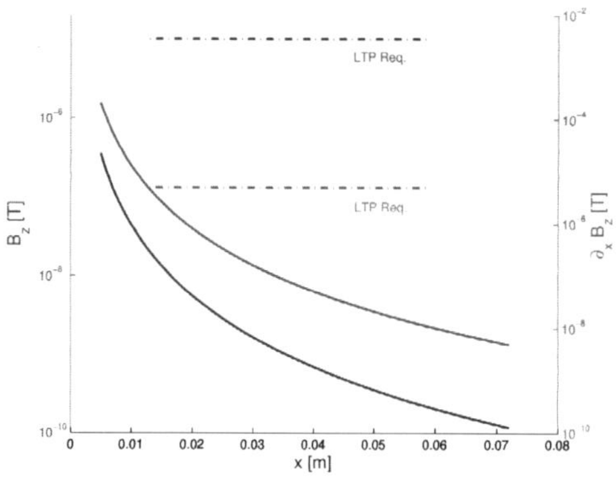

Figure 7. Magnetic field (blue) and magnetic field gradient (green) created by a *AMR*.

3. Controlled magnetic excitations: Coil current stability

Different current source topologies [5] have been tested at *LISA* Bandwidth, obtaining the best result with the *floating load current source*, although the *differential current source* topology could be a good option, within the requirements —see table 3.

Table 3. Noise density results for different current sources typologies.

Typology	1 mHz			0.1 mHz		
	$S_I^{1/2}$ $\left[\frac{\text{nA}}{\sqrt{\text{Hz}}}\right]$	$S_B^{1/2}(*)$ $\left[\frac{\text{pT}}{\sqrt{\text{Hz}}}\right]$	$S_{\partial B/\partial x}^{1/2}$ (*) $\left[\frac{\text{nT}}{\text{m}\sqrt{\text{Hz}}}\right]$	$S_I^{1/2}$ $\left[\frac{\text{nA}}{\sqrt{\text{Hz}}}\right]$	$S_B^{1/2}(*)$ $\left[\frac{\text{pT}}{\sqrt{\text{Hz}}}\right]$	$S_{\partial B/\partial x}^{1/2}(*)$ $\left[\frac{\text{nT}}{\text{m}\sqrt{\text{Hz}}}\right]$
Floating load	6	26.8	0.65	34	152	3.71
Differential	17	76	1.86	80	357.8	8.74
Howland	50	223.6	5.46	210	939.2	22.94

(*) To estimate $S_B^{1/2}$ and $S_{\partial B/\partial x}^{1/2}$ have been assumed a coil's mean radius of 56.5 mm and a distance *TM* center-Coil center of 85.5 mm (Actual *LTP* configuration)

Figure 8 shows the theoretical [6] and measured current stability in the *Floating load* topology, dominated by the $1/f$ noise. The magnetic field gradient noise density obtained with this current pump at *LISA* Bandwidth (0.1 mHz) is shown in figure 9, where x is the distance of the TM from the coil and ρ is the distance from the axis of the coil.

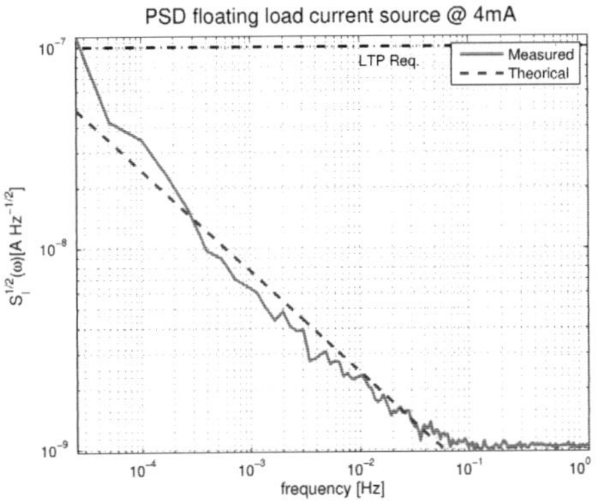

Figure 8. Current noise density of the *Floating load* current pump.

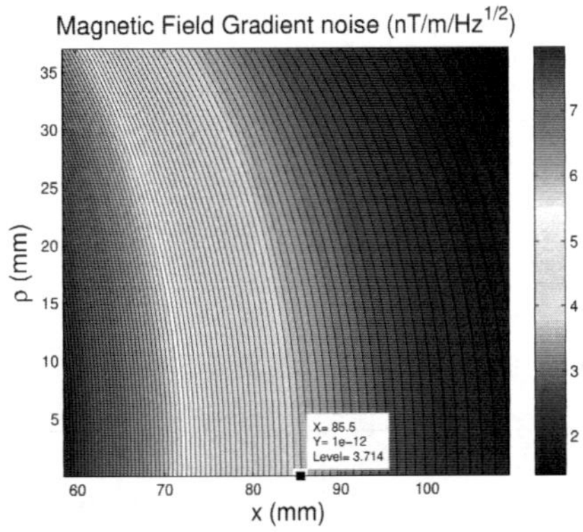

Figure 9. $S_{\partial B/\partial x}^{1/2}$: Magnetic field gradient noise. Label points out the *TM* centre.

4. Conclusions

AMR magnetometers sensors are smaller and present lower power consumption than fluxgate magnetometers, hence, higher spatial resolution with less power consumption is attainable. In addition, we have shown that the measurement noise levels requirement are also fulfilled when using *AMR* sensors in the *LTP* bandwidth and near to the requirement in the frequency of interest of *LISA*. Moreover, the very low magnetisation of the *AMR* sensors allows to place the sensors very close to the *TM*'s and, in consequence, an accurate measurement of the magnetic field in the *TM*'s can be made. For these reasons *AMR* magnetometers appear as a solution to the magnetic field measurement in the *LISA* mission. We also have shown that the *floating load current source* topology fulfils clearly the coil current stability requirements in the *LISA* bandwidth.

Acknowledgments
We wish to thank Núria Clos for her excellent work at the Serveis Científico-técnics of the Universitat de Barcelona and Ivan Lloro, Institut de Ciències de l'Espai, for his great help in managerial affairs. Support for this work came from Project No. ESP2004-01647 of Plan Nacional del Espacio of the Spanish Ministry of Education and Science (MEC).

References

[1] Vitale S 2005 Science Requirements and Top-level Architecture Definition for the *LISA* Technology Package (*LTP*) on Board *LISA* Pathfinder (*SMART-2*), report no. LTPA-UTN-ScRD-Iss003-Rev1
[2] Sanjuán J, Lobo A, Nofrarias M, Ramos-Castro J and Riu P J. 2007 Thermal diagnostics front-end electronics for *LISA* Pathfinder *Rev. Sci. Instrum.* 78 104904
[3] Hauser H, Stangl G, Fallmann W, Chabicovsky R and Riedling K 2000 *Proc. Workshop Preparation, Properties, and Applications of Thin Ferromagnetic Films (Viena)* pp 15-27
[4] Sanjuán J, Lobo A, Nofrarias M, Ramos-Castro J, Mateos I and Xirgu X 2008 Magnetic polarisation effects of temperature sensors and heaters in *LISA* Pathfinder *Rev. Sci. Instrum.* 79 084503
[5] Franco S 2002 *Design with Operational Amplifiers and Analog Integrated Circuits* (London: McGraw-Hill)
[6] Pallàs R and Webster JG 1999 *Sensors and Signal Conditioning* (New York: Wiley-Interscience)

Phase noise contribution of EOMs and HF cables

Simon Barke, Michael Tröbs, Benjamin Sheard, Gerhard Heinzel, Karsten Danzmann

Max Planck Institute for Gravitational Physics, Callinstr. 38, 30167 Hannover, Germany

E-mail: `simon.barke@aei.mpg.de`

Abstract. Two key components of LISA's inter-spacecraft clock tone transfer chain are electro-optic modulators (EOMs) and high-frequency (HF) cable assemblies. At modulation frequencies of 2 GHz, we characterized the excess phase noise of these components in the LISA frequency range (0.1 mHz to 1 Hz). The upper phase noise limit was found to be almost an order of magnitude better than required. In addition, phase dependencies on temperature were determined. The measured coefficients are within a few milliradians per Kelvin and thereby negligible due to the specified on-board temperature stability.

1. Introduction

The total setup phase noise detected by the LISA phasemeter must not exceed an equivalent spacecraft displacement of $1\,\frac{\text{pm}}{\sqrt{\text{Hz}}}$ [1, 2]. This translates to a phase noise requirement of $\Delta\phi_{1\text{pm}} < 6\,\frac{\mu\text{rad}}{\sqrt{\text{Hz}}}$. To reach this 1 pm requirement, an inter-spacecraft clock tone transfer chain is necessary to remove the non-negligible ultra-stable oscillator (USO) phase noise. In the current baseline design, the signal generated by the on-board USOs is up-converted, and superimposed by EOMs on each of the six laser beams (Fig. 1, left) [3, 4]. The main carrier signal is then transmitted, together with the modulation sidebands, to the receiving spacecraft (Fig. 1, right). Here they are heterodyned with a local laser beam. A photodiode now detects three beat notes: The carrier-carrier beat at a frequency up to 20 MHz and the two sideband-sideband beats. Since the sidebands were imposed onto the laser beams by a signal phase locked to the USO signal, the sideband-sideband beat notes contain information about the USO phase noises. Using this information the USO noise can be removed to acceptable levels [5].

1.1. The ancillary modulation error

The components of the clock transfer chain might add excess phase noise to the sidebands which is referred to as ancillary modulation error (AME). To obtain a modulation frequency $f_{\text{mod}} \approx 2\,\text{GHz}$, one multiplies the USO frequency $f_{\text{USO}} \approx 80\,\text{MHz}$ by the factor $\frac{f_{\text{mod}}}{f_{\text{USO}}}$, while the ADC is triggered directly by the USO. The magnitude of a phase shift in a signal being up-converted by a frequency multiplier is increased linearly by the multiplication factor [6]. Therefore—with respect to the design sensitivity—one can derive an AME requirement of

$$\Delta\phi_{\text{AME}} < \Delta\phi_{1\text{pm}} \frac{f_{\text{USO}} \cdot f_{\text{mod}}}{f_{\text{het}} \cdot f_{\text{USO}}} \sqrt{1 + \left(\frac{2.8 \cdot 10^{-3}\,\text{Hz}}{f}\right)^4} \approx 0.6 \cdot \sqrt{1 + \left(\frac{2.8 \cdot 10^{-3}\,\text{Hz}}{f}\right)^4}\,\frac{\text{mrad}}{\sqrt{\text{Hz}}}, \quad (1)$$

which increases as $1/f^2$ for frequencies below 2.8 mHz and is valid in the LISA frequency range.

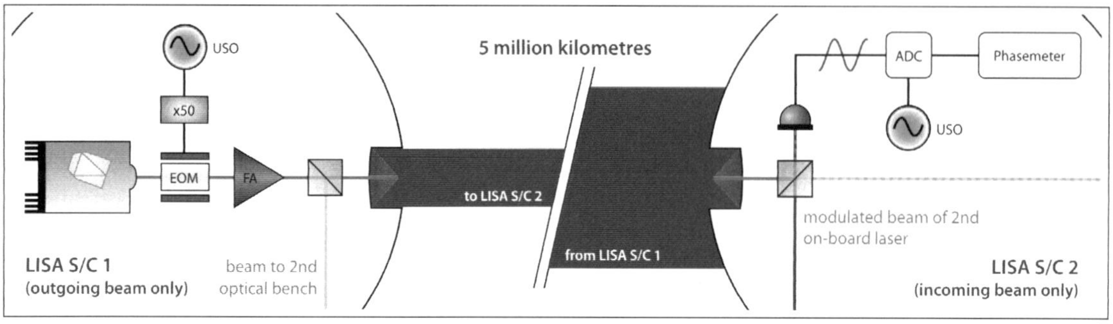

Figure 1. Inter-spacecraft clock transfer scheme: outgoing beam (left), incoming beam (right).

1.2. Phase characteristics of transfer chain components

Experiments are essential to characterize suitable components and to find devices phase stable enough to be used for the clock tone transfer chain. This paper presents measurements of the AME induced during the modulation process by the EOM or HF cable assemblies. EOM phase characteristics were investigated recently, but only for pairs of EOMs (differential processing of sideband pairs) and at modulation frequencies of 8 GHz [7]. The advantages of such high modulation frequencies are: (i) less sideband power is needed in order for clock calibration noise to be $\frac{1}{10}$ of the carrier-carrier shot-noise limit [8], and (ii) the relaxation upon the phase stability requirements is higher. The disadvantages, however, are a higher electrical power necessary for generating sidebands at higher frequencies and all common problems involved with HF electronics. We aim for a lower sideband frequency as proposed by [9].

2. Electro-optic modulator

For our sideband phase stability research we used a *Fibre-coupled Integrated Optical Phase Modulator* by Jenoptik. This EOM is polarization-maintaining and allows broadband phase modulation for frequencies up to 3 GHz. The electro-optic medium used is a waveguide made of magnesium oxide doped lithium niobate ($MgO{:}LiNbO_3$) which provides a high damage threshold. Single-mode fibres with FC/APC connectors are used to couple the light in and out with a typical overall insertion loss of 4 dB. The spectral bandwidth is specified as 1064 ± 60 nm. By connecting the EOM output to a fibre-coupled confocal scanning Fabry-Perot interferometer, we determined the modulation efficiency at a modulation frequency (f_{mod}) of $\sim$2 GHz to be $0.37\,\frac{\mathrm{rad}}{\mathrm{V}}$ which is equivalent to a half-wave voltage (the voltage necessary for a modulation index of $m = \pi$) of 8.6 V.

2.1. Potential phase noise sources

There are several effects in the ferroelectric substrate of the optical waveguides used in integrated EOMs which might add excess phase noise to the sidebands. A heat-induced expansion will change the length of the waveguide and thereby the optical path length, which leads to an unwanted phase shift. In addition, expansion of the waveguide substrate may lead to stress. The waveguide's refractive index is not constant over temperature which is also referred to as the thermo-optic effect. Due to the material's pyroelectricity, changes in temperature can also produce an electric potential (pyroelectric effect). This potential will then shift the phase of the light due to the linear electro-optic effect (Pockels effect) of lithium niobate [10]. Furthermore, for lithium niobate the refractive index decreases in response to higher optical powers which is referred to as the photorefractive effect [11].

 doi:10.1088/1742-6596/154/1/012006

2.2. Measurement setup

Fig. 2 shows a schematic of the setup used to measure the phase noise of a single EOM sideband. Two lasers at 1064 nm (L1 and L2) were offset phase locked to $f_{\mathrm{Ref}} = 2\,\mathrm{GHz} + 1.6\,\mathrm{kHz}$ with one beam (L1) sent through the EOM under test driven by a signal at $f_{\mathrm{Mod}} = 2\,\mathrm{GHz}$. We obtained a signal level of 18 dBm driving the EOM. This resulted in a modulation index of $m = 0.65$ or respectively 12% of the carrier power in a single sideband. Both laser beams were heterodyned at a beamsplitter and sent to a photodiode where different heterodyne frequencies were detected: the two carriers beating against each other; and the two sidebands of laser L1 beat against the carrier of laser L2. In a simplified description avoiding the spatial mode and polarization of the electrical fields, the two dectectable beat notes within the photodetector bandwith are: (i) for the carrier-carrier signal at 2 GHz+1.6 kHz

$$U_{\mathrm{HF}} \propto A_{\mathrm{L1}} J_0\left(m\right) A_{\mathrm{L2}} \cdot \cos\left(2\pi f_{\mathrm{Ref}} t\right), \tag{2}$$

and (ii) for the sideband-carrier signal at 1.6 kHz

$$U_{\mathrm{LF}} \propto A_{\mathrm{L1}} J_1\left(m\right) A_{\mathrm{L2}} \cdot \cos\left[2\pi\left(f_{\mathrm{Ref}} - f_{\mathrm{Mod}}\right) t - \phi_{\mathrm{EOM}-}\left(t\right)\right], \tag{3}$$

where $\phi_{\mathrm{EOM}-}(t)$ is the phase noise of the lower sideband. A_{L1} and A_{L2} are the electrical field amplitudes of the two laser beams (L1 and L2), m is the modulation index which is proportional to the amplitude of the signal driving the EOM ($m \propto A_{\mathrm{Mod}}$) and $J_n(m)$ is the Bessel function of the first kind, order n, for the value m. Since for the EOM phase characterization we used a low modulation index of $m \approx 0.6$ one can ignore the higher-order sidebands.

The carrier-carrier beat note (Eq. (2)) was passed to an electronic mixer and mixed down with the same signal driving the EOM. Considering only the low-pass filtered signal of the mixer output we thereby obtained another signal at 1.6 kHz:

$$U_{\mathrm{Mix}} \propto \frac{A_{\mathrm{L1}} J_0\left(m\right) A_{\mathrm{L2}} A_{\mathrm{Mod}}}{2} \cdot \cos\left[2\pi\left(f_{\mathrm{Ref}} - f_{\mathrm{Mod}}\right) t\right]. \tag{4}$$

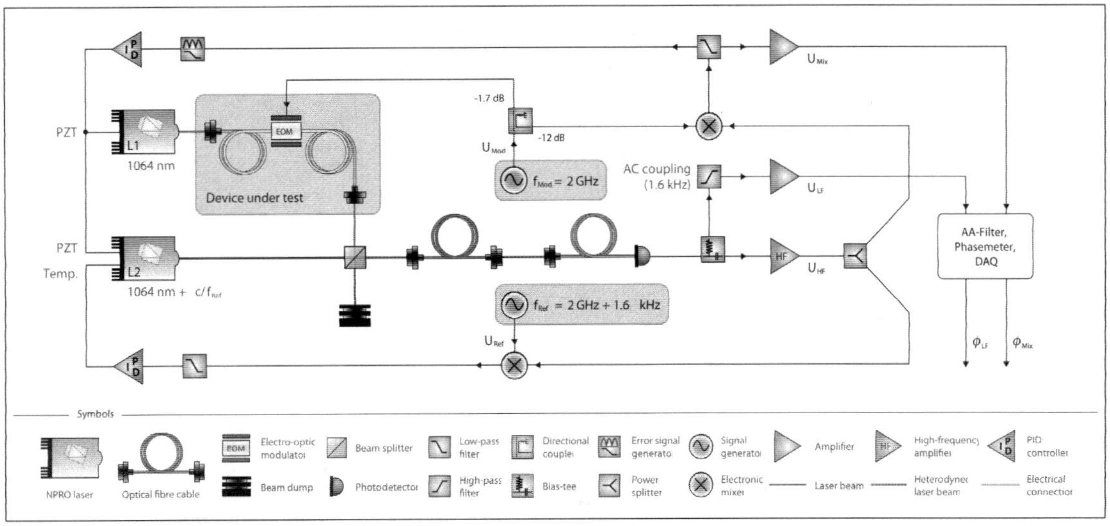

Figure 2. Setup for measuring the phase characteristics of a single EOM sideband at 2 GHz (schematic): Two lasers (one modulated by an EOM at $f_{\mathrm{Mod}} = 2\,\mathrm{GHz}$) offset phase locked to $f_{\mathrm{Ref}} = 2\,\mathrm{GHz} + 1.6\,\mathrm{kHz}$ were heterodyned and power stabilized with respect the amplitude of the 2 GHz beat note by a second control loop.

A DC voltage proportional to the amplitude of the $U_{\rm Mix}$ signal was passed to a loop filter which was connected to both laser diode current drivers. Hence we were able to stabilize the beat note power. The EOM phase noise measurement required the phase information of the two signals described by Eq. (3) and Eq. (4). A software phasemeter operating with the same principle as the LISA Technology Package (LTP) phasemeter was used for this purpose [12]. After determining the phase of both signals (where $\phi_{\rm Mix}$ is the phase of $U_{\rm Mix}$ and $\phi_{\rm LF}$ is the phase of $U_{\rm LF}$) we calculated the differential phase $\Delta\phi = \phi_{\rm Mix} - \phi_{\rm LF}$. The spectral phase noise densities of the differential phases were then calculated by the LPSD program [13].

One finds that the phase noise of the lower EOM sideband ($\phi_{\rm EOM-}(t)$) is observable in the differential phase $\Delta\phi$. Further phase noise induced by the frequency generators and the phase-locked loop (PLL) electronics are common-mode and thereby cancel in the differential phase. However, noise induced by cables, photodiodes, HF electronics and the EOM itself is not common-mode and therefore visible in the differential phase $\Delta\phi$. Thus all HF electronics, the lasers, optics and the EOM were isolated by extruded polystyrene boxes providing a low thermal conductivity of $0.03\,\frac{\rm W}{\rm m\cdot K}$ and thereby a temperature stability of about $0.1\,\frac{\rm K}{\sqrt{\rm Hz}}$ at $10^{-3}\,$Hz. Still the spectral phase noise density, calculated from a time series of $\Delta\phi$, represents an upper limit for the sideband phase noise induced by the EOM under test.

2.3. Measurement results

Fig. 3 shows the phase noise calculated from a time series of $\Delta\phi$ recorded over several hours. The upper limit of the EOM's phase noise (red solid curve) was better than the requirements over the whole frequency range. For comparison a measurement without amplitude stabilization is shown (red dashed curve). Here signal power fluctuations increased the phase noise by about an order of magnitude. The phasemeter sensitivity (blue curve) was well within the required level.

To measure the phase shift caused by a change in the EOM temperature, we constructed a two-part aluminium housing for the EOM with Peltier elements and a heat sink mounted on top. We then intentionally varied the EOM temperature in a range between 22 and 25 °C. Multiple temperature sensors were installed in the top and bottom parts. Fig. 4 shows a plot of the phase shift induced by the EOM versus its temperature. The measurement data is highlighted in red for the warming periods and in orange for the cooling periods. A linear fit is shown in blue which presents a correlation of $2.05 \pm 0.03\,\frac{\rm mrad}{\rm K}$.

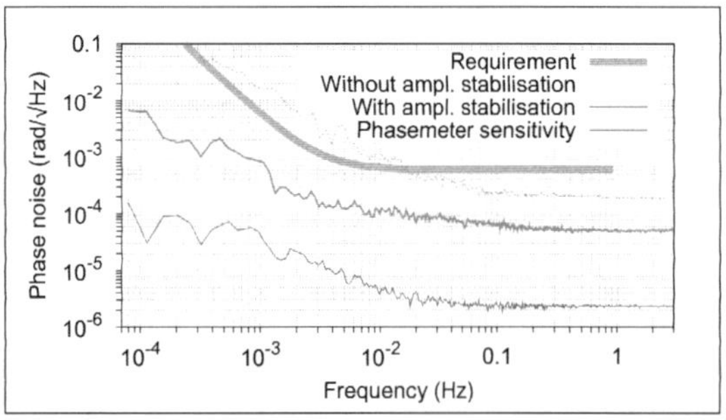

Figure 3. Upper Jenoptik EOM spectral phase noise density limit at 2 GHz modulation frequency.

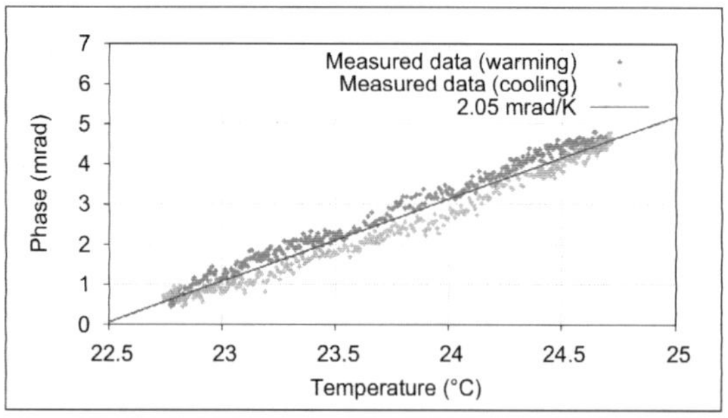

Figure 4. Differential phase versus EOM temperature measurement on Jenoptik EOM at 2 GHz modulation frequency.

3. HF cable assembly

In the EOM phase characteristics measurements discussed above, one cannot distinguish between EOM sideband phase noise added by the HF cable due to changes in temperature and the noise added by the EOM itself. Hence one needs to be sure that the cable is not adding a significant level of excess phase noise to a signal in the region of interest. In the following we concentrate on one representative cable assembly, a 1.22 m long micro-porous *Teflon* (PTFE) dielectric cable named *minibend LR* from Astrolab, USA. Its frequency range is up to 26.5 GHz with an attenuation of 0.93 $^{\mathrm{dB}}/_{\mathrm{m}}$ at 2 GHz.

3.1. Measurement setup

To measure the temperature-induced phase shift of a 2 GHz signal in an electrical cable, a split 2 GHz signal was sent through two identical cables and passed to mixers, each connected to a split 2 GHz + 1.6 kHz signal. Since the mixing process preserves any original phase shifts of the high-frequency signals, the low-pass filtered output of each mixer provided the phase information of the corresponding signal line and was sent to a phasemeter. The setup was divided into two parts. Each part was thermally isolated with an extruded polystyrene box. The first part, which contained the reference cable and all HF electronics, was kept at a constant temperature, while the temperature of the second part, which contained part of the cable under test, was periodically altered by a heating tape, connected to a remotely controlled power supply to alter the temperature of the cable under test between 25 and 45 °C.

3.2. Measurement results

Fig. 5 shows that the temperature-induced phase shift is not linear. Between 25 and 35 °C we observed a maximum slope of $1.2 \pm 0.2\,\frac{\mathrm{mrad}}{\mathrm{K\cdot m}}$. Between 35 and 45 °C the cable is much more phase stable with a maximum slope of $0.2 \pm 0.2\,\frac{\mathrm{mrad}}{\mathrm{K\cdot m}}$. Early measurements performed over a wider temperature range (tendential behaviour in Fig. 5, dashed curve) indicated a rapid phase shift at room temperature which is referred to as the *Teflon* knee [14].

The cable's long time phase noise is presented in Fig. 6. The blue curve shows the phasemeter sensitivity which is well within the phase stability requirement for 2 GHz shown in grey. The red curve represents the phase characteristics of the cable under test while it was exposed to laboratory's temperature fluctuations, while the orange curve is the phase noise when both cables were kept inside the thermal isolation box. The phase noise induced by the cable under test was in both cases better than the requirements.

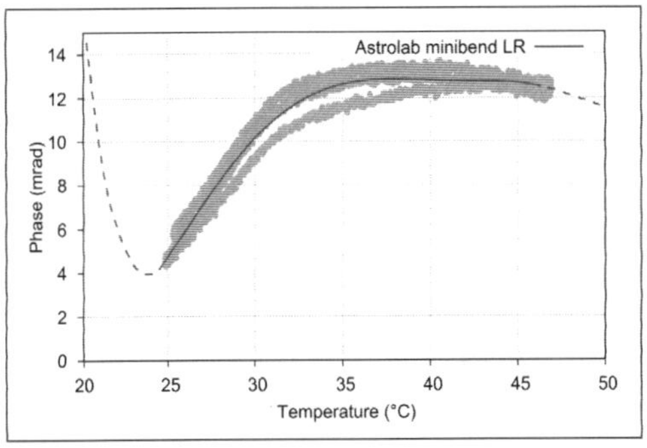

Figure 5. Differential phase versus temperature measurement for an Astrolab *minibend LR* cable, scaled to 1 m length.

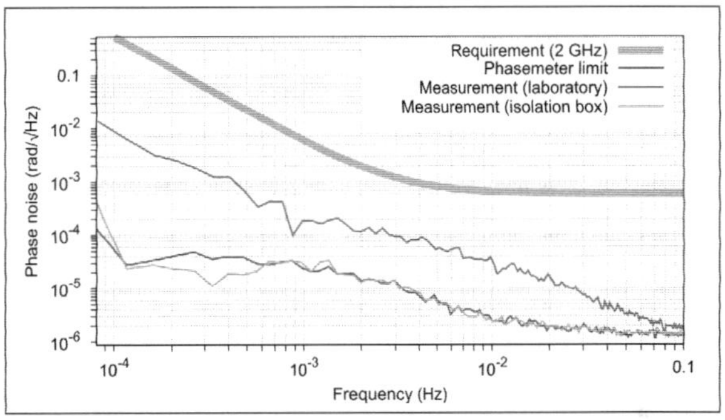

Figure 6. Spectral phase noise densities for an Astrolab *minibend LR* cable with and without thermal isolation box, scaled to 1 m length.

Journal of Physics: Conference Series **154** (2009) 012006 doi:10.1088/1742-6596/154/1/012006

4. Conclusions

We reported on the sideband phase characteristics of a Jenoptik fibre-coupled integrated EOM at modulation frequencies of 2 GHz in the LISA frequency range (1 mHz to 1 Hz). The upper phase noise limit was almost an order of magnitude better than the required $0.6\,\frac{\text{mrad}}{\sqrt{\text{Hz}}}$. The EOM's phase dependence on its temperature was measured to be $2.05 \pm 0.03\,\frac{\text{mrad}}{\text{K}}$. The designed temperature stability for LISA is $9.1 \cdot 10^{-8}\,\frac{\text{K}}{\sqrt{\text{Hz}}}$ at the optical bench and $10^{-5}\,\frac{\text{K}}{\sqrt{\text{Hz}}}$ at the gravity reference sensor (GRS) housing with an absolute temperature of 293 K [2]. Therefore, the given stability easily guarantees that the EOM fulfils the phase stability requirements with a large margin. One now needs to identify a space qualified version providing an equivalent performance.

Furthermore we showed that the phase noise induced to a 2 GHz signal by a representative HF cable—even when exposed to laboratory's temperature fluctuations—was well within the requirements. In addition, insight into the continuous dependency between temperature and phase of the HF cable under test was presented and we are about to perform measurements with a wider temperature range to map the phase unstable area around room temperature (*Teflon* knee).

Acknowledgments

We gratefully acknowledge support by Deutsches Zentrum für Luft- und Raumfahrt (DLR) (reference 50 OQ 0601).

References

[1] Faulks H et al 2000 *Study of the Laser Interferometer Space Antenna, Final Technical Report* (European Space Research and Technology Center)
[2] EADS Astrium 2007 personal communication
[3] LISA Study Team 1998 *LISA. Laser Interferometer Space Antenna for the detection and observation of gravitational waves. An international project in the field of Fundamental Physics in Space. Pre-Phase A report* (Garching, Germany: Max-Planck-Institut für Quantenoptik)
[4] Hellings R W 2001 *Phys. Rev. D* **64** 022002
[5] Tinto M et al 2002 *Phys. Rev. D* **65** 082003
[6] Couch II L W 1990 *Digital and Analog Communications Systems* (New York: Macmillan)
[7] Klipstein W et al 2006 *Laser Interferometer Space Antenna: 6th International LISA Symposium* (*American Institute of Physics Conference Series* vol 873) ed Merkovitz S M and Livas J C pp 312–318
[8] Folkner W et al 2004 *Presentation at the 5th International LISA Symposium*
[9] Tinto M et al 2008 *Classical and Quantum Gravity* **25** 015008 (12pp)
[10] Ruske J P 2004 *Integriert-optische Bauelementekonzepte zur Führung und Beeinflussung von Licht hoher Leistung und Photonenenergie* Habilitation thesis Friedrich-Schiller-Universität Jena Germany
[11] Steinberg S 1994 *Experimentelle Untersuchung optischer Phaseninstabilitäten an Streifenwellenleitern und elektrooptischen Phasenmodulatoren in LiNbO3 unter besonderer Berücksichtigung des Protonenaustausches* Ph.D. thesis Friedrich-Schiller-Universität Jena Germany
[12] Heinzel G et al 2004 *Classical and Quantum Gravity* **21** S581–S587
[13] Tröbs M et al 2006 *Measurement* **39** 120–129
[14] Micro Coax Application Notes 27 *Understanding Phase Versus Temperature Behavior: The Teflon Knee.*

Journal of Physics: Conference Series **154** (2009) 012007 doi:10.1088/1742-6596/154/1/012007

LTP – LISA Technology Package: Development Challenges of a Spaceborne Fundamental Physics Experiment

R Gerndt[1], for the entire LTP Team

Astrium Satellites GmbH, Claude-Dornier-Str., 88090 Immenstaad, Germany

e-mail: ruediger.gerndt@astrium.eads.net

Abstract. The LISA Technology Package (LTP) is the main payload onboard the LISA Pathfinder Spacecraft. The LTP Instrument together with the Drag-Free Attitude Control System (DFACS) and the respective LTP and DFACS operational software forms the LTP Experiment. It is completed by the FEEPs of the LPF spacecraft that are controlled by DFACS in order to control the spacecraft's attitude along with the experiment's needs. This article concentrates on aspects of the Industrial development of the LTP Instrument items and on essential performance issues of LTP. Examples of investigations on specific issue will highlight the kind of special problems to be solved for LTP in close cooperation with the Scientific Community.

1. Introduction

The LISA Technology Package (LTP) is the main payload onboard the European Space Agency's LISA Pathfinder Spacecraft (LPF). The development currently is on the turn from existing engineering models to building the final flight items. LPF is scheduled for launch in 2010.

LTP is a pre-cursor experiment that shall prove in space feasibility of the edge-of-technology techniques necessary to build the Laser Interferometer Space Antenna (LISA), a Gravitational Wave Observatory currently being under study by the European Space Agency (ESA) and by NASA together with space industry.

An overview on the LTP Mission and its science objectives has been provided by S Anza et al. [1]. The essentials of the Drag-Free Attitude Control System (DFACS) of the LTP Experiment have been summarized by, e.g., Fichter et al. [2]. DFACS will not be addressed specifically herein. Overviews on the LTP Instrument have been presented by R Gerndt et al. ([3] and [4]).

This paper is providing examples of specific challenges being encountered during development of the LTP Instrument. However, this snap shot is not meant to qualify techniques not mentioned here to be easily achievable. For all elements of LTP the development teams of the European LTP Consortium have had to exploit technical solutions of unusual precision for space borne instrumentation in order to comply with the demanding requirements put on LTP equipment and the LTP Mission.

---

[1] To whom any correspondence should be addressed and who herein is summarizing work done by a lot of LTP colleagues (see Acknowledgement Section).

2. LTP Instrument and its Mission

2.1. Instrument Overview

The LTP Instrument consists of two main functional subsystems, namely the Inertial Sensor Subsystem (ISS) and the Optical Metrology Subsystem (OMS) as shown in Figure 1, which are controlled by the Data & Diagnostics Subsystem (DDS).

The ISS is providing all technical means necessary to bring the two LTP Test Masses (TM) into orbit and then - steered by DFACS algorithms - to control the TM attitude and position through electrostatic actuation and suspension so that one TM is freely falling in direction towards the other.

The OMS serves as a high precision optical sensor of the differential movement of the two TM and of the movement of one of the TM with respect to the LPF Science Module (SCM). It is based on heterodyne Mach-Zender interferometry allowing for high precision measurements of TM position and attitude, e.g. intrinsically reaching the range of $6*10^{-12}$ m/Hz$^{1/2}$$*[1+(f/3mHz)^2]$ for 3 - 30 mHz in case of position sensing. For OMS details the reader is pointed to other publications like, e.g., those by G Heinzel et al. [5] and A F Garcia Marin [6].

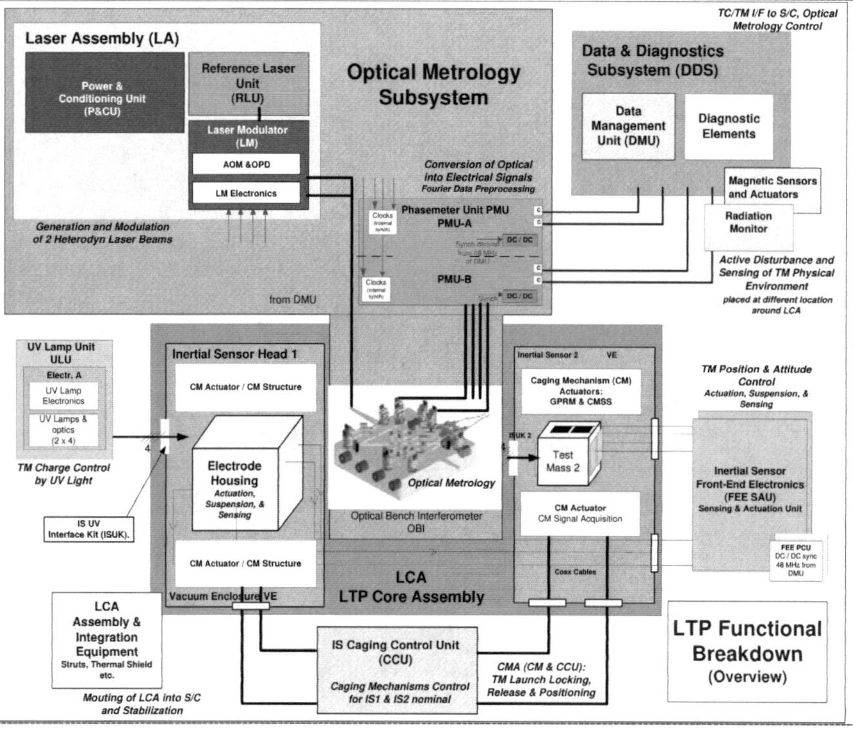

Figure 1. Functional Breakdown of LTP Instrument: The LTP Subsystems and Units

2.2. The Main Challenge of LTP: Launch Test Masses Safely and Keeping them Free-Floating

The essential task of LTP is to demonstrate that it is possible to launch a spacecraft without endangering the integrity of the Test Masses on-board and then to assure that on-station - despite of the artificial environment the TM are located in - they may be kept in a geodetic motion along their common coordinate axis x.

The acceleration noise level of less than 3×10^{-14} m s^{-2} Hz$^{-1/2}$ at 1 mHz to which the residual differential accelerations of the two TM along x need to be controlled is the main LTP mission requirement. Any time-varying process impacting directly on the TM movements or on the precision of the LTP electrostatic actuation and electrostatic and optical sensors need to be handled carefully in order to avoid disturbances of the geodetic motion beyond the level set by that prime mission requirement.

Any physical process may compromise the acceleration noise and thus quite a variety of field and direct forces need to be controlled by proper engineering of LTP and LPF items and their accommodation around the Test Masses. Figure 2 provides an overview on both, DC forces and AC forces, and the order of precision to which they need to be defined and controlled during the LTP mission in order to achieve the prime mission requirement.

Maximum Differential Acceleration Noise as a Measure for how free the Free Fall is:

$$S_a^{1/2}(f) \leq 3*10^{-14}\left[1+\left(\frac{f}{3mHz}\right)^2\right]\frac{m}{s^2}\frac{1}{\sqrt{Hz}}$$

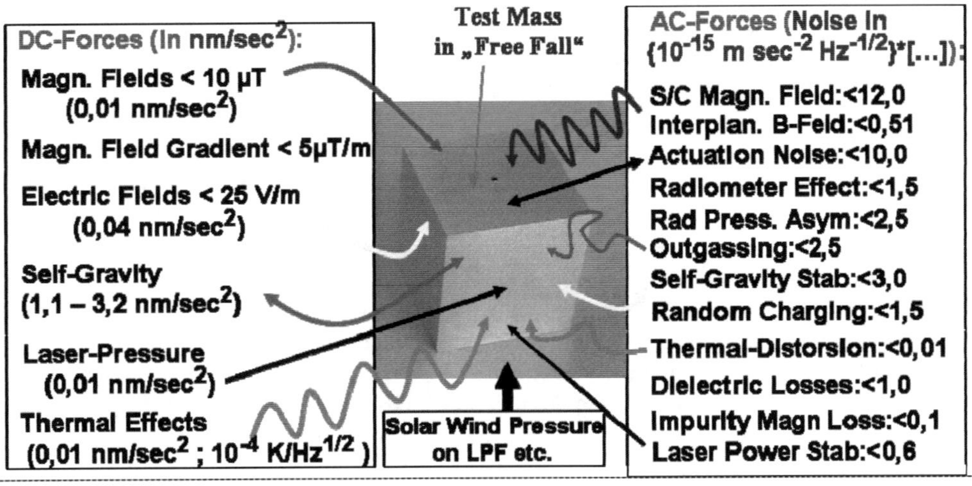

Figure 2. Forces impacting on Test Masses and Level of Control of these Forces necessary to Guarantee Sufficient Geodetic Motion of Test Masses

3. Technology Challenges Resulting from LTP Mission Task (Examples)

3.1. SCM Accommodation of LTP and Control of Self-Gravitation

Figure 3 shows the accommodation of the LTP elements on the LPF SCM. Besides the usual work to place the elements inside an spacecraft, locations of LTP and SCM items (and their internal mass distribution) need to be neatly defined during the design phase and the hands-on integration needs to be controlled with specific effort aiming at gaining the knowledge of the real mass distribution around the TMs. Based on the achieved as symmetrical as possible mass distribution, balance masses will be designed that internal to the ISHs will guarantee that the TMs finally will be located in the centre of the combined gravity field of all the spacecraft's mass surrounding them. This gravity balancing shall allow for keeping the residual gravitationally induced acceleration below about 1 to 3 nm/s^2 and to control down to $3*10^{-15} * [1+(f/3mHz)^2]$ m s^{-2} Hz$^{-1/2}$ the noise contribution due to gravitational field variations that in turn are caused, e.g., by thermal expansion of the SCM.

Still self-gravity will be one of the largest contributions to the differential acceleration noise and thus considerable effort needs to be spent in order to acquire precise knowledge on the mass distribution and the uncertainties about the as-built properties. As example, Figure 4 illustrates the complexity of items like the LTP Instrument harness. For defining the routing the gravitational balance needs had to be obeyed so that finally harness elements at SCM of cm-range length will be known in position to within several mm. This requires an accordingly precise pre-definition of the routing path and an integration process precise to an order not applied yet in space business.

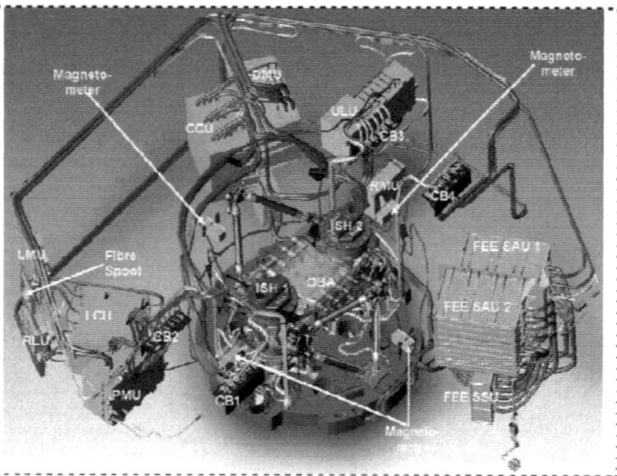

Figure 3. Accommodation of LTP on-board the LPF SCM.

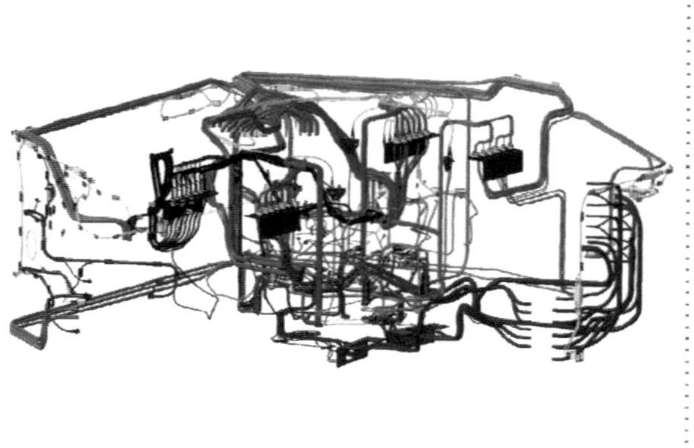

Figure 4. Routing of complex LTP Harness on SCM illustrating how extensive work necessary for controlling its gravitational impact on the TMs is.

3.2. Structural Integrity and Stability of the LTP Core Assembly (LCA)

The Optical Bench Assembly needs to support an Optical Bench made of Zerodur (OBI) centered between two heavy Inertial Sensor Heads (ISH). The Optical Bench Assembly then is integrated into the SCM by help of a set of struts providing a quasi-hyperstatic mounting of this set-up, called the LTP Core Assembly. For LCA design high launch load cases for LPF are to be encountered along with extremely tight structural stability and alignment requirements.

Alignment stabilities to be reached are in the order of 100 pm/Sqrt(Hz) for the IS heads distance and of 500 pm/Sqrt(Hz) and $0.1*10^{-6}$ rad/Sqrt(Hz) for position and attitude stability of OBI with respect to the Electrode Housings around the TM. The E2E performance simulation carried out to ensure a proper design of the LCA with respect to opto-dynamical stability and the self-gravity disturbance stability has been described in more detail by Brandt et al. [7].

Zerodur is the material of choice for the intermediate support structure due to the material's superior thermo-elastic stability. Application of Zerodur in space technology requires a good knowledge of the mechanical behavior of the challenging glass ceramic. The demanding Ultimate Physical Stress limit of 15 MPa including uncertainty factors needs to be guaranteed through specific design measures despite of the launch loads. The resulting design of an ultra-stable LCA structure is shown in Figure 5. Figure 6 shows the Structural Model of the LCA that was used to demonstrate that the LCA can survive the launch loads while keeping the internal alignment within the tolerances of several tens of microns respectively several mrad.

3.3. Launching Test Masses Safely and Releasing it into Free Fall

During launch quasi-static loads of about 50 g will result in forces of 1000 N acting on the TM. To safe-guard the TMs during launch the Caging Mechanism (CM) holds the TM with 8 fingers (Figure 7). A form-fit interface between fingers and TM ensures equal load per finger and avoids slippage of the 8 fingers on the TM that could damage its surface. 1-2 μm machining accuracy of fingers and TM corners are needed and technology tests have proven that such accuracies can be achieved despite of the fact that delicate materials are to be handled. To lay-out and to prove the design solution of the TM-to-finger interface extensive structural analyses had to be performed like such illustrated in Figure 8.

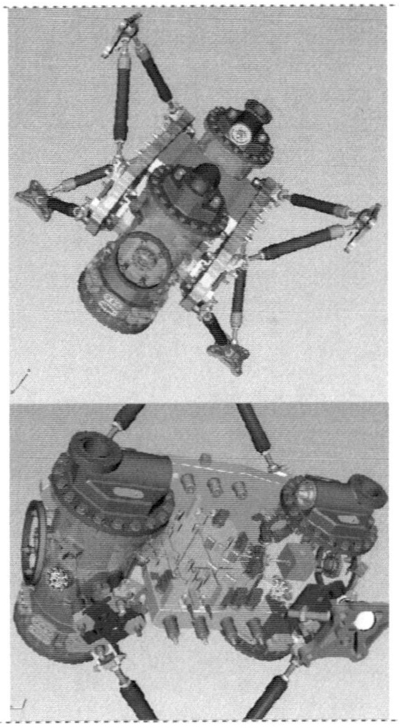

Figure 5. LTP Core Assembly Design: The LCA integrates the ISHs with the OBI and provides a very stable alignment between TMs and OBI.

Figure 6. The LCA Structural Model (upper picture) has been successfully tested. The lower picture shows the LCA integrated into the S/C Interface Structure.

The interface concept chosen after extensive studies has been successfully tested on a CM breadboard model and the design is currently facing Engineering Qualification Model testing.

Each of the fingers still has to carry a load of 125 N causing, together with other effects, quite high adhesion forces between TM and CM fingers. To overcome these adhesion forces without damaging the TM surfaces and to release the TM into free fall another part of the CM, the Grabbing, Positioning and Release Mechanism, is used. The challenges of this mechanism are as well demanding but cannot be discussed here.

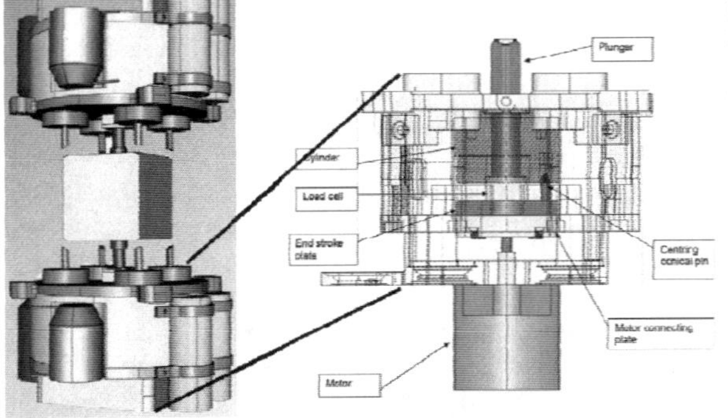

Figure 7. LTP Caging Mechanism showing the 8 fingers securing the TM during launch. On-station the plunger (blue central finger) will hold the TM and it will be used for releasing the TM.

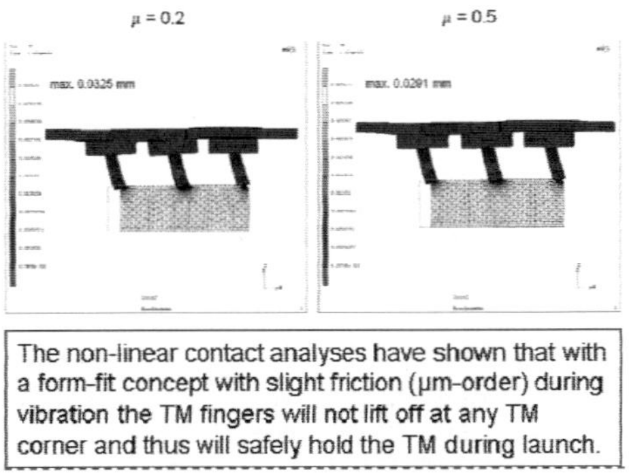

Figure 8. Structural analyses demonstrating that the finger-to-TM interface concept is capable of holding the TM during launch without TM slippage.

3.4. Finite-Element Electrostatic Analysis of the Inertial Sensor (IS)

Comprehensive finite-element electrostatic analysis of the Inertial Sensor's electrostatic fields within the Electrode Housing with MAXWELL 3D (and 2D) field simulation software have been performed to verify the approximations of the IS electrostatic modeling (used in performance analysis, simulation and on-board algorithms). This tool was used as well to confirm compatibility of operational scenarios with electrostatic need for issues like test mass release with resulting electrostatic impact of grabbing plunger inside the EH volume close to the TM. Another example is the investigation of the acceptability of TM corners - the specific form design of which is dedicated to the CM finger interface needs - for the electrostatic TM control process. As well TM surface patch effects for un-coated surfaces have been analyzed. Figure 9 illustrates the effort put into this performance issue of LTP.

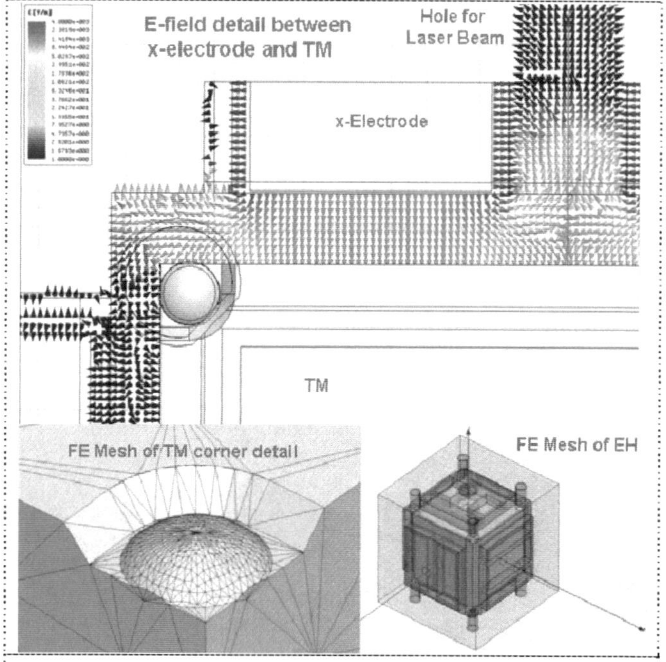

Figure 9. Finite-Element Electrostatic Analysis

4. Conclusion

The LTP Instrument consists of elements most of which are based on edge-of-technology designs that shall assure achieving the extreme level of performance of their functionality necessary to reach the LTP mission requirement on the Test Masses differential acceleration noise.

Extensive performance analyses and performance budget control show that the expected performance of the LTP Instrument is compliant with the mission requirement. The performance analysis is now more and more confirmed by incoming results from performance measurements performed with Engineering Models of the LTP configuration items, so that technical evidence is building up that the real hardware and software of the flight model can achieve the required performance.

Acknowledgments

The Author wishes to acknowledge and to thank for the outstanding efforts spent by all teams of the European LTP development consortium to realize the mission and the courage not to capitulate to the extreme technical challenges. Without the driving force and support of the European Space Agency and of the numerous National Space Agencies involved, the project could not be realized. Instead of listing the large number of valuable individual contributors which would desire to be mentioned the institutions they are with shall be named: University of Trento, Italy (PI); Albert-Einstein-Institut, Hannover, Germany (Co-PI); The Agencies: ESA, DLR (Germany), CNES (France), ASI (Italy), ICE (Spain), STFC/UK; LTP Consortium: Astrium GmbH, Germany (LTP Industrial Architect), with their Subcontractors: Kayser-Threde, Germany, Tesat, Germany, Thales Alenia Space Italy, and with the National Suppliers: Carlo Gavazzi (Italy), University of Glasgow (UK), University of Birmingham (UK), Imperial College London (UK), Oerlikon Space (Switzerland), SRON (The Netherlands), NTE (Spain).

References

[1] **S Anza, M Armano, E Balaguer, M Benedetti, C Boatella, P Bosetti, D Bortoluzzi, N Brandt, C Braxmaier, M Caldwell, L Carbone, A Cavalleri, A Ciccolella, I Cristofolini, M Cruise, M Da Lio, K Danzmann, D Desiderio, R Dolesi, N Dunbar, W Fichter, C Garcia, E Garcia-Berro, A F Garcia Marin, R Gerndt, A Gianolio, D Giardini, R Gruenagel, A Hammesfahr, G Heinzel, J Hough, D Hoyland, M Hueller, O Jennrich, U Johann, S Kemble, C Killow, D Kolbe, M Landgraf, A Lobo, V Lorizzo, D Mance, K Middleton, F Nappo, M Nofrarias, G Racca, J Ramos, D Robertson, M Sallusti, M Sandford, J Sanjuan, P Sarra, A Selig, D Shaul, D Smart, M Smit, L Stagnaro, T Sumner, C Tirabassi, S Tobin, S Vitale, V Wand, H Ward,W JWeber and P Zweifel,**
The LTP experiment on the LISA Pathfinder mission, Class. Quantum Grav. **22** (2005) S125–S138,

[2] Walter Fichter, Peter Gath, Stefano Vitale and Daniele Bortoluzzi
LPF drag-free control & system implications, Class. Quantum Grav. 22 (2005) S139–S148,

[3] **R Gerndt, W Fichter, N Brandt, D Gerardi, F Montemurro, A Schleicher, T Ziegler and U Johann**
LISA Technology Package System Design And Operations,
in: LASER INTERFEROMETER SPACE ANTENNA: 6th International LISA Symposium, Greenbelt, Maryland (USA), 19-23 June 2006, AIP Conference Proceedings, Volume 873, pp. 668-674 (2006), ISBN: 978-0-7354-0372-7, DOI: 10.1063/1.2405115

[4] R Gerndt
LTP System Design Synthesis Report, S2-ASD-RP-3009, Issue3, Astrium GmbH, 2007

[5] G Heinzel, J Bogenstahl, C Braxmaier, K Danzmann, A Garcia, F Guzman, J Hough, D Hoyland, O Jennrich, C Killow, D Robertson, Z Sodnik, F Steier, H Ward and V Wand,
Interferometry for the LISA Technology Package: An Update, J. Phys. Conf. Ser. 32 (2006), 132-136

[6] A F Garcia Marin, G Heinzel and K Danzmann
Interferometry for LISA and LISA Pathfinder, J. Physics: Conf. Ser. 66 (2007) 012052 (5pp)

[7] N Brandt, W Fichter, M Kersten, S Lucarelli and F Montemurro
LISA Pathfinder E2E performance simulation: optical and self-gravity stability analysis, Class. Quantum Grav. 22 (2005) S493-S499

Revised electrostatic model of the LISA Pathfinder inertial sensor

Nico Brandt[1,2] **and Walter Fichter**[2]

[1] Astrium GmbH, 88039 Friedrichshafen, Germany
[2] iFR, Universitaet Stuttgart, Pfaffenwaldring 7a, 70569 Stuttgart, Germany

E-mail: nico.brandt@astrium.eads.net

Abstract. A comprehensive electrostatic finite-element (FE) analysis of the LISA Pathfinder Inertial Sensor (IS) has been carried out at Astrium GmbH. Starting with a detailed geometrical model of the IS housing and test mass (TM) flight units, FE results were derived from multiple analyses runs applying the MAXWELL 3D field simulation software. The electrostatic forces and torques on the TM in 6 DoF, as well as all capacitances between the TM, the 18 electrodes, and the housing, have been extracted for different TM translations and rotations. The results of the FE analyses were expected to confirm the existing IS electrostatic model predictions for forces and torques used in performance analyses, simulations, and on-board algorithms. Major discrepancies were found, however, between the FE results and the model used so far. As expected, FE results give considerably larger capacitance values than the equivalent infinite non-parallel plate estimates. In contrast, however, the FE derived forces and torques are significantly lower compared to the analytic IS electrostatic model predictions. In this paper, these results are discussed in detail and the reasons for the deviations are elaborated. Based on these results, an adapted analytic IS electrostatic model is proposed that reflects the electrostatic forces, torques, and stiffness values in the LISA Pathfinder IS significantly more accurate.

1. Motivation

The electrostatic FE analysis of the LISA Pathfinder IS has been conducted for the following reasons: First, the results are used to verify the approximations of the current IS electrostatic modeling applied by performance analyses, simulations, and on-board algorithms. Second, operational scenarios, such as the TM release strategy, are confirmed by analyzing the impact of the retractable grabbing plunger in close vicinity of the TM. Third, caging finger and TM interface design trades are supported and patch effects due to non-coated surfaces are analyzed. In this paper, the focus is on the first topic; specifically on the electrostatic forces and torques exerted onto the TM. The FE derived results are compared with the existing infinite non-parallel plate based analytic IS electrostatic model. This model is currently applied throughout the LISA Pathfinder project [1, 2, 3, 4].

2. FE Analysis Tool, IS Model, and Settings

All FE analyses results presented in this paper have been generated with the field simulation software MAXWELL 3D [5]. In order to verify the time consuming three-dimensional simulation results, significantly faster solutions have been computed independently using the field solver of MAXWELL 2D for a simplified two-dimensional geometrical model of the IS. Both electrostatic

Journal of Physics: Conference Series **154** (2009) 012008 doi:10.1088/1742-6596/154/1/012008

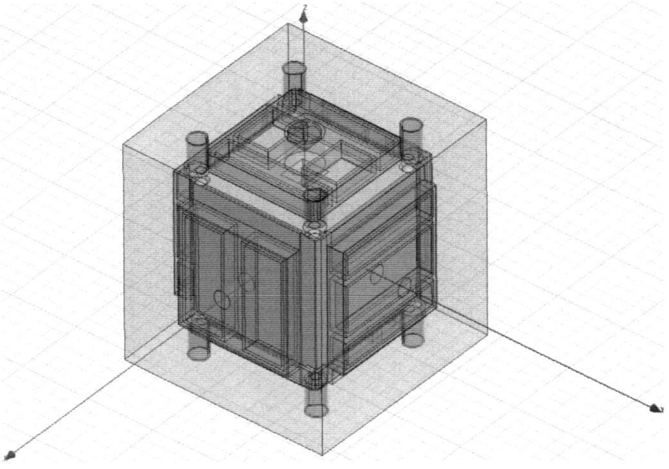

Figure 1. IS geometrical model in MAXWELL 3D. In the housing center the TM is depicted in yellow with its intricate corner features for the caging fingers and the pyramidal indention for the upper grabbing plunger.

field simulators provide a large amount of software specific settings that influence the surface approximation of the geometrical model, the convergence of the field solution, and the solution parameters. They allow for the direct calculation of the electrostatic forces, the torques, and the capacitance matrix for all specified conductors. The forces and torques are computed independently from the capacitances via the principal of virtual work; that is, by distorting the surface mesh.

A detailed three-dimensional geometrical model of the IS housing and the TM has been build from scratch in MAXWELL using current flight unit mechanical interface drawings [6]. Consequently, conversion artifacts from CAD model importing have been avoided. The model of the IS used for the 3D analyses is depicted in figure 1. It consists of the 18 electrodes, the guard ring and housing structures, the eight caging fingers, the two grabbing plungers, and the test mass with its indentions and corner features.

In order to find the best trade-off between accuracy and computation time, and to quantify the accuracy of each solution, a significant amount of time was spent on assessing the electrostatic field solution specific software settings. MAXWELL is equipped with an adaptive mesh generator that refines the mesh based on user defined convergence criteria. The local field error contribution to the total field energy was chosen as one of the stopping criteria to be smaller than 0.5%. At the same time, the surface approximation by the mesh was largely improved in order to cope with the intricate TM corner features. The electrostatic force or torque on the TM, depending on the conducted simulation run, were defined as additional error criteria in order to enable monitoring their convergence for each mesh refinement pass during the field solution computation.

The electrostatic forces and torques exerted onto the TM are generated by assigning corresponding realistic voltages on the 12 IS model actuation electrodes according to [2]. Multiple analyses runs have been carried out for varying TM translations ($\pm 200\,\mu$m) and rotations (± 2 mrad). For each of these runs, the forces and torques in 6 DoF, as well as all capacitances in-between the TM, the 18 electrodes, and the housing, were computed. Least-squares polynomials have been fitted to the electrostatic force, the torque, and capacitance data in order to allow for the accurate prediction of their gradients and second order derivatives within the model accuracies.

3. Results and Discussion

Major discrepancies have been identified between the FE analyses results and the IS electrostatic model predictions used so far.

Table 1. Finite-element vs. current analytic model capacitances (for nominal TM position and orientation). Analytic capacitances are derived from electrode dimensions and gaps using the infinite non-parallel plate capacitor formula according to [1, 2]. Analytic $C_{TM,H}$ is adopted from [3]. C_{TM} is computed as sum of all analytic capacitances.

Conductors		Capacitance [pF]		Error [%]
		FE	Analytic	
$C_{EL_{1-4},TM}$	(C_x)	1.18	1.09	7.8
$C_{EL_{5-8},TM}$	(C_y)	0.95	0.83	12.5
$C_{EL_{9-12},TM}$	(C_z)	0.70	0.61	13.0
$C_{EL_{13-14},TM}$	$(C_{y_{inj}})$	1.29	1.35	4.6
$C_{EL_{15-18},TM}$	$(C_{z_{inj}})$	0.41	0.43	3.8
$C_{TM,H}$		18.67	≈ 10.82	42.0
C_{TM}	(C_{tot})	34.21	≈ 25.34	25.9

Table 1 compares the FE derived capacitance values with the current infinite non-parallel plate estimates for nominal (zero) TM position and orientation. For the actuation electrode capacitances ($C_{EL_{1-12},TM}$), and the capacitance between TM and housing ($C_{TM,H}$), the FE results give considerably larger values than the equivalent analytic estimates. In contrast, the capacitances between injection electrodes and test mass ($C_{EL_{13-18},TM}$) are a little lower. In sum, however, the total capacitance (C_{TM}) between the TM and its surrounding is more than 25% larger than its current estimate.

These differences in capacitance were expected and are attributable to the fact that the infinite non-parallel plate approximations ignore fringing field effects at the electrode edges and, thus, underestimate the capacitances by neglecting the effective area increase. The magnitude of the differences matches very well the analytic predictions of capacitance correction formulas given in literature for Kelvin guard-ring capacitors [7]. Furthermore, the results are consistent qualitatively with previous findings using the ANSYS FE software for a less mature IS geometrical model [4]. Since the injection electrodes are further recessed behind the guard rings compared to the actuation electrodes, see figure 2, the infinite non-parallel plate model overestimates their capacitance. That is, despite neglecting the fringing field, the dominating effect in this case is the effective area decrease due to field lines closing onto the neighboring guard ring rather than the TM.

In contrast to the capacitances, the FE derived forces and torques are in general significantly lower compared to the analytic IS electrostatic model predictions. This is shown in table 2, where FE derived forces and torques at nominal TM position and orientation are compared to the existing infinite non-parallel plate based analytic model estimates. For the forces along x and y, as well as the torques around ϕ and θ that are generated by using the same actuation electrodes, the FE results give about 20% lower magnitudes. For the z-η actuation couple, the relative error is as big as 30%.

These large differences in forces and torques were not expected and are attributable to the following reasons: First, the current analytic IS electrostatic model is based on infinite non-parallel plate capacitor approximations for the capacitances (gradients) between actuation electrodes and TM. Second, the model considers *only* these gradients and neglects any additional capacitances. For example, the capacitance between each electrode and its surrounding housing is neglected, which is driving the identified differences as detailed in the following.

Table 2. Finite-element vs. current analytic model forces and torques (for nominal TM position and orientation). RMS voltages, which correspond to the maximum IS forces and torques, are applied to all actuation electrodes simultaneously. The TM voltage is kept at zero potential. Analytic forces and torques, and RMS actuation voltages, are derived from the analytic IS electrostatic model according to [1, 2].

DoF	Force [N]		Error [%]
	FE Analysis	Analytic Model	
x	2.12×10^{-9}	2.55×10^{-9}	20.2
y	3.59×10^{-9}	4.32×10^{-9}	20.3
z	5.49×10^{-9}	7.27×10^{-9}	32.5
	Torque [Nm]		
θ	1.55×10^{-11}	1.87×10^{-11}	20.5
η	1.19×10^{-11}	1.59×10^{-11}	33.2
ϕ	9.14×10^{-12}	1.10×10^{-11}	20.8

In table 3, the capacitances and their gradients relevant for the electrostatic force along x are shown. Again, FE derived results and their equivalent analytic estimates, where available, are compared for the nominal TM position and orientation. The electrostatic force is proportional to the gradient of the capacitances w.r.t. the force direction and the square of the potentials applied. Hence, the first capacitance gradient serves as a measure of each individual conductor's contribution to the overall force. The values clearly show that the capacitance gradient between actuation electrode and TM is the dominant force contributor. However, the gradient magnitude of the capacitance between each electrode and its surrounding housing ($C_{EL_{1-4},H}$) is in the order of 20% of the former with opposite sign, which coincides with the force error along x.

This result can be explained by the fact that the fringing fields at the actuation electrode edges change, depending on the test mass position altering the gap. The field lines will couple to their closest neighboring conductor. That is, by approximating the electric field lines to first order, depending on the shortest field line path either they close onto the guard ring or the TM. This effect is clearly visible in the electric field line plot, depicted in figure 2, which is

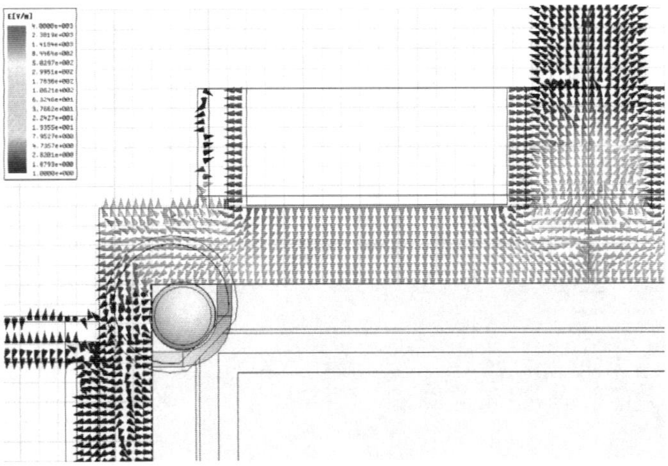

Figure 2. Electric field detail between electrode and TM in the x-y plane. Visible on top is a cut through an x-actuation electrode and the laser hole. To the left a cut through a y-injection electrode recessed behind the guard rings is shown. Clearly visible is the fringing field at the electrode edges.

Table 3. Finite-element vs. current analytic model capacitance gradients (for nominal TM position and orientation). Analytic capacitances and gradients are derived from electrode dimensions and gaps using the infinite non-parallel plate capacitor formula according to [1, 2].

Conductors	Capacitance [pF]		$\partial C_{i,j}/\partial x$ [pF/m]		$\partial^2 C_{i,j}/\partial x^2$ [pF/m^2]	
	FE	Analytic	FE	Analytic	FE	Analytic
$C_{EL_{1-4},TM}$	1.18	1.09	± 300	± 273	1.5×10^5	1.4×10^5
$C_{EL_{1-4},H}$	5.94	–	∓ 70	–	-1.6×10^3	–
C_{EL_1,EL_2} (C_{EL_3,EL_4})	2.18×10^{-4}	–	∓ 0.32	–	3.2×10^2	–
$C_{TM,H}$	18.67	≈ 10.82	0	0	8.2×10^5	–

generated from a MAXWELL analysis run. Thus, for a decreasing gap, the capacitance between the actuation electrodes and the TM is increased (positive gradient) while the capacitance between the actuation electrodes and the housing is decreased (negative gradient). The current analytic IS electrostatic model neglects the latter contribution. This explains the significantly larger forces and torques compared to the FE results. The in-between electrode capacitance gradient magnitude (C_{EL_i,EL_j}) is two to three orders of magnitude lower than the other two gradients; hence, it contributes only negligibly to the electrostatic force calculation.

The second derivatives in table 3 can be viewed as a measure for the stiffness (force gradient) contributions of each conductor. In contrast to the electrostatic force, the contribution of the electrode to housing capacitance to the electrostatic stiffness is small; it is about two orders of magnitude lower than the electrode to TM capacitance. The in-between electrode capacitance contribution, again, is three orders of magnitude smaller; thus, it can be safely neglected. The capacitance between TM and housing ($C_{TM,H}$), however, is significantly impacting the electrostatic stiffness. Its second derivative is about a factor of five larger than the electrode to TM capacitance contribution.

4. Improved Analytic IS Electrostatic Model

The inertial sensor electrostatic system can be described fundamentally by the schematic configuration of conductors and capacitances depicted in figure 3. As discussed above, in the current analytic IS electrostatic model the forces and torques exerted on the TM by the actuation electrodes are computed taking into account *only* the gradient of the electrode to test mass capacitance ($C_{EL_i,TM}$) w.r.t. the force direction. This results in a force and torque error of about 20–30% when compared to FE electrostatic analyses results. The main reason for this effect was shown to be the capacitance (gradient) between each electrode and its surrounding housing that has been neglected so far. As a consequence of this, an improved electrostatic force and torque model taking into account all capacitive gradients in the inertial sensor system is derived. Such a mathematical model in its most general form, derived from standard energetic considerations of the electrostatic field of conductors, is defined as

$$F_q = \frac{1}{2} \sum_{i=1}^{N} \sum_{j=i+1}^{N} \frac{\partial C_{i,j}}{\partial q} (V_i - V_j)^2 \tag{1}$$

Here, q is the generalized coordinate, V_i and V_j are any two conductor potentials in a system of N conductors, and $C_{i,j}$ is the corresponding capacitance between them.

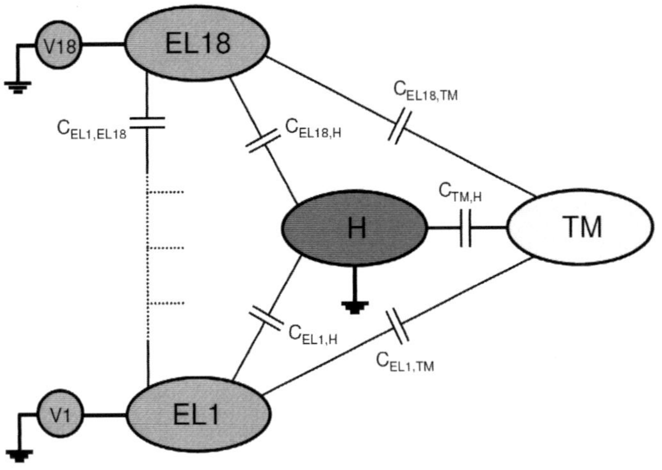

Figure 3. Schematic configuration of IS conductors and capacitances. The 18 electrodes with their respective voltage sources are denoted by EL1 to EL18, TM is the test mass, and H the grounded inertial sensor housing. All possible capacitances between the conductors are depicted.

Using the nomenclature of figure 3, the improved IS specific force and torque model can be derived to

$$F_q = \frac{1}{2}\sum_{i=1}^{18} \frac{\partial C_{EL_i,TM}}{\partial q} (V_i - V_{TM})^2 + \frac{1}{2}\frac{\partial C_{TM,H}}{\partial q} V_{TM}^2 \tag{2a}$$

$$+ \frac{1}{2}\sum_{i=1}^{18} \frac{\partial C_{EL_i,H}}{\partial q} V_i^2 \tag{2b}$$

$$+ \frac{1}{2}\sum_{i=1}^{18}\sum_{j=i+1}^{18} \frac{\partial C_{EL_i,EL_j}}{\partial q} (V_i - V_j)^2 \tag{2c}$$

Here, the term (2a) represents the existing IS electrostatic model, the term (2b) accounts for the non-negligible capacitance gradient between each electrode and its surrounding housing, and the last term (2c) represents the negligible contribution of the in-between electrode capacitance gradient. The test mass potential V_{TM} used in above equation is defined by

$$V_{TM} = \frac{\sum_{k=1}^{18} C_{EL_k,TM} V_k + Q_{TM}}{C_{tot}} \tag{3}$$

with Q_{TM} as the accumulated charge on the TM and $C_{tot} = \sum_{k=1}^{18} C_{EL_k,TM} + C_{TM,H}$.

Without taking into account the term (2b), using FE derived results for the electrode to TM capacitances (gradients), the force and torque error is about 30–50% when compared to FE electrostatic analyses results. The improved analytic IS electrostatic model, however, including the term (2b), reflects the FE derived forces and torques to better than 1% when using FE derived results for both the electrode to TM and electrode to housing capacitances (gradients).

References

[1] Brandt N, Hirth M, Fichter W, Schubert R, Warren C and Wealthy D 2008 *Experiment performance budget*, S2-ASD-TN-3036, LISA Pathfinder Project Documentation

[2] Schleicher A 2008 *Test mass actuation algorithm for DFACS*, S2-ASD-TN-2011, LISA Pathfinder Project Documentation

[3] Bortoluzzi D and Weber W 2005 *Test mass electrostatic suspension actuation stiffness and cross-talk matrix*, S2-UTN-TN-3009, LISA Pathfinder Project Documentation

[4] Shaul D N A, Araújo H M, Rochester G K, Sumner T J and Wass P J 2005 *Class. Quantum Grav.* **22** 297–309

[5] Maxwell 3D and 2D (v11.1.1) are registered trademarks of Ansoft Corporation, 225 West Station Square Drive, Suite 200, Pittsburgh, PA 15219, USA
[6] Thales Alenia Space 2008 *LISA EH replica EIDP*, S2-AAS-M-DP-3-0002, LISA Pathfinder Project Documentation
[7] Moon C and Sparks C M 1948 *J. Res. Natl. Bur. Stand.* **41** 497–507

Journal of Physics: Conference Series **154** (2009) 012009 doi:10.1088/1742-6596/154/1/012009

Principles, operations, and expected performance of the LISA Pathfinder charge management system

T Ziegler[1,2], W Fichter[2], M Schulte[3] and S Vitale[4]

[1] Astrium GmbH, 88039 Friedrichshafen, Germany
[2] iFR, Universität Stuttgart, Pfaffenwaldring 7a, 70569 Stuttgart, Germany
[3] Blackett Laboratory, Imperial College, Prince Consort Road, London SW7 2BZ, UK
[4] Department of Physics, University of Trento, 38050 Povo, Trento, Italy

E-mail: tobias.ziegler@astrium.eads.net

Abstract. The test masses of LISA Pathfinder are free flying and therefore not grounded to the spacecraft by a wire. Because of galactic cosmic rays, solar energetic particles, and unknown microscopic surface effects during initial test mass release, an unacceptable level of absolute charge might be present on the test masses. A charged test mass can endanger transition to high accuracy control modes which are required for science experiments. Furthermore, charged test masses introduce unwanted disturbance accelerations for example due to Coulomb interactions with surrounding conducting surfaces. The charge management system is designed to discharge the test masses up to a tolerable level of absolute charge such that the mission goal can be achieved. It is therefore an essential part of the experiments to be performed with the LISA Technology Package. The paper describes charge management tasks to be performed on board the spacecraft and summarizes the principles of charge measurement and discharge control. An overview of the experiment operations is given where the interconnection of operational charge management system modes and operational modes of the drag-free, suspension and attitude control system is considered. Simulated performance results are presented.

1. Introduction

LISA Pathfinder aims at the verification of differential free fall between the two test masses along one axis, with a required level of $3 \cdot 10^{-14}\,\mathrm{m/s^2/\sqrt{Hz}}$ in the measurement bandwidth between 1 mHz and 30 mHz [1]. This is accomplished by measuring the distance between the test masses along this *sensitive axis* with a precise optical metrology system. From this data, the acceleration can be obtained on ground. A drag-free control system [2] controls the spacecraft inertial attitude and stabilizes the test mass dynamics in such a way, that the non-gravitational acceleration along the sensitive axis is minimized. The primary science task is to measure and characterize all non-gravitational forces along the sensitive axis in a series of runs, each run consisting of one or more experiments.

2. Motivation for on-board charge management

The test masses within the gravitational reference sensors are free-floating. They are electrically isolated from the spacecraft and may therefore accumulate an unacceptable level of absolute charge. The test mass charging is due to galactic cosmic rays, solar energetic particle events and unknown microscopic surface effects which might occur during initial test mass release from the caging mechanism assembly. Even though the test mass net charging rate from energetic radiation is predicted to be positive [3], the test mass can also be negatively charged due to the unknown charge separation phenomenons during release. High absolute test mass charge levels impose functional and performance limitations on the experiment:

(i) functional limitations
  - mode transitions from *DFACS accelerometer mode* (a control system control mode to capture test masses after release) to high accuracy control modes with less actuation authority (used to perform science experiments) might be endangered

(ii) performance limitations
  - charged test masses introduce differential disturbance acceleration noise due to coupling with stray DC voltages, magnetic fields, and electrostatic actuation noise [4],
  - DC forces due to errors in electrostatic actuation algorithm which assumes that test mass charge is zero, and
  - increase actuation stiffness (in science modes a constant stiffness is assumed).

In order to keep this limitations sufficiently small, the test mass charge must be actively controlled on board the spacecraft. This is accomplished by means of the charge management system (CMS) software application and specific hardware equipment.

3. Provided functionality and required performance

The CMS provides the functionality to perform on board

(i) charge measurement without any discharge actuation,
(ii) fast discharge control to ensure the transition from accelerometer mode to high performance science control modes and to discharge a test mass in preparation of specific science runs,
(iii) continuous discharge control during the main science test run is carried out (the reason for this experiment is that in LISA continuous measurement and discharging is planned to be part of the baseline science control mode).

Fast discharge and continuous discharge are closed-loop control algorithms. However, their purpose and therefore their principle of operation is different. The goal of fast discharge is to quickly discharge a test mass up to a desired level; disregarding the cause of any disturbance noise. Robustness of the control algorithm is driving the design since it should work directly after test mass de-caging before the discharge performance can be calibrated in-flight. Fast discharge is terminated as soon as a user defined threshold value is reached. The main purpose of continuous discharge is to keep the test mass charge below a certain threshold without introducing considerable disturbance noise on the differential test mass acceleration. The performance requirements of the CMS functionalities are summarized in table 1. The

Table 1. Functionalities of the charge management system and required performance.

Functionality	Requirement
Fast discharge in DFACS accelerometer mode	$7.23 \cdot 10^7$ e
Fast discharge in DFACS science mode	$1.36 \cdot 10^6$ e
Continuous discharge in DFACS science mode	$1.00 \cdot 10^5$ e
Only charge measurement in DFACS science mode	$1.00 \cdot 10^4$ e

requirement on fast discharge in DFACS science mode ensures the test mass charge level to be

Journal of Physics: Conference Series **154** (2009) 012009 doi:10.1088/1742-6596/154/1/012009

below 10^7 unit charges for the duration of a 24 h science run. Notice that a charge level of 10^7 unit charges is allocated in the experiment performance budget. However, for the preparation of an individual test run, the desired discharge performance might be less stringent and fast discharge can be terminated earlier. The derivation of fast discharge requirements is reported in [5]. Charge measurement without discharge actuation does not necessarily impose a requirement on the on-board application. The test mass charge can also be obtained by means of ground based estimation algorithms. This implies that the required data for ground based estimation algorithms must be visible in telemetry.

4. Experiment operation scenarios

The attainable closed-loop discharge control performance is limited by the charge estimation performance which in turn depends on the currently active DFACS mode of operation. The mode link table depicted in figure 1 shows possible CMS-DFACS mode combinations. It becomes clear that fast discharge control can be done in any DFACS mode where the test mass is free-flying. Continuous discharge is only possible in DFACS science modes.

The CMS has been designed such that the required functionalities can be realized by means of a flexible on-board software structure. The CMS implementation is a set of functional units (rather than a set of operational modes) which can be configured/parameterized according to the required functionalities. A certain configuration/parameterization defines an CMS operational

DFACS Mode / CMS Mode	Accelerometer Mode					Normal Mode		Science Mode 1	Science Mode 2
	ACC1	ACC2	ACC3	ACC4	ACC5	Submode 1 IS-FEE WR	Submode 2 IS-FEE HR	(M3)	(M1)
Fast Discharge TM1 (CMS_FAST1)	**P1**	n.a.	**P1**	P4	P5	P6	P7	**P2**	P8
Fast Discharge TM2 (CMS_FAST2)	n.a.	**P1**	**P1**	P4	P5	P6	P7	**P2**	P8
Continuous Discharge TM1 (CMS_CONT1)	n.a.	n.a.	n.a.	n.a.	n.a.	n.a.	n.a.	**P3**	P9
Continuous Discharge TM2 (CMS_CONT2)	n.a.	n.a.	n.a.	n.a.	n.a.	n.a.	n.a.	**P3**	P9

Figure 1. CMS-DFACS mode link table with possible CMS-DFACS mode combinations.

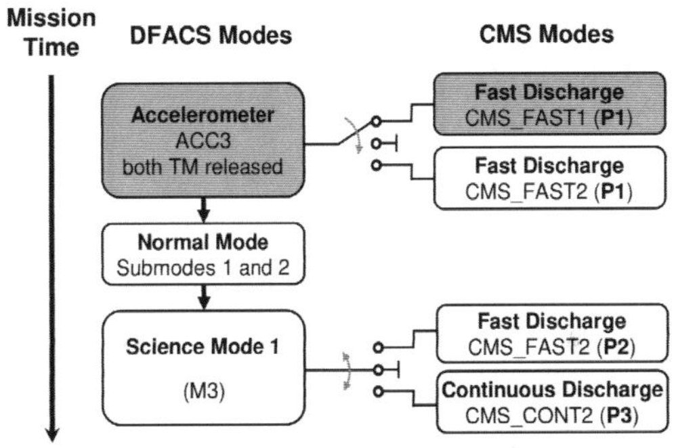

Figure 2. Example of charge management related experiment operations scenario.

mode. Configuration means the setting of configuration parameters (e.g. used test mass coordinate for charge measurement or UV lamps to be activated), whereas parameterization means the setting of any other parameter that influences the performance of an operational mode (e.g. frequency of dither voltage for charge measurement, controller gains, DC bias voltages on actuation electrodes, etc). The bold entries (**P1-P3**) indicate dedicated configurations/parameterizations of the functional units for baseline CMS-DFACS mode combinations. The entries P4-P9 are mode combinations which are in principle possible but not investigated as a baseline.

In principle all experiments (like the test of continuous discharge or the execution of fast discharge in the preparatory phase of other experiments) will be packed into the mission time. The complete definition of runs forms the Experiment Master Plan (EMP) of the mission. An example of a possible discharging operational scenario is illustrated in figure 2: Firstly, the test masses are discharged in accelerometer mode such that the transfer to DFACS science modes via the transition modes (*normal modes*) works smoothly. As soon as the DFACS science mode is in steady state, fast discharge of test mass 2 is performed such that the test mass charge is well below 10^7 e. This can be considered as a preparation of the following science run; which is

Journal of Physics: Conference Series **154** (2009) 012009 doi:10.1088/1742-6596/154/1/012009

the test of continuous discharge in this example.

The fundamental elements of closed-loop discharge control are charge estimation and discharge actuation. Their basic principles are summarized in the following.

5. Principle of discharge actuation

The test masses are discharged by using ultra-violet (UV) light via the photoelectric effect. Depending on the polarity of the test mass charge, the surface of the test mass itself or those of the electrodes and the electrode housing will be illuminated with UV light in order to release photoelectrons. The *charge management device* (CMD) hardware unit [6] generates the UV light and directs it to the gravitational reference sensors using feedthroughs that enter the shielding vacuum enclosures. In order to discharge a positively charged test mass, one of the feedthroughs pointed towards the electrode housing will be used; a negatively charged test mass will be discharged by using the feedthrough directed towards the test mass (see figure 3).

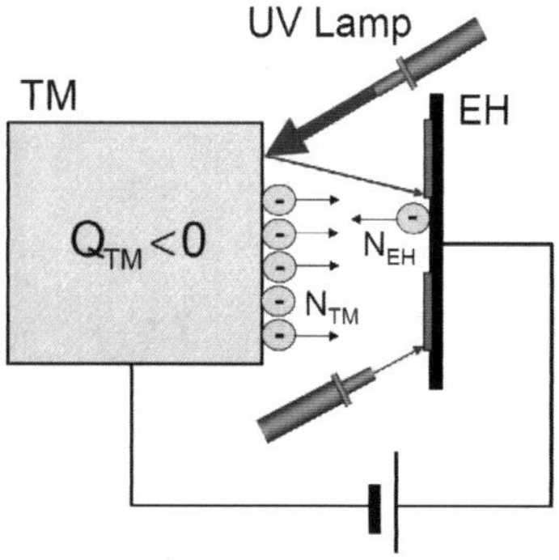

Figure 3. Principle of discharge actuation.

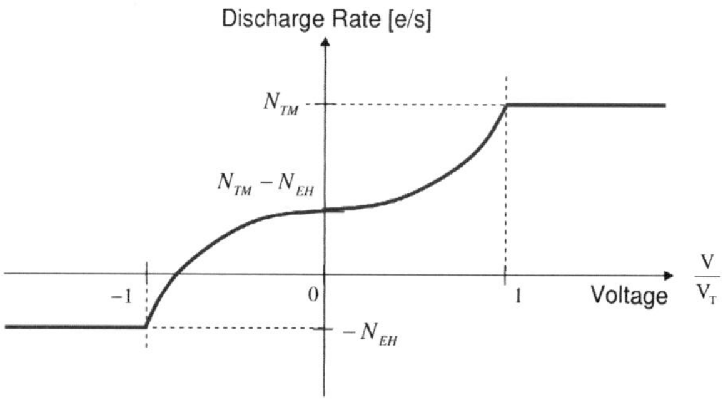

Figure 4. Dependency of discharge rate on voltage difference.

Primary illumination of the test mass should cause a dominant electron flow N_{TM} from the test mass (TM) to the electrode/housing surfaces (EH). As illustrated in figure 3, a significant amount of light is reflected off the primary surface, hitting opposing surfaces which causes an undesired electron flow N_{EH} from the EH to the TM (moreover, a powered lamp which is not actively used to support discharge actuation emits some light). A positive test mass discharge rate ($\frac{dQ_{TM}}{dt} > 0$) is obtained if $N_{TM} - N_{EH} > 0$. Thus, the dominant electron flow determines the sign of $\dot{Q}_{TM}$. The efficiency of photoelectron emission depends on the workfunction of the gold coated surfaces, their reflectivity, the quantum yield, and the amount of illumination. These parameters influence the effectiveness of the discharge, and need to be measured accurately in order to ensure a working design. A measurement campaign [7] to improve the knowledge of the flight test masses and electrode housing surface properties is currently carried out. The discharge performance depends not only on the emitted photoelectrons but also on the voltage difference between the test mass and the electrode/housing surfaces. The dependency is schematically illustrated in figure 4 and can be used to enhance/suppress the discharge. DC bias voltages are applied during fast discharge actuation.

The relation between UV lamp and voltage commands and the finally obtained discharge rate is non-linear, complex, and uncertain. For the design of the controllers and the actuation algorithms, a static actuation model has been derived which is a function of surface properties, CMD hardware parameters, and applied voltages.

6. Principle of charge measurement

Sinusoidal dither voltages are applied to a set of four actuation electrodes (see figure 5). If a charge is present on the test mass, a force or torque along the considered test mass DoF is generated [8]. Assuming equal electrodes and no DC bias voltages, the force/torque signal is proportional to the charge and has same frequency as the dither voltage. Such forces/torques can either be compensated with the suspension controllers (test mass charge information would be part of the controller signal) or it produces a test mass displacement. In the latter case, the resulting test mass motion along the corresponding degree of freedom can be measured using electrostatic or optical sensors. A recursive time-variant Kalman filter based on a model of the test mass dynamics and certain stochastic disturbances is used for on board charge estimation.

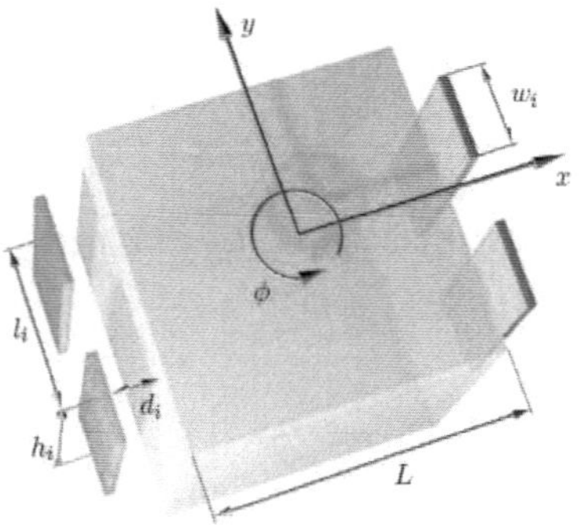

Figure 5. Test mass and x/φ-actuation electrodes.

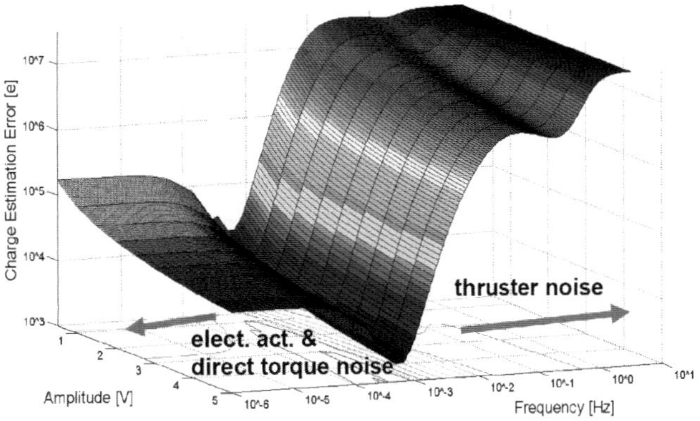

Figure 6. Standard deviation of expected charge error vs. test signal frequency and amplitude after estimation time of 3600 s.

The achievable estimation performance depends on factors like estimation time, stochastic force and torque disturbances acting on the test mass, measurement noise from electrostatic or optical displacement sensors, charging noise, etc. A detailed breakdown of all considered disturbances is given in [5]. The ideal sinusoidal dither voltages for charge estimation are determined by means of an optimal error covariance analysis. Figure 6 shows the obtained standard deviation of the charge estimation error as a function of dither frequency and amplitude. For an estimation time of 3600 s, the ideal dither voltage frequency is 0.88 mHz.

7. Performance Results

All baseline CMS-DFACS mode combinations have been tested with an end-to-end simulator. This simulator features (among others) a detailed inertial sensor model on voltage level for electrostatic actuation and sensing, an optical metrology model, models for environmental test mass charging and discharging using UV light, various disturbances models as well as an 18 DoF non-linear dynamics of the spacecraft and test masses. It also includes the on board computer software, including a top-level logic for experiment control, the DFACS and the CMS implementation.

Figure 7 shows the nominal performance of the continuous discharge control loop as well as the commanded UV lamp currents. In the performed test run, the transition to DFACS science mode (M3) is initiated after 10000 s; steady state is reached after 15000 s. At this time, the continuous discharge mode for test mass 2 is enabled via mission timeline. After a specified estimator settling time, the reference signal $Q_{TM} = 0$ is linearly faded in to avoid overshoots. The initial test mass charge at the start of the continuous discharge control mode is approximately 10^7 unit charges. As soon as the 8000 s reference signal fade-in is over, the test mass charge is kept below the requirement. The configuration/parameter settings are according to the derived baseline for continuous discharge control (CMS parameterization P3). The spectra of the differential test mass acceleration for two scenarios (with and without continuous discharge) have been computed in order to assess the impact on the main science data. The only apparent difference in the measurement bandwidth occurs at double dither voltage frequency; the influence is uncritical.

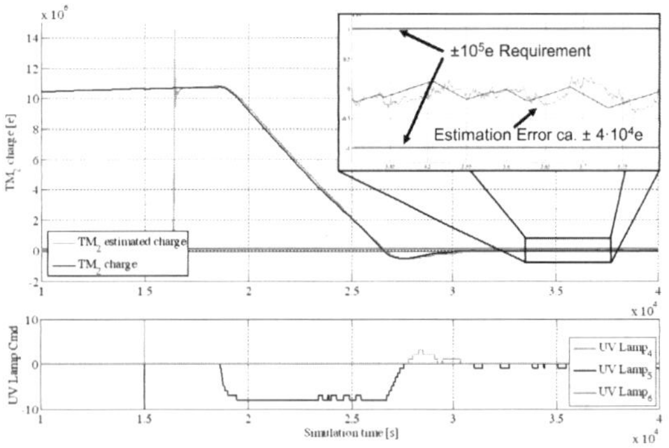

Figure 7. Nominal performance of continuous discharge control mode of operation.

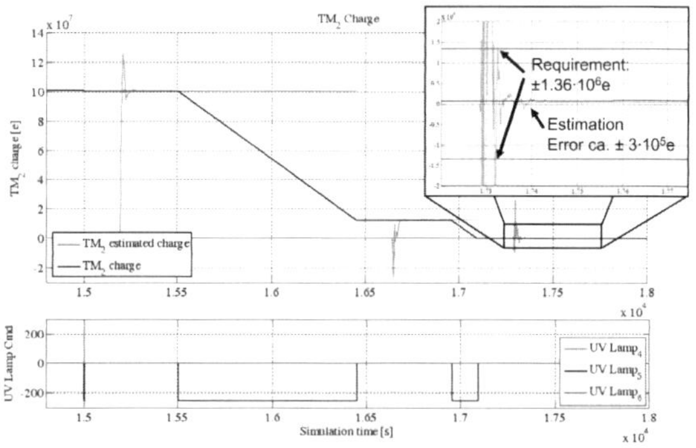

Figure 8. Nominal performance of fast discharge in DFACS science mode (M3).

Figure 8 shows the closed-loop performance of fast discharge performed in science mode (M3). Fast discharge is initiated after 15000 s via the experiment control logic. The initial test mass charge corresponds to approx. 10^8 unit charges. Fast discharge can be considered as discrete-time closed-loop control with variable stepsize. The estimation intervals are constant and set to 500 s; the actuation cycles are variable depending on the absolute level of test mass charge. After 15500 s the first discharge actuation cycle is performed. The discharge actuation is done with the maximum possible negative discharge rate which is obtained by applying appropriate DC bias voltages. After the second discharge actuation, the test mass charge has been controlled below the requirement which is verified by the third estimation interval. This is reported to the higher level experiment controller logic and fast discharge is automatically terminated. The fast discharge control algorithm can adapt the controller gains in order to cope with uncertainties in the discharge chain.

References

[1] Anza *et al* 2005 The LTP experiment on the LISA Pathfinder mission *Class. Quantum Grav.* **22** 125–38
[2] Fichter W, Schleicher A and Vitale S 2007 *Drag-Free Control Design with Cubic Test Masses* (Lasers, Clocks, and Drag-Free) ed H Dittus, C Lämmerzahl and S G Turyshevl (Berlin: Springer Verlag) pp 365–80
[3] Wass P J *et al* 2005 *Class. Quantum Grav.* **22** 311-17
[4] Brandt N *et al* 2008 *Experiment performance budget* S2-ASD-RP-3036, LPF project documentation
[5] Ziegler T and Bergner P 2008 *CMS design and analysis* S2-ASD-TN-2044, LPF project documentation
[6] Schulte M and Shaul D *et al* 2007 *CMD design synthesis report* S2-ICL-DDD-3001, LPF project documentation
[7] Schulte M *et al* 2008 Inertial sensor surface properties for LPF Pathfinder and their effect on test mass discharging *Class. Quantum Grav.* (submitted)
[8] Vitale S 2007 *Continuous charge measurement* S2-UTN-TN-3033, LPF project documentation

Journal of Physics: Conference Series **154** (2009) 012010 doi:10.1088/1742-6596/154/1/012010

Ground-based self-gravity tests for LISA Pathfinder and LISA

C Trenkel, C Warren and D Wealthy

Astrium Ltd, Gunnels Wood Road, Stevenage SG1 2AS, UK

E-mail: christian.trenkel@astrium.eads.net

Abstract. Gravitational coupling between the free-falling test masses and the surrounding spacecraft is one of the dominant noise sources for both LISA Pathfinder and LISA. At present, there are no plans to verify any of the self-gravity requirements by test, on the ground. Here, we explore the possibilities of conducting such tests, using a customised torsion balance. We discuss the main sources of systematic and statistical uncertainty present in such a set-up. Our preliminary assessment indicates that the sensitivity is sufficient to carry out meaningful self-gravity tests.

1. Introduction

Self-gravitational effects contribute to limit the free-fall quality of the suspended test masses inside drag-free spacecraft such as LISA Pathfinder (LPF) and LISA. The dominant gravitational interactions between spacecraft and free-floating test masses are:

- DC gravitational field – due to spacecraft imbalance
- Thermoelastic effects – modulation of DC gravitational field
- Spacecraft residual jitter – modulation of DC gravitational field
- Gravitational stiffness due to spacecraft

The first three are direct forces acting on the test mass, while the last one represents a parasitic stiffness, which can couple to the test mass via the position sensor noise and/or residual spacecraft motion.

As a result, in order to meet the free-fall quality requirements, requirements on the total DC gravitational field, and gravitational stiffness, have been derived [1], [2]. These requirements are shown in Table 1:

Quantity	LISA	LPF	Units
DC Gravitational Field	$5\times10^{-10} - 1\times10^{-9}$	1×10^{-9}	m/s^2
Gravitational Stiffness	$3\times10^{-8} - 1\times10^{-7}$	-5×10^{-7} $+7\times10^{-7}$	$1/s^2$

Table 1 LISA and LPF requirements on total DC gravitational field and stiffness

For LPF, which is in the advanced implementation phase, there has been a detailed apportioning of the contributions due to the Inertial Sensor (IS), the remainder of the LISA Technology Package (LTP), and the Spacecraft (SC) [2]:

Quantity	IS	LTP	SC	Units
DC Gravitational Field	$3.5x10^{-10}$	$4x10^{-10}$	$3.5x10^{-10}$	m/s^2
Gravitational Stiffness	$-1x10^{-7}$ $+2x10^{-7}$	$-2x10^{-7}$ $+3x10^{-7}$	$-2x10^{-7}$ $+3x10^{-7}$	$1/s^2$

Table 2 Apportioning of total DC gravitational field and stiffness requirements for LPF

It should be pointed out that gravitational effects make a significant contribution to the overall error budget, and that at present no experimental verification of the above requirements is envisaged. For a meaningful verification of these requirements, target sensitivity levels of $10^{-10}ms^{-2}$ and of $10^{-8}s^{-2}$ can be set for the DC gravitational field and gravitational stiffness, respectively.

The purpose of this paper is to explore the potential of verifying the above requirements to as large an extent as possible, using an experimental method based on a customised torsion balance.

2. Gravitational Requirement Verification

2.1. Experimental Concept

The basic experimental concept is schematically illustrated in Figure 1. The spacecraft (LISA or LPF) is placed on a turntable which also incorporates a translation stage, such that the spacecraft can be both rotated and translated.

A torsion balance of design very similar to the original Eötvös torsion balance is proposed as force sensor. It has a lower test mass, which is used as "near" or sensing mass, and a balancing or "far" test mass. The sensing test mass needs to be located on the rotation axis of the turntable, and the torsion balance as a whole has to be structurally decoupled from the spacecraft / turntable assembly.

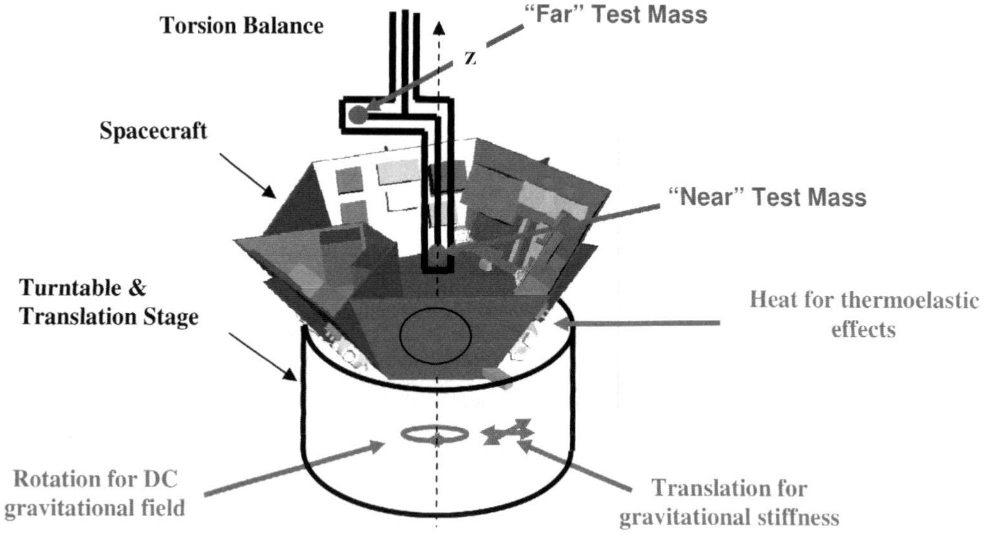

Figure 1 Schematic principle of experimental verification of gravitational requirements using a torsion balance

The positioning of the spacecraft on the turntable / translation stage will be such that the near test mass is at the point of interest, ie where gravitational requirements have to be verified.

In this setup, continuous rotation of the spacecraft will result in a modulated force projected onto the sensitive torsion balance direction, and can be used to measure the residual DC gravitational field. Linear oscillations of the spacecraft using the translation stage can in turn be used to determine the gravitational stiffness. Ideally these measurements would be carried out as late as possible during the AIT phase, such that as much hardware as possible participated.

As indicated in Figure 1, in principle one could even look at thermoelastic effects, by deliberately applying heat to the spacecraft. This possibility is not further explored here.

2.2. Possible Verifications

The schematic illustration above already shows that not all of the gravitational requirements will be directly verifiable using this method. It will be difficult to verify some requirements at the location of the test mass itself, in the presence of all the hardware. Neither the LPF nor the LISA spacecraft have been designed to allow access for a torsion balance.

Tthe requirements on the spacecraft could be verified by removing the payload. For LPF, it may be possible to remove an individual IS, in order to verify the LTP contribution.

In order to include as much flight hardware as late as possible in the AIT phase as possible, one could also measure the gravitational field and stiffness at one or more locations *outside* the IS. This would then allow to verify the effect of the entire mass distribution. The gravitational models could be used to predict fields and stiffness at these other locations, and an experimental confirmation of the predictions could then be taken as verification of the actual mass distribution on board the spacecraft.

3. Sensitivity Analysis

3.1. Acceleration sensitivity requirement

As discussed above, an acceleration sensitivity of $10^{-10}ms^{-2}$ is adequate for verifying DC fields, and if the spacecraft carries out linear oscillations of amplitude 1cm, then the stiffness sensitivity requirement of $10^{-8}s^{-2}$ is also equivalent to an acceleration sensitivity of $10^{-10}ms^{-2}$. This, then, will be the acceleration sensitivity required from the torsion balance, for all experiments. If we assume that a single measurement should not take more than 10^4s, we can derive a spectral density acceleration noise requirement of $\delta a \approx 10^{-8}\, ms^{-2}/\sqrt{Hz}$ for the torsion balance.

State-of-the-art torsion balances are performing close to their thermal noise limit ([3], [4]). Their acceleration sensitivity is approximately $\delta a \approx (3-10)\times 10^{-13}\, ms^{-2}/\sqrt{Hz}$. This is about 4 orders of magnitude better than what is required here. Nevertheless, these are compact, highly optimized torsion balances, and the design proposed above is very different.

The remainder of this paper assesses, qualitatively, the main sources of systematic and random uncertainty, in order to get some preliminary idea as to the feasibility of the proposed verification.

3.2. Random Noise Sources

The following random noise sources will in general be present:

- Thermal Noise (via dissipation processes)
- Magnetic Noise (coupling of torsion balance to random magnetic field fluctuations)
- Sensor Noise
- Seismic Vibration (excitation of torsion balance due to ground vibrations)
- Gravitational Noise (coupling of torsion balance to random gravitational field fluctuations)

A qualitative assessment shows that the last two may be significantly larger for the torsion balance proposed here than for the torsion balances referred to earlier ([3]and [4]), mainly because of its much larger size and also shape.

This qualitative assessment is summarized in Table 3:

Noise Source	Predicted noise levels compared to [3] and [4]	Comment
Thermal Noise	Same or slightly higher	Sources include intrinsic fibre material losses, residual gas damping and others. Intrinsic fibre losses could be higher if fibre diameter larger
Magnetic Noise	Same	Determined by torsion balance material properties, magnetic shielding and magnetic environment
Sensor Noise	Same	Same type of sensor (optical or capacitive) could be used, no reason to expect increased noise levels
Seismic Vibrations	Higher	Depends on seismic environment, torsion balance modes and damping, and torsion balance geometry. Large quadrupole will increase noise levels
Gravitational Noise	Higher	Depends on gravitational environment and torsion balance geometry. Large quadrupole will increase coupling

Table 3 Qualitative assessment of random noise sources

A quantitative analysis of the likely noise levels has not been carried out. However, given a margin of 4 orders of magnitude in random noise levels, we are confident that the sensitivity requirements could easily be met.

3.3. Systematic Noise Sources

During the experiments, the following systematic noise sources will in general be present:

- Mechanical – coupling between spacecraft motion and torsion balance suspension
- Magnetic – direct magnetic coupling between spacecraft, and/or turntable, and torsion balance
- Thermal – torsion balance may be sensitive to thermal fluctuations correlated with spacecraft motion
- Gravitational I – Turntable and / or translation stage may generate a gravitational field or stiffness
- Gravitational II – direct coupling to "far" torsion balance test mass which is not infinitely far

The first four effects are standard effects, and well-known procedures exist for dealing with them. However, the last effect turns out to be highly relevant in our case, and is investigated further in the next section. The problem is that the far mass will also couple gravitationally to the spacecraft, and the net torque on the torsion balance will not be just due to the quantities of interest.

3.4. Gravitational Coupling to "far" mass

The most obvious approach to deal with this problem is to place the far test mass as far away from the spacecraft as possible, but it is important to note that the LPF and LISA spacecraft have typical diameters between 2.5m and 3m. It can therefore be anticipated that dimensions of the torsion balance will also have to be of this order.

In order to make quantitative predictions, we have modeled the distribution of all discrete electronic units on the latest LISA spacecraft design [5].

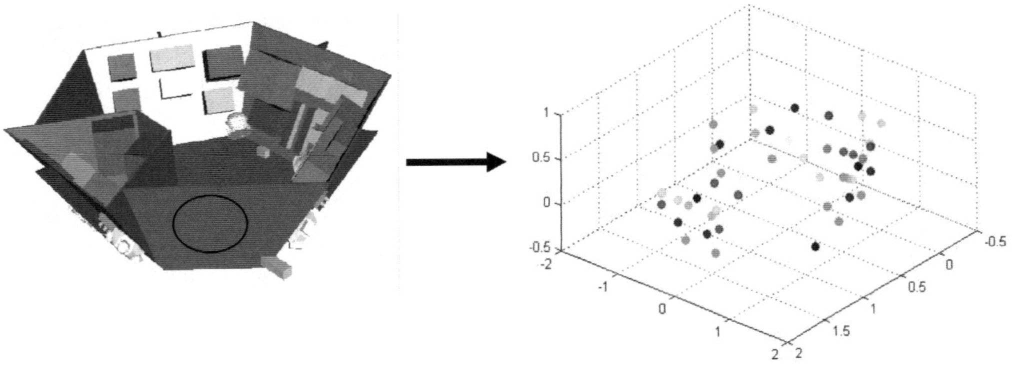

Figure 2 Simplified Gravitational Model of LISA Spacecraft – Electronic Units only

This model, shown in Figure 2, we have then used as "source" mass, and have calculated the total torque on a simple torsion balance of the design shown above. In order to see how far the far test mass had to be in order to reduce the systematic coupling to less than about $10^{-10}ms^{-2}$, we maintained a constant torsion balance beam length of 1m, and then varied the torsion balance height (effectively the vertical distance between the two test masses).

The systematic acceleration error, as a function of torsion balance height, is shown in Table 4:

Torsion Balance Height [m]	**Acceleration error [m/s^2]**
2	$4x10^{-10}$
3	$1.5x10^{-10}$
4	$6x10^{-11}$

Table 4 Acceleration measurement error due to coupling to "far" mass

As can be seen, a torsion balance height in excess of 3m would be required to reduce the unwanted coupling to less than $1x10^{-10}ms^{-2}$. We have also investigated the impact on stiffness measurements, and found this to be negligible. Compared to a spacecraft oscillation amplitude of 1cm, a test mass at a height of 2m is already far enough.

In principle, there might be other options to minimise the effect of the far test mass:

- Optimise torsion balance design
  - Increase beam length
  - Alternative torsion balance design and gravitational multipoles
- Conduct differential experiments using torsion balances of different heights

For this preliminary work, these alternatives have not been further investigated.

4. Conclusions and Future Work

We have shown that the self-gravity requirements existing for spacecraft such as LPF and LISA could be verified, to a large degree, in a ground based experiment using a customized torsion balance. These measurements could be used to verify the gravitational models and, crucially, also the hardware assembly and integration.

A qualitative assessment of the main random and systematic noise sources has been carried out, and the most critical sources have been identified. Although a detailed quantitative analysis has not been carried out, the sensitivity requirements on the torsion balance are very modest: even noise levels up to 4 orders of magnitude larger than those found in the best performing torsion balances would still be acceptable.

Therefore, even without this more rigorous analysis, our preliminary conclusion is that a meaningful experimental verification of self-gravity requirements for LPF and LISA on the ground is definitely possible. We have had preliminary contacts with Birmingham and Glasgow University, and would certainly be interested in taking this idea further. We also point out that a LPF Spacecraft mass dummy is available at Astrium UK, and could be used in any experimental demonstration of the method.

Acknowledgements

We would like to thank Neil Dunbar and also our colleagues at Astrium Germany for useful discussions.

References

[1] *LISA Requirement Breakdown*, Document LISA-ASD-TN-5001, issue 2.0, 2007

[2] *Science Requirements and Top-level Architecture Definition for the LTP on board LPF*, Document LTPA-UTN-ScRD-Iss003-Rev1, 2005

[3] Y.Su et al, *New tests of the universality of free fall*, Physical Review D, Volume 50, Number 6, page 3614 (1994)

[4] L.Carbone et al, *Upper limits on stray force noise for LISA*, Classical and Quantum Gravity, Volume 21, page S611 (2004)

[5] *LISA Spacecraft Design Description*, Document LISA-ASU-DD-4001, issue 1.3, 2008

LTP fibre injector qualification and status

J Bogenstahl, L Cunningham, E D Fitzsimons, J Hough, C J Killow, M Perreur-Lloyd, D Robertson, S Rowan and H Ward

Institute of Gravitational Research, Department of Physics and Astronomy, University of Glasgow, Glasgow G12 8QQ, UK

E-mail: j.bogenstahl@physics.gla.ac.uk

Abstract. This paper presents the current state of the LISA Technology Package (LTP) fibre injector qualification project in terms of vibration and shock tests. The fibre injector is a custom built part and therefore must undergo a full space qualification process. The mounting structure and method for sinusoidal vibration and random vibration tests as well as shock tests will be presented. Furthermore a proposal will be presented to use the fibre injector pair qualification model to build an optical prototype bench. The optical prototype bench is a full-scale model of the flight model. It will be used for development and rehearsal of all the assembly stages of the flight model and will provide an on-ground simulator for investigation as an updated engineering model.

1. Introduction

The LTP optical bench interferometer (OBI) will be built at the Institute of Gravitational Research (IGR), University of Glasgow. The aim of the OBI is to measure displacements of two test masses [1]. To do this it uses four heterodyne interferometers, two of which have beams reflected from the test masses. Figure 1 shows the optical layout created by OptoCAD [2] of the OBI. The optics have to be bonded [3] to the ultra stable bench to achieve the desired performance [4]. Silicate bonding is a technique which has been used as part of a previous space mission [5]. As part of the LTP OBI two customised ultra-stable fibre injectors are attached to the bench by monolithic joints. These fibre injectors have been developed and will be built at the IGR. The fibre injector consists of glass components which are either glued or silicate bonded together and an optical fibre which includes strain relief cladding. The design and construction process of the fibre injector are subject to space qualification. Vibration and shock tests are part of the qualification tests. It has to be demonstrated that the tests cause no significant changes of the optical performance of the fibre injectors. A fibre injector pair qualification model will be built to the same requirements using the same processes as the proto flight model. This fibre injector pair will be bonded to a representative prototype Zerodur® baseplate for the qualification process.

2. Fibre Injector Qualification

The fibre injector optical sub-assembly (FIOS) consists of the fibre with part of the cladding stripped which is attached to a ferrule and then glued into a hole in a fibre mounting cube. There is also a strain relief at the fibre input side. The fibre mounting cube is polished flat and silicate bonded onto the FIOS baseplate. The collimating lens is first glued into a fused silica U-groove and this assembly precision bonded to the FIOS baseplate. Finally, a polariser to maintain the correct polarisation is glued onto the FIOS baseplate. The FIOS baseplate will be silicate bonded onto a post which is also silicate bonded onto the OBI Zerodur® baseplate. A complete assembly of two FIOS and their mounting post is shown in Figure 1. To have a fully representative FIOS qualification process, a FIOS pair will be bonded onto a prototype Zerodur® baseplate of the same dimension as the OBI baseplate and at the same location. The Zerodur® baseplate will be used as reference baseplate for optical performance testing before and after the FIOS pair are subjected to environmental tests.

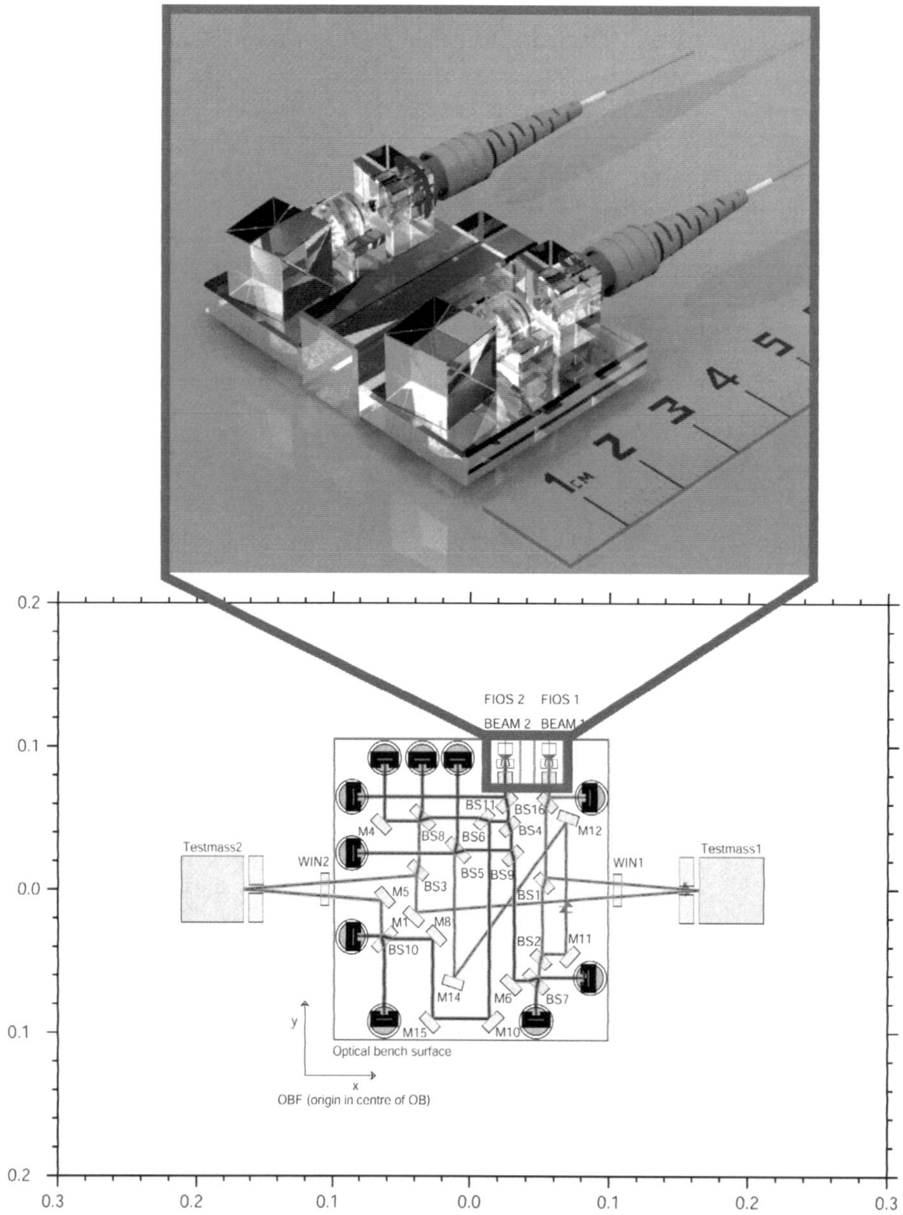

Figure 1. Optical layout of the LTP optical bench interferometer with a CAD rendering close-up of the FIOS pair

Journal of Physics: Conference Series **154** (2009) 012011 doi:10.1088/1742-6596/154/1/012011

2.1. *Qualification plan of vibration and shock tests*

The main stress for the optical bench assembly is during the launching process. Part of the vibration will be absorbed by the spacecraft and the loading structure. Nevertheless, the requirements for the LTP OBI and therefore for the FIOS qualification are demanding. The qualification process of the fibre injector involves sinusoidal tests, random tests and shock tests along all three axes. During the sinusoidal tests the FIOS will experience an acceleration of up to 40 g at frequencies up to 100 Hz. The random test loads are in the frequency range 100 - 500 Hz with 0.5 g^2/Hz with a RMS acceleration of 20 g. A shock response spectrum (Q=10) of 750 g at 10 kHz and 1.5 kHz and 20 g at 100 Hz has to be demonstrated. In the case of the LTP OBI environmental shaking tests the bench will be mounted using specially engineered inserts, however the prototype Zerodur® baseplate does not have inserts, therefore another way of mounting the baseplate has to be found.

2.2. *Mounting technique*

A suitable mounting structure is needed, which holds the prototype Zerodur® baseplate and can be easily mounted onto the vibration table. To mount a Zerodur® baseplate into this mounting structure a strong but temporary joint is needed. Waxes are commonly used to mount glass for polishing and cutting, therefore different waxes were tested in small shear tests, of which *Logitech Quartz* wax [8] seemed the most suitable for this application. Large scale tensile tests were carried out using two 10 cm diameter discs, which had been wax bonded together and were then pulled apart. The six test runs measured forces of 8.4 ± 1.2 kN. This equates to $\sim 10\,\mathrm{kg/cm^2}$, shown in Figure 2. These measurements show that the wax-bond would be strong enough to hold a Zerodur® baseplate under the toughest load of 40 g. But the behaviour of the wax-bond is unknown during vibration and shock tests, therefore a test run with a wax-bonded dummy load was carried out.

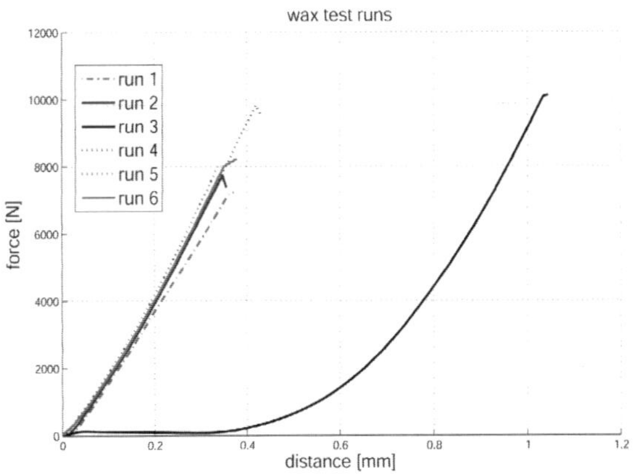

Figure 2. Graph of measured force of tensile tests

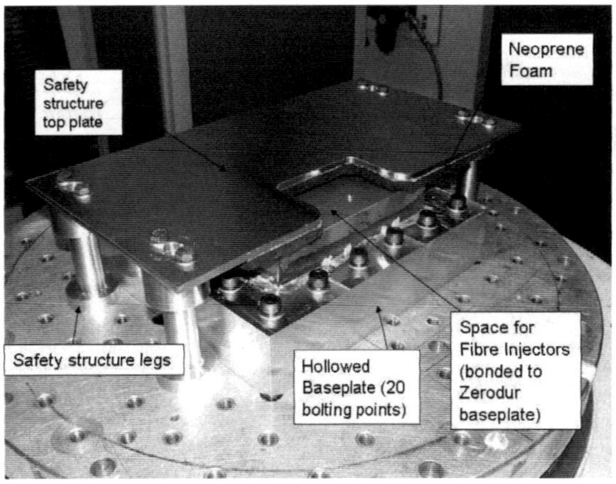

Figure 3. Picture of mounting structure on vibration table

Figure 3 shows the dummy Zerodur® load wax-bonded into the hollowed aluminium adapter plate and the safety structure top plate. Using finite element (FE) analysis [7], the hollowed adapter plate and safety structure were modelled. A hollowed adapter plate of 60 mm thickness, hollowed out to a depth of 20 mm to the size of the Zerodur® baseplate, has a first resonance mode over 2000 Hz. The safety structure has a first mode of 163 Hz, therefore the safety structure is placed over the Zerodur® baseplate with a gap and is never in contact with the Zerodur®

baseplate. The limiting factor for the thickness of the hollowed adapter plate is the mass - the vibration table has a mass limit of 20 kg for our vibration requirements.

2.3. Results of Pre-tests

The pre-tests of a dummy Zerodur® baseplate (without the FIOS pair) wax-bonded into the mounting hardware have been conducted and the results were successful. Figure 4 shows the graph of the sinusoidal vibration test in the horizontal direction. Three accelerometers were mounted onto the Zerodur® baseplate, the hollowed adapter plate and the top plate. It is easy to see that the acceleration of the Zerodur® baseplate and hollowed mount are similar. The acceleration of the top plate is increasing with frequency, due to the fact that the top plate resonance mode is lower and easier to excite, and the top plate shows in Figure 5 the first resonance mode as simulated at 163 Hz. The Zerodur® baseplate and the hollowed adapter plate are showing in Figure 5 the desired effect of behaving like a monolithic entity with the predicted first resonance mode at 2.1 kHz.

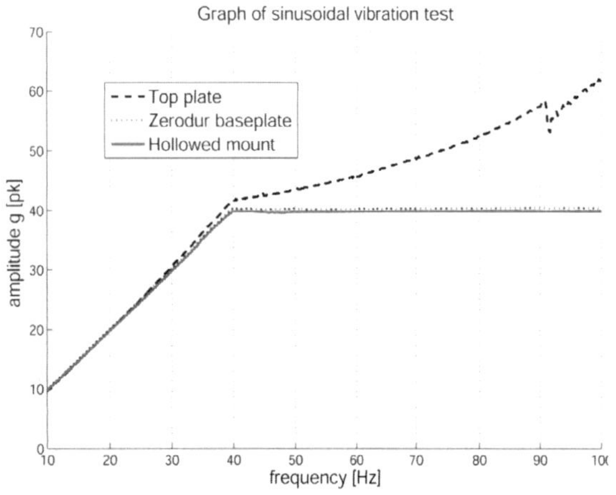

Figure 4. Graph of sinusoidal vibration test

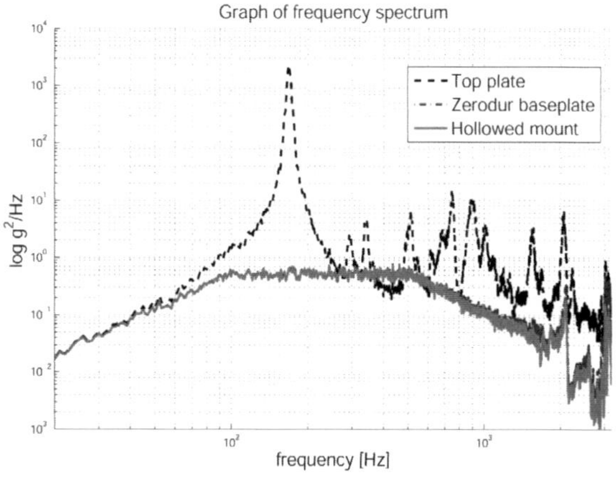

Figure 5. Graph of frequency spectrum

The shock tests were even more challenging than the vibration tests. The wax-bond was not renewed before the shock tests were carried out. During the pre-test, damaged areas of the wax-layer were detected after 6 shocks (2×3 shocks in vertical orientations). Afterwards three further shocks in horizontal orientation were carried out and two more in vertical orientation before the wax-layer and the Zerodur® baseplate became loose. The Zerodur® baseplate was left undamaged. As a summary, it can be said that the wax-bond fulfilled the requirements for a suitable joining technique, because after each shock the wax-layer could be examined and if necessary the wax-bond could be redone by heating the structure to re-form it. This was demonstrated by placing it onto a heat plate and re-melting the wax. The pre-tests of the mounting structure and the wax-bond technique was successful.

3. LTP OBI optical prototype bench

The aim of building an LTP OBI optical prototype bench is to have an exact optical copy of the flight model, including the same design of fibre injectors and therefore similar beam performance. In such a way, the optical prototype bench will be the replacement for the engineering model [9], and can be used for ground-based readout experiments. It can also be used as a dummy bench for flight model OBI performance procedure set-ups. Furthermore, the construction of the optical prototype bench provides practice for the assembly stages of the flight model and will provide feedback to prove the optical bench alignment tools at every construction step.

3.1. Realisation plan

Some optical components do not require high precision location and can be placed within $\sim 50\,\mu$m precision using templates. Some other optical components have to direct the beam to certain points, for example the beams have to hit the test masses accurately in the middle of the front face with a maximum error of $\sim 25\,\mu$m. Using micro precision tools the optical components can be manipulated. This will be done in real time having a beam which will be aimed onto a target consisting of a pair of calibrated quadrant photodiodes (CQP). The CQP will be used to steer the beam to a pre-calculated position of the beam vector and will also be used to measure the position of the final beam after each step, and in this way to provide feedback for the simulation program that generates the next corrected optical target point. The CQP consists of two quadrant photodiodes and a beamsplitter, which splits the beam into two paths with a pathlength difference of 44 cm, shown in Figure 6. The CQP is constructed on a stable Invar structure. Using two quadrant photodiodes the beam position can be detected using the first photodiode and the beam angle using the second photodiode. The readout precision of the quadrant photodiodes is better than $1\,\mu$m. Modulated light is used to reduce the sensitivity to DC offsets on the photodiodes and electronics. For calibration it is necessary to use a stable beam, therefore a FIOS engineering model was glued to a Zerodur® block. Using this stable beam the quadrant photodiode pair was calibrated by centring it to this beam and measuring the position of the stable Invar structure. This process was repeated in the same manner several times at different distances and angles with the non-changed stable beam. In such a way the quadrant photodiode pair was calibrated with a remaining uncertainty of $\pm 3\,\mu$m and $\pm 30\,\mu$rad per single measurement.

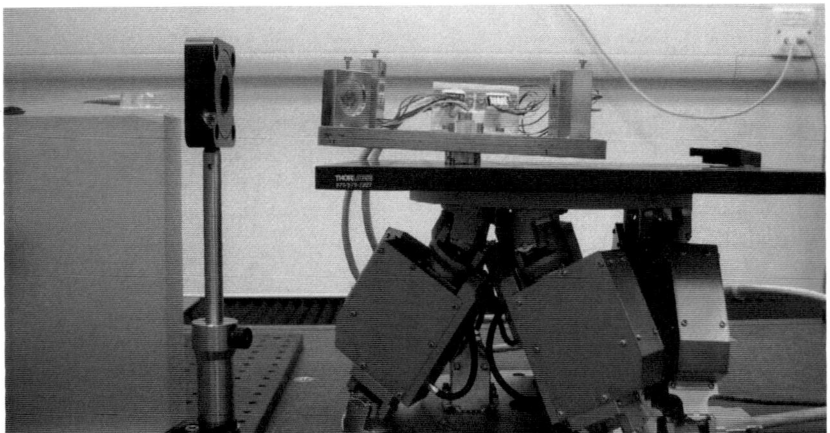

Figure 6. Calibration of the CQP using a stable beam from a FIOS engineering model. The CQP is aligned on a six-dimension micron-precision stage.

4. Conclusion and Acknowledgement

A test procedure to qualify the fibre injectors for LTP has been presented. The mounting hardware has been built and successfully demonstrated. The CQP fulfils the necessary specifications. We wish to acknowledge the support of STFC and the University of Glasgow. We would like to thank SELEX-Galileo (Edinburgh) for their assistance. Furthermore, we would like to thank the Albert Einstein Institute Hannover for their contribution to the LTP OBI optical prototype bench.

References

[1] S.Anza and the LTP Team 2005 Class. Quantum Grav. 22 S125-138
[2] Program written by Roland Schilling: http://www.rzg.mpg.de/~ros/OptoCad/
[3] E J Elliffe et al 2005 Class. Quantum Grav. 22 S257-S267
[4] Robertson D et al 2005 Class. Quantum Grav. 22 S15563
[5] Gwo D H 1998 *Proc. SPIE* 136 Vol.3435
[6] ASU RS 2031 General Design and Interface Requirements Specification (GDIR)
[7] See the information available on http://www.ansys.com/
[8] See the information available on http://www.logitech.uk.com/
[9] G Heinzel et al 2005 Class. Quantum Grav. 22 S149-S154

Journal of Physics: Conference Series **154** (2009) 012012 doi:10.1088/1742-6596/154/1/012012

Ground testing, with a four mass torsion pendulum facility, of an Optical-Read-Out for the LISA gravitational reference sensor

A Cavalleri[1], G Ciani[2], R De Rosa[3], L Di Fiore[3], R Dolesi[2], F Garufi[3], M Hueller[2], D Nicolodi[2], A La Rana[3], L Milano[3], D Tombolato[2], S Vitale[2], P J Wass[2], W J Weber[2]

1 INFN and Centro Fisica degli Stati Aggregati, 38100, Povo, (Trento), Italy

2 INFN and Dipartimento di Fisica, Università di Trento, 38100, Povo, Trento, Italy

3 INFN and Università degli studi di Napoli "Federico II", Via Cintia, 80125 Napoli, Italy.

E-mail: luciano.difiore@na.infn.it

Abstract. In the last few years the Lisa group in Napoli has developed an Optical Read-Out (ORO) system based on optical levers as an auxiliary and backup readout for the Gravitational Reference Sensor (GRS) of LISA. Bench-top measurements, with a rigid set-up have successfully proven that the ORO fits the requirements for sensitivity both in translational and rotational DOFs, exceeding the capacitive sensor performance in a wide range of frequencies. Last year an ORO system designed in Napoli in collaboration with the Trento LISA group, has been installed, as an auxiliary readout system, on the four mass torsion pendulum developed in Trento. In this paper we report on the testing of this ORO device and its performances in comparison with the capacitive one; we also outline further improvements and their advantages for the torsion pendulum facility performances.

1. Introduction

The GRS, also called inertial sensor, is one of the key elements of LISA [1]. It is a position sensor that measures, on six degrees of freedom (DOF), the position of each test mass (TM) with respect to the spacecraft. The GRS signals are used in the drag free control loops to force the spacecraft to follow the TM geodesic motion. The reference solution for LISA, also adopted for the LISA-Pathfinder technology demonstration mission, is a capacitive sensor, developed by the LISA group in Trento [2]. The Trento group has also developed torsion pendulum facilities for the ground testing of the prototypes and models of the GRS designed for the LISA-Pathfinder mission. In particular, a single mass and, more recently, a four mass torsion pendulum facilities have been developed. For a complete description of these devices refer to references [2] and [3].

As an alternative to GRS capacitive readout, optical readout (ORO) systems have also been proposed [4, 5, 6, 7, 8]. In particular, the LISA group in Napoli has proposed and developed an ORO system, based on optical levers and quadrant photo-diodes (QPD), that allows to improve the sensitivity of the GRS [5, 6, 9]. The main goal of this activity is not the replacement of the capacitive sensor (that will be tested on flight with LISA-Pathfinder) with an optical one, but the integration of the two in a single GRS system in such a way to get a back-up sensor in case the other fails after the launch of LISA, with consequent mission risk reduction. Furthermore, the ORO system can be more sensitive than the capacitive readout, and thus using ORO for satellite control on the Y and Z axes and θ rotation around the TM x axis, where there is no interferometric readout, could allow to relax the specification on cross-couplings in the drag-free control loops. This can be achieved adopting the ORO as the main sensor or using its signals for subtracting the extra noise of the capacitive readout.

For the capacitive readout, we assume as reference sensitivity 2 nm/Hz$^{1/2}$ in displacement and 200 nrad/Hz$^{1/2}$ in angle. For the ORO, a sensitivity performance exceeding the one of the capacitive readout system has been already demonstrated with bench top experiments on a rigid set-up in Napoli [6, 9].

Last year, in collaboration with the Trento group, a prototype of the ORO has been installed on the four mass torsion pendulum facility. The goal is twofold; the first one is to check the performance of the ORO in a situation that is as close as possible to free fall and to compare the noise with contemporary measurements with both ORO and capacitive readout, aiming to verify the improvement in sensitivity provided by the ORO. The torsion pendulum could also measure (or put upper limits to) any excess force noise created by the ORO. The second motivation is that the potentially more sensitive ORO can improve the performance of the torsion pendulum device as a facility for ground testing of the LISA-Pathfinder hardware.

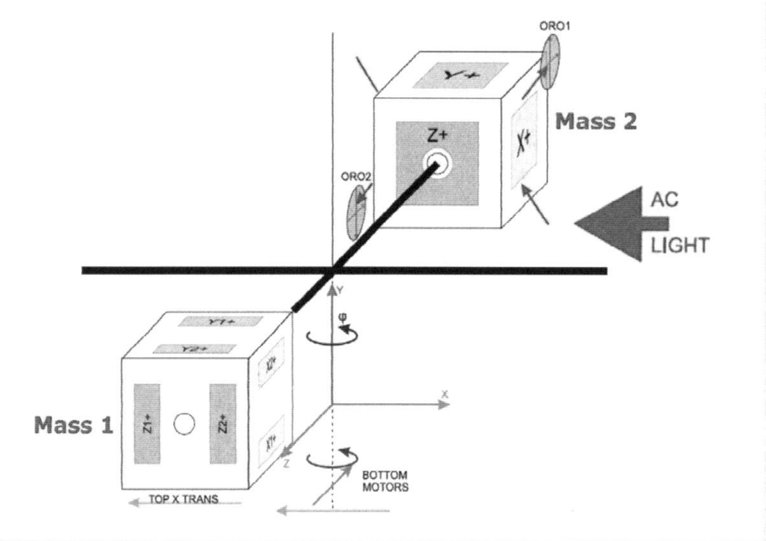

Figure 1. Schematic view of the sensors on the four-mass torsion pendulum. The EM, measuring all the six DOFs, is mounted around test mass 1. The STC (measuring two displacements, X and Z) and the ORO (measuring φ, η and X) are mounted on TM 2. An autocollimator, that uses a laser beam reflected by a mirror placed above the pendulum payload shaft, is used for calibration.

At the time of measurement the torsion pendulum was used for the testing of the engineering model (EM) of the LISA-Pathfinder GRS that was mounted around one of the four test masses. On the opposite test mass, there is another sensor, called stiffness compensator (STC) mainly used for improving readout sensitivity. Due to the electrode configuration, the STC is only sensitive to displacements in three DOFs, while the EM measures all the six DOFs. The ORO was mounted on the STC, in a configuration with two beams and two QPDs that allows the independent measurement of two angles and one displacement (φ, η and X). In particular, each QPD provides two signals (proportional to horizontal and vertical displacements of the spot across the detector area). The two vertical signals are used for separating pendulum η and X, while the horizontal ones provide independent measurements of φ (that we average for noise reduction). The light sources used for the ORO are super-luminescent LEDs (S-LED), coupled to single-mode optical fibers, operating at 820 nm (total power about 400 nW). There is also an autocollimator that provides a measurement of the rotation around the vertical axis, and is used for calibration of all the other signals. In figure 1 there is a scheme of the various sensors.

2. Noise measurement

In this section we report on the results of a measurement campaign where the ORO and the capacitive readout were operative at the same time. The aim is to compare the measurement of the TM position (both in displacement and angle) to the ones of the capacitive sensors (mainly the EM, that is the model for LISA).

In figure 2, we report the measurement for the rotation around the vertical axis (φ). As we can see, the ORO measurement (blue line) is more sensitive than the EM capacitive one (red line), confirming the results obtained in bench-top experiments [9]. The ORO sensitivity (~ $2\cdot10^{-8}$ rad/ $Hz^{1/2}$ above 10 mHz) corresponds to a beam displacement across the QPD of about $8\cdot10^{-10}$ m/ $Hz^{1/2}$. On the same plot, are shows the estimated pendulum thermal noise and the expected ORO noise (computed according to the bench-top results). It is worth noting that the EM φ signal reported in Fig.1 doesn't represent the best capacitive readout available on the torsion pendulum, that is instead obtained as the difference between the X displacement measured by EM and STC divided by the arm-length distance (between the two TM centers), and is a white noise currently at a level of about $2.5\cdot10^{-8}$ rad/$Hz^{1/2}$; anyhow, as long as we are interested in the ORO as a back-up solution for LISA, we are interested in comparing it with the performance of the EM alone.

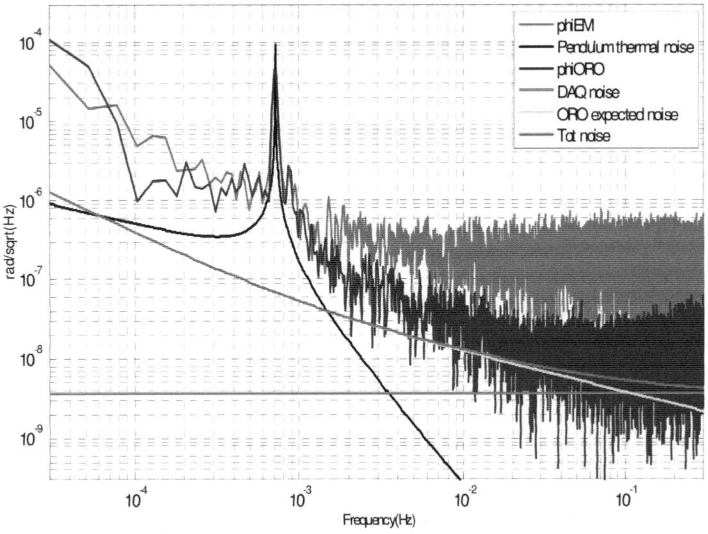

Figure 2. ORO measured φ sensitivity (blue line) compared to the one of the EM capacitive sensor (red). The EM sensitivity is limited to ~ $2\cdot10^{-7}$ rad/ $Hz^{1/2}$, while the ORO reaches $2\cdot10^{-8}$ rad/ $Hz^{1/2}$.

Journal of Physics: Conference Series **154** (2009) 012012 doi:10.1088/1742-6596/154/1/012012

In figure 3, it is shown the same measurement for η (rotation around the horizontal axis connecting the two TMs). In this case, the capacitive readout is limited by sensor noise (~ $3 \cdot 10^{-7}$ rad/ $Hz^{1/2}$), while the ORO one, in a large frequency interval, is limited by the residual TM motion and reaches $2 \cdot 10^{-8}$ rad/ $Hz^{1/2}$ in high frequency

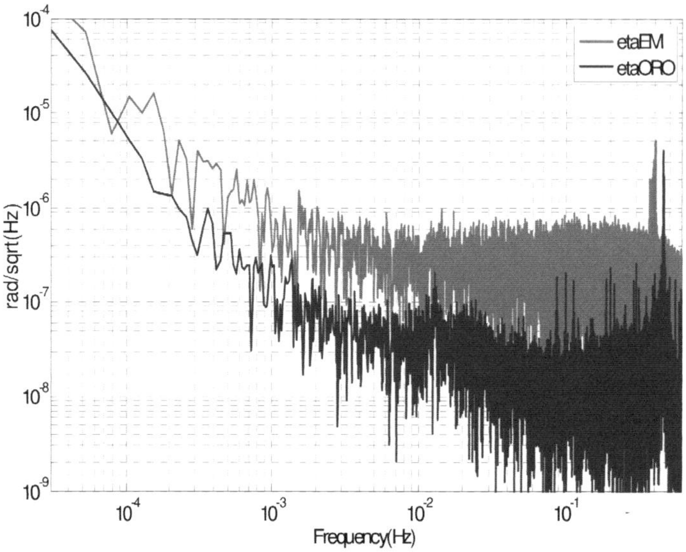

Figure 3. ORO measured η sensitivity (blue line) compared to the one of the EM capacitive sensor (red line). Also in this case, the EM sensitivity is limited to ~ $3 \cdot 10^{-7}$ rad/ Hz1/2, while the ORO reaches $2 \cdot 10^{-8}$ rad/ Hz1/2.

In fig. 4, the X signals (i.e. the measurement of the TM X position inside the sensor) of ORO (blue line), EM (red line) and STC (black line) are shown. All three sensors measure essentially the same signals, dominated by the residual simple pendulum swing mode motion and different readout performances are not appreciable The differential signals XSTC-XEM (orange) and XORO-XEM (magenta) are also shown. They are proportional to the pendulum torsion angle φ, and are dominated by the true pendulum torsion motion around and below the 0.7 mHz resonance and by the combined readout noise at higher frequencies. As both signal combinations show the same level of noise, we can conclude that the dominant term of the incoherent sum is coming from the EM (common to both combinations), and place an upper limit on the ORO noise level not to exceed the EM one (2 nm/$Hz^{1/2}$) above about 5 mHz.

3. The ORO as an auxiliary sensor for the torsion pendulum

Up to now, we dealt with the characterization of the ORO as a sensor for the LISA mission. The ORO is also interesting as a sensor for the torsion pendulum, if it can improve its performance as a ground testing facility for the LISA-Pathfinder hardware. Form this point of view, the most important measurement is the torque, that is used to place upper limits on the surface forces originating in the GRS (until now, such tests have studied the EM GRS used in this study, while tests on a prototype of the flight model design will begin in the near future. For this measurement, it is normally used the differential measurement of the two position sensor mounted on the opposite test masses (EM and STC), already shown in figure 4. This is, thanks to the rather long arm between the two test masses (20 cm), the best measurements of φ. In the same way, we can use the differential signal of ORO end EM.

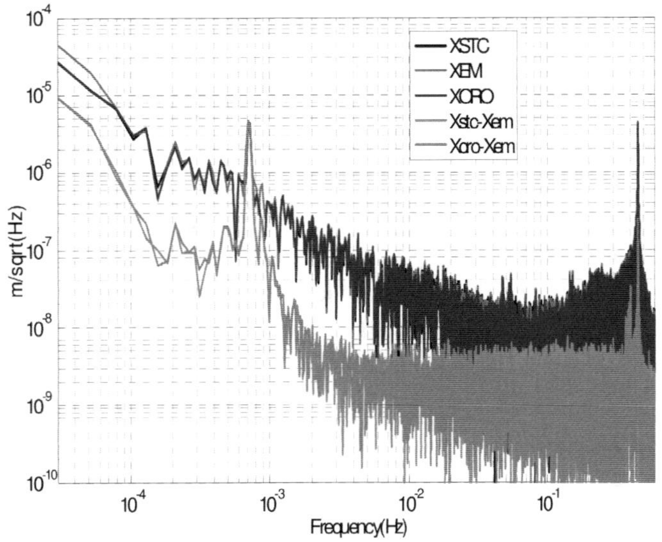

Figure 4 - Measured noise power spectrum of the ORO (blue) X readout, compared to the one of the capacitive sensors (EM in red, STC in black). The differential signals X_{STC}-X_{EM} (orange) and X_{ORO}-X_{EM} (magenta) are also shown. They allow to measure, in high frequency, the incoherent sum of the noise of sensor couples.

In figure 5 it is reported the residual force noise spectrum measured as the difference of TM displacements. The magenta and black lines represent the upper limit to the force noise obtained with ORO-EM and STC-EM combinations respectively. On the same graph, the limits posed by using the measurements performed on a single test mass with only the ORO (blue) and only the EM (red) are also shown.

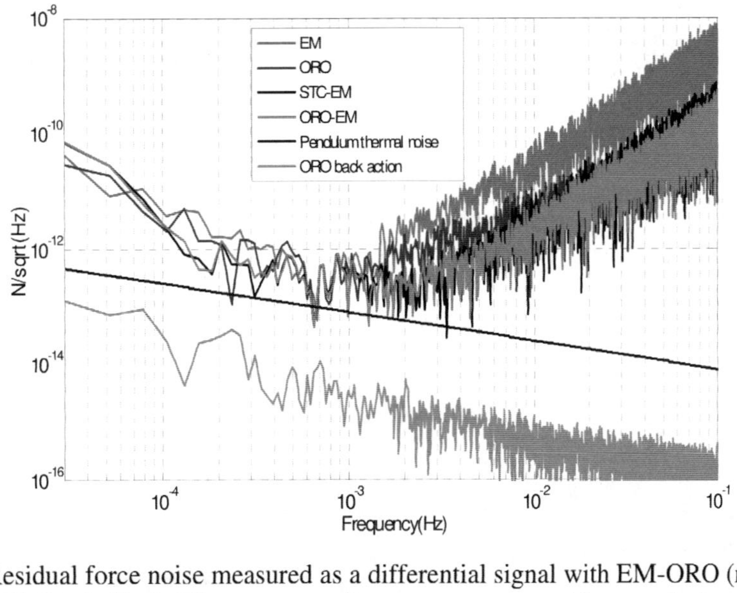

Figure 5. Residual force noise measured as a differential signal with EM-ORO (magenta line) and EM-STC (black line). The corresponding measurement with one single sensor is also shown: EM (red) ORO (blue). The orange line reproduces the ORO back-action, estimated form the measured S-LED power fluctuation.

As we can see, ORO and STC essentially allow reaching the same limit. This could be improved by inserting a second ORO on a different test mass so that the total noise of the differential measurement (presently limited by the capacitive sensors) will be limited by the incoherent noise of the two OROs. Other solutions, using a single ORO and multiple reflections, to directly measure φ, even if in a configuration not representative of the one proposed for LISA, are also being considered.

In figure 5 is also shown the upper limit to the ORO back action (orange), due to radiation pressure fluctuation, estimated from the measured power fluctuation of the light beams impinging on the QPDs. Clearly, the back action is not directly measurable because it is much below the torsion pendulum thermal noise. It could be only measured if artificially increased by modulating the S-LED's power.

4. Conclusion

We have tested the performance of an optical read-out system for the LISA GRS on the four-mass torsion pendulum facility operating in Trento. The results are encouraging, confirming that, as expected, the ORO shows a better sensitivity compared to the capacitive sensor in the interesting frequency band (this is directly measured for angles, while we only put upper limits for displacement).

We also verified that the ORO can already be used as e sensor for the torsion pendulum facility in place of one of the two capacitive sensors, obtaining the same sensitivity in force noise measurements.

Further upgrades of the ORO system, presently under definition, should permit to improve the performance of the pendulum as a ground testing facility for the LISA-Pathfinder hardware.

References

[1] Bender P L et al. 2000 Laser Interferometer Space Antenna for the detection and observation of gravitational waves. *ESA-SCI* **11**.

[2] Carbone L, Cavalleri A, Ciani G, Dolesi R, Hueller M, Tombolato D, et al. 2006 Torsion pendulum facility for direct force measurements of LISA GRS related disturbances. In: Merkovitz SM, Livas JC, editors. Laser Interferometer Space Antenna: 6th International LISA Symposium. *American Institute of Physics Conference Series* **873** 561–5.

[3] Cavalleri A, Carbone L, Ciani G, Doles R, Hueller M, Nicolodi D, Tombolato D, Wass P J, Weber W J and Vitale S Direct force measurements for testing the LISA Pathfinder gravitational reference sensor, 7th LISA symposium proceedings, *Class. Quantum Grav.* in press.

[4] Chiao M P, Dekens F and Abramovici A 2003 Optical displacement sensor (ODS): an inertial reference sensor candidate for LISA, *Proc. SPIE-Int. Soc. Opt. Eng.* **4856** 98.

[5] Acernese F, Calloni E, De Rosa R, Di Fiore L, Garcia L and Milano L 2004 *Class. Quantum Grav.* **21** S261–7

[6] Acernese F, Calloni E, De Rosa R, Di Fiore L and Milano L 2005 *Class. Quantum Grav.* **22** S279–85.

[7] Speake C C and Aston M S 2005. *Class. Quantum Grav.* **22**, S269-77.

[8] Sun K, Buchman S and Byer R, 2006 *J. Phys.: Conf. Ser.* **32,** 167-179.

[9] Acernese F, Calloni E, De Rosa R, Di Fiore L, Garufi F, La Rana A and Milano 2006 Some progress in the development of an optical readout system For The LISA gravitational reference sensor in: Merkovitz SM, Livas JC, editors. Laser Interferometer Space Antenna: 6th International LISA Symposium *American Institute of Physics Conference Series* **873** 339–43.

LISA Mission and System Architectures and Performances

Peter F. Gath, Dennis Weise, Hans-Reiner Schulte, Ulrich Johann

Astrium GmbH Satellites, 88039 Friedrichshafen, Germany

peter.gath@astrium.eads.net

Abstract. In the context of the LISA Mission Formulation Study, the LISA System was studied in detail and a new baseline architecture for the whole mission was established. This new baseline is the result of trade-offs on both, mission and system level. The paper gives an overview of the different mission scenarios and configurations that were studied in connection with their corresponding advantages and disadvantages as well as performance estimates. Differences in the required technologies and their influence on the overall performance budgets are highlighted for all configurations. For the selected baseline concept, a more detailed description of the configuration is given and open issues in the technologies involved are discussed.

1. Introduction

The starting point for the LISA Mission Formulation study was the design established in the 2000 Final Technical Report (FTR) [1]. During the course of the study, alternative designs were identified that offered a high potential for improvements in the overall payload system and performance.

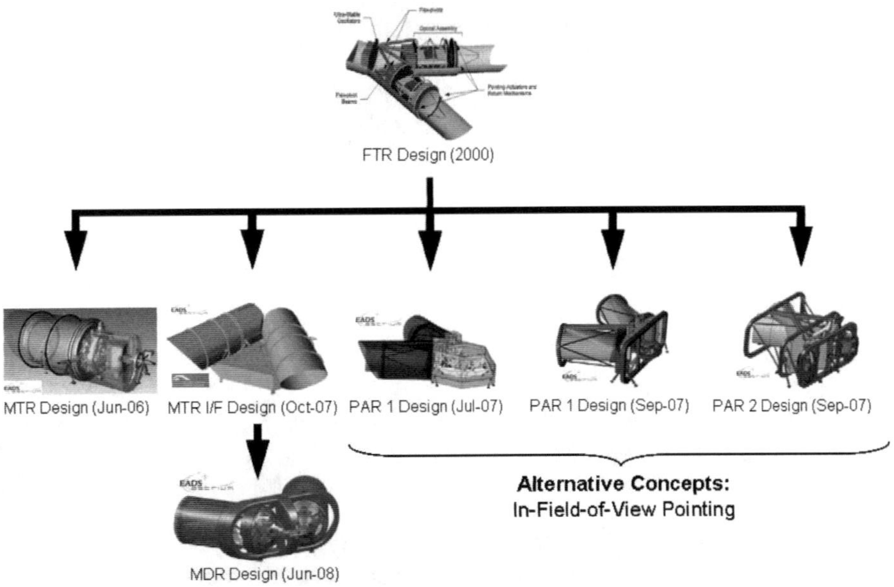

Figure 1: Different payload designs investigated.

Several trade-off studies have been performed on individual subsystems such as the propulsion system, launch stack configuration, and especially for the payload configuration itself. Among other trades, the main trade-offs performed for the payload are:

- telescope type
- Gravitational Reference Sensor (GRS) vacuum enclosure or open GRS
- optical assembly pointing vs. in-field pointing
- realization of point-ahead angle compensation
- strap-down / two step-interferometry
- Use of optical GRS readout for the drag-free and attitude control system

This paper focuses on the trade off between optical assembly pointing and in-field pointing. Figure 1 gives an overview about the investigated designs.

2. Mid-Term Review: MTR Design

By the time of the mid-term review of the study in June 2006, the established payload design baseline was based on optical assembly pointing, i.e. the whole optical assembly, consisting of telescope, optical bench, and GRS, is rotated around a pivot axis in order to compensate for the varying angle between two LISA interferometer arms. The design was an evolution of the original FTR baseline design and is depicted in Figure 2.

Already by the time of the MTR, an interesting alternative design concept was identified which avoids the articulation of the whole optical assembly. Instead of moving the whole assembly, a small scan mirror within the telescope is compensating the changes in the angle between two interferometer arms.

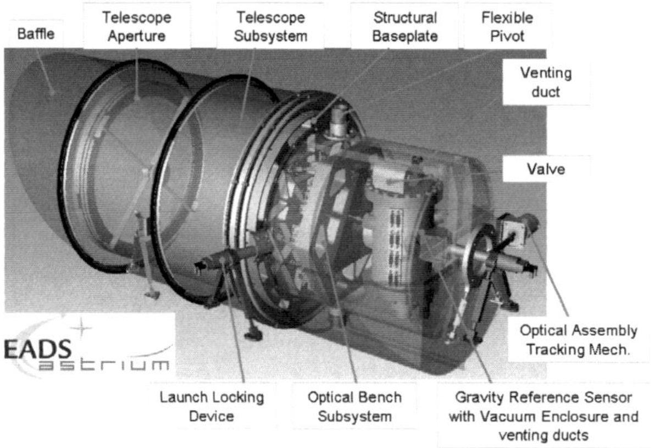

Figure 2: MTR design baseline

While this seems to be an attractive concept from the mechanical side, the requirements imposed by this approach on the optical metrology system and especially on the telescope are very challenging. This can easily be understood since the scan mirror is located directly in the interferometry path where pm-stabilities or at least a knowledge about pm changes in the LISA measurement bandwidth are required for the science measurement. When moving the complete optical assembly, several nm can be allowed since this only has an impact on the movement between the proof mass and the GRS housing.

3. Payload Architecture Review 1: PAR-1 Single Active GRS

The first in-field-pointing configuration investigated was a configuration with only one active GRS (Figure 3, left). For redundancy and reliability issues, a second, cold redundant GRS is installed in a geometrically less optimal position on each S/C. However, only one GRS is operated during science operation, i.e. a full drag-free control can be established in all lateral degrees of freedom of the proof mass and only suspension actuation around rotational degrees of freedom is required.

In order to successfully establish this payload concept, one key point was the development of a telescope design that allows full performance over a large field-of view of approx. ±0.5°. More details on this are given in [2].

Sizing efforts on the spacecraft and propulsion module showed that for the PAR-1 design, the launch stack mass becomes unacceptably high. Therefore, this particular configuration was not considered to be a reasonable candidate for a potential new baseline. An effort was taken in order to significantly reduce the overall envelope of the payload and a modified design of the single GRS configuration was introduced. This requires that the two GRS are moved together as close as possible, which requires either an open GRS system or a common vacuum enclosure for the two GRS. It was shown that the self-gravity issues can be handled by a proper placement of compensation masses.

The main benefit of this modified configuration illustrated in Figure 3 (right) is that it does not only reduce the payload envelope. By moving together the two GRS, also the science performance requirements can be met with all three interferometer arms in case of a failure of the nominal GRS.

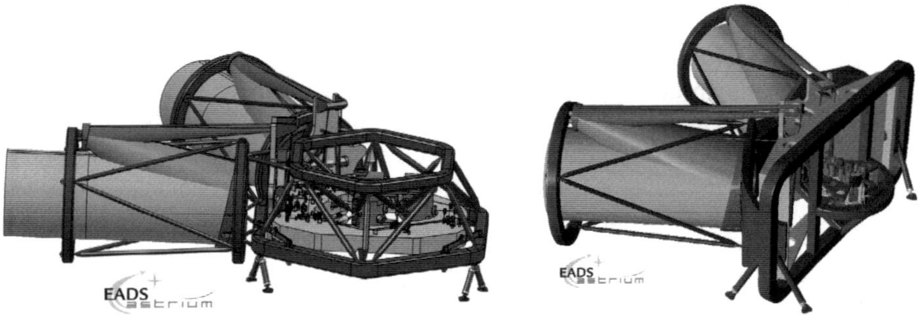

Figure 3: Single active GRS configuration with in-field pointing (left) and with full GRS redundance (right)

4. Payload Architecture Review 2: PAR-2 Two Active GRS

The second in-field pointing configuration investigated was a configuration with two active GRS, each GRS serving one interferometer arm. From a DFACS point of view, such a configuration is very similar to the MTR baseline, except that no telescopes need to be moved. The configuration is shown in Figure 4.

It turns out that by crossing the two lines of sight of the two telescopes, the most compact payload configuration can be achieved. However, as for all in-field pointing designs, there exists a very strong coupling between the subsystems in terms of alignment.

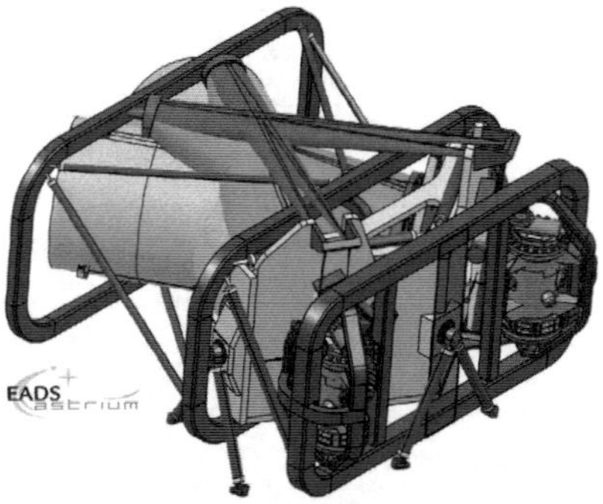

Figure 4: Two active GRS configuration with in-field pointing

5. MTR-Design with Interface Structure

For the trade-off study, also the MTR baseline was modified by introducing a common interface structure between the two separate optical assemblies and the spacecraft. This significantly simplifies the interface definition to the S/C and also improves the AIVT process. The design of this structure followed the same mechanical requirements as they were imposed on the in-field pointing configurations. The resulting configuration is depicted in Figure 5.

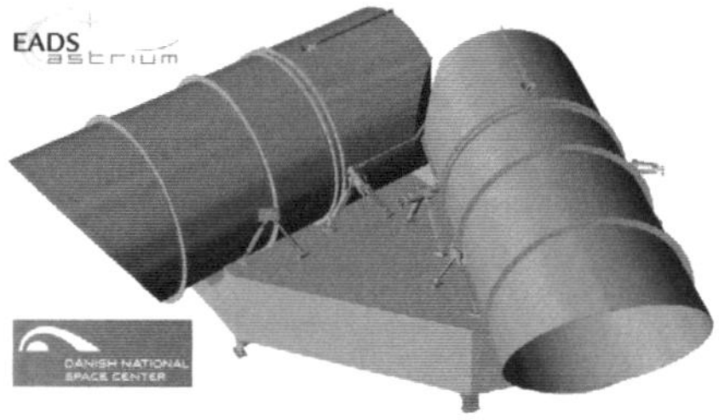

Figure 5: Moving optical assemblies with interface structure

6. Performance Comparison and Trade-Off

In terms of overall performance it turned out, that all configurations can meet the LISA science requirements. When compared with the MTR design, the PAR-2 configuration showed a slight increase of the noise at low frequencies. This is mainly due to a larger contribution from the path-length error, based on the assumptions made for thermal stability and its frequency dependency.

The single active proof-mass configurations show a slightly better performance at medium frequencies due to the full drag-free control in lateral degrees of freedom. However, the performance is still very close to the performance of the MTR configuration with moving optical assemblies.

A comparison of the configurations is shown in Figure 6.

The complete trade-off study took into account 71 different parameters evaluated for the different configurations. These also included estimates for the risks of technology developments, launch stack mass, power consumptions, number of required measurement channels etc. It turned out that all configurations show a similar overall "performance" when performing a weighted trade-off analysis.

Due to the very similar performance of all configurations, the MTR design was finally chosen as the baseline configuration for the remainder of the LISA Mission Formulation study. This decision was mainly driven by a reduced complexity of the optical system, especially the telescope, and the elimination of the scan mirror with a fairly large dynamic range and pm stability requirements from the optical metrology path. Especially this scan mirror was considered to be a very high risk for the in-field pointing designs.

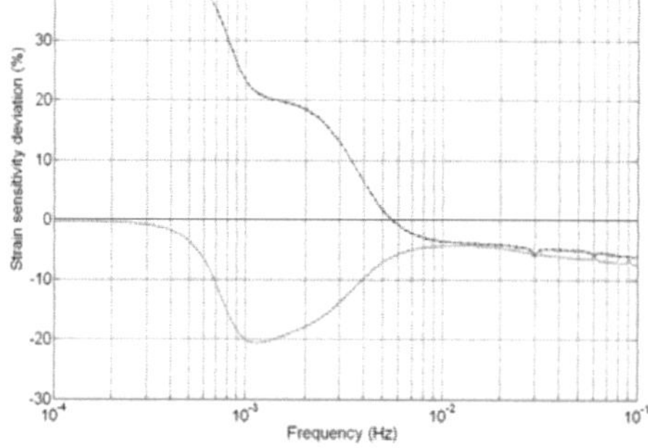

Figure 6: Comparison of strain sensitivity

Journal of Physics: Conference Series **154** (2009) 012013 doi:10.1088/1742-6596/154/1/012013

7. Mission Design Review: Moving Optical Assemblies

The baseline adopted for the mission design review was derived from the MTR design and includes many lessons learned during the study of the in-field pointing designs. In order to reduce the risk of stray-light, an off-axis telescope was also designed for this configuration, but with a smaller field-of-view which significantly simplifies the design. Also the overall envelop of the design could be improved by integrating the interface structure to the spacecraft directly into the payload design. This is illustrated in Figure 7.

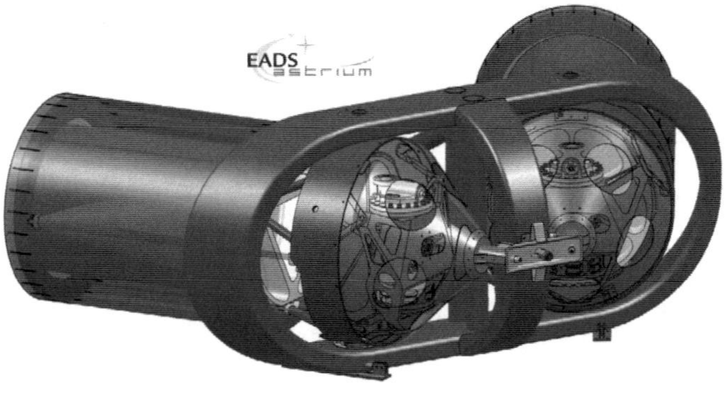

Figure 7: Moving optical assemblies – MDR Baseline

In order to significantly improve the telescope pointing knowledge and allow for ground calibrations of pointing sensitivities, an additional optical metrology system for the point ahead angle actuator is introduced in the design. The estimated jitter of this mechanism was limiting the pointing knowledge in the MTR design.

Another element introduced into the design is the optical truss which measures any path-length change in the telescope subsystem. While this was originally thought to be mandatory for achieving the performance in the in-field pointing designs, it turned out to be an excellent tool during ground testing and AIV. On orbit, it serves as a backup system in case of any unexpected path-length change effects, e.g. due to material outgassing, and thus increases the robustness of the mission. The optical bench design is shown in Figure 8.

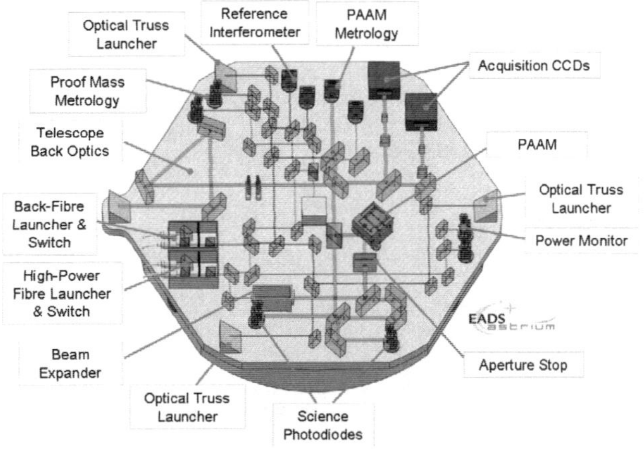

Figure 8: Optical bench – MDR baseline

8. Conclusions

Detailed trade-off studies were performed on all mission and system levels during the LISA Mission Formulation study. This paper describes the particular trade-off for different payload configurations. Lessons learned during the investigation of the alternative payload designs were included in the finally selected baseline design, i.e. a final iteration on the payload design was included after a baseline concept was selected. This new payload design is now complemented with the required specifications for a further refinement in the next phase of the LISA project.

9. Acknowledgements

This work was performed in the context of the LISA Mission Formulation Study, ESTEC Contract No. 18756/04/NL/HB and partially supported by DLR, Contract No. 50 OQ 0701.

The authors also acknowledge the work provided by the Danish National Space Center for the interface structure of the MTR design, namely the contributions from Hans Ulrik Nørgaard Nielsen and Niels Christian Jessen.

References

[1] Hammesfahr, A., et al., Study of the Laser Interferometer Space Antenna, Final Technical Report, Report No. LI-RP-DS-009, April 2000

[2] Weise, D., et al., Alternative Opto-Mechanical Architectures for the LISA Instrument, 7th LISA Symposium, 16-19 June 2008, Barcelona, Spain

Journal of Physics: Conference Series **154** (2009) 012014 doi:10.1088/1742-6596/154/1/012014

Control System Design for the Constellation Acquisition Phase of the LISA Mission

Francesca Cirillo and Peter F. Gath

Astrium GmbH Satellites, 88039 Friedrichshafen, Germany

E-mail: francesca.cirillo@astrium.eads.net, peter.gath@astrium.eads.net

Abstract. The objective of the constellation acquisition phase for the LISA mission is to establish the three laser links between the three spacecraft of the LISA constellation so that the interferometric measurements for the science experiment can commence. The laser beam acquisition for LISA is extremely challenging given the 5 million km distance between the spacecraft, the inherent limits of the attitude sensors accuracy, the orbit determination accuracy issues and the time required to phase-lock the incoming and outgoing laser signals. This paper presents the design of the control system for the acquisition phase of the LISA constellation: the acquisition operational procedure is outlined, guidance laws are defined together with the Gyro Mode attitude control principle, which implements a Kalman filter for disturbances rejection purposes. Constellation-wide non-linear simulations demonstrate that the LISA constellation acquisition phase is feasible by means of the proposed control strategy.

1. Introduction

The main goal of the LISA mission is to detect gravitational waves and characterize their sources. The underlying measurement principle is a laser interferometry system built up with three identical satellites, each carrying two telescopes for laser pointing and flying in a triangular constellation with an edge length of $5 \cdot 10^6$ km. Nevertheless, the interferometric measurements for the LISA science experiment are only possible once the three laser links between the three LISA spacecraft are acquired on the six quadrant photodiodes (QPD) (one for each optical bench, i.e. one for each telescope) so that the QPD can start to measure the arm-length variations with a few picometer accuracy and to operate as a highly accurate inertial wavefront sensing for the spacecraft and telescope attitude control.

The constellation acquisition phase for the LISA mission is strongly demanding given the distance between satellites and the laser beam pointing determination issue which is mainly related to the inherent limits of the attitude sensor accuracy and to the orbit determination uncertainties.

In fact, the acquisition uncertainty half-cone angle, i.e. the size of sky where the receiving spacecraft is expected to be located w.r.t. the sending spacecraft, is estimated to be 15.2 μrad and it has contributions from the navigation error, the laser beam pointing long-term drift due to the accuracy of the available attitude sensors and the alignment accuracy between star-tracker (STR), telescope line-of-sight (LOS) and outgoing laser direction. The laser beam half-cone angle at the receiving side, instead, spans 1.43 μrad being the full-width of the emitted beam at half-maximum reduced by the local laser beam pointing short-term drift due to the attitude sensor accuracy. Since the uncertainty cone is larger than the laser beam cone, the acquisition cannot be direct. Moreover, since the QPD field-of-view (FOV) is 1 μrad, and therefore smaller than the laser beam cone size, a fine acquisition strategy must be foreseen.

The attitude sensor accuracy issue rises from the fact that the QPD is not available until it receives a continuous laser signal. Therefore, a star tracker, together with an optical assembly tracking mechanism (OATM) sensor, must be used as auxiliary coarse attitude sensor when no links are acquired. A fine accuracy acquisition sensor (CCD) can then be used for laser pointing determination when the CCD itself has acquired a laser signal and under the condition that the local laser is off. The current baseline for sensor equipment is shown in table 1: in order to enhance the FOV and the resolution required to realize the measurement chain from one sensor to another, also a drastic increase in the read-out noise is produced.

Table 1. Attitude sensor field-of-view, resolution and noise.

	STR	**CCD**	**QPD**
Field-of-View	0.3 rad	155 μrad	1 μrad
Resolution	255 μrad/pixel	0.4 μrad/pixel	1 nrad/pixel
Read-Out Noise	29 μrad (RMS 3σ)	60 nrad (RMS 3σ)	750 prad/√Hz

The design of a dedicated acquisition control strategy was therefore mandatory in order to perform the constellation acquisition phase for the LISA mission. Such a strategy includes:

- The establishing of an acquisition operational procedure.
- The design of the control system: guidance laws, measurement processor and controller.
- The set up of a constellation-wide non-linear simulator to verify that the LISA constellation acquisition phase is feasible by means of the proposed control strategy.

2. Constellation Acquisition Operational Procedure

In order to fulfill the LISA constellation acquisition a dedicated *Laser Beam Acquisition Mode* must be used. Spacecraft are first placed in *Gyro Mode* so that the inertial sensor measurements of the test masses states are used as gyro-like measurements to filter, smoothen and propagate the STR data and the OATM sensor data for disturbances rejection purposes. After a transient period, spacecraft point at each other based on ground-provided ephemeris data, STR and OATM sensor measurements (*Steady State Mode*). The strategy adopted for the constellation acquisition is then sequential in the sense that it establishes one laser link at time. The process for the single-link acquisition is initiated from ground command and then accomplished autonomously as described in the following section.

2.1. *Single-Link Acquisition Process*

The single-link acquisition process is divided into three main phases: a signal acquisition on the CCD, a signal fine acquisition on the CCD and a signal acquisition on the QPD.

2.1.1. *Signal Acquisition on the CCD.* The goal of this acquisition step is to find the laser signal on the CCD of both spacecraft (S/C). The sending S/C turns on the laser and performs a scanning maneuver to cover the whole uncertainty cone while the receiving S/C holds its reference attitude based on ground-provided navigation data. At a certain time during the scanning, the receiving S/C detects a laser signal on its CCD, estimates the offset between the real and the ground-provided incoming laser signal, performs an attitude offset correction and switches on its laser. However, only after switching off its local laser at the end of the scanning maneuver, the sending S/C can detect a signal on its CCD and, analogously to the receiving S/C, apply the necessary attitude correction to its on-board reference attitude by steering its telescope to the updated reference direction. Both spacecraft along one arm of the constellation have now acquired a laser signal on their CCD but the laser transmission cannot be continuously operated yet, since the CCD is blind when the local laser is activated.

2.1.2. *Signal Fine Acquisition on the CCD.* Sending and receiving S/C perform a fine attitude correction by switching lasers on/off in order to bring the incoming laser signal to a reference position of the CCD within the FOV of the QPD. CCD measurements are used to support attitude control (*CCD Mode*). This sub-phase may be bypassed if *Gyro Mode* performances allow getting a signal on the CDD which is already within the QPD field-of-view.

2.1.3. *Signal Acquisition on the QPD.* In order to obtain the laser signal on the QPD of both S/C along one constellation arm, the receiving S/C performs a frequency scan by changing the frequency of its local laser in order to detect an interference signal on its QPD. Once the receiving S/C has found a beat signal, the sending S/C will also find a beat signal approximately 16.7 s later, i.e. the delay of any information travelling through the $5 \cdot 10^6$ km arms between the satellites. As for the fine acquisition, CCD data can be used to support attitude control depending on *Gyro Mode* performances.

2.2. Laser Link Acquisition Sequence

The single-link acquisition process must be repeated three times in order to establish the three laser links between the LISA spacecraft. The order in the link acquisition is arbitrary: the sequence illustrated in figure 1 has been selected just as exemplification. Note that the arrows indicate the direction of traveling light. Note also the configuration of the telescopes (figure 2): telescope 1 (T1) is actuated and can be rotated in the constellation plane, while telescope 2 (T2) is fixed to the S/C. The first laser link to be established is link L_3/L_3' where T1 of S/C1 defines the sending pointing direction and T2 of S/C2 defines the receiving pointing direction. Then, the acquisition of the link L_2/L_2' is performed with T2 of S/C1 as sender and T1 of S/C3 as receiver. Eventually, the laser link L_1/L_1' is acquired where T1 of S/C2 is the sending direction and T2 of S/C3 is the receiving direction.

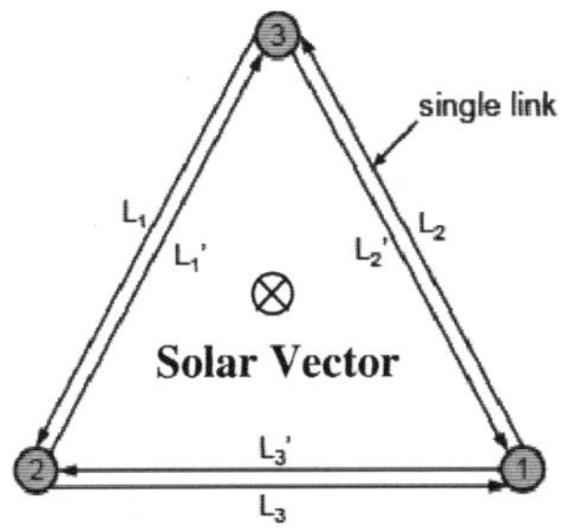

Figure 1. Laser links acquisition sequence.

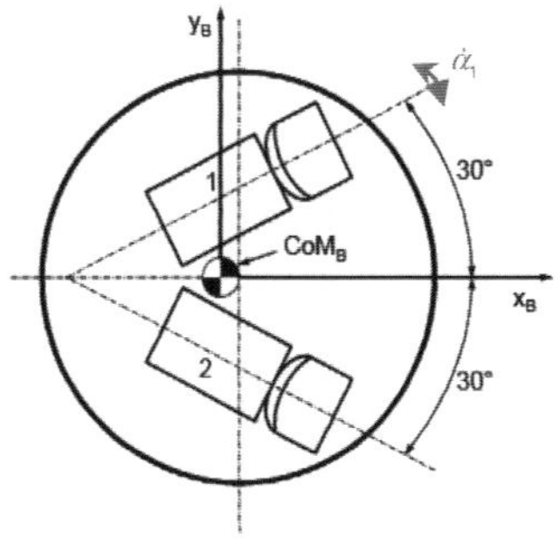

Figure 2. Telescope configuration.

3. Constellation Acquisition Control System

The key element to perform the constellation acquisition for the LISA mission stands in the design of the acquisition control system. This is divided into three parts: a guidance law design capable to drive the laser pointing during the acquisition, a measurement processor design necessary to reduce the disturbances on the telescope pointing due to the coarse attitude sensor noise and a controller design.

Three cases can be considered for attitude determination during the acquisition phase:

- No available fine attitude sensor measurements: attitude determination must be achieved by means of a measurement processor used to reject the sensor noise conveyed by the STR.
- One link of fine sensor data (CCD, QPD) available: fine sensor data can be used with or within the measurement processor to provide higher-accuracy attitude knowledge. The *CCD Mode* can be necessary to perform fine pointing and to achieve QPD acquisition while QPD data can be used to improve the acquisition of the second S/C arm once the first is acquired.

- Two links of fine sensor data (QPD) available: QPD data can be exclusively used to compute the telescope pointing as in *Science Mode*.

It has been conservatively assumed for the design of the control system that only the STR and the OATM sensor data are available during the whole duration of the constellation acquisition phase [1].

3.1. Guidance Law Design

Three different guidance laws are derived in order to drive the telescope pointing during the several constellation acquisition sub-phases.

The reference attitude law for the *Gyro Mode* is based on ground-provided ephemeris data, including point-ahead angle computation and compensation, where the point-ahead angle is the angle between the send and the receiving direction due to the relative motion of the S/C in the constellation.

The scanning law selected for the scanning maneuver consists in Archimedean spiral motion from the center, where is more likely to find the receiving S/C, to the outer boundary of the uncertainty cone, performed with constant tangential velocity. The parametric equations of the spiral motion are:

$$\begin{cases} \rho = \dfrac{k \cdot r}{\pi} \cdot \theta \\ \theta = \left(\dfrac{2 \cdot \pi \cdot a \cdot (t - t_0)}{k \cdot r} \right)^{1/2} \end{cases} \tag{1}$$

and additional conditions derived by imposing to cover the uncertainty cone in N+1 rounds are:

$$N = \left(\frac{R}{k \cdot r} - 1 \right) \cdot \frac{1}{2} \tag{2}$$

$$a \cdot \Delta t_{id} = 2\pi \cdot k \cdot r \cdot \left(N + 1 \right)^2 \tag{3}$$

where R is radius of the uncertainty cone (~ 76 km), r is the radius of the laser beam cone (~ 7.15 km), $k \in [0,1]$ is the spiral pitch, N+1 are the rounds required to cover the uncertainty cone, a is the constant tangential velocity, t_0 is the starting scanning time, t is the current time and Δt_{id} is the ideal time required to cover the uncertainty cone. Note that k and Δt_{id} are the only free parameters and must be tuned in order to guarantee a sufficient coverage of the uncertainty cone.

The guidance law selected for the attitude offset correction is a time-optimal slew maneuver with torque and angular velocity limits. Details on the governing equations can be found in [3], nevertheless it is important to point out that the maneuver is completely determined once the maximum torque delivered by the actuators and maximum angular velocity permitted by the sensor equipment or by the controller are selected.

3.2. Measurement Processor and Controller Design

Since the STR does not provide a good enough attitude reference, the inertial sensor (IS) is used to support the attitude control system over short time-periods. This control mode is addressed as *Gyro Mode* since it is based on the same principle, which is often used for spacecraft attitude determination in high accuracy pointing applications and which relies on the hybridization of STR and gyroscope measurement data. In analogy to the STR and gyros data fusion approach, the IS data are used as gyro-like measurements to filter, smoothen and propagate the STR data, supporting in this way the attitude control system over short-time periods. This can happen because the two signals have complementary error characteristics: the STR has a good long-term stability but suffers from short and medium-term noise, on the contrary the IS ("gyro") has a good short-term stability but suffers from medium and long-term drifts due to unknown or uncertain DC forces and torques acting on the test masses.

The way to implement the *Gyro Mode* is to use a state-augmented continuous-time steady-state Kalman filter opportunely tuned, whose detailed equations can be found in [1] and [4], that processes the measurement data coming from all the available sensors (STR, OATM sensor and IS unit) providing enhanced estimates of the S/C attitude, the T1 in-plane pointing angle and the test masses coordinates. The enhanced estimations are then fed in closed-loop to the controller in place of the sensor data. The controller used in acquisition mode has the very same structure of the drag-free and attitude control system (DFACS) used in *Science Mode*. Details about the DFACS can be found in [5].

4. Performances Analysis

The end-to-end (E2E) constellation non-linear simulator shown in figure 3 has been developed in order to analyze and validate the proposed control strategy in a time domain framework and, therefore, to demonstrate the feasibility of the LISA constellation acquisition.

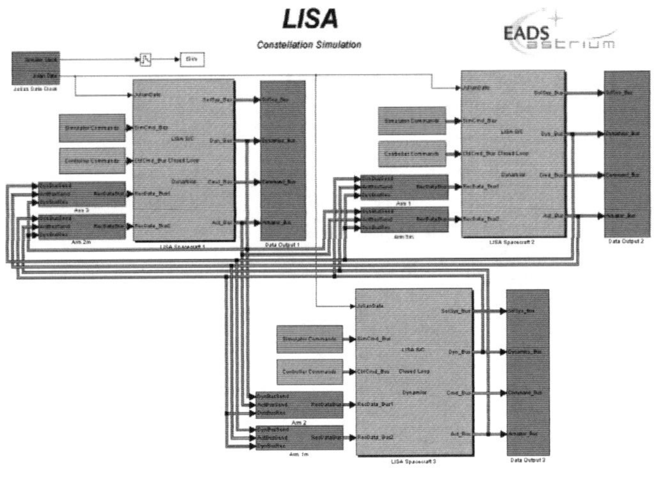

Figure 3. Main structure of the LISA E2E constellation simulator. It consists of three individual spacecraft running in parallel and connected via six arm delay models. Spacecraft can be independently configured by means of separate telecommanding blocks. The model includes the full rigid-body dynamics of the LISA constellation: non-linear dynamics with 57 degree-of-freedom including spacecraft, telescope and test masses states, reference attitude computation via Chebychev propagator algorithm with point-ahead angle compensation, DFACS control system, detailed actuator and sensor models.

Figures 3 and figure 4 show the time series of the telescope pointing performances during the acquisition of the laser link $L_3/L_3^{'}$.

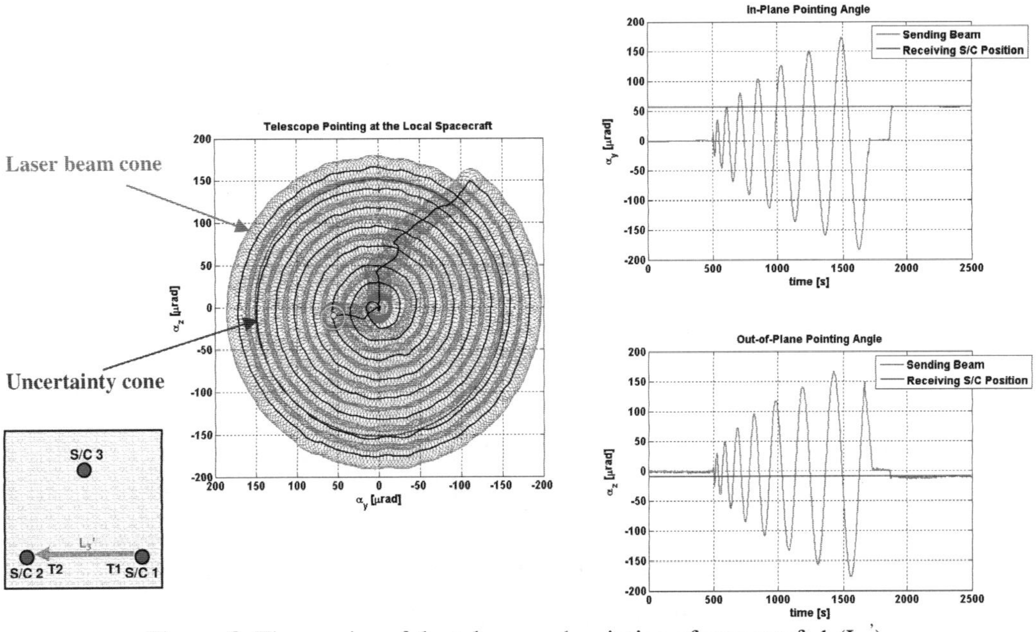

Figure 3. Time series of the telescope 1 pointing of spacecraft 1 ($L_3^{'}$)

Journal of Physics: Conference Series **154** (2009) 012014 doi:10.1088/1742-6596/154/1/012014

The use of the Kalman filter and the choice of an appropriate spiral pitch guarantee that the laser beam of spacecraft 1 is able to map the whole uncertainty cone with a coverage varying between 1.8 s and 8.6 s. The time required to complete the laser link acquisition is about 1400 s (ca. 24 minutes). The telescope pointing performances during the acquisition of the remaining two links are basically the same and are not shown for editorial reason (see [4] for details). The resulting total duration of the constellation acquisition phase is about 70 minutes, that is 100 minutes including margins.

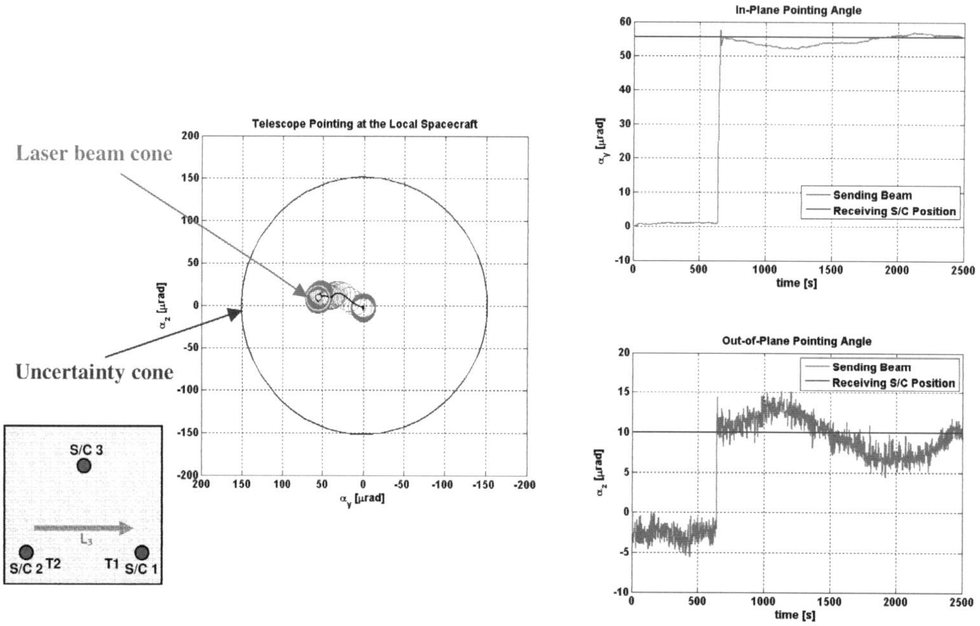

Figure 4. Time series of the telescope 2 pointing of spacecraft 2 (link L_3)

Figure 5 shows the complete time series of the telescope pointing errors from the starting of the *Gyro Mode* up to the end of the constellation acquisition phase.

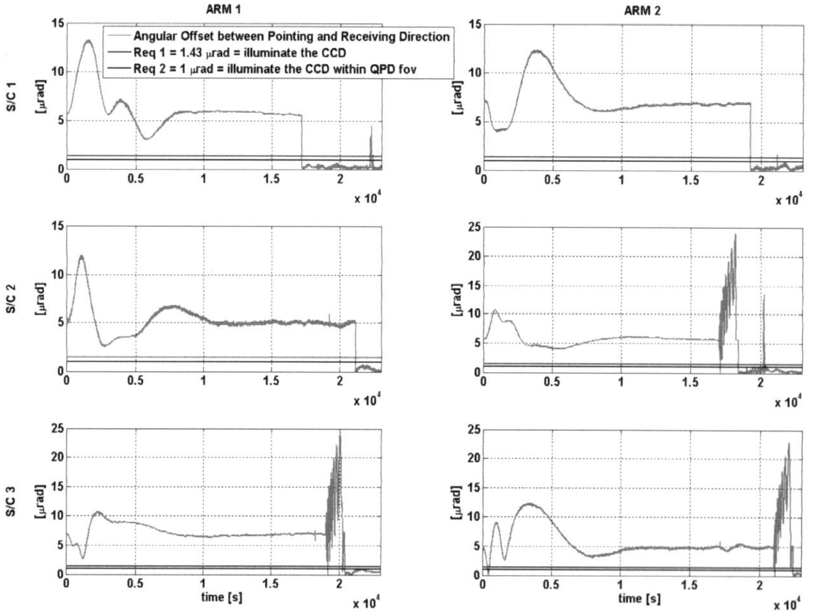

Figure 5. Time series of the six telescope pointing errors during LISA constellation acquisition

The angular offsets between the telescope pointing direction and the direction of the incoming laser beam for each arm of each satellite are shown. After a transient period of about 12000 s, the *Gyro Mode* achieves the steady state and non-zero offsets, larger than the laser beam cone size, affect all the constellation arms: no signal can be detected. The acquisition phase begins at 17000 s and ends about 5500 s later. The residual misalignment between the pointing direction and the laser beam incoming direction is lower than the laser beam cone size, i.e 1.43 μrad, and, moreover, lower than QPD field-of-view, i.e. 1 μrad, indicating that the *Gyro Mode* performance allow getting a signal on the CCD which is already within the FOV of the QPD.

5. Conclusions

A control strategy for the acquisition phase of the LISA constellation has been proposed: it includes the definition of an operational procedure, the design of a constellation acquisition control system and the development of an E2E constellation simulator.

Constellation-wide non-linear, fully coupled E2E simulations demonstrate the feasibility of the laser beam acquisition from initial scanning up to *Science Mode* for all constellation arms. The total duration of the acquisition phase results in about 70 min, i.e. 100 min including margins.

The acquisition phase is entirely performed in *Gyro Mode*. Time domain analyses indicate that *Gyro Mode* performances meet the requirement for a CCD acquisition directly in the QPD field-of-view in a timely manner. However, even if a fine acquisition is not strictly required, its implementation is advisable in order to guarantee larger safety margins for the QPD acquisition.

6. Acknowledgements

The present work was performed in the context of the LISA Mission Formulation Study, ESTEC Contract No. 18756/04/NL/HB. The authors acknowledge the fruitful and stimulating collaboration with the whole LISA mission formulation study team including ESA, NASA and the LISA science community.

References

[1] Cirillo F and Gath F P 2007 Constellation acquisition control for LISA *Astrium GmbH Tech Note*

[2] Maghami P G. Hyde T T and Kim J, 2005 An acquisition control for the laser interferometer space antenna *Class. Quantum Grav.* **22** 421-8

[3] Gath F P 2003 Time optimal slew maneuvers *Astrium GmbH Tech Note*

[4] Cirillo F 2007 Controller design for the acquisition phase of the LISA mission using a Kalman filter *Uni Pisa* etd-06202007-101808

[5] Gath F P, Schulte H R, Weise D and Johann U 2007 Drag free and attitude control system design for the LISA science mode *AIAA Guid., Nav. and Control Conf. and Exhibit (Hilton Head, South Carolina, 20-23 August 2007)* AIAA 2007-6731

Lisa Amplitude Modulation: A Study of the Angular Resolution of LISA for Monochromatic Gravitational Waves

G.Auger, H.Halloin, A.Petiteau and Eric Plagnol

APC, UMR 7164, Université Paris 7 Denis Diderot
10, rue Alice Domon et Léonie Duquet, 75025 Paris Cedex 13, France

E-mail: eric.plagnol@apc.univ-paris7.fr

J.Y.Vinet

ARTEMIS, Observatoire de la Côte d'Azur-C.N.R.S., 06304 Nice, France

Abstract. We present formulae for the amplitude modulation of the X,Y and Z TDI combinations for monochromatic gravitational waves and use these to study the LISA angular resolution in the case of large SNR. The angular resolution, $\Delta\Theta/(0.1 * SNR)$, is found to lie between 2° and 5° and is somewhat dependent on the ecliptic colatitude angle (β), on the polarisation $(h_+, h_\times)$ and much less on the ecliptic longitude angle (λ). Comparisons with other studies are presented. Future studies will treat the case of small SNR.

1. Introduction

The aim of this work is to provide a relatively simple method to extract the angular location and polarisation of a monochromatic gravitational wave (GW) of frequency f from the LISA TDI time series (X,Y and Z, for this study, A and E in a further development) and to estimate the precision with which this can be done. A description of the future LISA mission can be found in Danzmann et al. [1] and an explanation of the TDI (Time Delay Interferometry) laser noise supression method in Armstrong, Estabrook, and Tinto [2] and in Dhurandhar, Rajesh Nayak and Vinet [3] .

The method is based on an analytical formula of the amplitude modulation produced by these TDI generators. The formula is based on two assumptions: the LISA triangle has equal arms ($L_1 = L_2 = L_3 = L$) and the GWs have "low frequencies" ($f << L/c$). Although it is applied here to monochromatic GW sources, it can and will be generalised to GW emitted from coalescing black holes binaries.

The results presented here show the angular resolution obtained with this method, as a function of the source direction ($\vec{\omega}$) in the sky (defined by β : ecliptic colatitude and λ ecliptic longitude, or equivalently by $\theta = \pi/2 - \beta$ and $\varphi = \pi - \lambda$), of the polarisation $(h_+, h_\times)$ defined in the barycentric reference frame and of the signal to noise ratio (SNR).

As it is based only on the amplitude modulation, the precision of the measured direction (β,λ) will not be as good as the one obtained from methods which also include the phase modulation [4, 5], particularly at high frequencies ($> 10^{-3}Hz$). A study of how much precision is lost will be performed in the near future.

2. The Amplitude Modulation formula

Within the two approximations mentioned above, the amplitude modulation of the TDI X_i ($i = 1$ for X, $i = 2$ for Y and $i = 3$ for Z) is given by (see Petiteau [6], page 162, see also Cornish and Rubbo [7]):

$$A_{X_i}(t; \theta, \varphi, h_+, h_\times, f) = 4\pi f L \sin(2\pi f L) \left[h_+^2 F_{i+}^2 + h_\times^2 F_{i\times}^2\right]^{1/2}, \tag{1}$$

where F_{i+} and $F_{i\times}$ are the LISA antenna patterns defined by

$$F_{i+} = \left[\left(\vec{\theta} \cdot \vec{n}_{i+2}\right)^2 - \left(\vec{\phi} \cdot \vec{n}_{i+2}\right)^2\right] - \left[\left(\vec{\theta} \cdot \vec{n}_{i+1}\right)^2 - \left(\vec{\phi} \cdot \vec{n}_{i+1}\right)^2\right] \tag{2}$$

$$F_{i\times} = \left[2\left(\vec{\theta} \cdot \vec{n}_{i+2}\right) \cdot \left(\vec{\phi} \cdot \vec{n}_{i+2}\right)\right] - \left[2\left(\vec{\theta} \cdot \vec{n}_{i+1}\right) \cdot \left(\vec{\phi} \cdot \vec{n}_{i+1}\right)\right], \tag{3}$$

where n_i are the unit vectors of the arms of LISA and $\vec{\theta}$ and $\vec{\phi}$ are vectors defined by

$$\vec{\theta} = -\frac{\partial \vec{\omega}}{\partial \theta} = -\frac{\partial \vec{\omega}}{\partial \beta} \tag{4}$$

$$\vec{\phi} = \frac{-1}{\sin\theta} \frac{\partial \vec{\omega}}{\partial \phi} = \frac{-1}{\cos\beta} \frac{\partial \vec{\omega}}{\partial \lambda}, \tag{5}$$

and $\vec{\omega}$ is the unit vector of the direction defined by the angles β and λ.

3. Data Generation

A method to generate amplitude modulation data would be to use a LISA simulator (LISACode [8] in our case) to produce a year long (TDI) time series for a given source and instrumental noise function. Using this, one would split it into many (N_d) time windows of duration $\Delta T = 1$ year$/N_d$, obtain a measure of the amplitude (from a FFT analysis) and fit these N_d data points with equation (1).

Even with a fast simulator such as LISACode, generating all the necessary cases to cover the possible angular domain, polarisation values and signal to noise ratios would take a prohibitive amount of computing time.

We have therefore simplified this method in the following way. We use LISACode to generate the time series for the various angular locations and polarisation values *in the absence of instrumental noise*. From these time series, we extract the value of the amplitude at the central time of each of the N_d windows. To this amplitude, we add a noise vector by randomly choosing its 2-dimensional components from Gaussian distributions of variance σ, calculated for the requested SNR value (see eq. 11 below). The *measured* data corresponds then to the amplitude of the vector thus created. This method is correct as long as the power of the monochromatic signal is contained in only one frequency bin $\Delta f = 1/\Delta T$. The typical values of $N_d, \Delta T$ and Δf used in this work (31, 11.5 days and $10^{-6}Hz$) guarantee that this is generally the case. Otherwise the value of the noise has to be taken into account, as well as the number of bins over which the data is spread.

Figure 1 shows an example of data generated (black dots with error bars) for a SNR of 115 (same value as treated in ref. [4]) as well as the corresponding amplitude modulation (red line) as calculated from equation (1). The half size of the error bars corresponds to the value of σ. The title of the figure gives the fit values for the parameters and, in parenthesis, the exact values. Figure 2 shows a similar example for a SNR of 23. It is clear from this last figure that the probability distributions of the data are not Gaussian (for example, negative values are seen to be possible from the extension of the error bars). In fact, even on figure 1, some of the points that are close to zero will not have a Gaussian probability distribution. We will discuss this below.

Figure 1. Data (black dots with error bars) generated for a SNR=115. Angular location and polarisation of the source are indicated on the figure. The red line corresponds to a fit to the data.

Figure 2. Data (black dots with error bars) generated for a SNR=23. Angular location and polarisation of the source are indicated on the figure. The red line corresponds to a fit to the data.

4. Calculation of the experimental error (σ) for a given SNR

We assume here, that for frequencies close to f, the instrumental noise is represented by a white noise. Assuming a bandwidth limited white noise, of average value Γ_0 in the frequency domain $-B/2 < f < B/2$, and a signal S(f) (or S(t)), the Signal to Noise ratio (SNR) is defined by

$$SNR^2 = \int_{-\infty}^{+\infty} \frac{S^2(f')\delta(f'-f)}{\Gamma(f')} df' = \frac{1}{\Gamma_0} \int_0^T S^2(t)\, dt \ . \tag{6}$$

The frequency band B and time window width ΔT are related by

$$B = \frac{1}{\Delta T} = \frac{N_d}{T}, \tag{7}$$

where N_d is the number of time windows over a total time of T. The average power of the noise is defined by

$$\Gamma_0 = \frac{\sigma^2}{B} \ . \tag{8}$$

Converting from an integral to a discrete sum, one obtains

$$SNR^2 = \frac{B}{\sigma^2} \int_0^T S^2(t)\, dt \approx \frac{B}{\sigma^2} \sum_{i=1}^{N_d} S^2(t_i)\, \Delta T = \frac{1}{\sigma^2} \sum_{i=1}^{N_d} S^2(t_i) \ , \tag{9}$$

so that σ can thus be related to the SNR by

$$\sigma = \frac{1}{SNR}\sqrt{\sum_{i=1}^{N_d} S^2(t_i)} \ . \tag{10}$$

Note that, as three different signals (TDI X,Y and Z) will be analysed simultaneously, the (time independent) variance will be defined by

$$\sigma = \frac{1}{SNR}\sqrt{\sum_{i=1}^{N_d} (S_X^2(t_i) + S_Y^2(t_i) + S_Z^2(t_i))} \ . \tag{11}$$

An important point to notice is that, for a fixed SNR, the value of σ will depend on the value and number of data points and hence on the direction of the wave, on its polarisation and on the number of TDI combinations studied.

5. Data Analysis

Using these sets of data and the amplitude modulation equation (1), a χ^2-minimisation on parameters θ, ϕ, h_+ and $h_\times$ is performed.

This χ^2-minimisation produces a covariance matrix, $C_{P_jP_k}$ for the parameters P_j and P_k , from which the estimation of the angular resolution is calculated. The error on θ and ϕ are given by

$$\Delta\theta = C_{\theta\theta}^{1/2} \text{ and } \Delta\phi = C_{\phi\phi}^{1/2}, \tag{12}$$

and the error on the solid angle is defined by

$$\Delta\Omega = \sin(\theta)\ \Delta\theta\ \Delta\varphi \ . \tag{13}$$

The angular error on the direction of the wave is then defined as

$$\Delta\Theta = \sqrt{\frac{\Delta\Omega}{\pi}} \ . \tag{14}$$

This error can also be estimated from the Fisher Information Matrix (FIM) [9, 10, 11]. The Fisher matrix for a signal S(t) depending on parameters P_j and P_k and with errors identical for all data points, can be approximated in the monochromatic case by

$$J_{P_jP_k} = \frac{1}{\sigma^2}\sum_{i=1}^{N_d} \left(\partial_{P_j} S(t_i)\right)(\partial_{P_k} S(t_i)), \tag{15}$$

where $\partial_{P_j} S(t_i)$ is the derivative of the signal at time t with respect to parameter P_j .

For $P_j = P_k = \theta$, and using the above value of σ, one obtains

$$J_{\theta\theta} = SNR^2 \frac{\sum_{i=1}^{N_d} (\partial_\theta S_X(t_i))^2 + (\partial_\theta S_Y(t_i))^2 + (\partial_\theta S_Z(t_i))^2}{\sum_{i=1}^{N_d} (S_X^2(t_i) + S_Y^2(t_i) + S_Z^2(t_i))} \ . \tag{16}$$

The error ($\Delta_F\theta$) on θ obtained from the FIM is given by

$$\Delta_F\theta = \sqrt{(J^{-1})_{\theta\theta}} \ , \tag{17}$$

and the angular error on the direction of the wave is given by

$$\Delta_F\Theta = \sqrt{\frac{\sin(\theta)\ \Delta_F\theta\ \Delta_F\phi}{\pi}} \ . \tag{18}$$

Note that for all cases treated in this work $\sqrt{(J^{-1})_{\theta\theta}} \approx (J_{\theta\theta})^{-1/2}$, i.e. the FIM is almost diagonal.

Figure 3 shows the values of $\Delta\Theta$ (see eq. 14) for monochromatic waves (f=0.1 mHz), for a variety of polarisation (points with solid lines) and for a SNR of 115. The legend gives, from top to bottom, the polarisation values of the curves of decreasing $\Delta\Theta$. The dotted lines show the predictions of the Fisher matrix (see eq. 18).

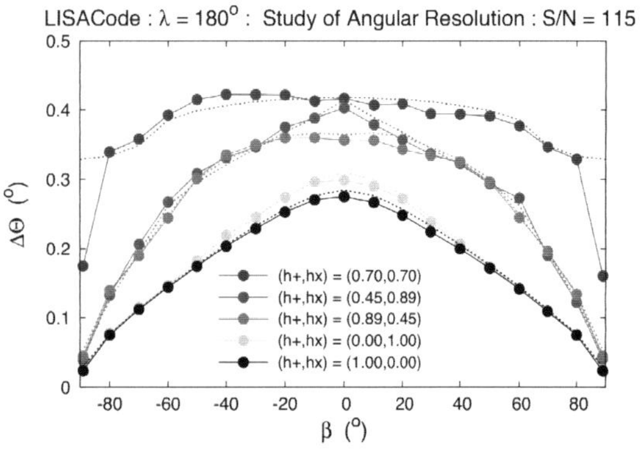

Figure 3. Variation of the angular resolution of LISA as a function of β and polarisation (h_+,$h_\times$) for $\lambda = 180°$ and $SNR = 115$. Note that the amplitude of the GW $(h_+^2 + h_\times^2)^{1/2}$ is kept fixed. The dotted lines give the predictions of the Fisher matrix (see text).

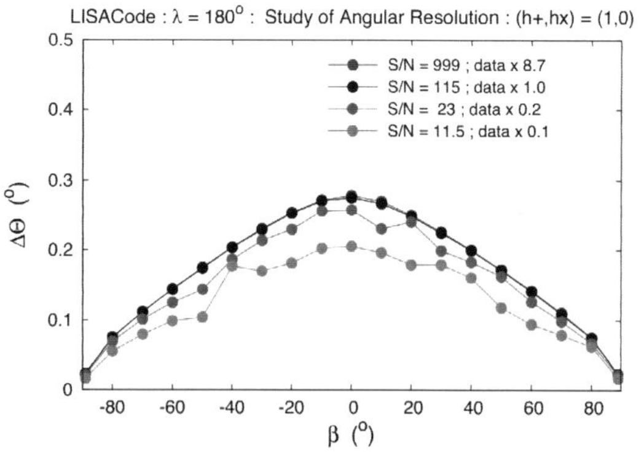

Figure 4. Variation of the angular resolution of LISA as a function of β and SNR for $\lambda = 180°$. Note that the polarisation is fixed at (h_+,$h_\times$) = (1,0).

From this figure one can observe that the angular resolution of LISA, for most of the angular range, lies between 0.2° and 0.5° for a SNR of 115 and hence possibly between 2° and 5° for a SNR close to 10, a value which could apply to effective observations by LISA. A clear effect due to the polarisation is also seen.

The predictions of the Fisher matrix (dashed line on these figures) follow very closely the results obtained from the minimisation of the χ^2 as is expected for large SNR values. The discrepancies observed for (h_+,$h_\times$) = (0.7,0.7) and $\beta = \pm 90°$ have been traced to numerical problems: it has been checked that, for very large SNR (i.e. 999), the resolutions follow precisely the values of the Fisher Matrix.

It has also been checked that because of the duration of the data acquisition (1 year) and of the use of different TDI combinations, the resolutions calculated depend very little of the ecliptic longitude λ. Hence, figure 3 maps most of the possible values the LISA angular resolution can take.

Figure 4 shows, for a polarisation (h_+,$h_\times$) = (1.,0.), the evolution of the resolution as a function of the SNR. One sees that the resolution scales very well with the SNR for SNR values above 100 but begins to deviate from this scaling for $SNR = 23$ and below. Why this product ($\Delta\Theta \times SNR$) decreases with SNR has yet to be understood.

Journal of Physics: Conference Series **154** (2009) 012015 doi:10.1088/1742-6596/154/1/012015

6. Comparison with other studies

As stated above, other studies have been performed on the angular resolution of LISA. Peterseim et al. [4] have treated the case of SNR = 115 with, as we understand, a unique TDI (or equivalent) combination. In this case, we deduce that this corresponds to an effective value of SNR divided by $\sqrt{3}$. Their results give (for $f = 3mHz$, $(h_+, h_\times) = (1,0)$, $\beta = 0, \lambda = 0$) a value of $\Delta\Omega = 16\ \mu$str compared to $70/\sqrt{3} \approx 23\mu$str obtained in our study (equivalent to $\Delta\Theta = 0.3°$). This difference is understandable because their method uses both the amplitude and the phase modulation and should therefore yield better results.

The study of Cutler [5] also makes use of the amplitude and phase modulation, but using two TDI combinations. His results, at low frequencies ($f = 10^{-4}Hz$) where the phase modulation should have little influence, give an average value of $\Delta\Omega(\times 2\pi) = 8 \times 10^{-2}$ str for a SNR of 10.

In our study, this should correspond to 79 μstr ($= 0.08 \times \sqrt{3} \times (10/115)^2/(2\pi)$) and is quite close to the values we obtain.

7. Summary and future work

Using realistic data generated from LISACode and a simple analysis method based on the study of the amplitude modulation of the LISA response for monochromatic GW, the angular resolution of LISA is shown to lie between $\Delta\Theta \times (SNR/115) = 0.1°$ and $0.5°$, somewhat dependent on the ecliptic colatitude angle (β), on the polarisation ($h_+, h_\times$) and much less on the ecliptic longitude angle (λ). Because this method uses only the amplitude modulation (and not the phase modulation), the results do not depend on the frequency of the GW, contrary to the cases for the other studies mentioned [5, 4]. These conclusions are valid, however, for large SNR because Gaussian data distributions are assumed. For small SNR (≤ 50), this is probably not the case and more exact methods have to be used. It is probable that the verification binary sources considered for LISA will fall into this class and such a study will be performed in the near future. We will also move from the X,Y and Z TDI combinations to A and E as these have zero covariance and are therefore better adapted to a standard statistical analysis.

References

[1] K.Danzmann et al., LISA - Laser Interferometer Space Antenna, Pre-Phase A Report, Max-Plansk-Institut fur Quantenoptic, Report MPQ 233, (1998).
[2] J. W. Armstrong, F. B. Estabrook, and Massimo Tinto. Time-delay interferometry for space-based gravitational wave searches. Astrophys. J., 527 :814826, December 1999.
[3] S. V. Dhurandhar, K. Rajesh Nayak and J.-Y. Vinet , Algebraic approach to time-delay data analysis for LISA. Phys. Rev. D 65, 102002 (2002).
[4] M.Peterseim, O.Jennrich, K.Danzmann and B.F.Schutz : CQG 14, 1507 (1997).
[5] C.Cutler Phys. Rev. D 57, 12, 7098 (1998).
[6] A.Petiteau, Ph.D. Thesis, University Paris 7, France, 30 June 2008.
[7] N.J.Cornish and Rubbo Phys. Rev. D 67, 022001 (2003).
[8] Antoine Petiteau et al. LISAcode : A scientic simulator of LISA. Phys. Rev. D 77, 023002 (2008).
[9] L.S.Finn Phys. Rev. D 46, 12, 5236 (1992).
[10] C.Cutler and E.Flanagan Phys. Rev. D 49, 6,2658 (1994).
[11] M.Vallisneri Phys. Rev. D 77, 042001 (2008).

Journal of Physics: Conference Series **154** (2009) 012016 doi:10.1088/1742-6596/154/1/012016

Lasers for LISA: overview and phase characteristics

M Tröbs[1], S Barke[1], J Möbius[2,3], M Engelbrecht[2,4], D Kracht[2], L d'Arcio[5], G Heinzel[1] and K Danzmann[1]

[1] AEI Hannover, (MPI für Gravitationsphysik und Leibniz Universität Hannover), Callinstr. 38, 30167 Hannover, Germany

[2] Laser Zentrum Hannover, Hollerithallee 8, 30419 Hannover, Germany

[3] present address: Viscom AG, Carl-Buderus-Straße 9-15, 30455 Hannover, Germany

[4] present address: Menlo Systems GmbH, Am Klopferspitz 19, 82152 Martinsried (Munich), Germany

[5] European Space Agency, postbus 299, 2200 AG Noordwijk, The Netherlands

E-mail: michael.troebs@aei.mpg.de

Abstract. We have investigated two alternative laser systems for the Laser Interferometer Space Antenna (LISA). One consisted of the laser of LISA's technology precursor LISA Pathfinder and a fiber amplifier originally designed for a laser communication terminal onboard TerraSar-X. The other consisted of a commercial fiber distributed feedback (DFB) laser seeding a fiber amplifier. We have shown that the TerraSar-X amplifier can emit more than 1 W without the onset of stimulated Brillouin scattering as required by LISA. We have measured power noise and frequency noise of the LISA Pathfinder laser (LPL) and the fiber laser. The fiber laser shows comparable or even lower power noise than the LPL. LISA will use electro-optical modulators (EOMs) between seed laser and amplifier for clock noise comparison between spacecraft. This scheme requires that the excess noise added by the amplifiers be negligible. We have investigated the phase characteristics of two fiber amplifiers emitting 1 W and found them to be compatible with the LISA requirement on amplifier differential phase noise.

1. Introduction

In comparison to LISA Pathfinder [1], the laser system for LISA [2; 3] must emit at least 30 times more power and include an electro-optical modulator for clock noise transfer between spacecraft. The integrated broadband optical modulators we have studied so far can only handle a fraction of the 1 W laser power required for LISA and must hence be followed by an optical amplifier. This amplifier must not change the phase of the sidetones imposed on the light by the EOM for clock noise transfer. In Section 2 we present current options for a LISA laser system and in Section 3 we report on phase characteristics measurements for two optical amplifiers.

2. Laser systems for LISA

LISA requires at least 1 W of laser output power and an electro-optical modulator is needed in the optical path to transmit clock information between spacecraft. The integrated EOMs known to us only handle less than 100 mW of optical power. Hence, a master oscillator power amplifier concept with the EOM before the amplifier seems necessary.

Nonplanar ring oscillators (NPROs) have proven as low-noise master oscillators [4; 5]. One option is to use the LISA Pathfinder laser as master oscillator for LISA. This laser is discussed in Section 2.1. A possible amplifier might be the one flying as part of a laser communication terminal [6; 7] on the TerraSar-X mission, launched on June 15th, 2007. This amplifier was designed for lower output powers than 1 W. We discuss our tests on the engineering model in Section 2.2. Another alternative for the LISA laser is to use an all-fiber system consisting of a distributed feedback (DFB) fiber laser with subsequent amplifier. Measurements on a commercially available fiber laser are presented in Section 2.3.

2.1. LISA technology package laser

The LISA Pathfinder laser (LPL) is a possible candidate for the LISA master oscillator. The LPL weighs 1 kg, has a 12x13 cm^2 footprint, and consumes 12 W of electrical power. The engineering model that was investigated emitted 14 mW of optical power, the flight model will emit 35 mW. The LPL consists of the laser head, a fiber-coupled pump module, and electronics. Power noise and frequency noise measurements are discussed in Section 2.4.

2.2. Space-qualified fiber amplifier

The TerraSar-X mission, launched in June 2007 includes a laser communication terminal [6; 7] with a master oscillator similar to the LPL, a broadband integrated optical modulator, and a two-stage polarization maintaining Ytterbium fiber amplifier.

Onboard TerraSar-X, the amplifier emits 300 mW. At higher output powers, nonlinear effects such as stimulated Brillouin scattering (SBS) might occur which would degrade the noise performance and ultimately limit the output power. We have investigated, if SBS occurs at 1 W of output power when seeded under conditions representative for LISA. We have seeded the amplifier with phase modulated input from an NPRO. Each sideband was spaced by 300 MHz from the carrier and contained 10% of the total power. As indication for the onset of SBS, we have used the amount of backscattered light detected by a monitor photodiode inside the amplifier. Figure 1 shows output power and backscattered light level as function of pump current. The onset of SBS is correlated with a significant slope increase of the backscattered light signal. We have not observed such a slope increase. We have extracted 1.2 W of output power without the onset of SBS. The measurements were performed in one of our laboratories where no special care was taken to minimize acoustic noise. The air conditioning was active during the measurements as well as fans from laboratory equipment such as laser diode drivers.

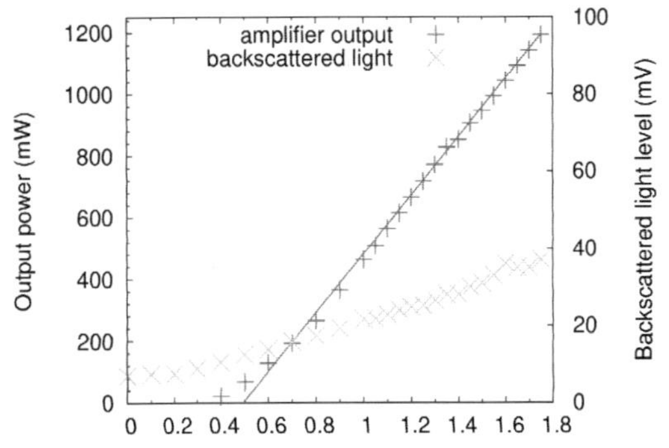

Figure 1. Measured output power and backscattered light level as function of pump current of amplifier with space heritage.

2.3. Master fiber oscillator with amplifier

As an alternative for the LISA laser we have investigated an Ytterbium-doped distributed-feedback fiber master oscillator power amplifier system (BoostiK by Koheras) emitting 1 W at

1064 nm. A fiber DFB laser was used as master oscillator followed by a polarization-maintaining two-stage fiber amplifier. The laser system had actuators for power and frequency tuning and a control loop was implemented to suppress the relaxation oscillation of the DFB laser. Power noise and frequency noise measurements are discussed in Section 2.4.

2.4. Power and frequency noise

Figure 2 shows measured relative power noise of the LPL and the fiber laser with amplifier. The output of the LPL was detected by a photodiode with transimpedance amplifier. For frequencies up to 200 Hz the photodiode signal was digitized using a computer, and the program lpsd [8; 9] was used to calculate spectral densities. For higher frequencies the photodiode signal was measured with a spectrum analyzer (Agilent 4396B). For the fiber laser a fraction of the output power was detected and for frequencies up to 2 Hz a computer and lpsd were used to produce the graph. For higher frequencies, the spectrum analyzer was used. None of the traces was limited by the measurement sensitivity. The relaxation oscillation of the LPL is visible at 600 kHz while the fiber laser used an internal control loop to suppress its relaxation oscillation.

Figure 3 shows the free-running frequency noise of the LPL and the fiber laser with amplifier. For comparison, the frequency noise of a commercial NPRO (Mephisto 500 by Innolight) emitting 300 mW is also shown. For frequencies up to 1 Hz the traces were obtained from beat measurements of the free-running lasers with a frequency-stabilized laser [10, pp. 71]. For higher frequencies, the lasers were frequency-locked to a stable Fabry-Perot cavity while actuator and error signals were recorded [10, pp. 111].

For Fourier frequencies below 0.2 Hz the frequency noise of the fiber laser is up to a factor of 10 above the frequency noise of the LPL. Between 0.2 Hz and 1 kHz, the frequency noise of the LPL is up to a factor of 30 above the noise of the fiber laser, and above 1 kHz the frequency noise of the fiber laser is up to three orders of magnitude higher than the noise of the LPL. Compared to the 300 mW NPRO, the LPL shows up to a factor of 20 higher frequency noise for Fourier frequencies above 10 mHz.

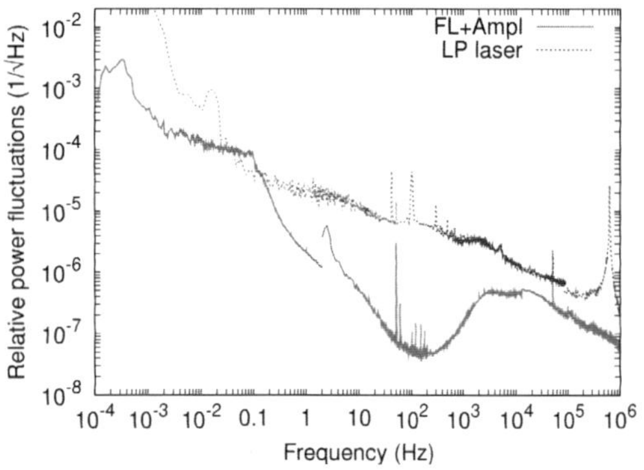

Figure 2. Measured power noise of fiber laser with amplifier and LISA Pathfinder laser.

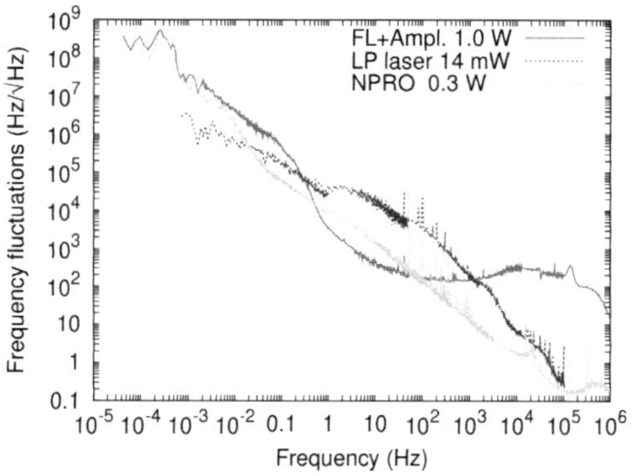

Figure 3. Measured free-running frequency noise of fiber laser with amplifier, LISA Pathfinder laser, and a commercial NPRO.

Due to the control loop suppressing the relaxation oscillation of the fiber laser, the measured relative power noise is not directly comparable to the noise of the LPL. However, for frequencies in the LISA measurement band from 0.1 mHz to 1 Hz the fiber laser shows comparable or lower noise than the LPL and is already within a factor of 20 within the stabilized power noise requirement for LISA. Using its power modulation input an external stabilization can be built to further reduce its power noise.

For Fourier frequencies below 1 kHz the free-running frequency noise of the fiber laser is comparable to or better than the free-running frequency noise of the LPL. For frequencies above 1 kHz the frequency noise of the fiber laser was significantly higher than the LPL frequency noise. This high-frequency frequency noise can impact the performance of the LISA phase measurement system due to aliasing. Its impact on the phase measurement system will need to be investigated in future work.

3. Clock noise transfer in optical amplifiers

We have investigated differential phase noise between carrier and sideband in two fiber amplifiers.

3.1. Potential noise sources

There are at least two possible sources of differential phase noise in fiber amplifiers:

(i) Nonlinear dispersion leads to differential phase noise when the seed laser frequency changes with respect to the amplifier gain profile. First measurements have been performed (not shown) that show a tiny effect, which is negligible for LISA.

(ii) Optical length changes in the amplifier, ΔL, due to e. g. changes in ambient temperature, pump power, or seed power lead to differential phase changes, $\Delta\phi$, of sidebands at f_{EOM} with respect to the carrier of

$$\Delta\phi = \frac{2\pi f_{\mathrm{EOM}}}{c}\Delta L, \qquad (1)$$

where c is the speed of light. We have measured the combined effect and discuss it in Section 3.4.

3.2. Optical amplifiers under test

We have investigated two amplifiers built at Laser Zentrum Hannover. Both used Ytterbium-doped fibers as gain medium pumped by laser diodes at 976 nm and emitted 1 W of optical power. One amplifier was pumped directly into the fiber core and consisted of two idential stages using 0.4 m of doped fiber each. The second amplifier used 3 m of polarization-maintaining double-clad fiber in a single stage and was pumped in the multi-mode cladding.

3.3. Differential phase noise measurements

Figure 4 shows the (simplified) schematic setup for measuring fiber amplifier differential phase noise. Light from two lasers (Mephisto 500 by Innolight) was superimposed and used to seed the fiber amplifier under test. Before and behind the amplifier, a fraction of the light was detected by fiber-coupled photodiodes (SIR5-FC by Thorlabs). The signal before the amplifier was split and one half used to offset-phase-lock both lasers to a difference frequency of 2 GHz+1.6 kHz. This was performed by an analog mixer with subsequent low-pass filter and loop filter. The other half of the photodiode signal and the signal behind the amplifier were mixed with a local oscillator at a slightly different frequency, namely 2 GHz. Both signals at 1.6 kHz were digitized by a computer and a phasemeter [11] implemented in software was used to determine the phase difference between both signals. The amplitudes of both 1.6 kHz signals were kept constant by control loops (not shown) acting on piezo-mounted tilt-mirrors in front of the fiber-coupled photodiodes.

3.4. Amplifier phase characteristics

Figure 5 shows measured differential phase noise between 2 GHz sideband and carrier at 1 W of amplifier output power for the core-pumped and the cladding-pumped fiber amplifiers. Both amplifiers meet the LISA requirements. The cladding-pumped amplifier showed slightly higher differential phase noise than the core-pumped amplifier. In an independent experiment, the

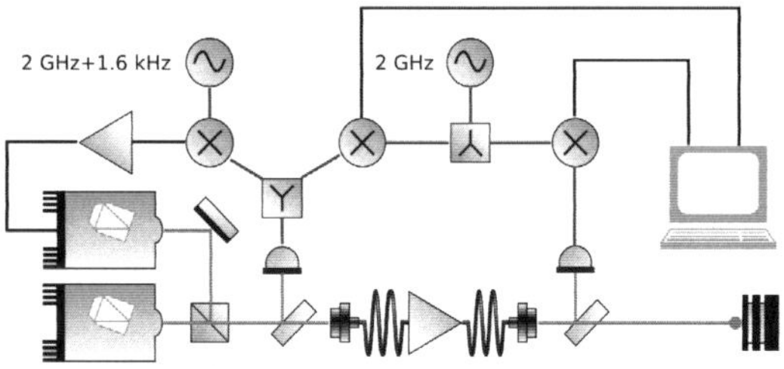

Figure 4. Schematic setup for fiber amplifier differential phase noise measurements.

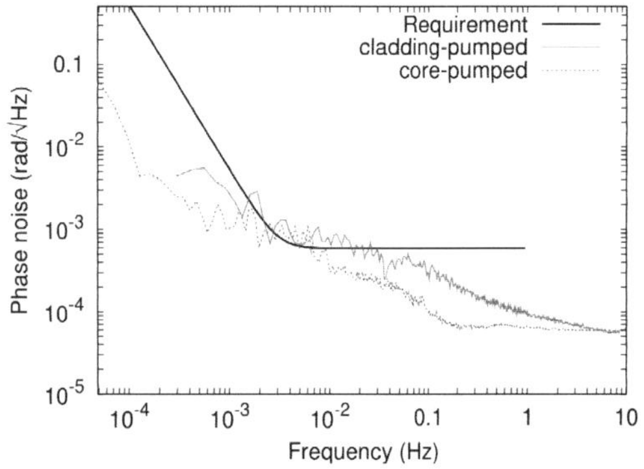

Figure 5. Measured differential phase noise of core-pumped and cladding-pumped fiber amplifier.

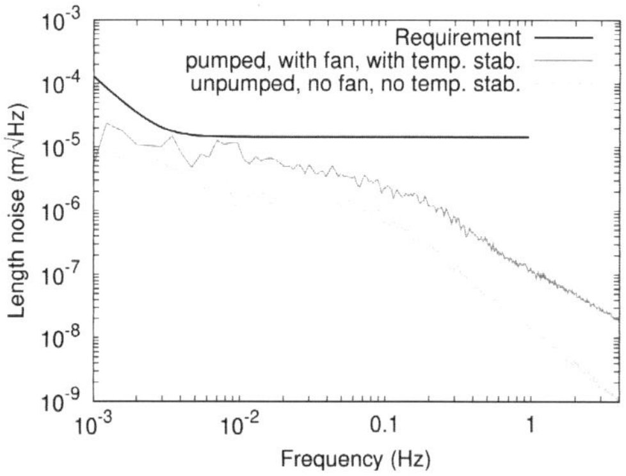

Figure 6. Measured length noise of cladding-pumped fiber amplifier.

cladding-pumped amplifier was seeded by a single laser and placed in one arm of a homodyne Mach-Zehnder interferometer with balanced detection. A mirror mounted on a piezo actuator was used to stabilize the interferometer armlength difference to mid-fringe. By reading the actuator signal the length fluctuations of the amplifier were measured. They are shown in Figure 6. The trace with the pumped amplifier shows higher noise than the trace of the unpumped amplifier, which proves that the measurement setup was sensitive enough. Length noise below the trace labelled "Requirement" leads to differential phase noise below the respective requirement. For Fourier frequencies around 10 mHz, the length fluctuations of the amplifier limit the differential phase noise. If lower differential phase noise shall be achieved, lower length fluctuations are required.

4. Conclusions

We have investigated two alternative laser systems for the Laser Interferometer Space Antenna (LISA) and we have measured the phase characteristics of two fiber amplifiers for clock noise comparison between spacecraft.

The laser of LISA's precursor LISA Pathfinder in combination with a fiber amplifier originally designed for a communication terminal and a commercial fiber distributed feedback (DFB) laser both delivered 1 W output power as required by LISA. Within the LISA measurement band from 0.1 mHz to 1 Hz the fiber laser showed power noise up to a factor of 20 above the requirement for the stabilized laser. Using the power actuator of the fiber laser further suppression of power noise should be possible to reach the stabilized LISA requirement.

In comparison to the LISA Pathfinder laser, the fiber laser showed significantly higher frequency noise for Fourier frequencies above 1 kHz, that can possibly impact the LISA phase measurement system. The impact of this frequency noise needs to be analysed in future work.

We have investigated differential phase noise between a carrier and a 2 GHz sideband in two different fiber amplifiers. Both amplifiers used Ytterbium-doped fibers as gain medium and emitted 1 W of output power. The amplifiers showed sufficiently low differential phase noise for clock-noise comparison between LISA spacecraft.

Acknowledgments

We thank Tesat Spacecom for providing the laser discussed in Section 2.1 and the amplifier discussed in Section 2.2. We gratefully acknowledge support by Deutsches Zentrum für Luft- und Raumfahrt (DLR) (references 50 OQ 0501 and 50 OQ 0601).

References

[1] McNamara P, Vitale S and Danzmann K 2008 *Class. Quantum Grav.* **25** 114034

[2] Danzmann K and Rüdiger A 2003 *Class. Quantum Grav.* **20** S1–S9

[3] Shaddock D A 2008 *Class. Quantum Grav.* **25** 114012

[4] Kane T J and Byer R L 1985 *Opt. Lett.* **10** 65–67

[5] Freitag I, Tünnermann A and Welling H 1995 *Opt. Commun.* **115** 511–515

[6] Lange R and Smutny B 2004 *Military Communications Conference, 2004. MILCOM 2004. IEEE* **1** 441–444

[7] Schwander T, Lange R, Kämpfner H and Smutny B 2004 *Proceedings of the 5th International Conference on Space Optics (ICSO 2004)* 419–421

[8] Tröbs M and Heinzel G 2006 *Measurement* **39** 120–129

[9] Tröbs M and Heinzel G 2009 *Measurement* **142** 170

[10] Tröbs M 2005 *Laser development and stabilization for the spaceborne interferometric gravitational wave detector LISA* Ph.D. thesis University of Hannover

[11] Heinzel G, Wand V, Garcia A, Jennrich O, Braxmaier C, Robertson D, Middleton K, Hoyland D, Rüdiger A, Schilling R, Johann U and Danzmann K 2004 *Class. Quantum Grav.* **21** S581–S587

Journal of Physics: Conference Series **154** (2009) 012017 doi:10.1088/1742-6596/154/1/012017

LISA phasemeter development: Advanced prototyping

Iouri Bykov, Juan José Esteban Delgado, Antonio Francisco García Marín, Gerhard Heinzel, Karsten Danzmann

Institute for Gravitational Physics of the Leibniz University Hannover
and Max Planck Institute for Gravitational Physics (Albert Einstein Institute), Germany

E-mail: Iouri.bykov@aei.mpg.de

Abstract.
We present the status of our investigations on the LISA Phasemeter. The new prototype is based on a custom-designed breadboard with four high-speed ADC and two DAC channels, extended readout capabilities and a large FPGA (field programmable gate array). The required main functionalities and performance of the prototype have been demonstrated in laboratory conditions.

1. Introduction

The main task to be implemented by the LISA Phase Measurement System (PMS) is the phase readout of the beat note coming from the photo-detectors on the optical benches. The required phase sensitivity is $2\pi \times 10^{-6}\,\mathrm{rad}/\sqrt{\mathrm{Hz}}$ on the frequency range from 0.1 mHz to 100 mHz. This corresponds to the detection of pathlength variations between the free floating test masses in the picometer range, required for the observation of gravitational waves. These accuracy in the phase measurements must be ensured over a frequency range of the heterodyne signal between 2 and 20 MHz. Apart from this, the PMS must be able to isolate and extract the phase from the various beat notes contained in the incoming photodiode signal. The most suitable phasemeter architecture, which can provide the required accuracy, is based on a so-called digital phase-locked loop (DPLL) [1, 2]. Once the phase readout has been performed, the back-end processing of the PMS must implement a high bandwidth laser phase-locking control system, as well as inter-spacecraft laser ranging, data communication and clock synchronization [4].

To achieve these goals, a new custom-designed prototype board has been recently manufactured taking as an starting point the experience from previous prototypes [3]. It consists of a multi-channel PMS with high-speed digital processing capabilities, high bandwidth analog input/output interfaces and full signal visibility.

2. Implementation of the digital phase-locked loop

Figure 1 shows a single channel of the developed phasemeter architecture. To avoid aliasing, the incoming signal is processed with an integrated analog LC-filter. After filtering, the signal is digitized in a low-noise analog-to-digital converter (ADC) at 50 MHz sampling rate and fed into the FPGA. Here, the phase detection begins with the multiplication of the incoming beat note with a sine signal generated by a numerically controlled oscillator (NCO), consisting of a lookup sine table (LUT), phase accumulator (PA) and a phase increment register (PIR). The result of the multiplication, after a low-pass filter (LPF), is proportional to the phase difference between the input signal and the NCO. Therefore, this result can be used as error signal in a control loop to adjust the phase of the NCO to the incoming beat note. This is done with a proportional-integral controller (PIC) that transforms the phase error in a feedback signal for the PIR.

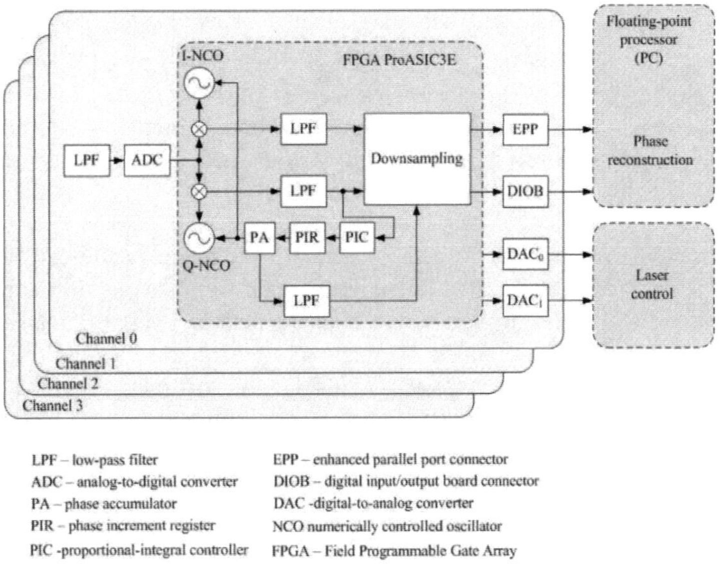

Figure 1. Block diagram of the phasemeter core design.

The primary results of the PMS are the values of the PIR and the PA registers, with correspond to frequency and phase of the input signal, respectively. Additionally, the filtered output signals of the I-LUT and Q-LUT branches can be used for phase measurement corrections. The downsampled data can be readout into a floating-point computing system (e.g. PC) through an integrated enhanced parallel port (EPP). It is also possible to employ a high-speed digital input/output board (DIOB) with direct memory access feature to test internal intermediate results of the PMS, thus providing extended opportunities for investigation of the system and efficient debugging.

3. Phase measurement technique

Figure 2 illustrates the detailed phase reconstruction of the DPLL algorithm presented previously in Section 2. The binary value stored in the phase accumulator (PA), consists of a fixed-point part A and fractional part B with a bit depth N. After multiplication by 2π and proper scaling, its results is the current phase value of the input signal in radians:

$$\varphi = 2A\pi + \frac{B}{2^N}. \tag{1}$$

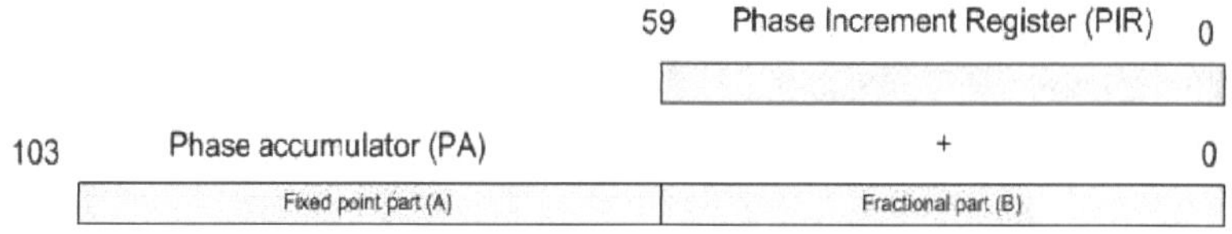

Figure 2. Phase reconstruction using phase accumulator.

4. Hardware Description

Figure 3 shows one of the prototypes of the design presented here that has been recently manufactured. This design implements four channels in one board and provides enhanced processing capabilities. For this purpose, ACTEL FPGAs are the most suitable choice due to their flexibility on design and space-flight compatibility. Our board supports up to 3 Million system gates with 147 high-performance inputs/outputs (I/Os). The software is designed in a very high speed integrated circuit hardware description languages (VHDL).

The digitization of the input beat note signals in the frequency range 2-20 MHz is performed by four A/D-converters (AD9446-100). The current prototype works with an unique internal quartz clock set at 50 MHz, but it is also possible to set an arbitrary clock by mean of an external port. Two D/A converters (AD9744) in combination with output filters are intended to be used for laser frequency stabilization purposes. The characteristics of the DAC enable also two important features for LISA: first, the control of different analog actuators in the LISA setup and second, the implementation of a laser-modulation unit with the use of an electro-optic modulator (EOM). This way, the breadboard can be used in a modulator/demodulator scheme to perform laser ranging [4]. Additionally, the four input channel can measure individually the phase of each quadrant of the photodiodes and deliver alignment signals for orientation control.

Figure 3. 4-channel prototype of LISA Phasemeter with extended signal processing and readout opportunities. EPP with output rate up to 1 MB/s. Digital input/output (DIO) with maximal rate of 2 GB/s.

5. Experimental results

In contrast to analog-based PMS implementations, a digital development makes it possible to update the firmware and eliminate the temperature drift. However, the digitization process suffers from quantization noise and subsequently the errors in the measurement need to be taken into consideration. In this section we present preliminary results in the assessment of the influence of these noise sources in our design.

Figure 4 presents the results of the first experiment, where the input signal was generated by an additional NCO implemented in the same FPGA as the PMS. The aim of the experiment was to verify the functionality and noise performance of the DPLL. The measurement shows a flat noise floor over the whole frequency range down to 1 mHz, limited only by the quantification noise of the ADC and the finite resolution inside the FPGA. The spikes at high frequencies are currently a subject of investigations, most probably being caused by suboptimal performance of the anti-aliasing filter applied to the output data before decimation.

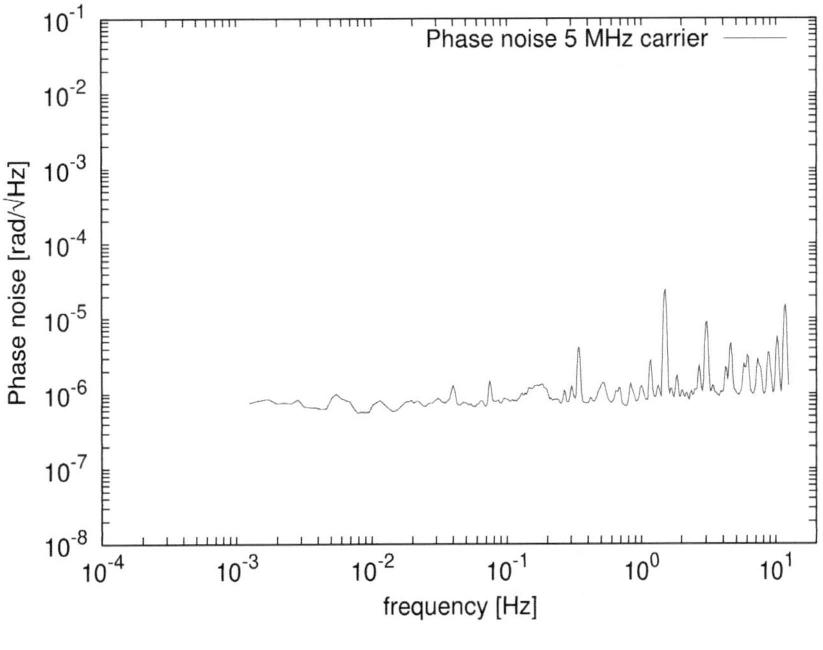

Figure 4. Noise performance of the phasemeter with 5 MHz input signal, generated by an additional NCO implemented in FPGA.

The next experiment was made in more realistic conditions: an external function generator, locked to the clock generator of the phasemeter, was used as input signal. In such conditions all the hardware was tested, including analog frontend and ADCs. The results of the experiment are shown in Figure 5. The noise behavior measured in this configuration is clearly limited by the performance of the external lock used by the PLL of the generator.

Another area of research concerns the back-end processing of the measured phase to perform ranging between the spacecraft: using the current breadboard, a delay-locked loop (DLL) system has been implemented. This system will enable to measure the light propagation time delay. For this purpose, the phase of laser carrier is modulated and a pseudo-random-noise codes (PRN)

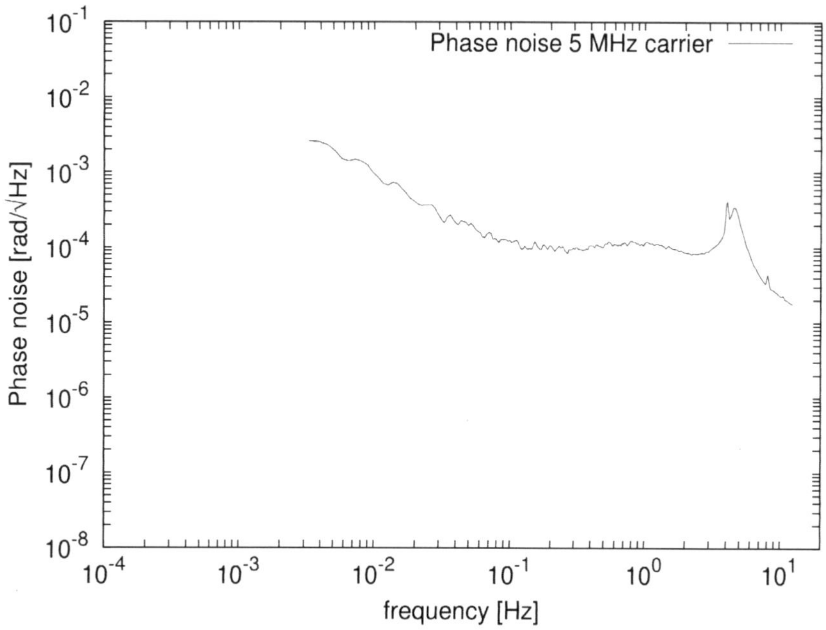

Figure 5. Noise performance of the phasemeter with 5MHz input signal, generated by an external function generator.

are transmitted to the remote satellites. The correlation properties of the PRN enable data communication and synchronization between different spacecraft[4].

6. Conclusions

A custom phasemeter prototype has been developed and its functionality demonstrated with electrical signals. The design aims to implement in a single platform the interferometric phase readout as core processing and a phase-locking control system, optical ranging, data transfer and clock synchronization as back end processing. The experimental results show that the new version of the phasemeter meets the requirement of $2\pi \times 10^{-6}\,\mathrm{rad}/\sqrt{\mathrm{Hz}}$ for the phase readout in the case of internally simulated input signals. We thus conclude that the current prototype can be expected to be a key component for the LISA interferometry test bench.

7. References

[1] Shaddock D, Ware B, Halverson P, Spero R, Klipstein B 2006 Overview of the LISA Phasemeter *AIP Conf Proc.* **873** 654-60

[2] Wand V, Guzman F, Heinzel G, Danzmann K 2006 LISA Phasemeter Development *AIP Conf Proc.* **873** 689-96

[3] Wand V 2007 Interferometry at low frequencies: Optical phase measurement for LISA and LISA Pathfinder. *Ph.D Thesis, University Hannover*

[4] Esteban J.J, Bykov I, García Marín A.F, Heinzel G, Danzmann K 2008 Optical ranging and data transfer development for LISA J. Phys.: Conf. Series. this issue

7th International LISA Symposium IOP Publishing
Journal of Physics: Conference Series **154** (2009) 012018 doi:10.1088/1742-6596/154/1/012018

Possible post-LISA science missions with gravitational reference sensors

P L Bender
JILA, University of Colorado and NIST, Boulder, CO 80309, USA

E-mail: pbender@jila.colorado.edu

Abstract. Among the many interesting possibilities for types of future missions that would benefit strongly from LISA and LISA Pathfinder technology development, three will be discussed. They are in the fields of fundamental physics, Earth science, and gravitational wave astronomy. The first is a mission to measure the gravitational time delay due to the Sun from a spacecraft near the L-1 point of the Earth-Sun system. It would require gravitational reference sensors (GRSs) with roughly 10^{-13} [10^{-6} Hz/f] m/s^2/$\sqrt{}$Hz performance at frequencies down to about 0.3 microHz. The second type of mission is future drag-free missions to measure time variations in the Earth's gravitational field. One example of such a mission will be described, with two satellites in the same polar orbit at about 300 km altitude. Changes in the roughly 50 km satellite separation would be measured with 10^{-14} or better accuracy, and spurious accelerations of the test masses in the GRS on each satellite would be the other main measurement accuracy limitation. The third mission is a possible moderately improved LISA follow-on mission aimed at being able to detect mergers of 10 solar mass black holes with IMBHs out to redshifts of about 10 in order to investigate the formation and growth of IMBHs in more detail than LISA will be able to achieve.

1. Introduction

There are several possible future missions that have been investigated at least briefly where very low levels of spurious accelerations of test masses are required. Thus the Gravitational Reference Sensor (GRS) results from the LISA Pathfinder mission and LISA will be of major importance for studies of such missions. In order to indicate the scientific opportunities involved and the technical requirements, three particular missions will be described in the following sections. All of these missions also require high-accuracy measurements of changes in light travel times between spacecraft.

2. L-1 gravitational time delay measurement

The gravitational time delay for electromagnetic waves passing near a massive body provides a well known test of general relativity [1]. For waves passing about 1.5 solar radii from the Sun, the round-trip delay can be about 240 microsec. If the measurements are made between two spacecraft that are far from the Sun, the delay depends mainly on the distance b of closest approach of the wave path to the Sun. The round-trip delay between the spacecraft is given to first order by [2]

$$\Delta\tau = \frac{2GM}{c^3}(1+\gamma)\ln\left(\frac{r_1 r_2}{b^3}\right).$$

Here M is the mass of the Sun, γ is the PPN "space curvature" parameter, and r_1 and r_2 are the distances from the spacecraft to the Sun.

The most accurate measurement of the gravitational time delay was reported in 2003 by Bertotti, Iess and Tortora [3]. It was based on high-accuracy microwave Doppler measurements between the Earth and the Cassini spacecraft when the line of sight passed close to the Sun. The value of gamma was determined to 2.3×10^{-5} from the measurements.

A number of proposals have been made for missions that could improve the accuracy for the time delay and thus gamma by a large factor. One possibility, presented at conferences in 2005 and 2006, was for optical measurements between a spacecraft near the L-1 point of the Earth-Sun system and a distant transponder spacecraft on a longer period orbit as the line of sight went past the Sun [4,5]. At that time, it was assumed that the time delay would be measured with respect to a cooled cesium hyperfine frequency clock on the L-1 spacecraft, and that the orbit for the distant spacecraft had a 1.5 year period.

The L-1 time delay mission possibility was looked at again recently [6]. It now appears reasonable to assume that the frequency standard in the L-1 spacecraft would be a cooled optical transition clock with $5\times10^{-15}/\sqrt{\tau}$ stability out to measurement times τ of 3×10^{6} seconds. As an example, this stability has been demonstrated in the laboratory over periods of a few thousand seconds for an optical lattice clock based on the 698 nm line of strontium-87 [7]. The time delay measurements would be made directly on the phase of a laser beam sent to and transponded by the distant spacecraft.

The other main change is that the distant spacecraft is now assumed to be in a 2.0 year period orbit, which gives solar conjunctions 1.0, 3.0, and 5.0 years after launch. In this case, the passages behind the Sun occur right around aphelion for the distant spacecraft, so that the spacecraft temperature is changing only slowly. In addition, the rate of motion of the line of sight with respect to the Sun is increased substantially to 1.9 solar radii per day, compared with 0.7 solar radii/day with the previously assumed orbit. This reduces the drag-free requirements on both spacecraft, as well as the clock stability requirement for a given accuracy for gamma.

For an L-1 time delay mission, the requirements on the spurious accelerations for the test masses in the GRSs in both spacecraft have to be very tight out to periods of about 3×10^{6} seconds. In simulations that have been done recently, the assumed requirement has been $1\times10^{-13}[10^{-6}\ \mathrm{Hz}/f]$ m/s$^2/\sqrt{\mathrm{Hz}}$ down to 0.3 microHz. Meeting such a requirement will require careful thermal design of the whole spacecraft, as well as the GRS. However, the LISA Pathfinder Mission and LISA plus careful thermal modeling can provide the necessary confidence that such requirements can be met.

Simulations are now being done to determine the accuracy for gamma that can be achieved, based on the new assumptions. One factor that is being investigated more carefully than in previous studies is how well orbit corrections for the two spacecraft can be determined simultaneously with gamma from the data. Preliminary simulations indicate that this won't be a substantial limitation, but confirmation of this is needed. If good accuracy for gamma can indeed be achieved, then the scientific value of such a time delay mission appears to be high.

3. Future drag-free missions for mapping time variations in the Earth's gravity field

The GRACE mission that has been flying for over 5 years is making much more accurate measurements of time variations in the Earth's gravity field than were possible previously. Two satellites are in the same nearly polar orbit, separated by about 200 km. Their altitude is roughly 450 km, in order to permit a mission lifetime of about 10 years. Each satellite contains an accelerometer built by ONERA in Paris, to permit correction during the analysis for non-gravitational accelerations. Changes in the satellite separation are measured with roughly micron/second accuracy by a two-frequency millimeter wave interferometric system.

The first US follow-on mission to GRACE is likely to be similar to GRACE, but possibly with a laser interferometric system added as an experiment in order to demonstrate improved capability for measuring changes in the satellite separation. However, the possibility of flying both satellites in a drag-free mode in other future missions has been widely discussed, and is being considered in a number of institutions. Whatever limitation due to spurious accelerations is present in GRACE data can be strongly reduced by drag-free operation. In addition, such missions can be flown at considerably lower altitudes, where the shortest wavelength spatial variations in the Earth's gravity field have not been attenuated as much by upward continuation. The combination of drag-free operation and laser measurements of the satellite separation is attractive, but the extra cost of drag-free operation has to be justified by the additional scientific benefits.

It unfortunately is not known at present how much of the reduction in spurious accelerations possible with the LISA Pathfinder GRS design could be taken advantage of in future GRACE follow-on missions. It takes about 25 days for the ground track of a GRACE-type pair of satellites to cover the Earth with roughly 1 degree separation of the tracks. Such close spacing of the ground tracks is highly desirable, since changes in the water storage in individual river basins is one of the sources of time variations in the gravity field that is of major scientific interest. During the 25 days, time variations in the geoid height anywhere on the Earth tend to get aliased in with the value at the measurement time at a particular place of interest. This non-local aliasing becomes a dominant limitation as the measurement accuracy is improved. The term "non-local" is included in order to distinguish this issue from aliasing due to time variations at a particular location of interest.

A number of studies are being carried out in order to try to understand the limitation due to non-local aliasing. In some of these, two or more pairs of satellites in polar orbits are assumed. An alternate approach [8], with one pair in a polar orbit and a second pair in a lower inclination orbit, will be discussed briefly below. To permit detailed studies, particular orbits have been chosen for what has been called a Dual-GRACE mission. Altitudes of about 310 km are assumed, along with 62.7 degree inclination for the second pair. A 79 revolution repeat ground track was chosen for the polar pair, and a 360 revolution repeat ground track for the lower inclination pair. The corresponding ground track repeat periods are about 5 days and 23 days.

The presumed advantage of the chosen orbits is that quite dense ground track coverage would be obtained at latitudes within about 29 degrees of the poles. During a 5 day period, the maximum spacing between ground track would be less than 115 km. Thus, in solutions that concentrate mainly on these high latitude regions, time variations with longer than 5 day periods would be resolved. These results then could be used in analyzing the measurements at lower latitudes. The high latitude parts of the 62.7 degree inclination pair ground tracks would provide improved information on the east-west variations in the geoid within 35 degrees of the poles, while the lower inclination of the pair would provide many more orbit crossings at low latitudes than with a second polar pair. Thus, for missions with two pairs of drag-free satellites, having one of the pairs in a non-polar orbit appears to be worth exploring. If the non-local aliasing problem can be reduced substantially by this or some other mission design choice, than the very low spurious acceleration level achievable with GRS type sensors can be taken advantage of in future Earth gravity variation missions.

4. Possible future LISA follow-on mission

It may seem very early to consider possible future gravitational wave missions for after LISA has been flown. However, suggestions have been made for several extremely challenging future missions aimed at detecting a primordial gravitational wave background in the frequency range from 0.1 to 10 Hz. Thus it seems desirable to consider what could be accomplished by a mission with less severe measurement requirements, and what new scientific objectives would justify such a mission.

One such suggestion is for a mission called the Advanced Laser Interferometer Antenna, or ALIA [9]. It would be like LISA, but with a factor 10 shorter arm lengths (500,000 km). The strain measurement accuracy would be a factor 30 better than for LISA. This corresponds to 6×10^{-14} m/√Hz above roughly 30 millihertz. Possible choices for design parameters are 1.0 meter telescope diameters and 30 Watts of laser power at 1.064 micron wavelength. Spurious acceleration levels below 3×10^{-16} m/s^2/√Hz down to about 1 millihertz also would be required.

LISA is expected to observe signals from a considerable number of mergers involving intermediate mass or larger black holes out to redshifts of about 10. The main purpose is to investigate the formation and growth of massive black holes. But LISA won't have enough sensitivity to see 10 solar mass black holes merge with larger ones, except at low redshifts. Since many more mergers of roughly 10 solar mass black holes with intermediate mass black holes (IMBHs) at large redshifts are expected, a mission that could observe such mergers would give much more detailed information on the initial formation and growth of the IMBHs.

In view of the above, there appears to be a strong scientific justification for a LISA follow-on mission with substantially improved sensitivity, but not extremely high sensitivity. Such a mission might be quite different than what is proposed as ALIA. However, the required limit on spurious accelerations is likely to be at least as stringent as for ALIA, and thus additional research beyond the GRS development for the LISA Pathfinder mission and for LISA is expected to be needed. It will be interesting to see if some of the objectives for observing BH-IMBH binary mergers out to large redshifts can be achieved with the intermediate level mission called pre-DEGIGO [10], which is being proposed for study in Japan. This apparently will depend mainly on the arm length and level of GRS performance that are adopted for pre-DECIGO.

Acknowledgments

It is a pleasure to thank particularly the following: Stefano Vitale, Bill Weber, and Rita Dolesi for information concerning the development of the GRS systems for LISA Pathfinder and LISA; Neil Ashby for having carried out most of the studies of the possible L-1 gravitational time delay mission; David Wiese and Steve Nerem for their studies of the limitations from temporal aliasing on studies of the Earth's gravity field; and Seiji Kawamura for information concerning studies in Japan of possible future gravitational wave missions.

References

[1] Will C M 2001 *The Confrontation between General Relativity and Experiment, in Living Reviews in Relativity* (2001-4), pp 1-97
[2] Will C M 1993 *Theory and Experiment in Gravitational Physics* (Cambridge University Press)
[3] Bertotti B, Iess L and Tortora P 2003 *Nature* 425 374-376
[4] Ashby N and Bender P 2005 *Conference on Lasers, Clocks and Drag-Free Control*, Bremen, Germany May 30-June 1 2005, in *Lasers, Clocks and Drag-Free Control, Astrophys & Space Science Library* V 349 (H Dittus, C Lammerzahl and S G Turyshev eds, Springer) pp. 219-230
[5] Bender P, Ashby N, Armstrong J W et al 2006 *Workshop on Advances in Precision Tests and Experimental Gravitation in Space*, Firenze Italy Sept. 28-30
[6] Bender P L, Ashby N, Dolesi R et al 2008 *Quantum to Cosmos III Workshop*, Warrenton, VA July 6-10 2008
[7] Ludlow A D, Zelevinsky T, Campbell G K et al 2008 *Science* 319 1805-1808
[8] Bender P L, Wiese D N and Nerem R S 2008 in *Proc. 3rd Int. Symp. on Formation Flying, Missions and Technologies*, 23-25 April 2008, Noordwijk, The Netherlands (K Fletcher ed) ESA SP-654
[9] Bender P L and Begelman M C 2005 in *Proc. 39th ESLAB Symposium, "Trends in Space Science and Cosmic Vision 2020"*, ESA SP-588, Noordwijk, 19-21 April 2005 (F Favata, J Sanz-Forcada, A Gimenez eds) pp 33-37
[10] Ando M 2008 *7th Int. LISA Symposium*, 16-20 June 2008, Barcelona, Spain

Journal of Physics: Conference Series **154** (2009) 012019 doi:10.1088/1742-6596/154/1/012019

Mass Properties Measurement for Drag-free Test Masses

John W. Conklin, Aaron Swank, Ke-Xun Sun, and Dan B. DeBra
Stanford University, Stanford, CA 94305, USA
E-mail: johnwc@stanford.edu

Abstract. Space-borne gravitational wave observatories like the Laser Interferometer Space Antenna (LISA) and those beyond, which may utilize a Modular Gravitational Reference Sensor (MGRS), greatly benefit from precise knowledge of the mass center location and moment of inertia tensor of the test mass prior to launch. The motion of the mass center of a drag-free test mass, which follows a pure geodesic, must be inferred from measurements of the surface. Therefore, knowledge of the mass center is critical for calibration of the cross-coupling between rotational and translational degrees of freedom. Together with the moment of inertia tensor, the mass center can also provide an estimate of the material density inhomogeneity to quadratic order, and the gravitational potential to second order, which improves modeling of self gravitation forces. These benefits, which are independent of the test mass shape, motivate the development of three new techniques for improving mass center and moment of inertia measurements beyond the current state of the art. A static pendulum is proposed to determine the mass center of a cubic test mass to $\sim 1\ \mu$m by measuring the equilibrium position with the cube in up to 24 different orientations relative to the pendulum platform. Measuring the natural frequency of a dynamic torsion pendulum can determine both the mass center and moment of inertia tensor of arbitrarily shaped objects to $\sim 5\ \mu$m and 1 part in $\sim 10^4$ respectively. The velocity modulation technique for measuring the mass center of a sphere has raised the bar in precision to ~ 150 nm, a factor of 20 improvement over the work presented at the LISA 6th symposium. This new technique involves rolling the sphere down a set of parallel rails to spectrally shift the mass center offset information to the rolling rate frequency, in order to avoid the $1/f$ noise that typically prevents other techniques from achieving precision below 1 μm.

1. Introduction & Motivation

A drag-free spacecraft [1] shields an internal free-floating test mass (TM) from external disturbances, while minimizing disturbances caused by the spacecraft itself. A gravitational reference sensor contains and shields the test mass, and senses its position, enabling a feedback control system to command thrusters to keep the spacecraft fixed with respect to the test mass. In principle, the test mass is then completely freed from non-gravitational disturbances, so that its mass center (MC) follows a perfect geodesic.

One exciting application of drag-free technology is space-borne gravitational wave detectors such as the joint NASA and ESA Laser Interferometer Space Antenna (LISA) [2–4]. LISA will require unprecedented ~ 10 pm level accuracy in the measurement of the distances between the mass centers of drag-free test masses housed in three spacecraft separated by 5 Gm. In addition to the challenging metrology, the disturbances applied to the test masses must be reduced four orders of magnitude below what has been demonstrated by Gravity Probe B [5], and over four orders of magnitude in frequency (0.1 mHz to 1 Hz). LISA will detect and observe the emission of gravitational radiation from massive black holes in the distant Universe and compact binaries within our own galaxy, and consists of three drag-free spacecraft in a heliocentric orbit. Beyond LISA is the proposed Big Bang Observer (BBO) [6; 7], which will detect primordial gravitational waves generated during the birth of the Universe. BBO hopes to achieve acceleration noise levels two additional orders of magnitude below that of LISA, but at higher frequencies (0.1 - 10 Hz).

LISA (and LISA Pathfinder [8]) will use two $\sim$ 40 mm cubic test masses per spacecraft, one aligned with each of the two remote spacecraft. The choice of cubes is based on the flight heritage of precision accelerometers manufactured by ONERA. Cubes also provide a flat surface for reflecting laser light. Alignment of the cubes is maintained by electrostatic actuation so that changes in their orientation do not mimic translation of the mass center.

The Modular Gravitational Reference Sensor (MGRS) is an alternate design for future gravitational missions, including gravitational wave detection [7; 9; 10]. The MGRS uses a single spinning sphere as the gravitational reference, no electrostatic suspension, and a relatively large gap ($\sim$ 20 mm) between the spacecraft and TM. All of these features make the MGRS especially sensitive below 1 mHz, where acceleration noise dominates the sensitivity limit [11].

Several sources of acceleration noise are related, in some way, to the density distribution of the test mass material. These include the translation of the TM mass center coupling with the orientation control, gravitational attraction with the spacecraft, and magnetic interactions [12]. We will discuss now how precise determination of the mass center (MC) and moment of inertia tensor of the test mass improves our ability to estimate the magnitudes of these effects

1.1. Cross-coupling of Test Mass Degrees of Freedom

The science signal for LISA, and almost any drag-free mission, is the position of the mass center of the test mass. In the case of LISA it is the variations in the distance between the mass centers of test masses housed in three drag-free spacecraft. Measurement of the mass center position must be inferred from measurements of the TM surface. Therefore, precise knowledge of the mass center of the test mass is crucial for achieving the desired science signal.

There are two general situations that can potentially confuse rotation of the test mass with translation of the mass center. The first is rotation of the TM about its mass center, without any real mass center motion. For a faceted test mass, if the direction of the distance measurement to the TM surface does not pass through the mass center, then rotation about the mass center will produce a false measurement of mass center position variations. The second case occurs when the TM orientation control rotates the TM about an axis not coincident with the mass center. This will inadvertently produce a force acting at the mass center, which corrupts the science measurement. Both of these systematic effects can be mitigated by incorporating accurate mass center information in the orientation control software and in the data reduction.

1.2. Density Inhomogeneity

Measurements of the mass, mass center and moment of inertia tensor can provide some information regarding the density distribution of the TM. To be more quantitative, let us choose a particular model for the true density distribution, $\rho(x,y,z)$,

$$\begin{aligned}\rho(x,y,z) \approx c_0 &+ c_1\,x + c_2\,y + c_3\,z + c_4\,x^2 + c_5\,y^2 + c_6\,z^2 \\ &+ c_7\,yz + c_8\,xz + c_9\,xy + \dots . \end{aligned} \tag{1}$$

This model is simply a polynomial expansion of the density distribution to quadratic order, including cross product terms.

Assume that the three mass properties are measured, which represent ten numbers: the mass, three components of the mass center location and six independent components of the moment of inertia tensor. These ten numbers directly specify the parameters c_i in the model for $\rho(x, y, z)$ which minimizes the error in density (true density distribution minus the quadratic model) squared and integrated over the TM volume. In other words the mass properties provide the best fit polynomial model, to quadratic order, to the true density in a least-squares sense.

Knowledge of the density inhomogeneities is vital for materials like gold-platinum, where spinodal decomposition during the material cooling process, can cause the gold and platinum to separate. A uniform mixture of the diamagnetic gold and paramagnetic platinum is required for a near-zero magnetic susceptibility. A low susceptibility reduces the TM to spacecraft stiffness and TM disturbances due to fluctuations in the magnetic field in the LISA frequency band.

1.3. Gravitational Self-attraction

The gravitational self-attraction force between the satellite and the TM is a dominant contribution to the disturbance budget. Although the attraction force can not be easily measured directly, the mass attraction formula through a second order expansion consists of the measurable quantities of mass, mass center, and moment of inertia about the mass center. Thus, the gravitational self-attraction force on the drag free reference due to the satellite can be indirectly measured. By computing the attraction force from the measured mass properties, variations in density as well as geometric variations will be included in the gravitational self-attraction calculation. As a result, the exact density distribution within the test mass need not necessarily be known or measured in order to determine the actual gravitational attraction properties.

2. Measuring Mass Center & Moment of Inertia

The LISA laboratory at Stanford University is exploring three methods of measuring the mass center and moment of inertia tensor of objects on the same scale as the LISA test masses.

2.1. Mass Center Measurement for a Cube with a Static Pendulum

The first technique is a static pendulum specifically designed to quickly and accurately determine the mass center of a cube. A schematic of the apparatus, which is currently in the development stages at Stanford, is shown in Figure 1 (a). Five wires are used to suspend a triangular platform. The position and orientation of the five wires are chosen so that they constrain all but one degree of freedom. The free motion of the pendulum, shown in Figure 1 (a) by the dashed lines, is primarily lateral translation, with a small amount of tilt. The pendulum platform has a triangular hole to accommodate a cubic test mass. Six gauge spheres are fixed to the edges of the triangular hole so that the cube is kinematically constrained to the platform.

The measurement is the equilibrium position of the pendulum, which depends on the mass center locations of both the pendulum platform and cubic TM. The equilibrium position can be measured using a laser interferometer, shown in Figure 1 (a), or by an optical shadow sensor. To separate the mass center of the platform from that of the cube, and to determine the mass center of the cube in all three dimensions, many measurements are taken, each time with the cube in a different orientation. Each time the orientation of the cube is changed, the cube's mass center is changed relative the mass center of the platform. There a total of 24 possible orientations of the cube on the pendulum platform. This provides more than enough measurements to determine the mass centers of both the pendulum platform and the cubic TM in all three directions, and provide an estimate of the measurement error using the post-fit residuals.

The static pendulum can potentially measure the MC location of the cube to better than 1 μm. This assessment is based on the sensitivity of the optical detection, which can be less than 10 nm for either an interferometric or a shadow sensor, and the geometry and masses of the the pendulum and cubic TM. In addition, the measurement procedure is easy since it only involves changing the orientation of the cube and measuring the resulting equilibrium position. The data

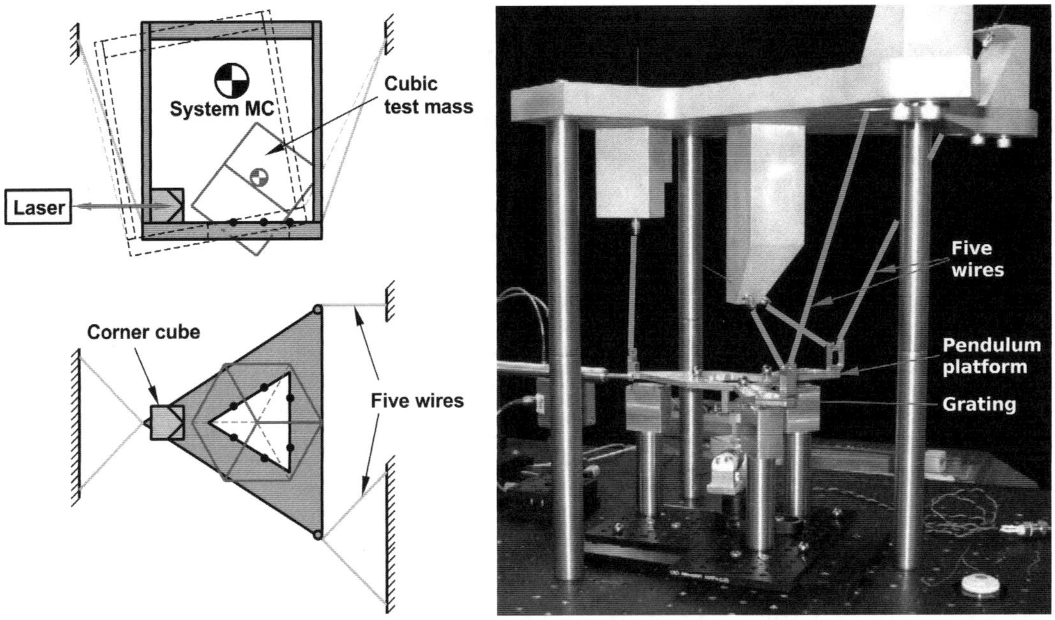

(a) (b)

Figure 1. (a) Elevation view (top) and plan view (bottom) of the static pendulum for measuring the mass center of a cubic test mass. (b) Schematic of the dynamic five-wire pendulum for measuring the moment of inertia tensor and mass center of arbitrary objects.

reduction entails the proper coordinate transformations for each of the cube orientations, and the proper weighting of the mass of the cube with respect to that of the platform.

2.2. Mass Center & Moment of Inertia Measurement with a Dynamic Pendulum

An apparatus which is designed to generate a pure rotation about an axis provides the ability to measure the mass center location and the moment of inertia. For example, in order to measure the moment of inertia, I, the object is rotated about an axis and the pendulum natural frequency, $\omega/2\pi$, is used to determine the radius of gyration, R_g, about that rotational axis:

$$R_g^2 = \frac{I}{m} = \frac{I_o + m_o r^2}{m} = \frac{k}{\omega^2} \tag{2}$$

Here, k is the torsion coefficient or stiffness constant of the pendulum and m is the total mass.

By measuring the natural frequency of rotation, the mass center offset from the geometric center of an object is also obtained. In Eq. (2), we replace the moment of inertia about the geometric center, I, by the moment of inertia about the objects mass center, I_o, plus the parallel axis theorem components of the object mass, m_o, times the square of the distance from the rotation center to the mass center, r^2. Therefore, placing the object on the pendulum far from the rotation axis amplifies the contribution due to the mass center offset from the geometric center, due to its quadratic dependence on r. By changing the orientation of the object with a fixed geometric location relative to the pendulum rotation center, the mass center offset is determined by measuring the change in the natural frequency, $\omega/2\pi$.

The difficulty in the measurements lies within the ability to generate pure rotation about one degree of freedom. Care must be taken to minimize the extra degrees of freedom in the system. In a five-wire torsion pendulum shown in Figure 1 (b), five wires are arranged to minimize translational degrees of freedom in order to reduce errors due to tilt and translation [13; 14]. By reducing the errors associated with generating a pure rotation, the five-wire pendulum has demonstrated a consistent frequency measurement to better than 10^{-5} Hz for a 2.5 Hz pendulum frequency. As show in Ref. [13], the precision on the frequency measurement results in a repeatable moment of inertia measurement on the order of a few parts in 10^4. For the mass center measurement, the five wire pendulum has demonstrated the ability to determine the mass center offset to within 5 μm.

2.3. The Velocity Modulation Technique for Mass Center Measurement
A new technique for measuring the mass center of a sphere to $\sim$ 100 nm, called velocity modulation, was presented in 2006 [15; 16]. Previous methods of determining the mass center, involving measurement of the sphere's natural pendulous period, are typically limited to MC offsets $\sim>$ 1 μm due to low frequency ($1/f$) instrumentation noise. In the velocity modulation technique, the sphere is rolled down a set of parallel rails to spectrally shift the MC offset information to the rolling rate frequency. The sphere's trajectory is measured optically by recording the times that the sphere crosses five optical gates.

The experimental apparatus is shown in Figure 2 (a). The two $\sim$ 1 m long steel rails are supported above a base plate mounted on an optical bench. The light from a single laser is broken into five beams, by five polarizing beam splitters, so that the beams cross the rails at right angles at the exact height of the sphere's geometric center. Measurement of the sphere's location on the rails is made by *pairs* adjacent photo-detectors mounted opposite to the beam splitters. The reflected laser beams and photo-detectors are space one diameter of the sphere apart so that exactly half of the incident light is occulted when the sphere is precisely between two detectors. The voltage produced by the pair of photo-detectors is differenced so that when the voltage difference is zero, the sphere is exactly between the two optical gates. Fluctuations in laser intensity are consequently eliminated in the differential signal because laser intensity variations are common to all detectors. The sphere is rolled three times, each time with a different initial phase, in order to separate the geometric irregularities in the rails, which are fixed in the laboratory, from the MC offset, which depends on the initial conditions. A Monte-Carlo parameter search is used to fit all relevant parameters in a simplified mechanics model to the measured data, including the optical gate locations, initial velocities and MC location.

The velocity modulation technique can also determine the MC of an arbitrarily shaped test mass (including a cube) to $\sim$ 1 μm by placing the TM inside a spherical fixture. To separate the MC of the TM from that of the spherical fixture, several measurements are made, each time changing the orientation of the TM with respect to the fixture. In this case, the measurement error is limited by the repeatability of the placement of the test mass inside the spherical fixture.

The precision of the velocity modulation technique is demonstrated by the repeatability of independent measurements. Figure 2 (b) shows four independent measurements of the mass center of a stainless steel gauge sphere. All four measurements are consistent within their uncertainties estimated by the RMS of the post-fit residuals. The mean result and one standard deviation error ellipse with maximum radius less than 150 nm are also shown in Figure 2 (b).

3. Summary
Future drag-free missions like LISA, with strict requirements for both metrology and disturbance reduction greatly benefit from precise determination of the mass center and moment of inertia tensor of its test masses prior to launch. This allows verification of several design requirements, including degree of freedom cross-coupling, gravitational disturbances and density inhomogeneity. Three techniques developed at Stanford for measuring these mass properties represent the state of the art, with accuracies listed in Table 1.

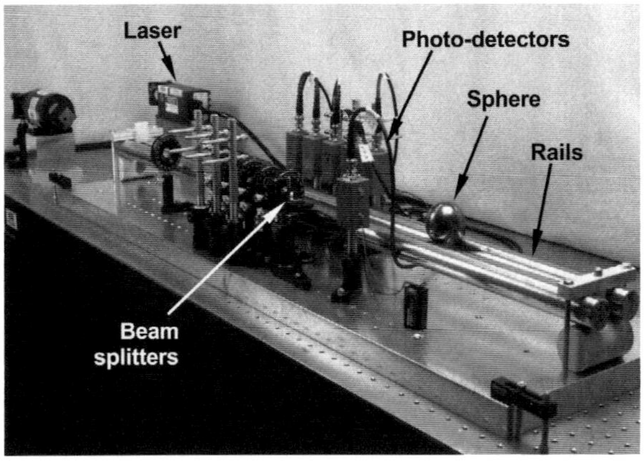

(a)

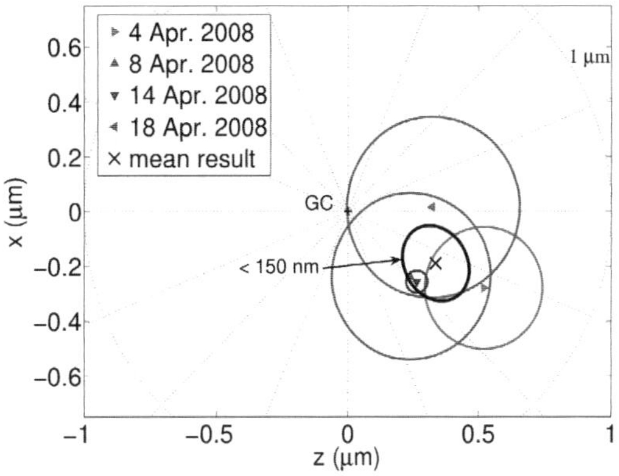

(b)

Figure 2. (a) Velocity modulation apparatus for mass center measurement. (b) Precision of mass center measurement demonstrated through the repeatability of four independent measurements. The mean result and error ellipse are shown in black.

Mass property	Technique	Accuracy
Mass center (any shape)	Static/dynamic pendulum or velocity modulation	$\sim 1\ \mu$m
Mass center (sphere)	Velocity modulation	~ 150 nm
Moment of inertia (any shape)	Dynamic pendulum	$\sim 10^{-4}$

Table 1. Mass center and moment of inertia measurement capability.

Acknowledgments
This work is supported by ROSES Proposal Beyond Einstein Foundation Science 06-BEFS06-56, NASA Grant No. NNX07AK65G

References

[1] DeBra D B 1997 *Classical and Quantum Gravity* **14** 1549–1555

[2] Danzmann K and Rüdiger A 2003 *Classical and Quantum Gravity* **20** S1–S9

[3] LISA Study Team 1998 LISA, Laser Interferometer Space Antenna for the detection and observation of gravitational waves LISA Pre Phase A Report, 233 Max-Plank-Institut für Quantenoptik Garching, Germany

[4] Jennrich O (ed) 2004 *Proceedings of the 5th International LISA Symposium and the 38th ESLAB Symposium* (*Classical and Quantum Gravity* vol 22) (Institute of Physics Publishing)

[5] Conklin J W 2008 *Proceedings of the Sixth International Conference on Gravitation and Cosmology* Journal of Physics: Conference Series

[6] Phinney E S *et al.* 2004 The Big Bang Observer: Direct detection of gravitational waves from the birth of the Universe to the present NASA OSS Vision Missions Program, Proposal VM03-0021-0021

[7] Sun K X, Allen G, Buchman S, DeBra D B and Byer R L 2004 *Proceedings of the 5th International LISA Symposium and the 38th ESLAB Symposium* (*Classical and Quantum Gravity* vol 22) pp S287–S296

[8] Anza S, Armano M, Balaguer E, Benedetti M, Boatella C, Bosetti P, Bortoluzzi D, Brandt N, Braxmaier C, Caldwell M, Carbone L, Cavalleri A, Ciccolella A, Cristofolini I, Cruise M, Da Lio M, Danzmann K, Desiderio D, Dolesi R, Dunbar N, Fichter W, Garcia C, Garcia-Berro E, Garcia Marin A F, Gerndt R, Gianolio A, Giardini D, Gruenagel R, Hammesfahr A, Heinzel G, Hough J, Hoyland D, Hueller M, Jennrich O, Johann U, Kemble S, Killow C, Kolbe D, Landgraf M, Lobo A, Lorizzo V, Mance D, Middleton K, Nappo F, Nofrarias M, Racca G, Ramos J, Robertson D, Sallusti M, Sandford M, Sanjuan J, Sarra P, Selig A, Shaul D, Smart D, Smit M, Stagnaro L, Sumner T, Tirabassi C, Tobin S, Vitale S, Wand V, Ward H, Weber W J and Zweifel P 2005 *Classical and Quantum Gravity* **22** 125–+

[9] Sun K X, Allen G, Buchman S, Byer R L, Conklin J W, DeBra D B, Gill D, Goh A, Higuchi S, Lu P, Robertson N and Swank A 2006 *2007 AAS/AAPT Joint Meeting, American Astronomical Society Meeting 209, #74.07* (*Bulletin of the American Astronomical Society* vol 38) p 991

[10] Sun K X, Allen G, Buchman S, Byer R L, Conklin J W, DeBra D B, Gill D, Goh A, Higuchi S, Lu P, Robertson N A and Swank A J 2006 *Laser Interferometer Space Antenna: 6th International LISA Symposium* (*American Institute of Physics Conference Series* vol 873) pp 515–521

[11] Gerardi D, Allen G, Conklin J W, Sun K X, DeBra D B, Buchman S, Gath P, Byer R and Johann U Achieving Disturbance Reduction for Future Drag-Free Missions to be submitted

[12] Schumaker B L 2003 *Classical and Quantum Gravity* **20** S239–S253

[13] Swank A J, Hardham C, Sun K X and Debra D 2006 (ASPE)

[14] Swank A J, Sun K X and DeBra D 2006 *Laser Interferometer Space Antenna: 6th International LISA Symposium* (*American Institute of Physics Conference Series* vol 873) pp 588–592

[15] Conklin J W, Sun K X and DeBra D B 2006 *Laser Interferometer Space Antenna: 6th International LISA Symposium* (*American Institute of Physics Conference Series* vol 873) pp 566–570

[16] Conklin J W, Sun K X and DeBra D B 2006 *Proceedings of the 21st Annual Meeting of the American Society for Precision Engineering, Monterey, CA* American Society of Precision Engineers (Raleigh, NC)

Journal of Physics: Conference Series **154** (2009) 012020 doi:10.1088/1742-6596/154/1/012020

Analog phase lock between two lasers at LISA power levels

Christian Diekmann, Frank Steier, Benjamin Sheard, Gerhard Heinzel and Karsten Danzmann

Max-Planck-Institute for Gravitational Physics, Callinstr. 38, D-30167 Hannover, Germany

E-mail: Christian.Diekmann@aei.mpg.de

Abstract. This paper presents the implementation of an analog optical phase-locked-loop with an offset frequency of about 20 MHz between two lasers, where the detected light powers were of the order of 31 pW and 200 μW. The goal of this setup was the design and characterization of a photodiode transimpedance amplifier for application in LISA. By application of a transimpedance amplifier designed to have low noise and low power consumption, the phase noise between the two lasers was a factor of two above the shot noise limit down to 60 mHz. The achievable phase sensitivity depends ultimately on the available power of the highly attenuated master laser and on the input current noise of the transimpedance amplifier of the photodetector. The limiting noise source below 60 mHz was the analog phase measurement system that was used in this experiment. A digital phase measurement system that is currently under development at the AEI will be used in the near future. Its application should improve the sensitivity.

1. Introduction

LISA will measure the distance fluctuations on the order of $12\,\mathrm{pm}/\sqrt{\mathrm{Hz}}$ between 5×10^9 m distant free-falling test masses in the frequency range between 1 Hz and 3 mHz, with a decreasing sensitivity below 3 mHz with f^{-2} mainly due to test mass acceleration noise [1].

The length measurements will be done interferometrically and due to the large beam divergence between the satellites only a small fraction of the transmitted power is detected at the remote satellite. The transmitted laser power, the wavelength of the light, the telescope diameter and the losses in the optical chain influence the amount of detected power which finally yield a detectable light power on one of the photodetectors that are used for the science measurement of about 100 pW [2].

The limit for the measurement sensitivity which mainly depends on the detectable power at the remote satellite is approximately $9\,\mathrm{pm}/\sqrt{\mathrm{Hz}}$ and is a consequence from the shot noise introduced by the lasers at the photodiode [2].

The phase information of the incoming light must be transferred to another laser, the so-called slave laser. In the case of LISA this will be done by an optical phase-locked-loop (OPLL) with a maximum offset frequency of 20 MHz which depends on the time-varying Doppler shifts that are induced by the changing velocities of the satellites [3, 4].

The incoming beam interferes with the local laser. Their beat note is detected by a photodetector, which might introduce additional phase noise on one hand due to electronic noise of the photodetector and on the other hand due to the available laser power from the slave laser on the photodiode [3, 5].

The more laser power from the slave laser is used, the lower is the influence of the electronic noise of the transimpedance amplifier (TA). The limits for the usage of slave laser power are set by two facts: The transmitted power to the remote satellite must be as high as possible in order to minimize the shot noise, and the deposed energy on the LISA optical bench must be as small as possible in order to avoid thermal gradients. The amount of power of the slave laser that is used for the OPLL on the LISA optical bench will be 1 mW.
With these constraints given, a suitable photocurrent transimpedance amplifier was designed and characterized in terms of its phase sensitivity.
In Section 2 the experimental setup for characterization of the phase noise performance is described. Section 3 deals with the analytic description of the significant noise sources, their tracing through the phase read-out chain and in Section 4 the experimental results from the analog OPLL are summarized.

2. Experimental setup

Figure 1a shows a simplified schematic of the experimental setup. For the OPLL two identical Nd:YAG lasers (Innolight Mephisto 500) were used which offer the possibility to control their frequency via piezo-electric and temperature actuators on the laser crystal. Both laser beams were attenuated before they were fed into the interferometer where the phaselock was performed. These attenuators were designed to provide output beams with minimum power fluctuations and to avoid stray light on the beat note that is used for the phase-lock. Thus, out-coupling mirrors were chosen for the attenuators set up separately from the interferometer described below. The attenuated beams were then guided via fibers onto a Zerodur optical bench, where they were brought to interference at a beam splitter. The attenuators were adjusted in such a way that the available power of the lasers at one output of the power beam splitter were approximately 31 pW and 200 μW, respectively. The weak beam power was estimated from the beat note amplitude combined with an estimate for the contrast. The slave laser power was determined from the DC photocurrent. The interference signal at one output was then used to produce the error signal for the OPLL. Photodetectors with dedicated transimpedance amplifiers were used to measure the beat note.
In contrast to the LISA photodetectors, single element photodiodes were used instead of quadrant photodiodes. Therefore, the available power from both lasers had to be adjusted. For the photodetector characterization the power of the master laser was chosen to be slightly higher compared to LISA which results in a lower limit for the achievable phase noise. Thus, other noise sources like the TA input current noise could be seen more clearly.
The phase read-out is done via a DC read-out scheme. The beat note was mixed with a reference frequency using a mixer (Mini Circuits TUF-3H). When both signals have approximately the same frequency, the output signal of the mixer contains information about the phase difference between both signals and an additional signal at the sum of both frequencies. The sum frequency was suppressed by an active low-pass filter, consisting of two second order Tschebyscheff low-pass filters which were implemented using an LT1028 operational amplifier. The input offset voltage of the operational amplifier was compensated using a composite amplifier design with an AD8628 in the feedback chain [6]. The mixer, the low-pass and the Data acquisition system (DAQ) used to sample the low-pass output signals represent the phase measurement system (PMS).
The error signal was passed through a PI controller before being fed back to the slave laser via its piezo and temperature tuning inputs. Thereby, the phase difference between the beat note and the reference signal is kept close to zero. Further information on the setup can be found in [5].
Figure 1b shows a schematic of the OPLL where the significant sources and their coupling nodes have been highlighted. The important noise sources for the development of the photodetectors

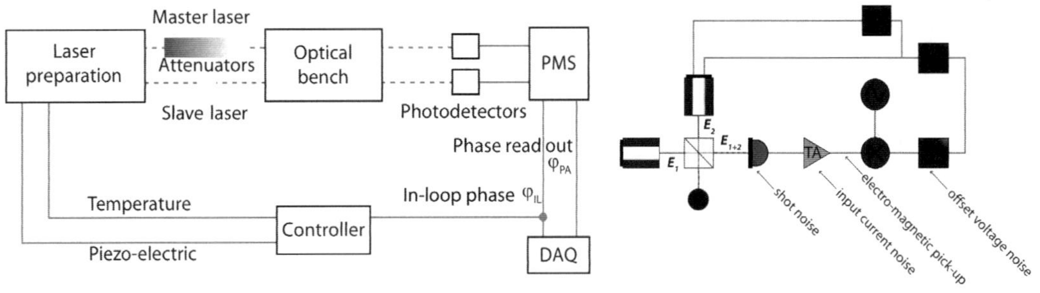

(a) Experimental setup of weak light OPLL (b) OPLL noise sources

Figure 1: Analog OPLL with offset frequency - Experimental setup and possible noise sources

are the input current noise of the transimpedance amplifier and the electro-magnetic pick-up of the electrical components of the OPLL [3].
The input current noise can be kept as low as possible by an optimized TA design. Due to the limited gain-bandwidth product, this design is a trade-off between high gain that reduces the effect of additional noises introduced by the following stages and the photodetector transimpedance bandwidth.

3. Photodetector design

In Figure 1b the electro-magnetic field E at the output of the power beam splitter is a interference signal of the electro-magnetic fields $E_{\mathrm{ML,SL}}$ at the inputs of the power beam splitter [7]. This interference signal E can simply be converted into a power

$$P = P_{\mathrm{ML}} + P_{\mathrm{SL}} + 2C\sqrt{P_{\mathrm{ML}}P_{\mathrm{SL}}}\cos\left(\Delta\omega t + \Delta\phi\right). \tag{1}$$

$P_{\mathrm{ML,SL}}$ denotes the regarding power from the master and the slave laser after the beam splitter, $\Delta\omega$ and $\Delta\phi$ describe the frequency and the phase difference between both laser beams and C the contrast of the beat note. The current i_{P} produced by the photodiode can be calculated by multiplication of Eq. 1 with the photodiode efficiency, ρ. The Perkin Elmer C30619G photodiode that was used in our experiment had an efficiency specified to be 0.68A/W [8].

$$\begin{aligned} i_{\mathrm{P}} &= \rho\left(P_{\mathrm{ML}} + P_{\mathrm{SL}} + 2C\sqrt{P_{\mathrm{ML}}P_{\mathrm{SL}}}\cos\left(\Delta\omega t + \Delta\phi\right)\right) & (2)\\ &= 0.14\,\mathrm{mA} + 21.7\,\mathrm{pA} + 0.1\,\mu\mathrm{A}\cos\left(\Delta\omega \mathrm{t} + \Delta\phi\right) & (3) \end{aligned}$$

For the next processing steps this photocurrent must be converted by a transimpedance amplifier into a voltage [9, 10]. When both the AC and the DC photocurrent are amplified, the DC term limits the amplification to a factor of about 10^5 when an operational amplifier with a typical output voltage range of $\pm 15\,\mathrm{V}$ is used. In this case the amplitude of the beat note would only be 10 mV which is not desirable because electro-magnetic pick-up and the noise contributions from following stages become more significant. In addition to this the mixer in the PMS needs a higher beat note amplitude to function properly. For this reasons the AC and the DC part of the photocurrent are decoupled to maximize the amplification in the first stage.
One major noise source is the input current noise of the TA. The shot noise i_{SN} at the output of the photodiode is dominated by the slave laser power because it is much higher than the master laser power [11]:

$$i_{\mathrm{SN}} \cong = \sqrt{2\mathrm{e}\rho\mathrm{P}_{\mathrm{SL}}}, \tag{4}$$

where e is the elementary charge. In our case the slave laser power received by the photodiode is 200 μW and, therefore, the shot noise spectral density of the photocurrent is 6.69 pA/$\sqrt{\text{Hz}}$. This current noise is amplified and converted into a voltage by the transimpedance amplifier. In the analog PMS, the voltage noise is multiplied as well as the beat note with the reference frequency. Thus the shot noise from the photodiode leads to a voltage noise at the output of the analog PMS which finally will limit the phase measurement.

The achievable phase sensitivity at the output of the PMS can be derived by knowing the properties of an analog PMS and tracing the noise from the photodetector through the phase read-out chain [12]. If the input current noise of the transimpedance is neglected and only the shot noise of the photodiode is taken into account, this assumption leads to the following equation for the observed phase noise

$$\delta\widetilde{\phi}_{\text{SN}} = \sqrt{\frac{\text{e}}{C^2 \rho P_{\text{ML}}}}. \tag{5}$$

The achievable phase noise depends in the ideal case only on the available power P_{ML} of the master laser.

Now a closer look at the input current noise of the transimpedance amplifier is taken. An ordinary transimpedance amplifier consists of an operational amplifier and a resistor [9]. The input current noise of an operational amplifier like the low noise amplifier LMH6624 is approximately 2 pA/$\sqrt{\text{Hz}}$ at 20 MHz plus the current noise from the transimpedance resistor, the so-called Johnson noise. The current noise for a resistor, i_{JN}, with resistance, R, is given by

$$i_{\text{JN}} = \sqrt{\frac{4\text{k}_\text{B}\text{T}}{R}} \tag{6}$$

where k_B denotes the Boltzmann factor and T the temperature. For a 2 kΩ resistor Eq. 6 yields approximately 2.8 pA/$\sqrt{\text{Hz}}$.

The equivalent input current noise for the transimpedance amplifier is of the same order as the shot noise from the slave laser and hence, cannot be neglected when designing a dedicated photodetector.

The two noise source of the TA are quadratically added and yield the additional input current noise, i_{dn}. The input current noise is uncorrelated to the shot noise of the slave laser and if both noise sources are kept in mind the calculations yield the following equation for the achievable phase noise:

$$\delta\widehat{\phi}_{\text{SN}} = \sqrt{\frac{\text{e}}{C^2 \rho P_{\text{ML}}} + \frac{i_{\text{dn}}^2}{2C^2 \rho^2 P_{\text{ML}} P_{\text{SL}}}}. \tag{7}$$

Assuming realistic values for the parameters one can calculate the achievable phase noise for a TA. If the detectable power from the master laser is 31 pW and the power from the slave laser is 200 μW, this yields a phase noise of 8.28×10^{-5} rad/$\sqrt{\text{Hz}}$ in the ideal case. If one additionally takes the unavoidable input current noise (i_{dn}) of about 3 pA/$\sqrt{\text{Hz}}$ into account, the phase noise increases to 9.4×10^{-5} rad/$\sqrt{\text{Hz}}$.

That is why the development of a fast and low noise TA design was one major issue for the LISA interferometry. The TA consisted of a cascode plus a common collector stage offers a high gain-bandwidth product [9][13][14]. It simultaneously decreases the Johnson noise of the feedback resistor since a higher resistance value can be used. Another advantage of the transistor design is the lower power consumption in comparison to operational amplifiers.

The AC part of the photodetector shown in Figure 2 has a transimpedance gain for the first stage of 30 kΩ (2 $\times$ 15kΩ) and a voltage amplifier with a gain of 20 dB in the second stage. This

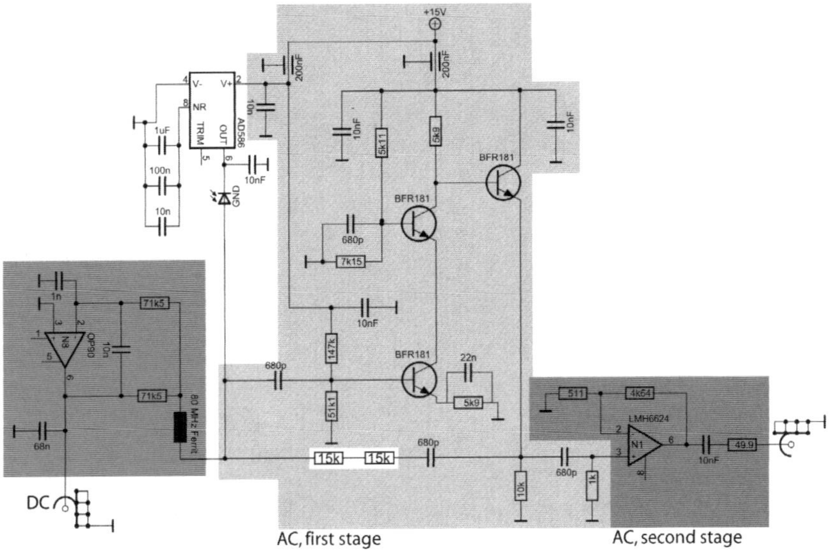

Figure 2: Schematic of the transimpedance amplifier. The AC amplifier uses only discrete transistor stages

yields an amplification factor of 1.5×10^5 [V/A] which includes the attenuation due to impedance matching between photodetector output and PMS input. The amplitude of the 20 MHz beat note at this stage was 15 mV. The amplitude was still too small for the analog PMS, which requires 0.5 and 2 V_{pp} input signals in order to achieve a sufficiently low phase noise. For this reason an additional amplifier with a gain of 50 was added, yielding an amplitude of 0.75 V at the input of the PMS. The introduced photodetector design was finally used for the OPLL at LISA power levels.

4. Results

The phase read-out sensitivity was measured by placing identical photodetectors at both outputs of one recombination beam splitter. One photodetector was used for the OPLL and the other was used to measure the out-of-loop phase noise of the photodetector plus the PMS. Figure 3a shows a comparison of two measurements of the phase read-out.

One important noise source was identified to be electro-magnetic pick-up. Its effect could be reduced by including better shielding of the photodetectors and the PMS and by using high frequency transformers (Mini Circuits T1-6T+) in the high frequency chains. Figure 3a shows that with these improvements a shot noise limited phase read-out could be achieved down to 60 mHz. The spectra were calculated using an improved FFT algorithm, called LPSD[15].

In order to identify the impact of the PMS on the noise performance, one photodetector signal was used as input for both the in-loop and the out-of-loop measurement paths. Thus, one output channel of the PMS was used for locking the two lasers and the other output of the PMS was measured by the DAQ. The spectrum of this measurement is shown as the green curve in Figure 3b indicates that the noise of the PMS is limiting the phase read-out below 60 mHz.

The light blue noise curve shows the noise of the phase read-out without any beat note for comparison. For this measurement only the slave laser beam was directed onto the photodiode. The shot noise introduced by the second laser on the photodiode can be calculated and converted into phase noise at the output of the PMS. The theoretical value is $1.1 \times 10^{-4}\,\mathrm{rad}/\sqrt{\mathrm{Hz}}$ and is almost identical with the measured value. A comparison of the measurement to the theoretical

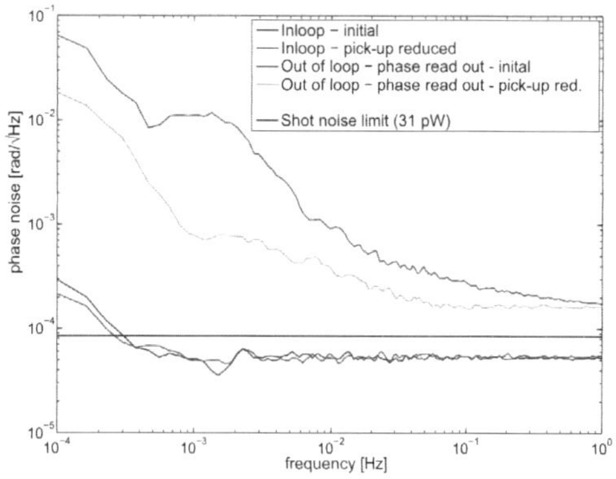

(a) Improved electro-magnetic shielding

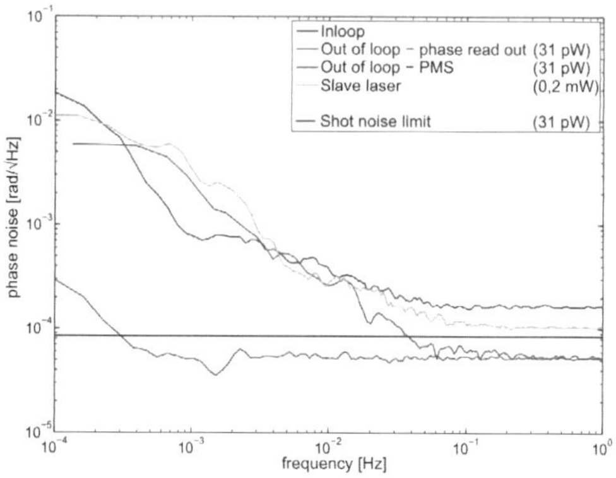

(b) Current phase noise sensitivity

Figure 3: Phase measurements with analog OPLL at LISA power levels

value gives information about residual unknown noise sources in the experiment which are still under investigation.

Acknowledgements
We gratefully acknowledge support by the Deutsches Zentrum für Luft- und Raumfahrt (DLR) (reference 50 OQ 0601).

Bibliography

[1] Danzmann K and Rüdiger A 2003 *Classical Quantum Gravity* **20** 1–9
[2] McNamara P W 2005 *Classical Quantum Gravity* **22** 243–247
[3] Huntington E 2005 *Optics Communications* **250** 178–190
[4] Dhurandhar S 2005 *Classical Quantum Gravity* **22** 481–487
[5] Steier F 2008 *Interferometry techniques for spaceborne gravitational wave detectors* Ph.D. thesis Institute for gravitational physics, University of Hannover
[6] Williams J 1986 Technical note 21 - composite amplifier Tech. rep. Linear Technology
[7] Bergmann and Schfer 2002 *Optik - Wellen- und Teilchenoptik* (de Gruyter)
[8] Datasheet http://optoelectronics.perkinelmer.com/content/datasheets/dts_c30619.pdf [date: 2008/10/29]
[9] Horowitz P and Hill W 1989 *The art of electronics - second edition* (Cambridge University Press)
[10] Mancini R 2002 *Op Amps For Everyone - Design Reference* (Elsevier LTD)
[11] Schottky W 1918 *Annalen der Physik* **57**
[12] Diekmann C 2008 *Phasenstabilisierung und -auslesung für LISA* Dipl. thesis Institute for gravitational physics, University of Hannover
[13] El-Diwany M 1981 *IEE Proc.* **128** 299–305
[14] Hobbs P 2000 *Building electro-optical systems - making it all work* (John Wiley and Sons,Inc.)
[15] Troebs M and Heinzel G 2006 *Measurement* **39** 120–129

Journal of Physics: Conference Series **154** (2009) 012021 doi:10.1088/1742-6596/154/1/012021

Current LISA Spacecraft Design

S. M. Merkowitz, K. E. Castellucci, S. V. Depalo, J. A. Generie, P. G. Maghami, and H. L. Peabody

NASA Goddard Space Flight Center, Greenbelt, MD 20771, USA

E-mail: Stephen.M.Merkowitz@nasa.gov

Abstract. The Laser Interferometer Space Antenna (LISA) mission, a space based gravitational wave detector, uses laser metrology to measure distance fluctuations between proof masses aboard three spacecraft. LISA is unique from a mission design perspective in that the three spacecraft and their associated operations form one distributed science instrument, unlike more conventional missions where an instrument is a component of an individual spacecraft. The design of the LISA spacecraft is also tightly coupled to the design and requirements of the scientific payload; for this reason it is often referred to as a "sciencecraft." Here we describe some of the unique features of the LISA spacecraft design that help create the quiet environment necessary for gravitational wave observations.

1. Introduction

The Laser Interferometer Space Antenna (LISA) mission will be sensitive to gravitational waves from astrophysical sources such as black hole, white dwarf, and neutron star binaries. LISA is comprised of three identical spacecraft separated by 5 million kilometers nominally forming an equilateral triangle. Each spacecraft (S/C) encompasses two freely floating proof masses that are the references for the gravitational wave measurement. Each leg of the triangle acts as a single arm of an interferometer that is used to measure any change in the distance between the distant proof masses.

The current design of LISA requires that all spurious accelerations of the proof masses be reduced to a level below $3\text{x}10^{-15}(1+(\text{f}/8\text{mHz})^4+(0.1\text{mHz}/\text{f})^2)^{1/2}$ $\text{m/s}^2/\sqrt{\text{Hz}}$ within the measurement bandwidth. A detailed noise budget was developed that set the required level of all known disturbance effects [1,2]. There are two types of contributors to this noise: effects that act directly on a proof mass (such as magnetic field fluctuations, voltage and charge effects, and thermally driven gas and radiation effects), and effects that couple S/C motion (caused by fluctuations in the forces acting on the spacecraft, such as thruster noise and variations in the solar photon pressure) to a proof mass through spatial gradients of the static force fields (gravitational, electric, etc.).

The design of the LISA S/C is driven by the need to keep the proof masses as undisturbed as possible by providing a quiet and stable environment. All the elements used for this purpose are grouped together into what is called the Disturbance Reduction System (DRS). The DRS consists of a set of sensors, actuators, and the control laws that bind them together to meet the disturbance and pointing requirements during LISA science operations. It also includes design features, such as thermal and magnetic shields, that are implemented to keep disturbances from reaching the proof masses. In this paper, we discuss how the current design of the LISA S/C supports the requirements of the DRS by creating the necessary environment for the gravitational wave observations.

2. Mechanical Design

The primary S/C structure consists of honeycomb panel upper and lower decks with aluminum alloy exterior sidewalls and a thermally isolated honeycomb panel solar array deck. Each of the three S/C will be nested inside of a propulsion module with the three propulsion modules stacked into a column inside the launch vehicle payload fairing. The stack design was chosen to carry the launch loads through the propulsion modules' outer shells, thereby isolating the S/C from the direct launch loads. A separation system, located on the bottom deck will serve to jettison the S/C from the propulsion module once the S/C has reached its operational orbit.

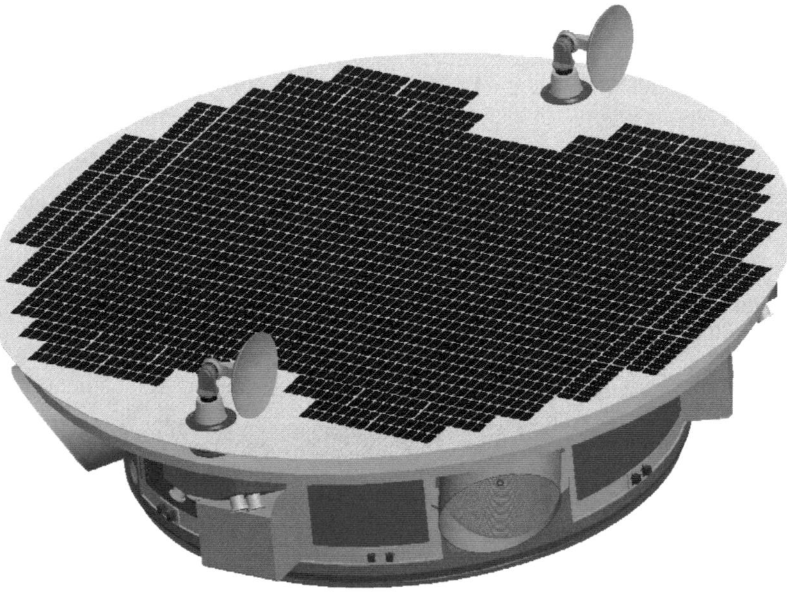

Figure 1. Top view of the LISA Spacecraft.

The S/C bus structure is designed seamlessly with the scientific payload. The optical assemblies are contained within thermal shrouds and incorporate the use of flex-pivots for rotational motion. The electronic boxes are mounted directly onto the bottom deck panel and the exterior wall. The current design philosophy to isolate the payload and bus electronics from solar heat input and to reject extraneous electrical waste heat into space means that mounting of any electronic components to the top deck panel is avoided.

In addition to housing the payload, the S/C bus structure must also provide mounting accommodations for the high gain and omni antennas, sun sensors, star trackers, and micronewton thrusters on the outside of the main structure as seen in figures 1 and 2.

All initial integration activities will occur with the solar array and top deck components removed, including the panels themselves. The solar array and top deck components will be installed in the last steps of assembly. In order to accommodate removal and replacement of components, maintenance or repairs after the S/C is completely assembled, access panels are provided at six locations around the circumference of the bus structure.

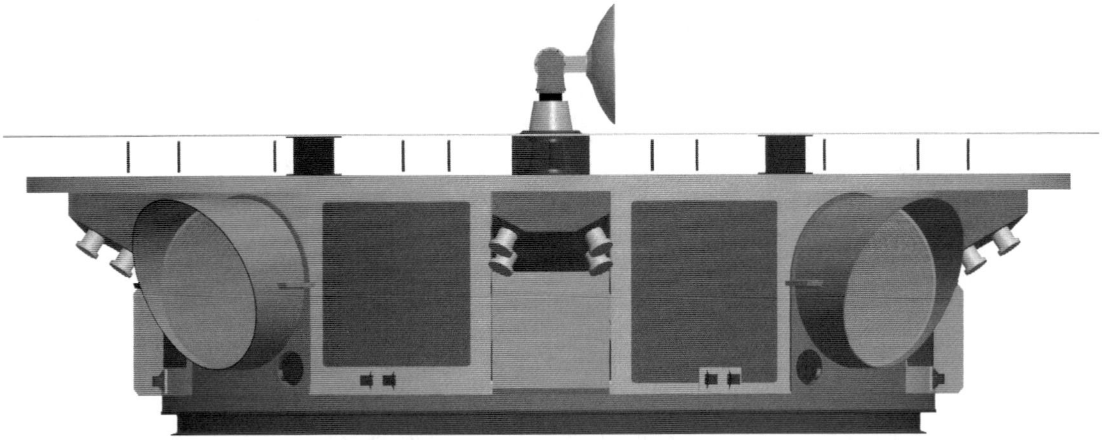

Figure 2. Side view of the LISA Spacecraft.

3. Thermal Design

The LISA thermal design benefits greatly from the stable thermal environment provided by the operational orbits, as the S/C operate far from the earth where eclipse, albedo, and planetary heating effects are negligible. The orientations of the S/C are held at a constant 30° angle between the S/C normal and the solar vector, providing for a stable solar heat input. All electrical components are power stabilized to provide constant power dissipations, with switching operations minimized, particularly within the LISA measurement bandwidth. The shape and design of the S/C also provides for further thermal stability [3].

The LISA S/C thermal design has two primary functions: protect the payload and mission critical components from the harsh thermal environment of space throughout the mission, and maintain thermal stability within the LISA measurement bandwidth during the science phase of the mission. Survival of the instrumentation during the cruise phase is achieved through the use of heaters and multi-layer-insulation on the S/C and propulsion module. Active thermal control is applied using thermostats and thermistors located inside the electronics boxes during the cruise phase. During the science phase, the survival heaters are turned off, being supplemented by electronics box waste heat. Current thermal analysis indicates that the required thermal stability during the science phase can be achieved without the use of active temperature control through the use of successive thermal shields [4].

The first layer of thermal isolation is provided by the solar array deck. The deck will be made of aluminum or composite honeycomb. The sun facing facesheet of the deck contains either gold or vapor deposited aluminum coated optical surface reflectors and solar cells. The solar array deck is isolated from the bus structure using low conductivity flexures. The top surface of the top deck panel is gold coated to minimize absorption of any radiation from the bottom surface of the solar array deck. A high emissivity coating around the bottom outer edge of the solar array will facilitate heat rejection to space. This isolation results in less than 1% of incident solar energy being transferred to the structure.

A second layer of isolation is provided by the payload shield, which is gold coated on the inside and outside surfaces in order to minimize the absorption of any radiated energy (e.g. electronic box radiated heat). Low conductivity standoffs are also placed between the payload shield and the bottom plate to reduce the conduction path.

The last layer of payload thermal isolation is provided by an internal shield, which is also gold coated on both sides and includes low conductivity mounts. This helps to shield the optical bench from any disturbances or gradients in the payload shield.

Finally, the cylindrical exterior wall and bottom deck of the S/C function as radiators to reject electronics waste heat. The area and coatings of these surfaces are optimized to keep the temperature inside the S/C within the operating range of the electronics without the need for additional heaters.

4. Self-Gravity

The gravitational field at the proof masses must be kept as small, flat, and as stable as possible. The static gravitational field at the proof masses must be kept below 5×10^{-10} m/s^2 along the measurement axes. In meeting this requirement, the amount of static force that must be compensated for by the gravitational reference sensor electrodes will be minimized. This is important for two reasons: first, the force fluctuations generated by the applied compensating electric field are proportional to the total force applied by that field. Fields in both the measurement axes and the other degrees of freedom are important as some of the force fluctuations from other directions will leak into the measurement axis through cross-couplings. Second, the compensating electric field creates a virtual spring between the S/C and the proof mass. The residual motion of the S/C will couple through this stiffness causing an acceleration disturbance to the proof mass [2].

The gradient of the gravitational field at the proof mass locations must be kept below 3×10^{-8} s^{-2}. The gravity gradient creates a virtual spring between the S/C and the proof mass. The residual motion of the S/C will couple through this stiffness causing an acceleration disturbance to the proof mass. Finally, fluctuating distortions of the S/C will change the gravitational field. These distortions must be minimized such that their acceleration disturbance to the proof masses is kept below 5×10^{-16} m/s^2/$\sqrt{\text{Hz}}$.

These self-gravity requirements are met through careful mass balancing of the spacecraft components and the minimization of moving parts. The use of items such as deployed solar arrays are ruled out due to both their uncertain positioning and the gravitational field fluctuations due to thermal distortions. The S/C structure is also designed to minimize thermal distortions and excitable vibrational modes.

The self-gravity requirements flow down to set requirements on the knowledge of the mass properties and placement of all hardware in the LISA S/C. The S/C can be divided into zones where all items within a zone have the same knowledge requirements. This does not set the accuracy needed in producing each part; rather it sets the accuracy needed in measuring and identifying the part after it is manufactured. In other words, it defines how well the part must be weighed, measured, and placed within the S/C. A detailed discussion of self-gravity analysis and balancing can be found in references [5-7].

5. Magnetic Environment

The magnetic properties of the proof mass and characteristics of the magnetic field contribute to several effects throughout the DRS error budget. The leading effects are: the interaction between the fluctuating interplanetary magnetic field with the S/C magnetic gradient, fluctuations from the magnetic gradient induced current dissipations (Eddy current damping), and a fluctuating S/C magnetic gradient [2]. The current budget value for the magnetic gradient at the proof mass is 5×10^{-6} T/m. Magnetic gradient fluctuations must be kept below 2.5×10^{-8} T/m/$\sqrt{\text{Hz}}$ at 0.1 mHz at the proof mass locations.

The magnetic field inside the spacecraft is driven by the magnetically hot components. Magnetic parts are generally avoided in the LISA design, however it is not cost effective to completely eliminate them all. All magnetic items are tracked so that a magnetic budget can be maintained. The budget magnetic field should be easily met by placing all the magnetic components far from the proof masses and including a modest amount of magnetic shielding and compensation.

A number of design rules are available to minimize magnetic fields. The obvious solution is to minimize the use of magnetic material. Unavoidable permanent magnets can be compensated for with oppositely oriented duplicates such as a cold spare. The remaining permanent magnets can be shielded using high permeability foil magnetic materials such as METGLAS or VITROVAC. Twisted pair is used for all wiring. Current loops in the power system is eliminated by using a star or single point grounding. No primary supply currents are allowed to flow through the S/C structure.

The solar array can be a very significant source of stray magnetic fields due to the large currents, however, its linear geometry makes it straightforward to cancel out by correct placement of forward and return interconnections using a technique called "backwiring." In backwiring the return wire from each string of solar cell modules is returned directly underneath the modules in that particular string and carefully routed along a line just behind the centerline of the modules. Each string and module of the string is self-canceling and does not depend on the magnetic field of an adjacent module or string for cancellation. If a module fails in flight the current in both the string and the return drop to zero simultaneously leaving no uncompensated currents in the array.

6. Attitude and Drag Free Control

The LISA attitude and drag-free control system is responsible for ensuring that the residual acceleration of the two proof masses fall below the LISA sensitivity requirements by providing tight pointing and translational control of the LISA spacecraft and its proof masses. Stringent requirements are placed on the rotational (8 nrad/√Hz) and translational dynamics ($1.5\times(1+(f/8\text{mHz})^4)^{1/2}$ nm/√Hz) of each spacecraft to ensure that the proper sensitivity for science measurements can be achieved. This is implemented by performing a number of control functions [8,9].

Along the two sensitive axes (telescopes' lines of sight), the S/C is controlled around the proof-masses such that all residual accelerations in the sensitive axes are minimized. In addition, the S/C follows the average out-of-plane motion of the proof masses in order to compensate for the solar radiation pressure. The control error signals are obtained from the optical and electrostatic readouts of the proof masses. The very sensitive optical proof mass metrology is used on the two sensitive axes in order to reduce the sensor noise in the generated force signals to the micronewton propulsion system. The drag-free control covers all 3 translational degrees of freedom of the S/C.

During science mode, the S/C attitude control is performed by feeding back the information from inertial wavefront sensing of the incoming laser light to the micronewton propulsion system. Since two telescopes are on-board and mounted at an angle of nominally 60 deg with respect to each other, inertial wavefront sensing provides a total of 4 tip and tilt error angles. By applying the corresponding geometric relations, these angles can be used to determine the complete S/C attitude error to align the telescopes with respect to the incoming laser beams as well as the angular error between the two telescopes.

Due to orbital mechanics, the angle between the two interferometer arms is constantly changing, therefore, this angle must be controlled as well. The information from the inertial wavefront sensing is used again to determine the angular error between the two telescopes. The measured angular error is used to generate a feedback signal for the optical assembly actuators. Note that one of the two optical assemblies will remain in a fixed position while the second one will be constantly actuated at 10 Hz (either assembly can be actuated, providing redundancy).

The constellation of three S/C rotates about the constellation center on a once per year period. Consequently, the three S/C are always moving relative to each other. Because of this motion and the 5 million km separation between S/C, the outgoing beam must be pointed ahead of the incoming beam in order to be directed towards where the distant S/C will be when the light arrives. This is accomplished by a point-ahead mirror that can be actuated. The mirror is actuated out-of-plane only, since that component of the point-ahead angle exhibits large variations throughout the year. The in-plane component shows a small variation about a fixed bias, which may be accommodated by a pre-fixed tilt of the mirror. The mirror pointing can be performed by a simple lookup table as the angles are slowly varying and are sufficiently predictable.

LISA requires micronewton thrusters to provide the fine spacecraft attitude and position control for drag free flight and beam pointing to the distant spacecrafts. The thrusters are operated continuously during science operations with their thrust levels set by the disturbance reduction system control loops. Three different thruster technologies are nearing flight readiness for LISA Pathfinder and have demonstrated the LISA thrust and thrust noise requirements: the colloid micronewton thruster (CMNT) made by Busek Co. in Boston [10], the indium needle field emission electric propulsion (In-FEEP) thruster made by ARC Seibersdorf in Austria [11], and cesium slit FEEP (Cs-FEEP) made by ALTA S.p.A. in Italy [12].

At least six micronewton thrusters on each spacecraft must be operating continually during science operations for the entire LISA mission. Enough consumables are carried for the entire extended mission. With sunlight photon pressure as the largest external force acting on the spacecraft, the micronewton thrusters must produce on the order of 10 μN of thrust with better than a 0.1 μN resolution during science measurements. Furthermore, over the LISA science measurement bandwidth, thrust and thrust noise must be stable and within the error limitations of the DRS over the entire mission, < 0.1 $\mu N/\sqrt{Hz}$ (open loop) at the high end of the measurement band. Brief periods of higher thrust, >30 μN, may be required during tip-off recovery, constellation acquisition, and safe-mode operations.

7. Summary

The LISA S/C and scientific payload are tightly coupled into one seamless design in order to optimize the sensitivity to gravitational waves. While some of the LISA requirements are unique among spacecraft design (such as self gravity), the design principles and methodologies are familiar. Modeling and analysis of the current LISA S/C design indicates that these requirements can be met through standard engineering practices without the need for new technologies. For example, the self-gravity requirements can be met through careful tracking of the mass of all components. The level of precision may be more than a traditional space mission, but not beyond standard metrological techniques. With careful design and strong systems engineering, the LISA S/C will be ready to support the exciting astrophysical measurements from the first space based gravitational wave observations.

References

[1] R. T. Stebbins *et al.*, Class. Quant. Grav. **21**, S653 (2004).
[2] S. M. Merkowitz, in Proceedings of the 6th International LISA Symposium, edited by S. M. Merkowitz and J. C. Livas (American Institute of Physics Conference Proceedings, New York, 2006).
[3] H. Peabody and S. Merkowitz, Class. Quant. Grav. **22**, S403 (2005).
[4] H. Peabody and S. M. Merkowitz, in Proceedings of the 6th International LISA Symposium, edited by S. M. Merkowitz and J. C. Livas (American Institute of Physics Conference Proceedings, New York, 2006).
[5] A. M. Gopstein *et al.*, in Proceedings of the 6th International LISA Symposium, edited by S. M. Merkowitz and J. C. Livas (American Institute of Physics Conference Proceedings, New York, 2006).
[6] S. M. Merkowitz et al., Class. Quant. Grav. 22, S395 (2005).
[7] M. Armano et al., Class. Quant. Grav. 22, S501 (2005).
[8] P. Maghami *et al.*, in Proceedings of the 18th International Symposium on Space Flight Mechanics, (2004).
[9] P. G. Maghami and T. Hyde, Class. Quant. Grav. 20, S273 (2003).
[10] J. K. Ziemer *et al.*, in Proceedings of the 6th International LISA Symposium, edited by S. M. Merkowitz and J. C. Livas (American Institute of Physics Conference Proceedings, New York, 2006).
[11] M. Tajmar *et al.*, in 42nd AIAA/ASME/SAE/ASEE Joint Propulsion Conference and Exhibit, (2006). AIAA-2006-4826
[12] L. Biagioni *et al.*, in 41st AIAA/ASME/SAE/ASEE Joint Propulsion Conference and Exhibit, (2005). AIAA-2005-4261

Journal of Physics: Conference Series **154** (2009) 012022 doi:10.1088/1742-6596/154/1/012022

Measurement of the non-reciprocal phase noise of a polarization maintaining single-mode optical fiber

Roland Fleddermann[1], Frank Steier[1], Michael Tröbs[1], Johanna Bogenstahl[2], Christian Killow[2], Gerhard Heinzel[1] and Karsten Danzmann[1]

1 Albert-Einstein-Institute Hanover, Max Planck Institute for Gravitational Physics and Leibniz University Hanover, Callinstr. 38, D-30167 Hannover, Germany

2 Department of Physics and Astronomy, University of Glasgow, Glasgow G12 8QQ, UK

E-mail: `Roland.Fleddermann@aei.mpg.de`

Abstract. Polarization maintaining single-mode optical fibers are key components in the interferometry of the Laser Interferometer Space Antenna (LISA). LISA's measurement principle relies on the availability of space qualified fibers of this type which influence the phase of light with a wavelength of 1064 nm passing in opposite directions through them with differences smaller than $6\,\mu\mathrm{rad}/\sqrt{\mathrm{Hz}}$. We present a measurement scheme suitable to sense these non-reciprocal phase changes, as well as results obtained using this setup on samples of commercially available fibers. The experimental setup for the fiber characterization consists of a quasi-monolithic interferometer which constitutes a representative cut-out of the local interferometry on-board LISA concerning the fiber. Several noise sources are identified and improvements to the setup are presented to overcome them. The noise level achieved using this setup is between approximately $40\,\mu\mathrm{rad}/\sqrt{\mathrm{Hz}}$ and $400\,\mu\mathrm{rad}/\sqrt{\mathrm{Hz}}$ in the frequency range between 1 mHz and 1 Hz. It is also verified that this noise level is limited by the setup and not introduced by the fiber.

1. Introduction

The Laser Interferometer Space Antenna (LISA) [1–3] requires optical fibers as key components for the mission. These fibers are foreseen in the current LISA baseline design, since a flexible optical connection within the interferometric path is desired. The angle between the satellites deviates periodically during one year from its nominal value of 60° by up to 1.5° while the constellation moves around the Sun [4, 5]. Therefore, the direction of the light beams sent out from the spacecraft has to be changed. This is to be accomplished by changing the angle between two individual optical benches inside each satellite. Fibers are the most promising candidates for the task of connecting the two moving optical benches.

Currently there are several experiments going on in research institutes all over the world to measure the reciprocity of optical fibers, among them the University of Florida, University of Glasgow and the Jet Propulsion Laboratory.

In the application on-board LISA non-reciprocal phase changes in the fiber cannot be distinguished from distance changes between the test masses whose position is to be measured with picometer sensitivity. Hence we measure the non-reciprocal phase noise of such fibers in order to verify that this noise source does not spoil the measurement performance. Reciprocal noise is expected as well, but this does not have to be taken into account, because it can

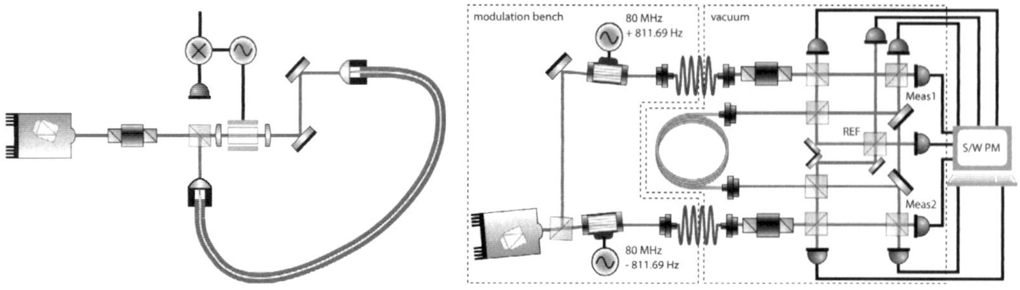

(a) Sagnac setup (b) LISA-like setup

Figure 1: Two different optical setups used for the measurement of the non-reciprocal phase noise of optical fibers.

be subtracted by a technique called Time Delay Interferometry (TDI) [6, 7] during data post processing.

Non-reciprocal phase noise must be smaller than the required phase sensitivity of LISA which is $6\,\mu\text{rad}/\sqrt{\text{Hz}}$. To accomplish the task of measuring this phase noise a measurement setup with an inherent non-reciprocity of about one order of magnitude below this level is required.

2. Experimental setup

Two alternative optical setups have been evaluated for the measurement of the non-reciprocal phase noise.

At first it was believed that the easiest way is to measure the non-reciprocal phase noise using a Sagnac interferometer. A Sagnac interferometer is to first order exclusively sensitive to non-reciprocal phase changes [8, 9]. Therefore, it seemed to be an ideal tool for the characterization of the fiber.

This setup has the advantage of being relatively simple. Furthermore, due to the reciprocity of the setup, coupling of path length fluctuations and laser frequency noise to the phase are largely eliminated. The main disadvantage of the setup is that a heterodyne phase readout is not possible and hence additional components are needed to improve the sensitivity.

The alternative is a setup resembling the situation on-board a LISA spacecraft. It has the advantage that a heterodyne measurement can be performed. An even bigger advantage is that the interferometer layout is a representative copy of the fiber interferometry on-board LISA which offers the opportunity to characterize not only the fiber, but also other important aspects that are due to the interferometer design. Disadvantages are that the setup is more complex and that path length and laser frequency changes couple into the non-reciprocal phase observed.

First experiments were conducted using the Sagnac setup shown in Figure 1a. Here, the light of a Nd:YAG laser (Innolight Mephisto 500) is split by a beamsplitter. Both beams are then sent through a fiber such that after passing through the fiber they interfere at the same beamsplitter where the beam was split. An electro-optical modulator (EOM) is introduced close to one end of the fiber to allow for a non-reciprocal modulation which results in a better sensitivity to small non-reciprocal phase changes.

Unfortunately it turned out that the sensitivity of this setup was limited to about $5\,\text{mrad}/\sqrt{\text{Hz}}$ to $10\,\text{mrad}/\sqrt{\text{Hz}}$ in the frequency range between 1 mHz and 100 mHz and that this noise level could not be improved although a lot of noise hunting was done. Therefore, the Sagnac setup is not detailed here further and the following discussions are focused on the LISA-like setup.

Figure 1b shows the basic LISA-like setup. In this setup, the light of one Nd:YAG laser is split into two beams by a power beamsplitter. Both fractions are then sent through acousto-optical modulators (AOMs) working at slightly different frequencies, namely 80 MHz + 811.69 Hz and 80 MHz - 811.69 Hz, resulting in a frequency difference between the two output beams of 1.623 kHz. These beams are launched into the optical bench containing the interferometers, which are thus used in a heterodyne scheme. This part of the setup is called modulation bench.

The optical bench itself consists of three individual interferometers which are used in combination to measure the non-reciprocal phase changes in the fiber under test.

In the reference interferometer, labeled REF in Figure 1b, both beams are directly routed to the interference beamsplitter. The measured phase of this interferometer is used for subtraction of the pathlength changes on the modulation bench which are common-mode in all interferometers. In the first measurement interferometer, Meas1, the first beam is traveling directly to the interference beamsplitter, while the second beam passes through the fiber on its way to this point. The situation in the second measurement interferometer, Meas2, is exactly opposite. Here, the second beam travels directly to the beamsplitter, while the first beam passes through the fiber.

Thus the reciprocal phase changes are measured in the two measurement interferometers, Meas1 and Meas2, in opposite directions. The non-reciprocity can now be found by comparison of the changes in the phases measured in both interferometers. Photodiodes at the corresponding beamsplitters sense the beat signal at the heterodyne frequency of 1.623 kHz. Its phase is measured with a software phasemeter implemented on a PC like the one described in [10, 11].

3. Noise hunting

Initial measurements using the LISA-like setup described in the previous section were performed using an interferometer based on an aluminum baseplate. These measurements showed a relatively high non-reciprocal noise level. This can be seen in the corresponding trace in Figure 2. The non-reciprocal noise has a magnitude of about $100\,\mathrm{mrad}/\sqrt{\mathrm{Hz}}$, decreasing with 1/f to higher frequencies.

In the following paragraphs the noise hunting that was performed on the setup is described stepwise. The topic under investigation is emphasized at the beginning of each paragraph.

Quasi-monolithic setup Investigations soon suggested that the non-reciprocity was introduced by length changes in the part of the setup where the two beams sensing the fiber do not travel through the same path. Non-homogenous thermal expansion of the baseplate is most likely to be the limiting noise source.

To overcome these limitations a revised version of the interferometer has been devised which is optimized for the application in a Zerodur® baseplate interferometer. This material has been chosen because of its very low coefficient of thermal expansion of approximately 10^{-7}/K. Small modifications that were necessary for the construction of a quasi-monolithic interferometer have been added to the setup in comparison to the one based on aluminum.

The interferometer was built using the hydroxide-catalysis bonding technique [12–14]. Due to the challenging alignment of the interferometer, some components (BS6, BS7 and M||) were added to the initial setup from Figure 1b for simplification of construction.

The most important change compared to the initial setup is an additional null-measurement path via BS5 and BS6 that is confined to the Zerodur® baseplate and thus allows for characterization of the instrument sensitivity without the fiber.

The results of the first measurement using this setup are represented by the corresponding trace in Figure 2. It is evident that only a very small improvement with respect to the aluminum setup is achieved. Therefore, further investigations have been undertaken to identify the limiting noise source.

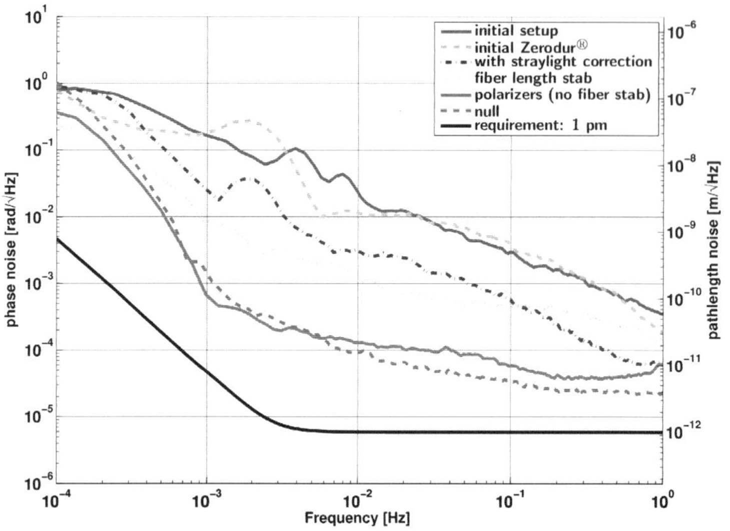

Figure 2: Non-reciprocal path length noise spectral densities observed using different improved versions of the LISA-like setup for the fiber characterization.

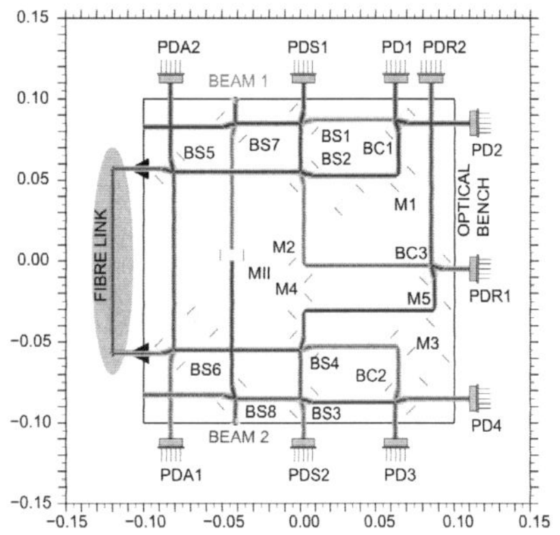

(a) Schematic created using OptoCad [15]

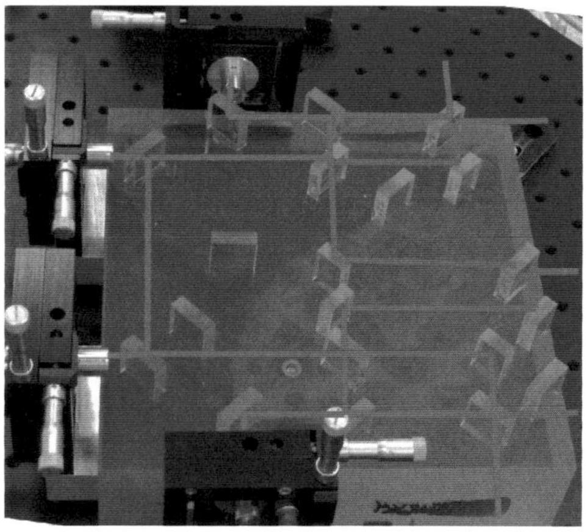

(b) Actual setup

Figure 3: LISA-like setup optimized for Zerodur® baseplate

Stray light It was apparent, that some light was reflected by either the fiber under test or the fiber coupler assemblies. This back-reflected light cannot be blocked because it travels along the same beam axis or is even in the same spatial mode as the counter propagating beam when it originates from a reflection occurring inside the fiber. The reflected light led to a spurious interference signal at the two measurement interferometers. This is exactly what happens at the fiber link on-board LISA: The beam coming from the fiber interferes with the beam entering it at the point where the entering beam is back-reflected.

Therefore, a stray light correction scheme was applied to remove this influence from the measured phase. The principle of this correction is based on phase measurement at both outputs of the interference beamsplitter [16]. It relies on the fact that the phase difference of the interference signals is 180° between both beamsplitter outputs while scattered light which has interfered in another place of the setup will have zero phase difference in both outputs.

This scheme was implemented in data post processing. It led to a significant reduction of the observed non-reciprocal phase noise as can be seen by comparing the trace in Figure 2 which represents the phase noise observed with scattered light correction to the trace which is not corrected for scattered light. The correction leads to a reduction of non-reciprocal noise by about one order of magnitude. This shows, that a significant amount of the noise observed at that point could be attributed to back-reflected light. Therefore, it was decided to use the stray light correction for all further measurements.

This is very valuable information for the design of the LISA optical benches. The fiber link part of the current baseline design of the LISA optical benches closely resembles the setup that was used here, so it might be necessary to implement a similar correction for LISA.

Fiber length stabilization In a next step a fiber length stabilization was implemented. This seemed necessary as observations showed that the phase observed at both ends of the fiber changed by up to hundred radians during one night. It was feared that this high reciprocal phase noise would couple to the observed non-reciprocal phase noise of about 100 µrad due to a limited "common-mode rejection" of the fiber.

The fiber length stabilization was implemented by gluing the fiber to a ring piezo [17]. The piezo was driven by a feedback control loop that used the difference between the phase at one end of the fiber and the phase of the reference interferometer as error signal.

The results of this measurement are represented by the respective trace in Figure 2. A comparison to the previous trace without fiber length stabilization shows that the fiber length stabilization reduced the non-reciprocal noise at frequencies below 100 mHz, while for frequencies above, the noise is slightly amplified. This led to the assumption that the method of stabilization had an effect on polarization maintaining properties of the fiber. This would then lead to a change in the output polarization of the fiber which in turn would appear to be non-reciprocal phase changes due to changing content of light in the interference signal.

Polarization It was therefore decided to add polarizers in front of both fiber ends. If the input light polarization axis does not match the fiber axis, light propagates through the fiber in both perpendicular polarization states. As the fiber is made deliberately birefringent in order to be polarization maintaining, the light in the two polarization states will generally experience different phase shifts. A change in the fiber axis or in the refractive index of one axis through external influences like e.g. pressure or temperature will therefore lead to phase changes at the output of the fiber which are not necessarily reciprocal.

The results of this measurement are found in the corresponding trace in Figure 2. It is obvious that the introduction of the polarizers led to a significant reduction in observed non-reciprocal phase noise. The noise curve is about one order of magnitude below the one recorded without the polarizers. It is worth noting that this result was achieved even without stabilizing the length of the fiber.

This noise observed using polarizers in the setup is also very close to the noise that was observed when the fiber was removed and a null-measurement was performed. This measurement was done to monitor the inherent non-reciprocity of the measurement setup. From the fact that the noise curve without the fiber is identical to the noise curve with the fiber in place within the limits of variations between repeated measurements, it can be deduced that no significant additional noise is introduced by the fiber but that instead the currently observed noise level is limited by the setup. This means that to overcome this limit a redesign of the setup might be necessary. Investigation of the critical parameters is currently underway.

4. Conclusions and outlook

We presented a possible setup for the investigation of non-reciprocal fiber phase noise as well as first results obtained using this setup. The results do not yet meet the requirements but they are very close to them in the frequency range between 100 mHz and 1 Hz.It has also been shown that this phase noise is inherent to the setup and that no extra non-reciprocal phase noise is introduced by the fibers under investigation.

To achieve the current sensitivity, a technique has been successfully employed to subtract the phase noise that is caused by reflections. This led to a significant reduction of the observed non-reciprocal phase noise. Another significant improvement of the sensitivity was achieved by fixing the polarization state and aligning it to the fibers. Due to the similarity of the setup that was used here to the LISA setup this is valuable information for the LISA optical bench design.

Further investigations are required to identify the limiting noise source of the current setup. Several external influences will have to be studied in more detail. Depending on the results a redesign of the interferometer setup will be necessary.

This redesigned interferometer might also offer the opportunity to test more realistic space qualified fiber launchers with respect to their influence on the non-reciprocal phase noise.

Acknowledgments

We gratefully acknowledge support by Deutsches Zentrum für Luft- und Raumfahrt (DLR) (reference 50 OQ 0601).

References

[1] Hammesfahr A 2001 *Class. Quantum Grav.* **18** 4045–4051

[2] Bender P *et al.* 1998 LISA. Laser Interferometer Space Antenna for the detection and observation of gravitational waves. An international project in the field of Fundamental Physics in Space. Pre-Phase A report Tech. rep. Max-Planck-Institut für Quantenoptik Garching, Germany

[3] Danzmann K and Rüdiger A 2003 *Class. Quantum Grav.* **20** 1–9

[4] Hughes S P 2005 *Advances in Astronautical Sciences* vol 111 ed Culp R D and Jolly S D p 61

[5] Guangyu L, Zhaohua Y, Heinzel G, Rüdiger A, Jennrich O, Li W, Xia Y, Fei Z and Haibin Z 2008 *International Journal of Modern Physics D* **17** 1021–1042

[6] Tinto M, Shaddock D A, Sylvestre J and Armstrong J W 2003 *Physical Review D* **67** 122003

[7] Shaddock D A, Ware B, Spero R E and Vallisneri M 2004 *Phys. Rev. D* **70** 081101

[8] Arditty H J and Lefèvre H 1981 *Optics Letters* **6** 401–403

[9] Arditty H J and Lefèvre H 1982 *Fiber-Optic Rotation Sensors and related Technologies* ed Ezekiel S and Arditty H (Springer Verlag New York) pp 44–51

[10] Heinzel G *et al.* 2004 *Classical and Quantum Gravity* **21** 581–587 URL `http://stacks.iop.org/0264-9381/21/S581`

[11] Wand V, Guzmán F, Heinzel G and Danzmann K 2006 *AIP Conf. Proc.* **873**(1) 689–696 URL `http://link.aip.org/link/?APCPCS/873/689/1`

[12] Gwo D H 1998 *Cryogenic Optical Systems and Instruments VIII* vol 3435 ed Heaney J B and Burriesci L G (SPIE) pp 136–142 URL `http://link.aip.org/link/?PSI/3435/136/1`

[13] Elliffe E J *et al.* 2005 *Classical and Quantum Gravity* **22** S257–S267 URL `http://stacks.iop.org/0264-9381/22/S257`

[14] Reid S, Cagnoli G, Elliffe E, Faller J, Hough J, Martin I and Rowan S 2007 *Physics Letters A* **363**(5-6) 341–345

[15] Schilling R 2002 Optocad internal note

[16] Shaddock D 2007 Scattered light phase error cancellation by balanced detection private communication

[17] García Marín A F 2007 *Minimisation of optical pathlength noise for the detection of gravitational waves with the spaceborne laser interferometer LISA and LISA Pathfinder* Dissertation Gottfried Wilhelm Leibniz Universität Hannover

Laser interferometer for spaceborne mapping of the Earth's gravity field

Marina Dehne, Felipe Guzmán Cervantes, Benjamin Sheard, Gerhard Heinzel and Karsten Danzmann

Albert-Einstein-Institute, Max Planck Institute for Gravitational Physics and Leibniz University, Callinstraße 38, D-30167 Hannover, Germany

E-mail: marina.dehne@aei.mpg.de

Abstract. The Gravity Recovery and Climate Experiment (GRACE) is one of the present missions to map the Earth's gravity field. The aim of a GRACE follow-on mission is to map the gravitational field of the Earth with higher resolution over at least 6 years. This should lead to a deeper insight into geophysical processes of the Earth's system. One suggested detector for this purpose consists of two identical spacecraft carrying drag-free test masses in a low Earth orbit at an altitude of the order of 300 km, following each other with a distance on the order of 50 to 100 km. Changes in the Earth's gravity field will induce distance fluctuations between two test masses on separate spacecraft. These variations in the frequency range 1 to 100 mHz are to be monitored by a laser interferometer with nanometer precision. We present preliminary results of a heterodyne interferometer configuration using polarising optics, demonstrating the required phase sensitivity.

1. Introduction

Various geophysical processes generate gravity anomalies with extensive spatial variations over the surface of the Earth. The resulting gravity field is known as the long-term average (or mean) gravity field. Measurement of these gravity anomalies provides, for example, a better understanding of the structure of the solid Earth. Shorter-term mass fluctuations like the variation in water content of the Earth's crust are known as the time-variable gravity field. It helps, among other things, to study the global sea level changes or the polar ice sheet balance. These changes have a significant impact on relevant climatic issues.

The Gravity Recovery and Climate Experiment (GRACE) was successfully launched in 2002 to monitor these changes [1]. GRACE consists of two identical satellites, one 220 km ahead of the other in the same orbit at an altitude of approximately 500 km. Temporal and spatial changes in the Earth's gravity field cause small variations in the inter-spacecraft separation, which is measured in order to determine the Earth's gravity field. The relative distance and velocity changes are measured with a microwave interferometer in the K-Band with an accuracy of 1 µm. The information gained about the global gravity field from GRACE raised an interest in developing a follow-on mission with better performance to improve our knowledge of e.g. the cryosphere and hydrological or atmospheric phenomena affecting the Earth's system. To achieve this, the metrology system for the distance measurement between the satellites needs to be improved, for example, by using laser satellite-to-satellite interferometry (SSI) [2–5] and it is beneficial to reduce the inter-satellite distance to improve spatial resolution [6]. Furthermore,

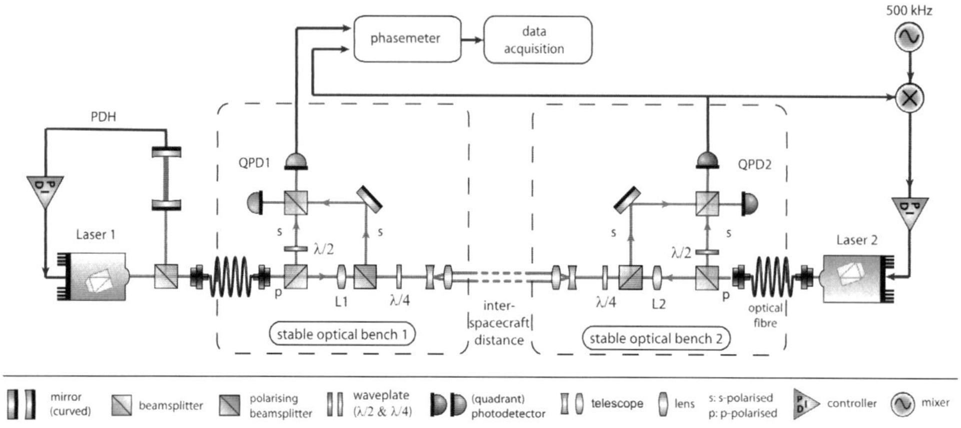

Figure 1: Baseline design for the SSI: a symmetric system of two heterodyne Mach-Zehnder interferometers with lasers and detectors on both spacecraft.

it is also desirable to choose a lower altitude since the gravity field diminishes with increasing height. The disadvantage of a lower orbit is the significant atmospheric drag which must then be compensated. Therefore, a suitable drag-free control system needs to be developed. In order to provide a constant thermal environment and to avoid sunlight radiation coupling onto the optical axis between the two satellites, a sun synchronous near-circular orbit would be suitable.

One possible GRACE follow-on mission can benefit from the current developments for the joint ESA-NASA space-based gravitational-wave detector "Laser Interferometer Space Antenna" (LISA) [7] and its precursor mission LISA Pathfinder (LPF) [8, 9], such as a precise drag-free technology and the interferometric readout. Interferometric length measurements in the frequency band of interest from 1 to 100 mHz, that meet the required phase sensitivity, have been demonstrated with the LPF configuration [10]. Our goal is to show that the following proposed measurement system containing polarising optics can be implemented with the proposed pathlength sensitivity of a possible GRACE follow-on mission.

2. Measurement system

The relative velocity between the spacecraft induces a Doppler shift in the interferometer beatnote. This effect together with the large variations in the inter-spacecraft separation make heterodyne interferometry ideally suitable as the pathlength readout scheme. A symmetric active system with lasers and detectors on both spacecraft would be appropriate, since this enables to maintain the heterodyne frequency away from DC. We assume from an early strawman design that the required interferometer sensitivity is $2.5\,\mathrm{nm}/\sqrt{\mathrm{Hz}}$ between 10 mHz to 100 mHz, increasing with $1/f$ between 10 and 1 mHz. The round-trip Doppler shift introduces variations in the interferometer beatnote up to a few 100 kHz. A suitable heterodyne frequency is, therefore, between several hundred kHz and a few MHz, with the lower limit given by the maximal Doppler shift and the required control bandwidth of the offset phase lock. The upper limit is given by technical considerations concerning the photodiodes and the phasemeter. The preliminary baseline is a heterodyne frequency on the order of 500 kHz.

2.1. Polarising optics in heterodyne interferometer

For the heterodyne interferometer, we have selected a polarising layout as preliminary baseline (Fig. 1). We decided to separate the incoming and the outgoing beam on each optical bench with

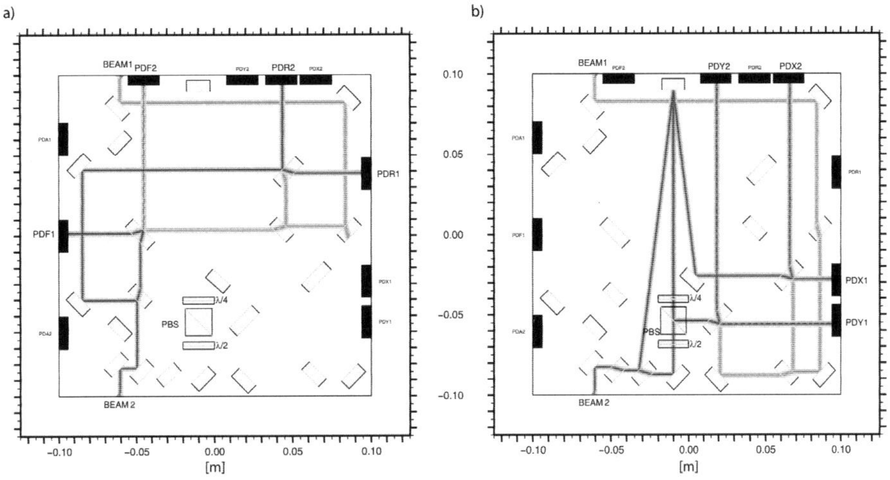

Figure 2: OptoCad models of the laser interferometer for a comparison between polarising and non-polarising optics: a) reference and frequency noise interferometer (unequal armlenghts) b) non-polarising and polarising interferometer.

polarising optics for the following reasons: A non-polarising setup simplifies the optical path but results in a significant loss of incoming light. This light must be removed with a Faraday isolator in order to avoid laser instabilities. By contrast, with a polarising layout one has the freedom to select the optimal local oscillator level and to utilise all the light power. On the other hand, thermal sensitivity of polarising components being used in transmission, are able to modify the optical pathlength, the exinction ratio and the polarising plane. For these reasons, the influence of polarising optics on the interferometer sensitivity is currently under investigation.

In the proposed layout shown in Fig. 1 the outgoing p-polarised beam is transmitted through the polarising beamsplitter (PBS). Therefore, the incoming light must be s-polarised to be reflected by the PBS. This could be realised by adding two quarter wave plates to the optical path. The two beams in each Mach-Zehnder interferometer are finally recombined at a beamsplitter. Apart from the interferometers, two telescopes are included for both transmitting and receiving the laser beams between the two spacecraft. The lens (L1 or L2 as illustrated in Fig. 1) in the outgoing beam path can be used to adjust the divergence of the outgoing beam because it does not affect the beam of the receive path.

Alignment control signals can be obtained by the technique of differential wavefront sensing (DWS) [11]. For this purpose, we intend to use four-quadrant photodiodes (QPD) since longitudinal information can be ascertained over the sum of the quadrants. The obtained DWS signals contain the information about the angle between the two interfering wavefronts.

Before implementing the setup illustrated in Fig. 1, an optical bench containing four heterodyne Mach-Zehnder interferometers has been designed for a trade-off study between polarising and non-polarising optics in interferometry. The four interferometers with equal number of transmissions and reflections presented in Fig. 2 can be described as follows:

(i) *Reference interferometer*: gives a reference phase containing all environmental noise contributions such as mechanical and thermal fluctuations, which occur outside the stable optical bench and are common to all four interferometers (Fig. 2 a).

(ii) *Frequency noise interferometer* with intentionally unequal armlengths: senses laser frequency noise. Its output signal can be used for active frequency stabilisation of the laser (Fig. 2 a).

(iii) *Non-polarising interferometer*: to be used for a relative length measurement on the bench (Fig. 2 b).

(iv) *Polarising interferometer*: to be applied for the same purpose as (iii). The beam path of these two interferometers (Fig. 2b) has been designed to cover a similar area on the bench, in order to reduce effects of thermal gradients on the length measurement.

2.2. Frequency-stabilised laser system

We expect a strong coupling of laser frequency noise $\widetilde{\delta\nu}$ into the interferometric readout of the phase due to the large inter-spacecraft separation L of about 50 km. To keep the optical pathlength noise $\widetilde{\delta s}$ under the allocated budget of $1\,\mathrm{nm}/\sqrt{\mathrm{Hz}}$, the laser has to have a laser frequency stability of

$$\widetilde{\delta\nu} \approx 3\frac{\mathrm{Hz}}{\sqrt{\mathrm{Hz}}} \cdot \left[\frac{\widetilde{\delta s}}{1\,\mathrm{nm}/\sqrt{\mathrm{Hz}}}\right] \cdot \left[\frac{2L}{100\,\mathrm{km}}\right]^{-1}, \tag{1}$$

which follows from $\frac{\widetilde{\delta\nu}}{\nu} = \frac{\widetilde{\delta s}}{\Delta L}$. Laser frequency stabilities near this level have already been demonstrated using Fabry-Perot cavities [12] or transition in molecular iodine [13]. Because of the frequency noise coupling via the large pathlength difference, both laser frequencies must be stabilised. In our proposed configuration the first laser will be locked to a reference cavity using the Pound-Drever-Hall (PDH) scheme [14] while the second laser will be phase-locked to the incoming light from the remote spacecraft with an offset on the order of the desired nominal heterodyne frequency [15]. A number of experiments have already demonstrated the two techniques described here with the performance required for the LISA mission, which is also sufficient for the SSI.

2.3. Phase readout scheme

The proposed readout scheme measures the relative pathlength changes δL between the satellites using heterodyne interferometry. These variations translate into phase fluctuations $\widetilde{\delta\phi}$ of a sinusoidal beatnote at the heterodyne frequency. Pathlength changes of one wavelength measured by photodetectors correspond to a phase variation of one cycle or 2π rad. The purpose of the phase measurement system (PMS) is to faithfully extract the phase of the photocurrent generated in the photodetectors without limiting the sensitivity of the length measurement by adding significant extra noise. In LISA Pathfinder a Discrete Fourier Transform (DFT) method has been chosen as baseline [16], which is ideal if the signals have nearly constant frequency.

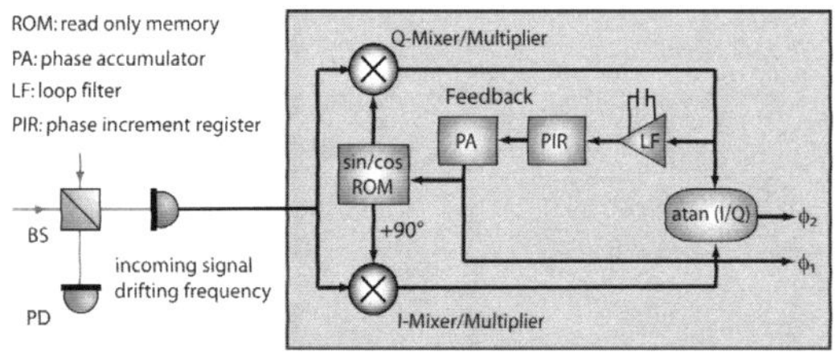

Figure 3: Phase measurement method: digital PLL as it is currently planned to be implemented on LISA.

In contrast to this, the measurement scheme for the interferometric phase readout chosen for LISA is a digital *phase locked loop* (PLL) that tracks the signal frequency and phase [17]. Figure 3 shows a schematic of such a phase readout system. Here, the photodiode signal is sampled by an analog-to-digital converter, and then multiplied with a sine and cosine signal of correct frequency. The output of the multiplication is used to track the signal's frequency and to compute the phase change. Since it is ideal for a variable heterodyne frequency, the digital PLL method has been chosen as preliminary baseline for a possible GRACE follow-on mission. Additionally, we benefit a lot from the ongoing LISA study. The algorithm will have to be implemented in hardware to make the phase readout fast and reliable. Hardware development based on Field Programmable Gate Arrays (FPGA) as main technology platform seems to be suitable. The requirements for the SSI phase measurement system are considerably relaxed compared to the LISA requirements [7]. The noise budget for the phase measurement is only $1\,\mathrm{mrad}/\sqrt{\mathrm{Hz}}$ compared to $6\,\mu\mathrm{rad}/\sqrt{\mathrm{Hz}}$ for LISA.

3. Performance of the laboratory breadboard

For breadboarding, we used standard laboratory laser-diode-pumped, monolithic Nd:YAG lasers (*Innolight Mephisto lasers,* 500 mW @ 1064 nm) [18]. These lasers generate single frequency, linear-polarised light with low intrinsic noise. In order to compare the behaviour of polarising and non-polarising optics in interferometry, we have set up three interferometers on an aluminium baseplate. The resulting optical scheme is shown in Fig. 4. All interferometers are divided into the two following functional parts:

- *The modulation bench* provides the beam preparation. The laser beam is split into two parts by a beamsplitter. The two beams are then modulated each by acousto-optical modulators (AOMs) driven at approximately 80 MHz with a frequency difference f_{het} of 1.6 kHz. The frequency shifted beams are coupled into optical fibres and injected into the optical bench.
- *The optical bench* contains the three interferometers. The polarising and non-polarising interferometers, with equal armlengths on the optical bench, perform the length measurements. In order to assess their performance, the obtained phases are to be compared to the phase reference obtained from the remaining non-polarising interferometer.

The experimental breadboarding has been performed with a digital phasemeter implemented on a PC as in LTP [16]. The heterodyne signals measured at the photodiodes are digitized and acquired by a computer. In order to extract the phase information, a DFT-Algorithm is applied as software post-processing of the data. This LTP based readout has been demonstrated to have sufficient sensitivity to measure. Therefore, it is used to assess and compare the opto-mechanical performance of our breadboard. In order to reduce coupling of acoustic and thermal effects into the phase readout, the measurement was conducted in a vacuum environment.

The noise measurements obtained using this system are presented in Fig. 5. The graph shows the measured interferometric phase noise as a linear spectral density in comparison to the total pathlength noise budget of SSI and LISA. We reach a phase readout sensitivity of $1\times10^{-2}\,\mathrm{rad}/\sqrt{\mathrm{Hz}}$ that corresponds to $1.7\times10^{-9}\,\mathrm{m}/\sqrt{\mathrm{Hz}}$ longitudinal fluctuations. One can see that the required longitudinal sensitivity for the SSI interferometer has been demonstrated with this simple setup. Thus, a polarising interferometer can reach the required length stability. The

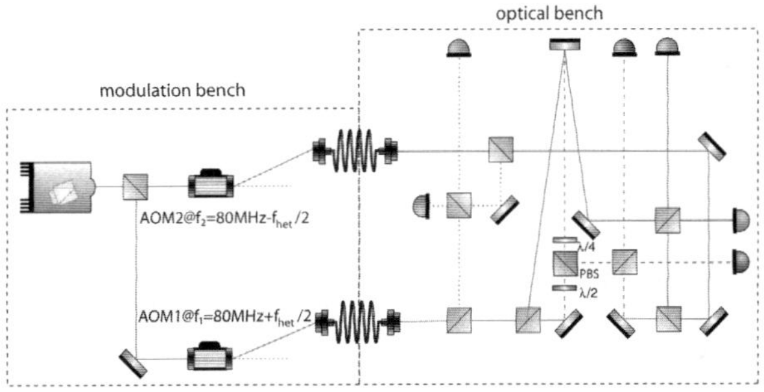

Figure 4: Schematic of the current setup. Left: the modulation bench. Right: the optical bench containing the reference (dotted), non-polarising (dashed) and polarising (solid) interferometer.

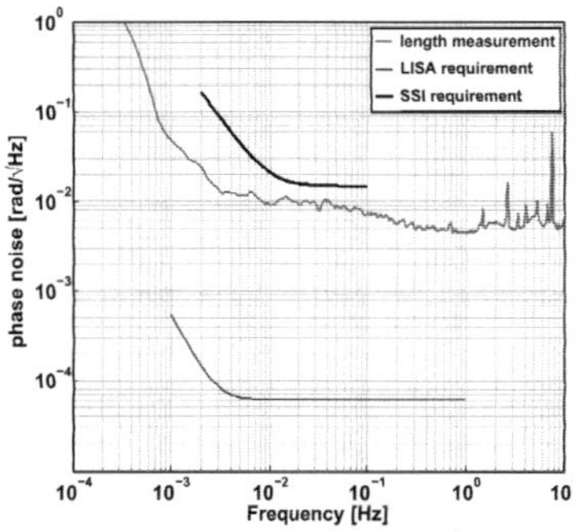

Figure 5: First pathlength noise measurements in a vacuum environment in comparison to the total pathlength noise budget of SSI and LISA.

sensitivity can be significantly increased by improving the mechanical and thermal stability of the optical bench, for example, by hydroxide-catalysis bonding the optical components onto an ultra-stable baseplate made of Zerodur®or Clearceram®, among other materials [19]. Clearceram®, as well as Zerodur®, is a material with low coefficient of thermal expansion of about 10^{-7}. Thus, it could be possible to further improve the length stability for the SSI and to even reach the longitudinal sensitivity requirements for LISA.

4. Conclusions and Outlook

First results of our investigations on a laser interferometer for spaceborne mapping of the Earth's gravity field are presented. Laboratory activities have demonstrated the feasibility of heterodyne interferometry containing polarising optics as metrology system for a possible GRACE follow-on mission. First performance measurements on the laboratory prototype, set up on a metal breadboard, show that we are able to reach the required longitudinal sensitivity of $2.5\,\mathrm{nm}/\sqrt{\mathrm{Hz}}$ for SSI. To obtain a significant improvement of the thermal and mechanical stability of the interferometer, we will bond the designed optical bench presented in Fig. 2 onto an ultra-stable glass-ceramic baseplate made of Clearceram®-HS. This leads to a more reliable and stable optical setup, suitble for a space application.

We are also planning to operate this bonded optical bench with a heterodyne frequency of 500 kHz. We have developed a modified version of our current phasemeter, which is able to operate at this frequency. However, it will be necessary to adapt the hardware phasemeter to their frequency, in order to track the Doppler shift of the beatnote in the measurement.

This effort will be expanded as part of the newly founded Center for Quantum Engineering and Space-Time Research (QUEST).

Acknowledgments

We gratefully acknowledge support by Deutsches Zentrum für Luft- und Raumfahrt (DLR) (reference FKZ 50 OQ 0501).

References

[1] Tapley B, Bettadpur S, Watkins M and Reigber C 2004 *Geophys. Res. Lett.* **31** L09607
[2] Bender P, Nerem R and Wahr J 2003 *Space Science Rev.* **108** 385–392
[3] Stephens M, Craig R, Leitch J, Pierce R, Nerem R, Bender P and Loomis B 2004 *4th annual Earth Science Technology Conference (ESTC)*
[4] Nagano S, Hosokawa M, Kunimori H, Yoshino T, Kawamura S, Ohkawa M and Sato T 2005 *Rev. of Scientific Instr.* **76** 124501
[5] Pierce R, Leitch J, Stephens M, Bender P and Nerem R 2008 *Appl. Opt.* **47** 5007–5019
[6] Leitch J *et al.* 2005 *Conference on Lasers and Electro-Optics/Quantum Electronics and Laser Science and Photonic Applications Systems Technologies* 1754–1756
[7] LISA study team 1998 Pre-Phase A Report (MPQ report 233)
[8] Vitale S *et al.* 2002 *Nucl. Phys. B (Proc. Suppl.)* **110** 209–216
[9] Anza S *et al.* 2005 *Class. Quantum Grav.* **22** 125–138
[10] Guzmán Cervantes F, Steier F, Wanner G, Heinzel G and Danzmann K 2008 *Appl. Phys. B* **90** 395–400
[11] Morrison E, Meers B J, Robertson D I and Ward H 1994 *Appl. Opt.* **33** 5037–5040 and 5041–5049
[12] Tröbs M 2005 *Laser development and charaterization for the spaceborne interferometric gravitational wave detector LISA* Ph.D. thesis university Hannover
[13] Leonhardt V and Camp J B 2006 *Appl. Opt.* **45** 4142–4146
[14] Drever R, Hall J, Kowalski F, Hough J, Ford G, Munley A and Ward H 1983 *Appl. Phys. B* **31** 97–105
[15] Diekmann C, Steier F, Sheard B, Heinzel G and Danzmann K 2009 *IOP Conference Proceedings of the 7th LISA Symposium (in prep.)*
[16] Heinzel G *et al.* 2004 *Class. Quantum Grav.* **21** 581–587
[17] Wand V, Guzmán F, Heinzel G and Danzmann K 2006 *AIP Conf. Proc.* **873**(1) 689–696
[18] Kane T J and Byer R L 1985 *Opt. Lett.* **10** 65–67
[19] Elliffe E J, Bogenstahl J, Deshpande A, Hough J, Killow C, Reid S, Robertson D, Rowan S, Ward H and Cagnoli G 2005 *Class. and Quant. Gravity* **22** S257–S267

Journal of Physics: Conference Series **154** (2009) 012024 doi:10.1088/1742-6596/154/1/012024

Implementation of Armlocking with a delay of 1 second in the presence of Doppler shifts

V. Wand[1], **Y. Yu**[1], **S. Mitryk**[1], **D. Sweeney**[1], **A. Preston**[1], **D. Tanner**[1], **G. Mueller**[1], **J. I. Thorpe**[2], **J. Livas**[2]

[1]Department of Physics, University of Florida, PO Box 118440, Gainesville, FL 32611, USA

[2]NASA Goddard Space Flight Center, Code 663 Bldg. 2, Greenbelt, MD 20771, USA

E-mail: `vwand@phys.ufl.edu`

Abstract. LISA relies on several techniques to reduce the initial laser frequency noise in order to achieve an interferometric length measurement with an accuracy of $\approx 10\,\mathrm{pm}/\sqrt{\mathrm{Hz}}$. LISA will use ultra-stable reference cavities as a first step to reduce the laser frequency noise. In a second step the frequency will be stabilized to the LISA arms which provide a better reference in the frequency band of interest. We present experimental results demonstrating Arm locking with LISA-like light travel times and Doppler shifts. We also integrated this system with a LISA-like pre-stabilization system using our ultra-stable cavities. The addition of realistic Doppler shifts led to further refinements of the arm locking controllers compared to the controller architecture discussed in the past. A first experimental result of the new controller is also presented.

1. Introduction

The joint ESA-NASA mission LISA (Laser Interferometer Space Antenna) is being designed to observe gravitational waves in the low frequency regime from 10^{-4} to $10^{-1}\,\mathrm{Hz}$ ([1]). LISA consists of 3 identical spacecraft (S/C) forming an equilateral triangle with an armlength of $5 \times 10^9\,\mathrm{m}$. The constellation orbits the sun trailing the earth by a distance of $50 \times 10^9\,\mathrm{m}$. Free falling test masses onboard of each spacecraft define the arms of a Michelson-like interferometer. Gravitational waves will change the lengths of the arms. These changes will be detected with a precision of about $10\,\mathrm{pm}/\sqrt{\mathrm{Hz}}$ by combining the individual interferometer outputs. The orbital dynamics of the constellation will cause mismatches of up to 50000 km between the arms. This mismatch requires a reduction and common mode rejection of laser frequency noise by 12 orders of magnitude with respect to its initial noise level ([2]). The current LISA baseline considers two methods to actively reduce the laser frequency noise. In a first step, the laser frequency will be stabilized to an ultra-stable optical cavity ([1],[3]). In a second step, the laser frequency will be stabilized using the LISA arms as references. This technique is called arm locking ([4]). In addition to these active frequency stabilization steps, data postprocessing algorithms (TDI - time delay interferometry ([5],[6],[7])) are used to synthesize a near equal arm interferometer to cancel the remaining laser frequency noise.

In this paper we will discuss the current status of our arm locking experiments. The motivation for arm locking is that the LISA arms provide a much better reference ($\Delta\mathrm{L}/\mathrm{L} = 10^{-21}$) in the LISA band compared to the ultra-stable cavities ($\Delta\mathrm{L}/\mathrm{L} = 10^{-13}$).

As of now, arm locking has been analyzed in several configurations ([4]) under the assumption that the Doppler shift can be subtracted from the interferometer signal. So far, experimental demonstrations have been limited to short 'light travel times' of a few μs without any Doppler shifts ([8],[9]). In addition, the incorporation of the pre-stabilization to the optical cavity has not been demonstrated experimentally although several designs have been discussed in the past. In this paper, we report on the first experimental demonstrations of arm locking with realistic Doppler shifts, an integration with a pre-stabilization system and light travel times of one second.

2. Arm locking

In an arm locked configuration, one laser on one of the S/C will act as the master laser. Every other laser is either directly or indirectly frequency offset phase locked with a known heterodyne frequency ω_H to this laser. In this configuration, the lasers on the far S/C act as transponders for this master laser. The beat signal between the transponded field and the master laser on the master S/C is then:

$$\mathrm{P}(t) = A\cos\left[(\omega_D + \omega_H)\,t + \varphi_1(t) - \varphi_1(t-\tau)\right] \tag{1}$$

where $\varphi_1(t)$ is the phase noise of the pre-stabilized laser, $\tau \approx 33\,\mathrm{s}$ is the two way propagation time between the spacecraft, and ω_D is the round-trip Doppler shift between two S/C. The phase of $\mathrm{P}(t)$ is then measured with a phasemeter. The phasemeter consists of a numerical controlled oscillator (NCO) which is phase locked to the signal. The NCO actuator signal of the digital control loop is then used to regenerate the phase of $\mathrm{P}(t)$ with respect to the initial frequency ω_{NCO} of the NCO:

$$\mathrm{S}(t) = (\omega_D + \omega_H)\,t + \varphi_1(t) - \varphi_1(t-\tau) - \omega_{NCO}t \tag{2}$$

In an idealized arm locking system the NCO frequency can be set to $\omega_{NCO} = \omega_D + \omega_H$. Thus $\mathrm{S}(t)$ can be written as:

$$\mathrm{S}(t) = \varphi_1(t) - \varphi_1(t-\tau). \tag{3}$$

Laplace transforming 3, the frequency response of the interferometer is given by ([4]):

$$\mathrm{G}(s) = \frac{\mathrm{S}(s)}{\varphi_1(s)} = 1 - e^{-s\tau} = 2\,\mathrm{i}\sin\left(\frac{\omega\tau}{2}\right)\mathrm{e}^{\frac{-\mathrm{i}\,\omega\tau}{2}}. \tag{4}$$

Figure 1 shows the Bode plot of the interferometer's transfer function for a delay of τ=1 s. Thus $\mathrm{G}(s)$ represents a transfer function of a sensor, susceptible to frequency changes and can be used for an active stabilization. Since the sensitivity rolls down with f to DC for $f < 1/\tau$, the sensor is insensitive at DC as well as at the Fourier frequencies $f_n = n/\tau$. In addition the phase changes abruptly at all f_n's from $-90°$ to $+90°$. Consequently, the controller has to roll off with less than $1/s$ to keep sufficient phase margin close to these frequencies ([4],[10]). In our experiments, we use combinations of pole-zero pairs to approximate a $1/\sqrt{s}$ slope in the controller to maintain a phase margin of approximately 45° in the entire system. This slope limits the low frequency gain above f_1 since the system has additional bandwidth limitations caused by additional delays and phase shifts in the sensor, controller, and actuator.

Our initial arm locking experiment uses a free running voltage controlled oscillator (VCO) as laser representative which is an realistic approximation in terms of frequency noise. Figure 2 shows the general setup. The free running VCO at 6 MHz is demodulated with an 8 MHz signal generated by a NCO. The 2 MHz beat signal is then split into two parts. The first part is sent to an electronic phase delay (EPD) unit ([11]) where it is delayed by a time $\tau \approx 1$ s and where its frequency is shifted by a Doppler frequency of -1 MHz. This delayed and frequency shifted

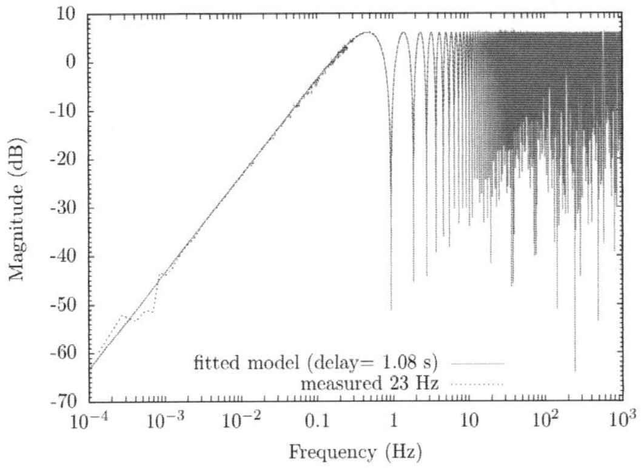

Figure 1. Transfer function of the arm locking sensor for an EPD delay of approximately τ=1 s. The measured transfer function (dashed curve) is in good accordance with the theoretical model (solid).

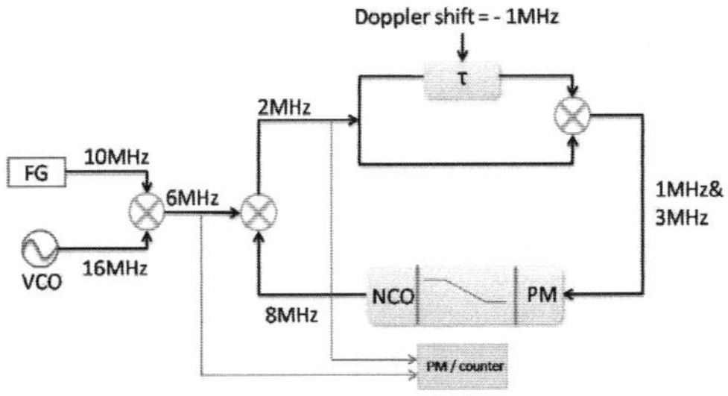

Figure 2. Setup of Armlocking implementation. The VCO signal with the frequency $\omega_0 = 6$ MHz represents a Laser beat note affected by phase noise $\varphi(t)$.

signal is then mixed in an analog mixer with the second non-delayed part of the 2 MHz signal. This is our interferometer signal which is identical to the LISA signal S(t) if we lump ω_D and ω_H together into the one Doppler frequency.

An unique feature of this configuration is that it allows us to measure the noise suppression of the arm locking controller by comparing the phase noise of the VCO with the phase noise of the 2 MHz beat signal. The phase of S(t) is then measured with the phasemeter (PM) ([12],[13]). The measured phase is the error signal which is then fed into the controller which then adjusts the frequency of the NCO. The closed loop will lock the difference frequency between the NCO and the VCO to the interferometer. The NCO will track the VCO frequency noise except at the frequencies where the interferometer response vanishes.

3. Arm Locking with perfect Doppler knowledge

In a first series of experiments we matched the initial NCO frequency in the arm locking PM to the Doppler shift added in the EPD unit. This corresponds to a prefect knowledge of the Doppler shift which will not be the case in LISA. The delay time in the EPD unit was set to approximately 1 s. Figure 1 shows the theoretical (solid curve) and the measured transfer function (dashed) of the arm locking sensor (interferometer). The delay time was fitted here by a least square fit estimating τ to be 1.08 s. The actuator signal to the NCO of the phasemeter was then filtered by the arm locking controller. The arm locking controller provided a slope of $1/\sqrt{f}$ at frequencies above $1/\tau$ and four integrator stages with corner frequencies at 0.1 Hz to boost the low frequency gain. This signal was then used to control the frequency of an NCO at the output of the controller.

In order to verify the performance of the arm locking loop, the initial phase/frequency noise of the VCO was measured as well as the phase/frequency noise of the stabilized beat signal. The linear spectral densities (LSD) of the VCO and the beat signal are shown in Figure 3. At frequencies given by integer multiples of the inverse delay time τ, the phase noise cannot be suppressed but is slightly enhanced (solid curve). Below about 0.1 Hz the integrators start to roll up the gain and suppress the noise further. Thus, the phase noise of the locked signal is significantly suppressed: The loop reaches zero gain[1] at a few hundred Hz (unity gain frequency)

[1] Strictly spoken the loop gain has several zero crossings (two near each spike). We refere here to the zero crossing

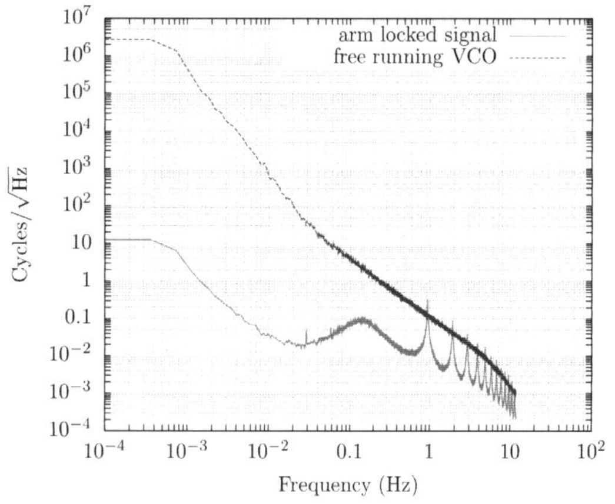

Figure 3. The free running VCO phase noise (dashed curve) and phase noise level by enabling arm locking (solid curve).

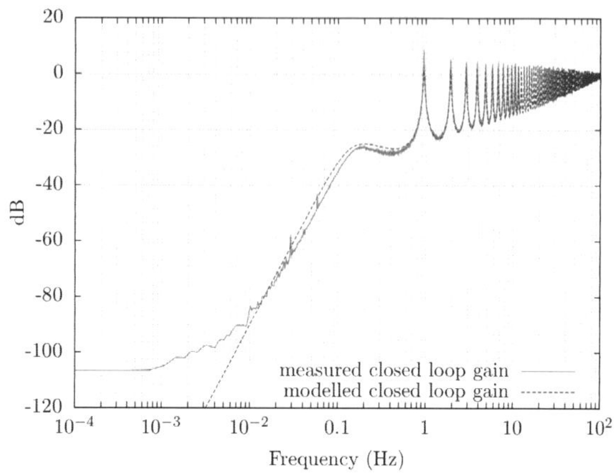

Figure 4. Modelled and measured closed loop gain (CLG).

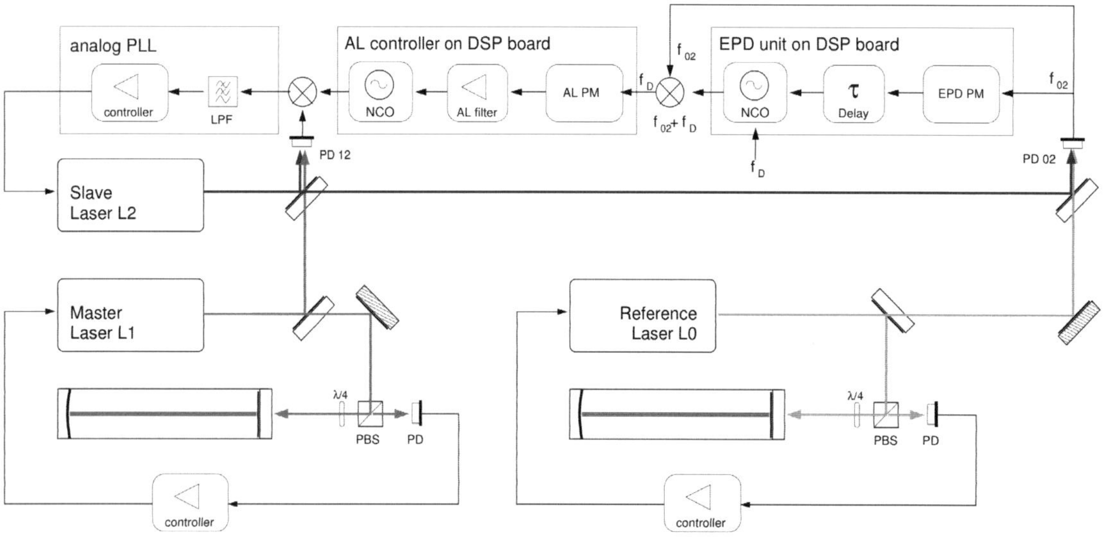

Figure 5. Electro optical setup of the laser PLL model.

rolling up to 1 Hz with $1/\sqrt{f}$ until a f^{-3} slope at approximately 0.5 Hz ensuring high gain at DC. Figure 4 shows the measured and expected closed loop gain of the system. They agree at all frequencies above about 10 mHz. The maximum noise suppression achieved is of the order of 10^5.

4. Integration with Pre-stabilization

Figure 5 illustrates the principle of our laboratory setup of a PLL-based implementation of arm locking and pre-stabilization. The master laser (L1) is pre-stabilized to a cavity. The slave laser (L2) is offset phase locked to the master laser such that its phase noise φ_2 follows the phase noise of the master φ_1 within the bandwidth of the PLL (about 40 kHz in our setup). The offset

of the envelope of the OLG magnitude

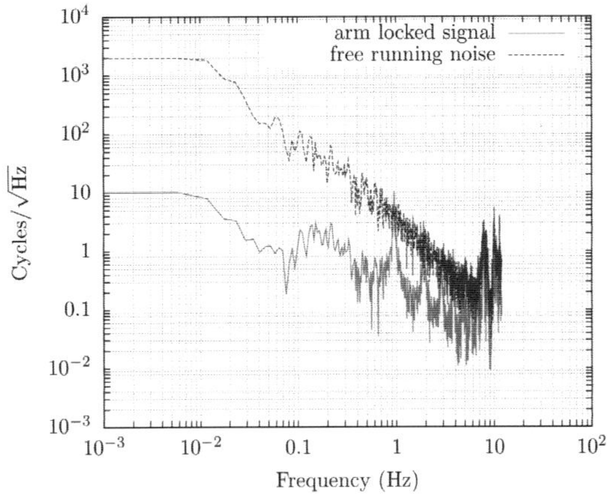

Figure 6. The upper graph is the phase noise obtained from the beat note L0-L1. The spectrum of the phase noise of the stabilized signal L0-L2 is shown as lower graph.

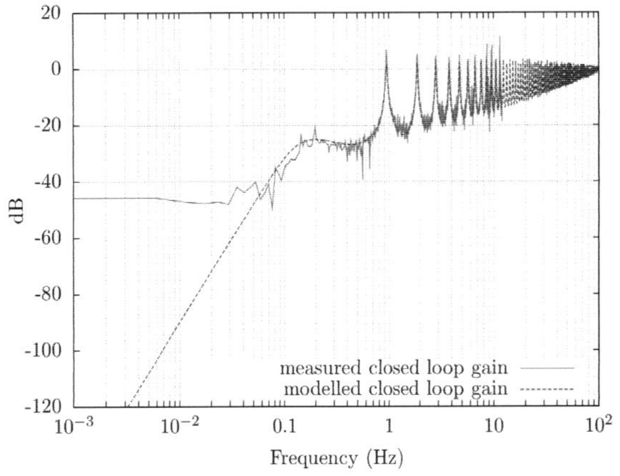

Figure 7. Modelled and measured closed loop gain. The plot shows a maximum suppression reached of about 100.

frequency is determined by a NCO which is part of the arm locking controller. The frequency of L2 can be controlled by changing the operation frequency of the NCO. In LISA, the beam of L2 propagates to the far spacecraft and the transponded copy, delayed by about 30 s, interferes with the prompt beam. Obviousely, it is impossible to set up an optical delay line with a comparable delay in an optics lab. Instead of an optical delay we use an electronic delay generated by the EPD unit. In order to provide an RF signal processable for the EPD unit, L2 is mixed with and an additional stabilized reference laser (L0). This signal is then electronically delayed and mixed with the instantaneous beat signal.

The obtained error signal is fed into the arm locking controller discussed above. The filtered data is used to tune the NCO to compensate frequency fluctuations of L2 with respect to L0. In other words, the phase noise φ_{02} of the beat note L0-L2 is being measured by mixing its delayed copy (1 s delay) with the prompt signal and is being suppressed by the feedback of the arm locking filter controlling the offset frequency between L2 and L1. Figure 6 shows the initial phase noise between L0 and L1 as well as the phase noise of the arm locked beat note L0-L2. The noise suppression reached at 10 mHz is about 100. Figure 7 shows the modelled and measured close loop gain of the system.

5. Conclusion and outlook

We reported on the first successful demonstration of arm locking using LISA-like signal travel times and Doppler shifts. We characterized our system first using a simplified electronic system with compensated Doppler shifts and achieved 5 orders of magnitude noise suppression in the mHz band. In addition, we also succeeded in incorporating arm locking in an existing laser pre-stabilization system. These experimental demonstrations show that single arm arm locking is now well understood and can be used to reduce the laser frequency noise in LISA significantly. This would further reduce the requirements on the phasemeter and TDI. We also analyzed arm locking with imperfect Doppler compensation and introduced a solution by rolling down the low frequency gain again. The experimental setup for a verification of this method is currently in implementation. Over the next months we will start to test dual arm locking, study the impact of the uncertainties and temporal changes in the Doppler shifts on dual arm locking, and integrate dual arm locking with the existing laser pre-stabilization system to demonstrate arm locking in a full LISA-like setup.

[1] P.L. Bender, K. Danzmann and the LISA Study Team, Laser Interferometer Space Antenna for the detection and observation of gravitational waves: Pre-phase A report, 2nd Edition, Max-Plank-Institut für Quantenoptik, MPQ233.

[2] Daniel A. Shaddock, Massimo Tinto, Frank B. Estabrook, and J. W. Armstrong. Data combinations accounting for LISA spacecraft motion. *Phys. Rev. D*, 68(6), 2003.

[3] R. Cruz et al. The LISA benchtop simulator at the university of florida. *Class. Quantum Grav.*, 23:761–767, 2006.

[4] B. Sheard, M. Gray, D. McClelland, and D. Shaddock. Laser frequency stabilization by locking to a LISA arm. *Phys. Lett. A*, 320:9–21, 2003.

[5] Massimo Tinto, Daniel A. Shaddock, Julien Sylvestre, and J. W. Armstrong. Implementation of time-delay interferometry for LISA. *Phys. Rev. D*, 67, 2003.

[6] Massimo Tinto and J. W. Armstrong. Cancellation of laser noise in an unequal-arm interferometer detector of gravitational radiation. *Phys. Rev. D*, 59, 1999.

[7] M. Tinto, F. B. Estabrook, and J.W. Armstrong. Time-delay interferometry for LISA. *Phys. Rev. D*, 65(082003), April 2002.

[8] Antonio F García Marín, Gerhard Heinzel, Roland Schilling, Albrecht Rüdiger, Vinzenz Wand, Frank Steier, Felipe Guzmán Cervantes, Andreas Weidner, Oliver Jennrich, Francisco J Meca Meca, and K Danzmann. Phase locking to a LISA arm: first results on a hardware model. *Class. Quantum Grav.*, 22(10):235–S242, 2005.

[9] B. Sheard, M. Gray, D. Shaddock, and D. McClelland. Laser frequency noise suppression by arm-locking in LISA: progress towards a bench-top demonstration . *Class. Quantum Grav.*, 22(10):221–226, 2005.

[10] J. I. Thorpe and G. Müller. Experimental verification of armlocking for LISA using electronic phase delay. *Phys. Lett. A*, 342:199–204, 2005.

[11] James Ira Thorpe, Rachel Jean Cruz, Shannon Sankar, and Guido Mueller. Electronic phase delay; a first step towards a bench-top model of lisa. *Class. Quantum Grav.*, 22(10):227–S234, 2005.

[12] D. Shaddock, B. Ware, P. G. Halverson, R. E. Spero, and W. Klipstein. Overview of the LISA phasemeter. *AIP Conf. Proc.*, 873:654–660, 2006.

[13] V. Wand, F. Guzman, G. Heinzel, and K. Danzmann. LISA phasemeter development. *AIP Conf. Proc.*, 873:689–696, 2006.

Journal of Physics: Conference Series **154** (2009) 012025 doi:10.1088/1742-6596/154/1/012025

Optical ranging and data transfer development for LISA

Juan José Esteban, Iouri Bykov, Antonio Francisco García Marín, Gerhard Heinzel, Karsten Danzmann.

Max-Planck-Institut für Gravitationsphysik (Albert-Einstein-Institut) and Universität Hannover, Germany

E-mail: juan.jose.esteban@aei.mpg.de

Abstract.
In order to implement optical inter-spacecraft ranging and data transfer for LISA, the carrier of the laser link must be phase modulated with pseudo-randon noise sidebands. The data acquisition and delay estimation are then implemented in the phasemeter as back end processing. This work presents a proposed demodulation scheme with submeter ranging accuracy and several kilobytes data rate. Its functionality is demonstrated both in a software simulation and in a FPGA-based hardware implementation.

1. Introduction

The Laser Interferometer Space Antenna is an international space project to detect and observe gravitational waves on the frequency range from 0.1 mHz to 100 mHz. LISA is a cluster of three satellites forming an equilateral triangle and separated by five million kilometers communicated via three bidirectional laser links. The primary quantity to be measured is the pathlength variation between the free-floating test masses onboard two different spacecraft by means of heterodyne interferometry with a required sensitivity of $1\,\mathrm{pm}/\sqrt{\mathrm{Hz}}$. It must be taken into account that, due to relative movements between the spacecraft, the distance of the interferometer arm is expected to change approximately $\pm1\%$ over one year orbit ($\simeq 50.000\,\mathrm{km}$). As consequence, the resulting unequal arm configuration of LISA turns laser frequency into excess noise added to the measured phase, which could spoil the scientific performance of the mission. One of the three methods planned to reduce the influence of laser frequency noise is time-delay interferometry (TDI). This consists in post-processing the data on ground to make LISA work virtually like an equal arm interferometer, suppressing the frequency noise contribution up to seven orders of magnitude. However, TDI is based on the knowledge of the absolute distances between the satellites with meters resolution and hence any error in this ranging measurement will turn into an error in the TDI post-processing scheme. To achieve the mentioned accuracy, a pseudo-random noise (PRN) sequence can be modulated on the phase of the laser link. The PRN sequence in the remote laser phase can be measured by the local phasemeter system (PMS) [1, 2] and its correlation with the local PRN can be performed as phasemeter back-end processing, delivering the light travel time τ between the satellites.

2. Architecture and system modelling

Each of the six lasers in the LISA constellation has to be phase-modulated with a different PRN codes that minimise the correlation with the other five for any given delay and with himself for delays different of zero. The set of six PRN sequences implemented shown in Figure 1 have been obtained by numerical optimisation and with selected correlation properties. This way, the correlation peak serves as a timestamp if the start of the PRN is synchronized with the clock of the remote spacecraft.

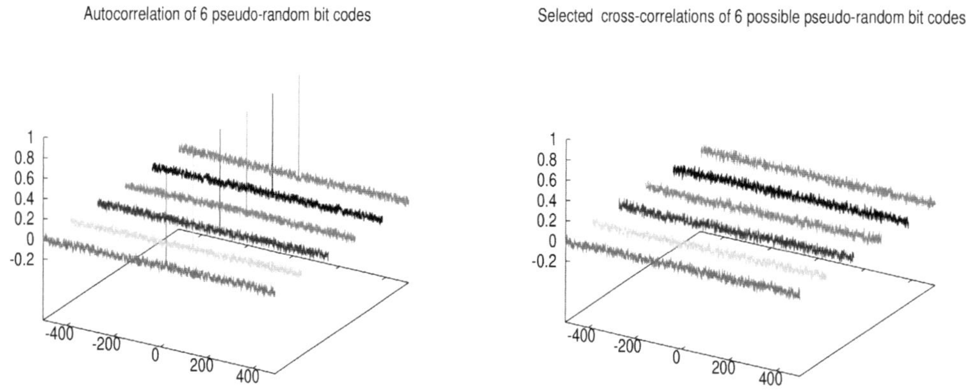

Figure 1. Autocorrelation and crosscorrelation between a possible set of pseudocodes

The proposed scheme uses a digital delay-locked loop (DLL) architecture[3] to track the correlation peak (see Figure 2). The delay-locked loop is based on the correlation of the input signal with three versions of the same reference PRN: a punctual, an early and a late. The early and late version are delayed by one chip[1] between each other and the punctual version is remained equal than the PRN transmitted. The punctual correlator is responsible for the data recovering and peak detection whereas the early and late correlators provide the tracking between the incoming and local PRN. The difference between early and late correlators can be used as error signal in a control loop to update the delay of the code generator to the input signal.

[1] One code bit of PRN is also called as chip

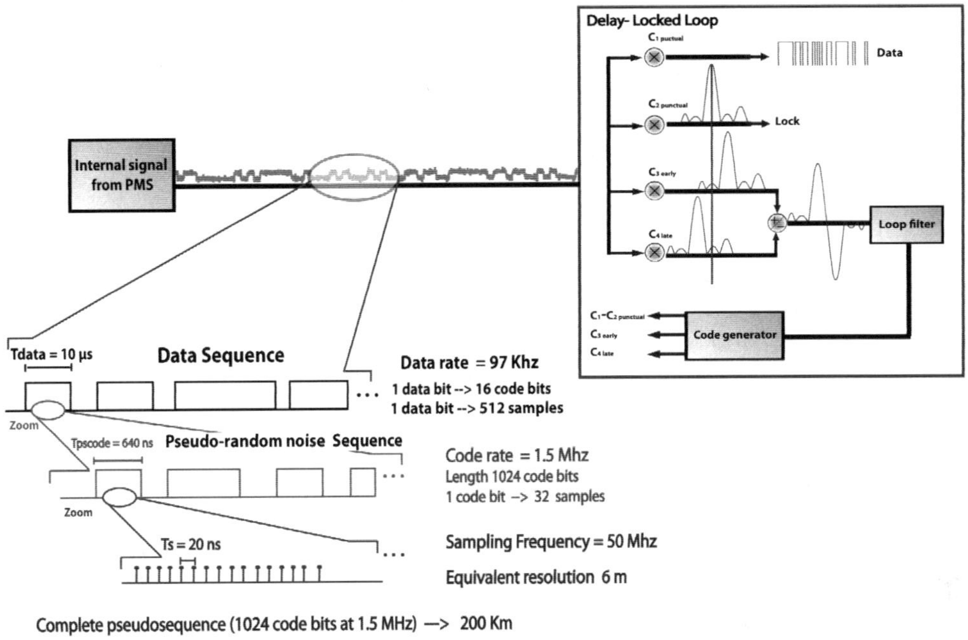

Figure 2. General block diagram of the delay-lock loop implemented in the same board than the phasemeter and design parameters for DLL implemented in the simulation and FPGA prototype

3. Simulation and hardware implementation

In order to demonstrate the functionality of the DLL, a software simulation was developed previously to the implementation on FPGA-based hardware. The design parameters for both of them are also shown in Figure 2: the bottom time series represents the phasemeter output sampled at 50 MHz. A pseudo-random noise (PRN) sequence is encoded on this phase signal at a chipping rate of 1.5 MHz, as shown in the middle time series. The PRN code-length of 1024 results in a code repetition every 200 km armlength, and the chipping rate is selected to optimise the ranging accuracy and robustness. The top part of the figure shows the data sequence, encoded on top of the PRN at 97 kHz.

Figure 3 summarises the current implementation running on the phasemeter board[4]. The processing unit is provided by a single FPGA (field programmable gate array) and the input signal to the DLL is internally synthesized. The DLL features two possibles working modes:

- **Acquisition mode:** In this mode, the delay between the incoming signal and the local PRN is estimated with one chip resolution (T_c =640 ns) without any previous information. To this end, the incoming signal is correlated during a complete period of the local PNR ($\simeq$ 1.5 kHz) using the punctual branch. The process must be repeated for all possible delay values. The resulting delay after this stage can be written as ($\tau = N \cdot T_c$), with N going from 0 to 1024.
- **Tracking mode:** Once the PRN chip of the incoming code has been identified, the early and late branches are used to correlate the incoming signal with a partial PRN-sequence (256 chips for the current implementation). This way, the delay between the incoming and local PRN-sequences can be estimated with nanoseconds accuracy.

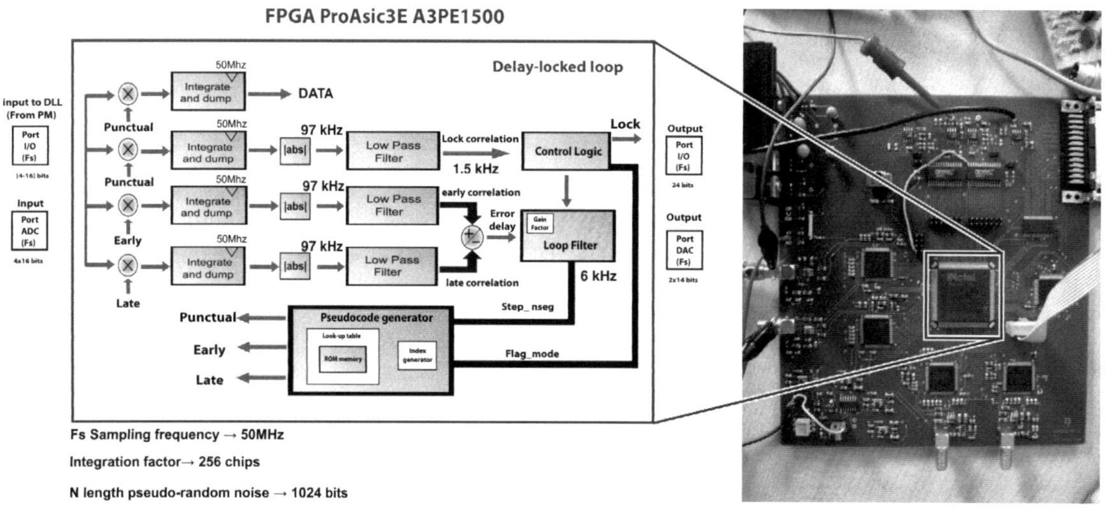

Figure 3. Implementation of the DLL in the same prototype board used for PMS development

The main characteristics of the used board can be summarised as follows:

- ProAsic3E Actel FPGA with 1.5 Million system gates with 147 high-performance inputs/outputs (I/Os)
- Four A/D-converters (AD9446-100) with 16-bit resolution
- Two D/A-converters (AD9744-210) with 14-bit resolution
- Enhanced parallel port (EPP) with output rate up to 1 MB/s
- High-speed PCI digital input/output (DIO) board with direct memory access features and maximal rate of 2 GB/s

4. Results

The results of the implementation of the proposed design in a software simulation are presented in Figure 4. These results have been verified in the FPGA target described in Section 3. The two top time series represent the input phase signal fed to the DLL with different levels of phase noise. In the left part of the graph, the input phase noise has been obtained assuming a normalized white noise of $A_{\mathrm{N}} = 0.2\,\mathrm{rad_{rms}}$ and a modulation index of 0.14 rad (for 1% power in side-band). This leads to a noise amplitude of $A_{\mathrm{LISA}} = 0.2\times 0.14 = 0.029\,\mathrm{rad_{rms}}$ and consequently a equivalent phase noise of $\tilde{\varphi}_{\mathrm{LISA}} = A_{\mathrm{LISA}}/\sqrt{f_{\mathrm{s}}} = 4\,\mu\mathrm{rad}/\sqrt{\mathrm{Hz}}$. In the left time series, the PRN sequence is clearly visible, in contrast with the right hand one which presents an increasing of $40\,\mu\mathrm{rad}/\sqrt{\mathrm{Hz}}$ in the equivalent phase noise. The bottom part of Figure 4 represents the error signal in the tracking mode corresponding to each upper input signal, respectively. The correlation value is given by the substraction of early and late correlators at measurement rate of $\simeq 6\,\mathrm{kHz}$ (integration time of 16 data bits, corresponding to 256 PRN chips). The system is able to match the correlation properties of the incoming signal and remains locked even in the presence of the described phase noise.

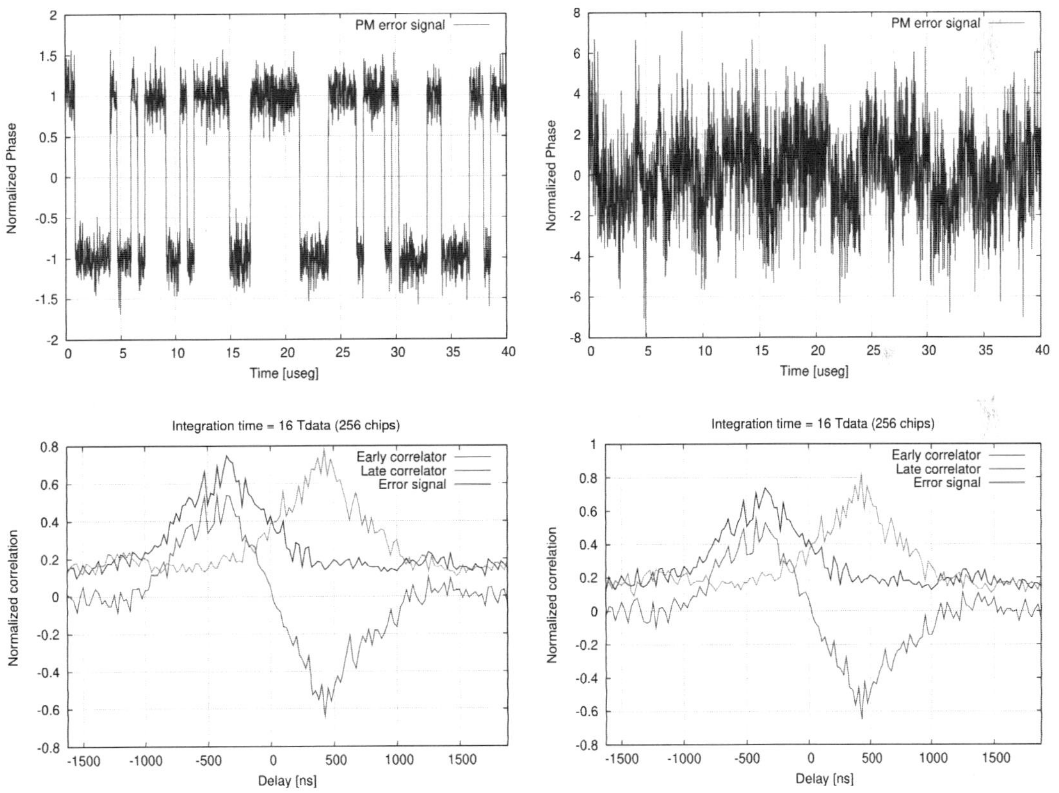

Figure 4. Noise impact on discriminator function for different amplitudes of additive gaussian noise, $\simeq 4\,\mu\mathrm{rad}/\sqrt{\mathrm{Hz}}$ and $\simeq 40\,\mu\mathrm{rad}/\sqrt{\mathrm{Hz}}$ respectively. Top: two time series of PRN with additive gaussian noise of amplitude. Bottom: resulting tracking error in presence of additive gaussian noise. Design parameters of the DLL: Integration time $T_{\mathrm{i}} = 16$ data bits $= 256$ chips ($\frac{1}{T_{\mathrm{i}}} \approx 6\,\mathrm{kHz}$).

Journal of Physics: Conference Series **154** (2009) 012025 doi:10.1088/1742-6596/154/1/012025

5. Conclusions

This paper presents an architecture based an a DLL for the implementation of optical ranging and inter-spacecraft data transmission for LISA. The proposed design can be implemented as back-end processing of the phasemeter. Its functionality has been tested in a software simulation and in a hardware implementation on the same board as our present phasemeter design.

The next steps include the evaluation of the performance with external phase signals and the integration, together with the phasemeter, in a full optical ranging experiment.

6. References

[1] Shaddock D, Ware B, Halverson P, Spero R, Klipstein B 2006 Overview of the LISA Phasemeter *AIP Conf Proc.* **873** 654-60

[2] Wand V, Guzman F, Heinzel G, Danzmann K 2006 LISA Phasemeter Development *AIP Conf Proc.* **873** 689-96

[3] Misra P and Enge P 2nd Ed Global positioning system, signals, measurements, and performance *Ganga-Jamura Press*

[4] Bykov I, Esteban J.J, García Marín A.F, Heinzel G, Danzmann K 2008 LISA phasemeter development: Advanced prototyping J. Phys.: Conf. Series. this issue

Modular Gravitational Reference Sensor Development

Ke-Xun Sun, Saps Buchman, Robert Byer, Dan DeBra, John Goebel*, Graham Allen, John W. Conklin, Domenico Gerardi, Sei Higuchi, Nick Leindecker, Patrick Lu, Aaron Swank, Edgar Torres, and Martin Trittler

Hansen Experimental Physics Laboratory, Stanford University, CA 94305, USA
*Code RET, NASA Ames Research Center, Moffett Field, CA 94035, USA

kxsun@stanford.edu

Abstract. The Modular Gravitational Reference Sensor (MGRS) is targeted as a next generation core instrument for both space gravitational wave detection and an array of other precision gravitational experiments in space. The objectives of the NASA funded program are to gain a system perspective of the MGRS, to develop key component technologies, and to establish important test platforms. Our original program was very aggressive in proposing ten areas of research and development. Significant advancements have been made in these areas, and we have met or exceeded the goals for the program set in 2007-2008. Additionally, we have initiated research projects for innovative technologies beyond the original plan. In this paper we will give a balanced overview of progress in MGRS technologies: the two layer sensing and control scheme, trade-off studies of GRS configurations, multiple optical sensor signal processing, optical displacement and angular sensors, differential optical shadow sensing, diffractive optics, proof mass center of mass and moment of inertia measurement, UV LED charge management, proof mass fabrication, thermal control and sensor development, characterization for various proof mass shapes, and alternative charge manage techniques.

1. Introduction

The Modular Gravitational Reference Sensor (MGRS) [1-2] represents the next generation of technology for space gravitational wave detection and other space-borne gravitational precision experiments. MGRS will employ new technologies to achieve higher performance required for missions beyond LISA. MGRS also has a simpler structure and thus lower cost. We will use all-optical sensing to achieve near-zero stiffness. We will use two-layer interferometric and shadow sensing to maximize the fidelity of science measurement and to ease drag-free control. We will use a spherical proof mass to establish a full 3-dimemsional geodesic reference. We will use a larger gap between proof mass and housing to minimize the disturbances. We will take advantage the latest semiconductor technologies by using UV LEDs for charge management. We have demonstrated the exceptional radiation hardness and long operation life time of an UV LED. We are developing ground testing platforms for characterization of the mass center location and moment of inertia of the proof mass. We have designed and are building thermal environment control with microkelvin stability. Our program, funded by NASA, has more than ten technical areas. We have made significant advancements, and met or exceeded the aggressive goals for the effective funding period of June 2007 to March 2008. In addition, we have initiated research projects beyond the original plan. In this paper we will provide an overview of our progress.

2. MGRS Technologies

2.1. System Technologies

2.1.1. *Sphere and cube GRS overview (Collaboration with EADS Astrium)* [2] Collaborating with EADS Astrium, we have conducted a high level review [2] of the state of the art of the LISA GRS. We also started a trade study of GRS configurations with cubic proof masses or with a single spherical proof mass (MGRS). The results shows that the performance of the cubic proof mass configurations are adequate for the LPF baseline requirement. Further, the MGRS with a spherical proof mass holds promise as a future GRS, because of its simpler structure and higher performance at low frequencies.

2.1.2. *Sphere and cube trade-off studies: Noise tree (Collaboration with EADS Astrium)*[14] Stanford and EADS Astrium collaboratively studied the acceleration noise trees for the MGRS, and compared the results with the cubic GRS. The results show that the noise from environmental disturbances for the MGRS and cubic GRS are 8.5×10^{-16} m/s^2 and 10.5×10^{-16}m/s^2, respectively, and the stiffness induced noises are 0.4×10^{-16} m/s^2, and 2.8×10^{-16} m/s^2, respectively. Based on these first estimates, the spherical MGRS has ~28% lower total acceleration noise as a result of its large gap, full optical sensing, and no electrostatic actuation. The MGRS further enables simpler drag-free control.

2.1.3. *Two-layer sensing and control* [8] We have devised a two-layer optical sensing scheme, in which the science measurement is accomplished via picometer precision interferometric sensors, and the drag-free control signal is mainly obtained via large dynamic range shadow sensors. There is a substantial overlap region between the two sensing layers.

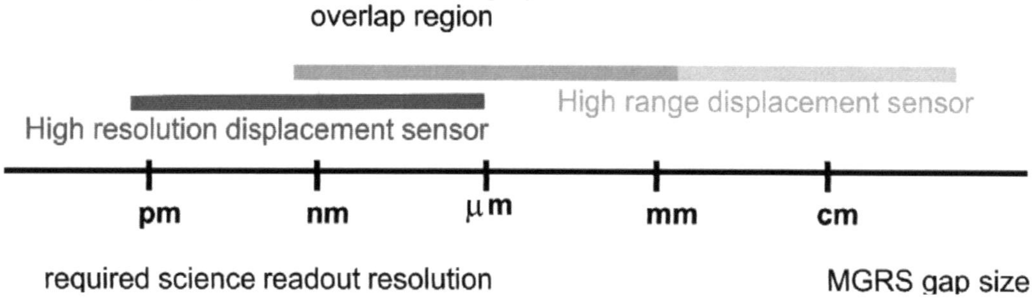

Figure 1. The two-layer sensing and control scheme. High precision laser interferometric sensors provide the science signal. High dynamic range optical shadow sensors provide the drag-free control signal. The two layers overlap in 0.1 nm-1 μm region.

2.1.4. *Mass center determination using multiple optical sensors* [3, 7, 8, 12, 13, 17]
We have undertaken an extensive set of analytical [12] and numerical [13] analyses to test the determination of the mass center position of a spinning sphere using a set of optical displacement sensors. The simulation demonstrates that we can accurately separate the proof-mass surface figure (~ 300 nm RMS) from its mass center motion (~ 30 nm RMS) using a numerical fit to the spin frequency and its harmonics. The simulation includes realistic system parameters such as sensor noise and residual drag-free error. The mass center displacement precision is better than 3 pm/Hz$^{1/2}$. This firmly establishes that a set of optical displacement sensors can be offer high precision and high system reliability.

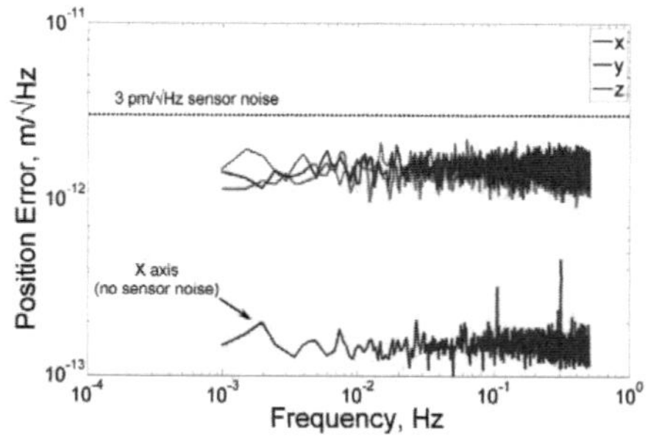

Figure 2: Simulation showing picometer determination of the mass center position, using multiple sensors. Head on sensors achieve even lower noise level (lower trace).

2.2. Optical Displacement Sensing [7, 8]

We have completed construction of a new test platform and vacuum chamber for the interferometric optical displacement sensor, consisting of a 900 line/mm grating, and a mirror surface simulating the proof mass. The cavity length is set at ~ 2 cm, resembling the gap size. The displacement signal is read out using a Pound-Drever-Hall RF modulation scheme. Preliminary testing in air on the new platform shows a noise level of 5 pm/√Hz at 1 Hz. The test platform will move into a vacuum chamber for improved performance.

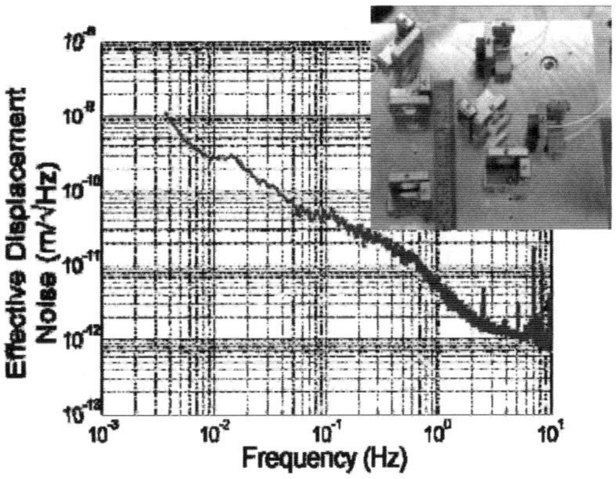

Figure 3. Preliminary results of the optical displacement sensor.

2.3. Differential Optical Shadow Sensing [7, 8, 19]

The Differential Optical Shadow Sensor (DOSS) scheme, shown in Figure 4, cancels laser intensity noise, while doubling the proof mass displacement signal. We have designed and constructed an optical shadow sensing test platform, on which the proof mass can be displaced with nanometer precision using a PZT driven flexure structure. We reduced the noise effects of electronics, electromagnetic interference, air flow and temperature, and achieved ~ 2 nm/Hz$^{1/2}$ at 2 Hz. The dynamic range is 1-3 mm, limited by the photodetector diameter. Our experience shows that optical shadow sensing is more robust than the interferometric sensors thanks to lower requirements in alignment and fringe counting. The method has been successfully applied to measurement of the mass center offset.

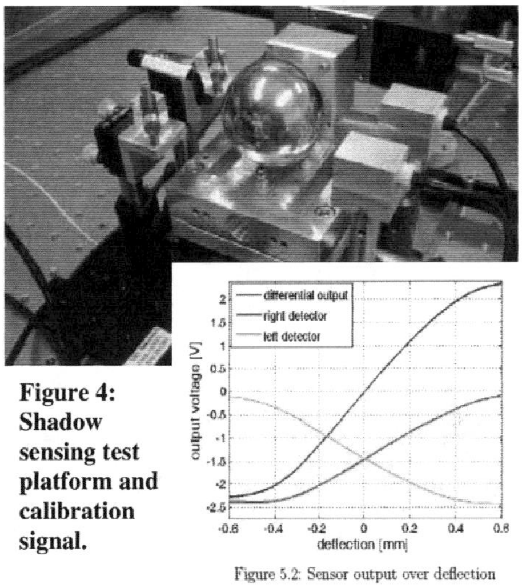

Figure 4: Shadow sensing test platform and calibration signal.

2.4. Optical Angular Sensing (Collaboration with Jet Propulsion Laboratory) [7, 8, 21]

We have improved the grating angular sensor by lowering noise and expanding the dynamic range. We have also constructed a vacuum enclosure for the entire grating angular sensor assembly. The photodetector and amplifier circuits now can receive higher laser power without saturation. With a mere working distance of 6 cm, and with an input laser power of 14 mW, we have observed an angular sensitivity of ~ 0.2 nrad/Hz$^{1/2}$ using the symmetric grating angular sensor. At low frequencies, we have achieved 1-2 nrad/Hz$^{1/2}$ at 1 Hz. The angular sensor will be applicable for both MGRS and space telescope steering.

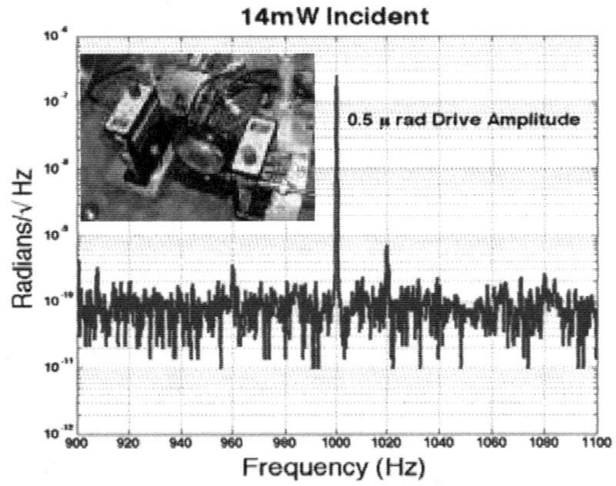

Figure 5. Measurement results from the grating angular sensor. The noise floor was below 2×10^{-10} rad/Hz$^{1/2}$. Inset: Grating angular sensor in vacuum chamber.

2.5. Diffractive Optics (Collaboration with Lawrence Livermore National Lab.) [18, 20]

We have expanded our diffractive optics work to characterize some LLNL gratings that may have applications in external interferometry, which requires high diffraction efficiency. We constructed grating cavities with extensive work in improving alignment, mechanical stability, interferometric calibrating the PZT actuation, and mode matching. Our highest observed finesse so far was 1002 ± 2.5, or a grating diffraction efficiency of 99.577 ± 0.002%. For higher power beam splitting applications, we tested the thermal characteristics of the dielectric gratings by illuminating a sample with up to 34.5 W of 1064 nm light in a 1.5 mm diameter spot without observing significant wavefront distortion with a Shack-Hartmann wavefront sensor.

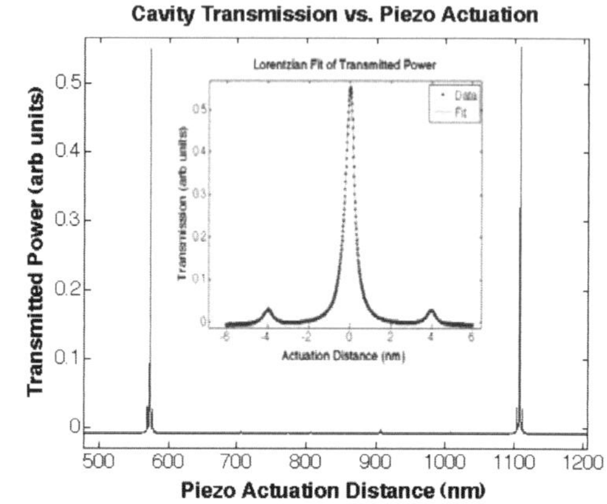

Figure 6. Grating cavity transmission versus cavity scan. Finesse of 1002 is demonstrated, representing diffraction efficiency of 99.577%. Inset: Blow up of the resonance portion.

2.6. Laser Frequency Stabilization Using a Grating Angular Sensor [7, 9]

The typical scheme for laser frequency stabilization utilizes a resonant optical cavity. Ambiguity in absolute frequency may occur due to periodicity in the cavity spectra. We propose to use the grating angular sensor as a non-resonant, robust laser frequency stabilizer. The grating angular sensor takes advantage of grating angular magnification and beam projection compression, thus exhibiting a high angular sensitivity of 0.1 nrad/Hz$^{1/2}$. The laser frequency deviation to produce such a small angle is ~ 300 kHz. Frequency stability at this level is sufficient for many practical applications, such as the absolute frequency indicator for LISA. In addition, the grating stabilizer will have a simpler structure and easier mechanical alignment.

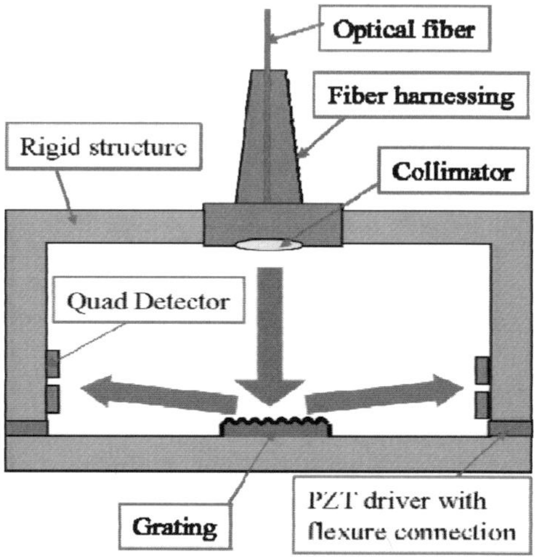

Figure 7. Structure of the grating laser frequency stabilizer

2.7. Center of Mass Measurement [7, 8, 22]

A new approach for measuring the mass center location of a spherical proof mass has been demonstrated to 150 nm precision. Knowledge of the mass center of a drag-free test mass is critical for calibrating the cross-coupling between rotational and translational degrees of freedom, and for inferring density inhomogeneities in the test mass material. In the past year, improvements in precision have come from the damping of mechanical vibrations due to the sphere rolling, careful isolation of the electronics to remove systematics caused by magnetic field fluctuations, shielding from air convection currents, and improving the detectors and data acquisition system design.

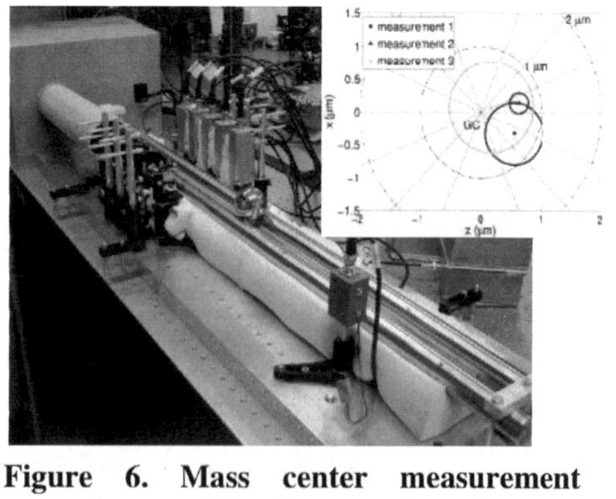

Figure 6. Mass center measurement apparatus, with improved vibration isolation and electronic noise reduction. Inset: Measurement of the mass center. Standard deviation < 150 nm.

2.8. Moment of Inertia and Self Gravity Attraction [16]

The five-wire torsion pendulum apparatus for moment of inertia measurements has been improved. A new version was designed and fabricated using SolidWorks CAD modelling software and a CNC milling machine. The improved design has a better mass balance and better tension distribution, resulting in reduced translational motion. In addition, a quad photo-detector was added to the grating angular sensor for higher sensitivity. The pendulum now can operate with higher spectral purity or signal to noise ratio. The amplitude spectral density plot shows a clean peak at the natural frequency around 3 Hz with error sources due to translation shifted above the measurement band.

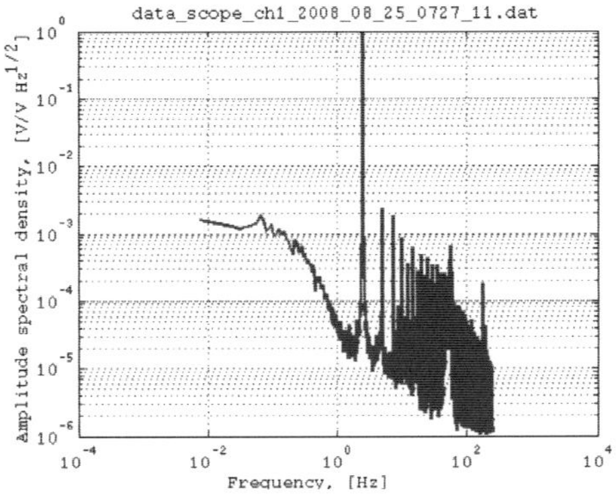

Figure 8. Spectrum from improved five wire torsion pendulum.

2.9. Proof Mass and Housing Fabrication [6][7][14]

Our efforts have focused on making a spherical test mass with preferred principal axis, accomplished using a TM with internally hollowed out portions and a spherical outer surface. For proof mass moment of inertia difference $\Delta I \geq 0.1I$ and a spin frequency > 10 Hz, the polhode frequency is > 1 Hz, above the science band of 0.1 mHz – 1 Hz. Fabrication of such a TM has been demonstrated using 50 mm diameter brass spheres with an average density loss < 20% [14]. Methods of lapping and polishing these spheres is under development with an ultimate goal of achieving an out-of-roundness < 100 nm.

Figure 9. Demonstration of assembly and lapping and polishing of 8 brass spheres with $\Delta I/I \sim 0.1$

2.10. Cubic Proof Mass Characterization [22]

In addition to developing the MGRS, our program is designed to help the LISA baseline configuration. We have designed several methods of measuring the mass center of a cubical proof mass. One technique would use our sphere mass center determination equipment with an internal kinematic mount for the cube inside a sphere which would have its mass center separated from the spherical holder mass center by reversal. A second method would use a torsion pendulum like apparatus shown in Fig. 10, but the cube would be offset a maximum distance from the center of rotation. The pendulum natural frequency depends quadratically on the distance between the mass center of the cube and the rotation axis. Mounted with its 4 major diagonals vertical there are 24 orientations of the cube in a triangular kinematic mount. Therefore, measuring the natural frequency with the cube in these various orientations allows us to accurately measure the cube's mass center. A third would develop a static pendulum with elastic hinges neutralized by a high mass center. A laser interferometer measures the neutral position of the pendulum. Reversal of the cubic mass on the triangular pendulum platform separates the mass center of the cube from that of the platform.

Figure 10. Five wire pendulum for proof mass characterization

2.11. UV LED Charge Management System [3][4][6][7][23]

We have continued UV LED power and spectral lifetime tests. The UV LED has now been operated more than 15,000 hours without significant power drop. The spectral shift is measured to be ~ 1 nm towards shorter wavelengths, which actually enhances photoelectric effects. Another power stability test has been initiated in a vacuum chamber several months ago. We have demonstrated that the UV LED is far superior to mercury lamps in reliability. Thus a UV LED based AC charge management system developed at Stanford should be the first choice for LISA and other high precision space flights requiring charge control.

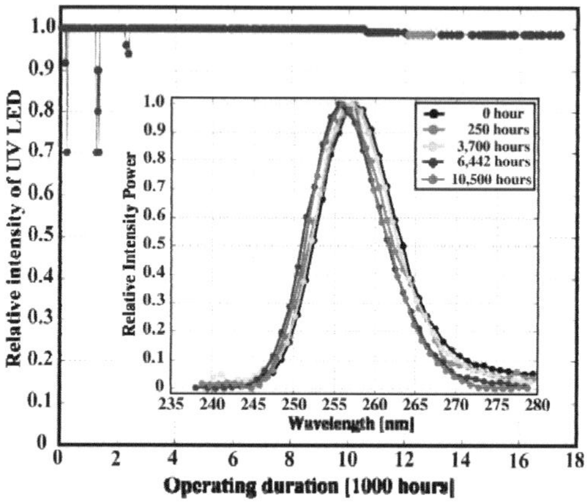

Figure 11. UV LED power and spectral stability.

2.12. UV LED Radiation Hardness Test [23]

We have conducted large dose radiation hardness tests using an accelerator source for 63 MeV protons. For proton fluence from 10^{10} to 10^{12} protons/cm^2, there was no significant power drop for UV LED light output at 255 nm wavelength. The UV emission spectrum also remains the same. This level of radiation test exceeded 100 years of radiation dose in the deep space LISA orbit. Therefore, we have demonstrated the extreme radiation hardness of UV LED. The combination of the successful tests in power lifetime, spectral stability, and radiation hardness have proven that UV LED should be primary choice for the charge management system for LISA and other high precision space flights.

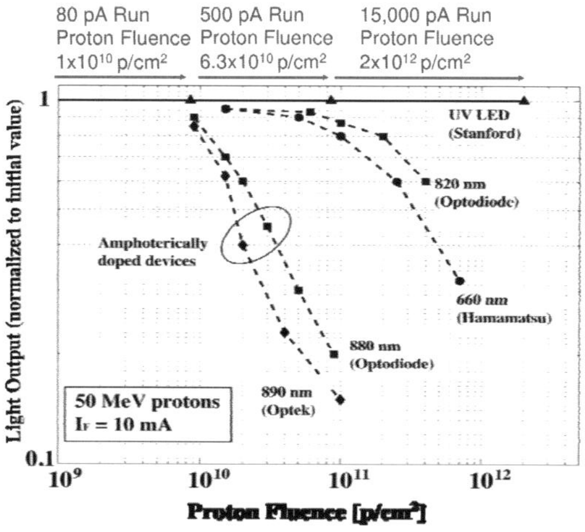

Figure 11. UV LED radiation hardness test. UV LED output power vs. proton fluence. Note: Data points other than UV LED are from Johnston et al. IEEE Trans. Nucl. Sci. p.2500 vol. 47 (2000).

2.13. Alternative Charge Management Scheme [11]

We proposed using ions and electrons of energy 1 eV–10 eV for neutralizing the charges on the non-conducting or isolated proof mass, which is possible for future GRS. By alternatively directing beams of positive and negative charges towards the mirror surfaces, we ensure the neutralization of the total charge as well as the equalization of the surface charge distribution. This method is compatible with operation in high vacuum, does not require measuring the potential of the mirrors and is expected not to damage sensitive optical surfaces.

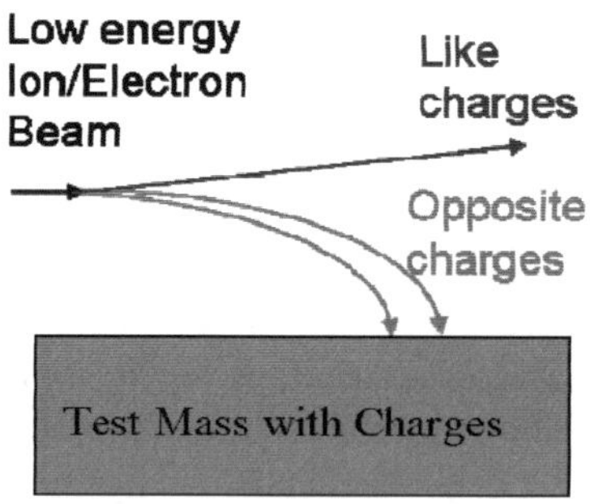

Figure 12. Charged particle neutralization for non-conducting surfaces.

2.14. Thermal Control [3] [24]
We are developing combined passive and active thermal control system with the goal of achieving sub microkelvin temperature stability and uniformity over a bench size volume. For the active control, we have developed a model predictive control (MPC) scheme, which will provide temperature controllability down to sub-microkelvin over the LISA science band. The very bottom of the curve in the Fig. 13 corresponds to the power spectral density of the new control law. For the passive control, we are designing a new thermal enclosure with multilayer structure with alternative conducting and insulating layers which insures the temperature uniformity and eases the burden on the active control. The upgraded thermal enclosure will be an important test facility for MGRS development. The stability of the new chamber is as low as 4 mK/Hz$^{1/2}$ at 1 mHz, which is a factor of 5 improvement over the insulation only enclosure. The uniformity improvement is expected to be even more significant.

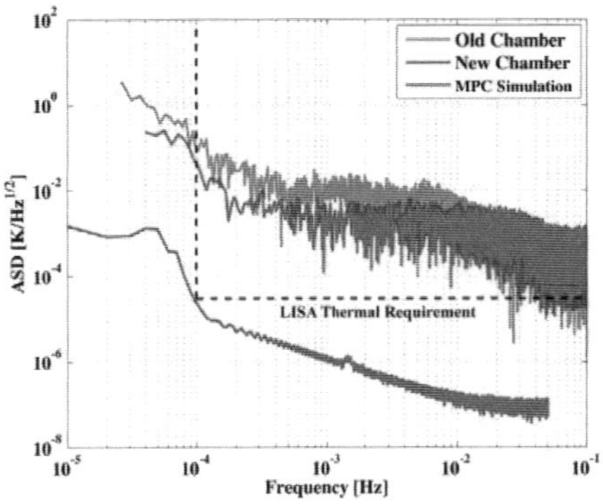

Figure 13. Results from model predictive controller model.

Figure 14. Newly constructed multi-layer thermal chamber.

2.15. Temperature Sensor
We studied temperature sensor stability and resolution of multiple thermistors. Four thermistors were mounted to a common thermal block with adjustable temperature, and connected to Wheatstone bridges. Temperature cycles of ~ 1 mHz were applied during typical periods of 72 hours. Thermocouples were used for coarse readout calibration. Temperature data shows that the correlations between thermistor readouts are typically greater than 98%. Therefore, multiple thermistors can be used to enhance signal to noise ratio and redundancy, while maintaining adequate consistency.

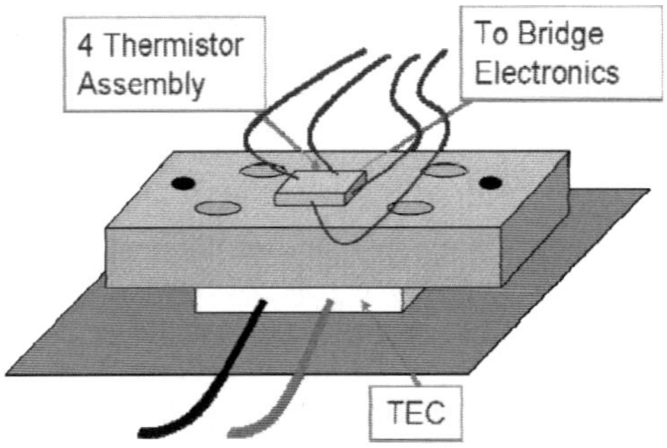

Figure 15. Thermistor, thermal couple (not shown), and TEC assembly for temperature sensor test.

3. Education and Science Outreach
The Stanford MGRS program has educated graduate, undergraduate, and high school students from diverse science and engineering backgrounds.

4. Conclusion
Our MGRS research has made significant progress in many key areas, in both enhancing the performance of the experiments and inventing new technologies. MGRS will contribute future precision gravitational space measurements.

Acknowledgements
This research was partially supported by NASA Beyond Einstein Foundation Science, Grant NNX07AK65G, for "Modular Gravitational Reference Sensor for Space Gravitational Wave Detection". We gratefully acknowledge the collaboration from Lawrence Livermore National Laboratory. The grating angular sensor research was funded by Jet Propulsion Laboratory.

References

[1] Ke-Xun Sun, G. Allen, S. Buchman, D. DeBra, and R. L. Byer, "Advanced gravitational reference sensor for high precision space interferometers", Class. Quantum Grav. 22 (10), S287 (2005).

[2] Ke-Xun Sun, Ulrich Johann (EADS Astrium), Dan B. DeBra, Sasha Buchman, and Robert L. Byer, "LISA Gravitational Reference Sensors," Journal of Physics CS 60 272–275 (2007).

[3] K.-X. Sun, S. Buchman, R. L. Byer, G. Allen, J. W. Conklin, D. DeBra, S. Higuchi, N. Leindecker, P. Lu, A. Swank, M. Trittler, "Technologies for an Advanced Modular Gravitational Reference Sensor," 18th International Conference on General Relativity and Gravitation, Sydney, Australia (2007).

[4] K.-X. Sun, N. Leindecker, S. Higuchi, S. Buchman, R. L. Byer, J. Hines, J. Goebel, E. Agasid, "Development of UV LED Based AC Charge Management Systems For Gravitational Wave Detectors," 7th Edoardo Amaldi Conference on Gravitational Waves, Sydney, (2007).

[5] Sasha Buchman, "Overview of GP-B Charging Issues" 2007 Workshop on Charging Issues in Experimental Gravity, Massachusetts Institute of Technology, July 26-27, (2007).

[6] Ke-Xun Sun, "UV LED AC Charge Management System for LISA and LIGO," Workshop on Charging Issues in Experimental Gravity, MIT, July 26-27, (2007).

[7] K.-X. Sun, S. Buchman, R. L. Byer, G. Allen, J. W. Conklin, D. DeBra, S. Higuchi, N. Leindecker, P. Lu, A. Swank, E. Torres, M. Trittler, "Technology Advances for LISA GRS and MGRS," LISA Science and Technology Team (LIST) Working Group 3 (DRS) Meeting, ESTEC, September (2007).

[8] Ke-Xun Sun for Stanford Team, "Progress in Optical Measurements for Science Signal and Spacecraft Control," LIST WG 2 (Interferometry) Meeting, ESTEC, September (2007)

[9] Ke-Xun Sun, Patrick Lu, and Robert Byer, "Laser Frequency Stabilization Using Diffractive Angular Sensors, Laser Science," Laser Science XXIII / OSA Frontier in Optics, San Jose, September (2007).

[10] S. Higuchi, D. B. DeBra, and S. Rock, "Sub-microkelvin Precision Thermal Control System Using Model Predictive Algorithm for Laser Interferometer Space Antenna (LISA) Ground Verification Facility", presented at the 22nd Annual ASPE Meeting, Dallas, Texas, October 14-19, (2007).

[11] S. Buchman, R. L Byer, D. Gill, N. A. Robertson, and K-X Sun, "Charge neutralization in vacuum for non-conducting, and isolated objects using directed low-energy electron and ion beams," Class. Quantum Grav. 25 035004 (2008).

[12] J. W. Conklin, G. Allen, K.-X. Sun, D. B DeBra, "Determination of Spherical Test Mass Kinematics with a Modular Gravitational Reference Sensor", AIAA J. Guidance, Control & Dynamics (2008).

[13] G. Allen, J. W. Conklin, K.-X. Sun, D. B. DeBra, R. L. Byer "Mass Center Position Determination of a Spinning Sphere as part of a Modular Gravitational Reference Sensor," to be submitted to AIAA Journal of Guidance, Control and Dynamics (2008).

[14] D. Gerardi, G. Allen, J. W. Conklin, K.-X. Sun, D. DeBra, S. Buchman, P. Gath, R. Byer, U. Johann, "Achieving Disturbance Reduction for Future Drag-Free Missions," to be submitted to CQG (2008).

[15] J. W. Conklin, D. Clark, M. Dreissigacker, M. Dolphin, M. Trittler, M. Ulman, and D. DeBra, "Fabrication of a Spherical Test Mass with a Preferred Principal Axis," in preparation (2008).

[16] Ke-Xun Sun, Saps Buchman, Graham Allen, Robert Byer, John Conklin, Dan DeBra, Sei Higuchi, Nick Leindecker, Patrick Lu, Aaron Swank, Edgar Torres, Martin Trittler, "Advances in Modular Gravitation Reference Sensor (MGRS) Technologies," 37th COSPAR, Montreal (2008).

[17] John W. Conklin, Graham Allen, Ke-Xun Sun, Saps Buchman, Robert L. Byer and Dan B. DeBra, "Modeling and Simulation of a Spinning Spherical Test Mass for Modular Gravitational Reference Sensor," 37th COSPAR, Montreal (2008).

[18] P. Lu, K.-X. Sun, R. L. Byer et al "Characterization of large size, high efficiency dielectric gratings," to be submitted to Optics Letters (2008).

[19] K.-X. Sun, M. Trittler, J. W. Conklin, R. L. Byer, "Differential Optical Shadow Sensing (DOSS) for LISA and MGRS Applications", 7th LISA Symposium (2008) (J. Physics C. This issue.)

[20] K-X. Sun, P. Lu, and R. L. Byer, "Characterization of High Efficiency Dielectric Gratings for Formation Flight Interferometry", *ibid*

[21] K.-X. Sun, P. Lu, and R. L. Byer ,"Grating Angular Sensor for LISA and MGRS Applications", *ibid*

[22] J. W. Conklin, A. Swank, K.-X. Sun, and B. DeBra, "Mass Properties Measurement for Drag-free Test Masses" *ibid*

[23] K.-X. Sun, N. Leindecker, S. Higuchi, S. Buchman, J. Goebel, R. L. Byer, "UV LED Operation Lifetime and Radiation Hardness Qualification for Space Flight", *ibid*

[24] S. Higuchi, K.-X. Sun, D. B. DeBra, S. Buchman, R. L. Byer, "Design of a Highly Stable and Uniform Thermal Test Facility for MGRS Development", *ibid*

Journal of Physics: Conference Series **154** (2009) 012027 doi:10.1088/1742-6596/154/1/012027

Recent LISA studies at the University of Colorado

M Nickerson, E Ames, and P L Bender

JILA, University of Colorado and NIST, Boulder, CO 80309-0440, USA

Email: nickermj@jila.colorado.edu

Abstract. We report on LISA experimental projects being pursued at JILA. Our focus is on the design and testing of a flight-compatible laser stabilization reference cavity. This is a dual cylinder ULE cavity, designed to provide high thermal and thermo-mechanical isolation in the millihertz frequency regime of interest to LISA. A modification of this hard-mounted design may allow for use in space without the need for clamping during launch. Progress so far consists of initial design, performance estimates, and construction. Simple thermal model calculations on the design indicate a thermal attenuation of 10^6 at 1 mHz, corresponding to a cavity strain of $3*10^{-16}$/rtHz for a 0.01 K/rtHz stability of the mounting surface. Finite element analysis indicates cavity strain attenuation of $5*10^7$ or better due to thermo-mechanical effects in the surrounding environment, and low sensitivity to vibration along the cavity axis. Setup and testing of two identical cavities and a laser-locking test system is ongoing. Another project was recently concluded, testing the low-frequency stability of commercial voltage references. Voltage reference performance is relevant to the stability of electrically applied forces on the LISA proof masses, and commercial references do not have well characterized noise in the sub-Hz regime. Our measurements confirmed that the best commercial reference was the AD587LN, with a typical noise of 2.1±0.6 *ppm*/rtHz at 0.1 mHz, in a temperature-stabilized environment of ~10mK/rtHz. This agrees closely with prior work by other groups.

1. Introduction

Considering highly stable laser reference cavities used in optical frequency standards, most designs have some sort of soft suspension. For use in space, such designs would need to be clamped down for launch. A fixed-mount cavity design obviates the need for a clamp system, reducing the overall complexity. Prior work has indicated that fixed, center supported cavities can be competitive with standard suspension designs [1]. Building on this work, we have developed a dual-cylinder cavity design that improves thermal isolation, while providing a center support to minimize vibrational disturbances along the axis. The aim of this design is to demonstrate the possibility of a flight-compatible cavity that provides high stability in the millihertz regime of interest to LISA. Progress so far in this project has consisted of the design and construction of two dual-cylinder cavities, and their support structures and vacuum chambers. The cavity design and performance estimates are presented.

2. Reference cavity design

2.1. Cavity construction

The laser reference cavity itself consists of two concentric cylinders and a connecting support disk, formed of a single contiguous piece of ULE. Both cylinders are 100 mm in length; the inner cylinder has a 10 mm ID and 30 mm OD, and the outer cylinder has a 50 mm ID and 70 mm OD. The 5 mm disk supporting the inner cylinder is vertically centered. A cutaway view of the cavity is shown in figure 1.

The inner cylinder forms the Fabry-Perot spacer, with optically contacted mirrors. It is centrally supported by the connecting disk, which provides symmetry along the cavity axis. For any acceleration along the axis, the change in length caused by the elastic deformation of one half of the cavity is compensated by a correspondingly opposite change in length of the other half. This provides a common-mode rejection of along-axis disturbances, which is the motivation for this support scheme. Thus, the outer cylinder can be firmly mounted to a lab surface, and the effect of any vibration of the surface will be strongly reduced.

The outer cylinder provides support for the connecting disk, and creates a stable environment for the inner cylinder. Any thermal changes must diffuse through the outer cylinder before affecting the Fabry-Perot cavity. Additionally, it serves as a mechanically isolating stage. ULE endcaps are epoxied to each end of the outer cylinder, to further isolate the inner cylinder from thermal disturbances.

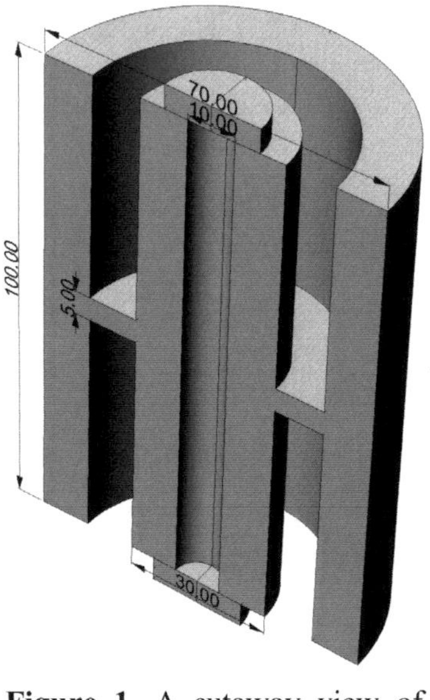

Figure 1. A cutaway view of the laser reference cavity. The inner cylinder forms the Fabry-Perot spacer.

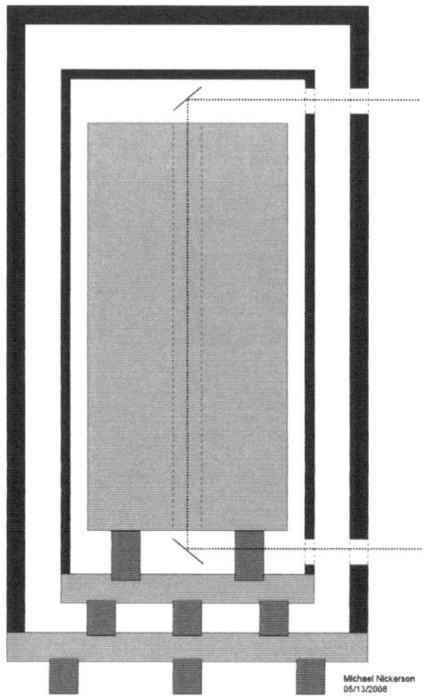

Figure 2. A diagram of the reference cavity and isolation platform. The lowermost legs rest on the vacuum chamber floor.

2.2. Cavity mounting

The laser reference cavity is mounted on a 55 mm high isolation platform, designed to provide thermal and thermo-mechanical isolation. The platform consists of three stages of 10 mm diameter ULE legs, separated by 10 mm thick Zerodur disks. The legs are attached to the disks and the lower cavity endcap with epoxy, and the lowermost set of legs rests on the floor of a vacuum chamber. Two aluminum thermal shields rest on the Zerodur disks, providing radiation shielding for the reference cavity. A diagram of this mounting scheme is shown in figure 2.

The three-stage platform results in significant thermal decoupling from the floor of the vacuum chamber, improving the thermal stability of the cavity. The series of legs also provides thermo-mechanical isolation, flexing to reduce thermally-induced strain changes in the vacuum chamber floor. The consistent use of low-expansion materials minimizes the thermo-mechanical contribution of the support structure itself.

3. Reference cavity performance estimates

3.1. Vibration isolation

Vibration performance estimates were developed using finite element analysis. To determine the cavity strain due to vibration, the cavity, endcaps, and upper mount legs were modeled with accurate dimensions and material properties. The upper mount legs were placed on 1 cm^2 blocks, which were given an upward acceleration. A high resolution FEA was then run, simulating the response of the system. An example result is shown in figure 3. Cavity strain was determined by averaging displacement over the inner cylinder's mirror surfaces.

Results of this finite element analysis indicate that the lengthwise strain on the Fabry-Perot cavity will be under $2*10^{-12}$ for a 1 m/s^2 acceleration along the cavity axis. This agrees well with preliminary calculations that were performed by T Rosenband of NIST.

Prior to design finalization, Rosenband also ran a series of similar tests to determine the effect of altering the vertical position of the connecting disk. As a result of this, the final design offsets the disk from the center by 190 microns. More detailed simulations agree with Rosenband's result to within manufacturing tolerances; the cavity strain due to vertical acceleration is minimized with a disk offset of ~0.20 mm.

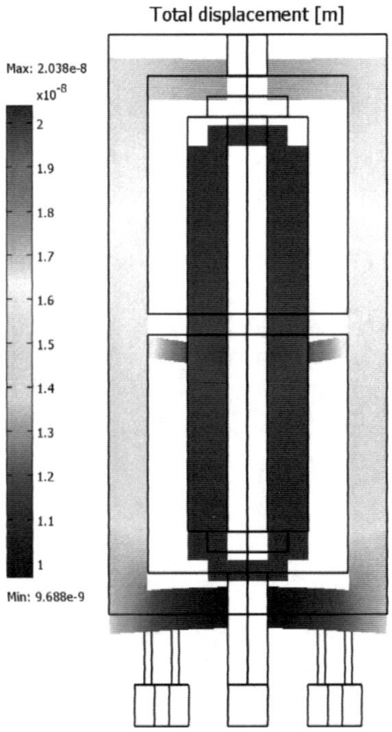

Figure 3. A slice of the laser cavity and upper stage mount, showing the relative displacement for a 1g vertical acceleration. The cavity length strain here is $1.4*10^{-11}$.

3.2. Thermo-mechanical isolation

Performance estimates of the thermo-mechanical isolation provided by the mounting system and outer cylinder were also determined using finite element analysis. Due to computation resource limitations, this was done in two stages, and only provides a lower limit on the isolation. The three-stage mounting system was first modelled without the cavity. A radial displacement was applied to the feet of the lowermost legs, simulating thermal expansion of the vacuum chamber floor. Running the FEA then determined the radial displacement at the base of the uppermost legs of the mounting system. A new system consisting of the uppermost legs and full cavity was then modelled. The radial displacement determined from the first simulation was applied to the base of the upper legs, and the cavity response simulated.

The results of this pair of simulations indicate that the cavity strain will be less than $2*10^{-8}$ that of any thermally induced strain at the base of the mounting system, i.e. a strain attenuation of $5*10^7$ or better from the base of the mounting system to the cavity length. Actual performance is likely to be better than this due to several factors. The simulation overestimates the movement of the uppermost legs, as the stiffness of the cavity is not present when determining the first stage simulation. An additional strain-attenuation factor of ~20 may also be present, due to the epoxy connections between the legs, disks, and endcap. Preliminary simulations indicate that there is a factor of ~2.5 strain attenuation across each pair of connections due to the presence of a 0.2 mm thick layer of epoxy. This has not been tested extensively, but indicates that the epoxy connections will have some beneficial contribution.

3.3. Thermal isolation

Estimates of the thermal isolation were developed using a simple thermal time constant approach. Assuming that the surfaces of the cavity are uncoated ULE, the thermal time constant for the inner cylinder responding to uniform changes in temperature of the outer cylinder is roughly 800 s. There are comparable contributions from radiation and conduction. At 1 mHz, this gives an attenuation factor of 5.

Assuming an emissivity of 0.05 for the aluminum thermal shields, and a high emissivity of the vacuum chamber, the time constants due to radiative coupling are estimated to be 25000 s from the vacuum chamber wall to the outer radiation shield, 60000 s between the thermal shields, and 58000 s from the inner thermal shield to the outer ULE cylinder. At 1 mHz, the total thermal attenuation through this path is $2*10^7$.

The conduction path from the vacuum chamber to the outer cylinder is considered as three steps in series. From the vacuum chamber to the lower disk through 3 legs has an estimated time constant of 3800 s; from the lower disk to the upper disk through 3 legs is 2400 s; and from the upper disk to the center of the outer cylinder through 3 legs and half of the outer cylinder is 31000 s. Thus the overall attenuation of the conductive path at 1 mHz is $2.2*10^5$. This is the dominant transfer mechanism by two orders of magnitude, so the radiative transfer path can be neglected.

The product of the attenuation factors from the conduction path to the outer cylinder and then between the two cylinders is $1.1*10^6$. For a thermal fluctuation on the floor of the vacuum chamber of 10 mK/rtHz, and the known $3*10^{-8}$/K ULE thermal expansion coefficient, the estimated cavity strain is only $3.3*10^{-16}$/rtHz at 1 mHz. The model used to make this estimate is crude, but the results are encouraging.

4. Voltage reference stability

One limit to the stability of electrically applied forces on the LISA proof masses is the noise of onboard voltage references in the relevant frequency range. Commercial voltage references do not have well characterized noise at low frequencies, and thus warrant investigation. We recently concluded a low-frequency performance investigation of a variety of references. The results from this study supplement earlier work done by Heinzel et al. in 2006 [2].

Noise spectra were obtained by running a pair of voltage references through a differential amplifier, and recording this output for analysis. Moderate temperature control (~10mK/rtHz) was applied to the references and amplifiers. Of the potential external noise sources, the differential amplifier was found to have under a 7% contribution to the overall noise. The measurement circuit and power supply were found to have similarly low noise contributions. Temperature correlation was also determined to be under 10^{-6} V/K, negligible in this case.

Many voltage references were tested, but four performed significantly better than the rest, and were selected for more detailed examination: the AD587LN, LT1021, MAX6126, and LT1236 references. Table 1 and figure 5 present the results of this analysis, giving the stability and a typical spectrum for the four chosen references. The best performer was the AD587LN, with a relative stability of 2.1±0.6 *ppm*/rtHz at 0.1 mHz [3]. These results are consistent with those obtained by Heinzel et al [2].

Table 1. Relative noise levels of several commercial voltage references at 0.1 mHz.

Voltage Reference	Output Voltage (V)	Noise Level (*ppm*/rtHz)
AD587LN	10	2.1 ± 0.6
LT1021	5	7.5 ± 1.1
MAX6162	5	5.3 ± 1.5
LT1236	10	6.6 ± 2.2

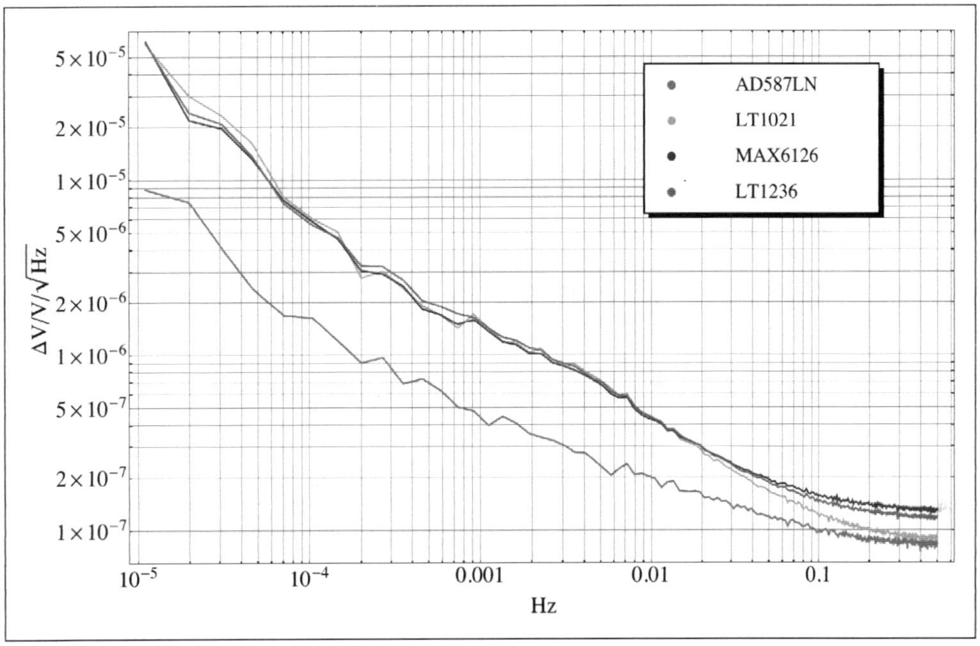

Figure 5. Typical noise spectrum of the AD587LN (lower line), LT1021, MAX6126, and LT1236 voltage references. The vertical axis is fractional noise level per rtHz.

Acknowledgements
Funding for this work was provided by NASA under NASA agreements NNG05GH33G, NGT-50451, NNG06GB73G, and NNX08AB62B. It is a pleasure to thank the following: T. Rosenband for preliminary cavity performance calculations during the design phase; J. L. Hall for insight regarding cavity design and performance, technical guidance during construction, and assistance with the laser setup; M. Notcutt and R. Lalezari for advice on the cavity design; C. D. Hamley for his contributions to the early parts of this work; and D. B. Pinegar for information on low-noise voltage references.

References

[1] M. Notcutt, L.-S. Ma, J. Ye, and J. L. Hall, Opt. Lett. 30, 1815 (2005).
[2] Heinzel et al., AIP Conf. Proc. 873, Laser Interferometer Space Antenna, 6th Int. LISA Symp., S. Merkowitz, J. Livas, 291-296 (2006).
[3] Ellery Ames, Characterization of Voltage Reference Noise for the Laser Interferometer Space Antenna, Honors Thesis, Univ. of Colorado (2008).

UV LED Operation Lifetime and Radiation Hardness Qualification for Space Flights

Ke-Xun Sun, Nick Leindecker, Sei Higuchi, John Goebel*, Sasha Buchman, and Robert L. Byer

Hansen Experimental Physics Laboratory, Stanford University, CA 94305, USA
*Code RET, NASA Ames Research Center, Moffet Field, CA 94035, USA

kxsun@stanford.edu

Abstract. We report measurements of ultraviolet light emitting diode (UV LED) performance under conditions simulating operation in an orbiting satellite. UV LED light output maintained within less than 3% observational uncertainty, over more than 19,000 hours of operation in a nitrogen atmosphere, and over 8,000 hours operation at a pressure of less than 10^{-7} torr vacuum. In addition, irradiation with 63 MeV protons to a total fluence of 2×10^{12} protons/cm^2 does not degrade the UV light output. Spectrally, the emissive center-wavelength and spectral shape are unchanged after proton irradiation within the precision of our measurement. These results qualify the UV LED operation lifetime and radiation hardness for space flights

1. Introduction

A radiation-hard, high-reliability ultraviolet (UV) light emitting diode (LED) will be of great value to future satellite missions requiring charge management technology employed in drag-free flights.

In drag-free control systems, the inertial reference of a satellite is a free-falling mass. The proof mass housing or satellite body shields this proof mass from external forces other than gravity. However, this isolation from external forces is poor if the housing itself exerts significant forces on the proof mass. Cosmic ray caused net electric charge on the proof mass will couple undesirable disturbance forces Minimization of surface differences of electrical potentials between the proof-mass and housing is thus crucial for optimal drag free operation.

Through Au surface photoelectron generation, UV light at ~255 nm renders an effective tool for charge management of the proof-mass and housing system. The recent breakthrough development of the deep UV LED based on gallium nitride (GaN) provides an attractive source. Compared with the traditional mercury vapor discharge lamps, the UV LED is lighter in weight, more compact in volume, lower in power consumption, and faster in time response. These advantages are more substantial when a UV source is required on a satellite where power and mass are at a premium. We have demonstrated successful AC charge management using a modulated UV LED in 2005 [1-4]. Since then, several groups have demonstrated similar UV LED utilization in precision experiments.

For deployment in extended duration space flights, the long term power and spectral stabilities, and radiation hardness of the UV LEDs need further verification. The operating lifetime of visible light LEDs have been quoted as long as 100,000 hours by industrial vendors. GaN materials and devices have been suggested insensitive to lattice defects and radiation [5]. UV LEDs emerged very recently, and there have been no prior tests for radiation hardness.

We report experimental results that conclusively demonstrate the long term power and spectral stabilities, and radiation hardness of GaN/AlGaN based UV LEDs. In the ongoing measurements at Stanford, we have reliably operated UV LED under the AC charge management working parameters for more than 19,000 hours in nitrogen and more than 8,000 hours under vacuum while observing no significant power loss and spectral emission wavelength shift. We have further demonstrated radiation hardness of a UV LED with 63 MeV fast proton irradiation doses up to a fluence of 2×10^{12} protons/cm^2. Our experimental results indicate that the reliability of UV LEDs exceeds that of gas lamps, and therefore UV LEDs are appropriate for extended space missions such as LISA, and beyond.

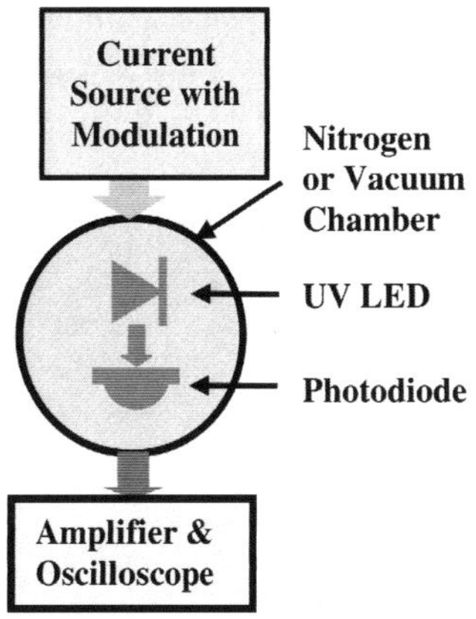

Figure 1. Experimental setup for UV LED operational lifetime tests.

2. UV LED Operation Lifetime Measurements

2.1. Long term power stability for operation in nitrogen

The experimental setup for UV LED operation life time test in a nitrogen purged chamber is shown in Fig. 1. It remains the same as presented at LISA 6th symposium [2]. The nitrogen atmosphere was chosen for more rigorous testing in a space-like environment, as there have been claims that oxygen may "clean" the surface contaminated with molecules disassociated by UV light.

The devices being tested is a nominally 255 nm deep UV LED (UVTOP255) available from Sensor Electronic Technology. The UVTOP LED is packaged with a flat UV grade fused silica window. A UV enhanced detector (OSI UV-20) is used to monitor the UV LED light output continuously. The detector is operated with zero bias in a transimpedance configuration for reduced dark current and lower temperature sensitivity. The UV LED is driven with ~10% duty cycle pulses of ~2 mA at a rate of 1 kHz. These settings represent those used for AC charge management experiment, in which phased light and bias voltage modulations direct electron movements [1].

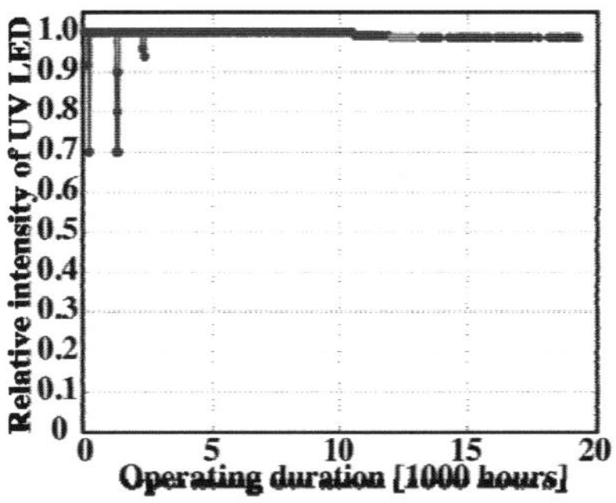

Figure 2. UV LED operational lifetime test in nitrogen reaches more than 19,000 hours

As shown in Fig. 2, the UV LED light output power level is stable for more than 19,000 hours, with an uncertainty of 3% or less. A few early hiccups were due to temperature fluctuation during the test. The minute power drop later was possibly caused by lab move, during which the test apparatus was disassembled and assembled again.

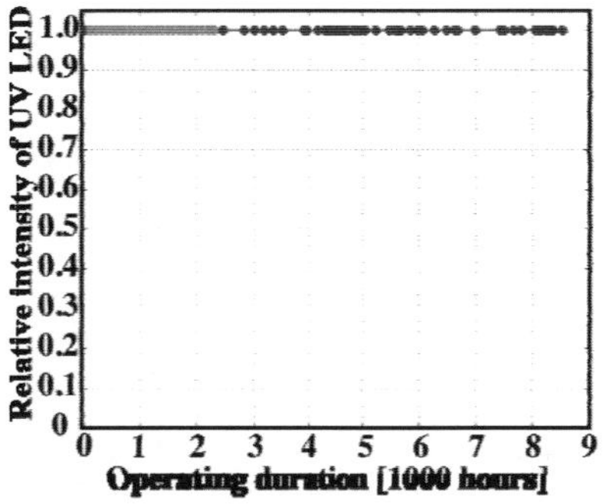

Figure 3. UV LED operational lifetime test in vacuum reaches more than 8.000 hours.

2.2. Long term power stability for operation in vacuum

To fully simulate space vacuum environment, we have placed a UV LED and a UV detector in an ion-pumped vacuum chamber with a pressure maintained at lower than 10^{-7} torr. The UV LED is packaged with a ball lens which substantially collimates the light output. The driving current was modulated at 20 kHz, using an improved electronics box built in-house. Figure 3 shows the ball-lens UV LED light output level during the test. The UV LED has been operated for more than 8,000 hours in vacuum while maintaining its initial power level within 3% measurement uncertainty. The UV LED in the high vacuum environment shows that the UV LED is on track for excellent long-term power maintenance.

2.3. Long-term spectral stability

The UV LED emission spectral stability is another important parameter for successful charge management. If the emission spectrum shifts to longer wavelengths, quantum efficiency for photoelectron generation will be reduced.

For the flat window UV LED maintained in the nitrogen-purged chamber, the UV LED emission spectrum was measured at several points during the test. The UV LED was removed from the chamber for these measurements. The UV spectrum was recorded using a grating UV monochromator (Resonance Research) coupled through UV transmissive 400 μm optical fiber to a photoconductive detector biased at -15V.

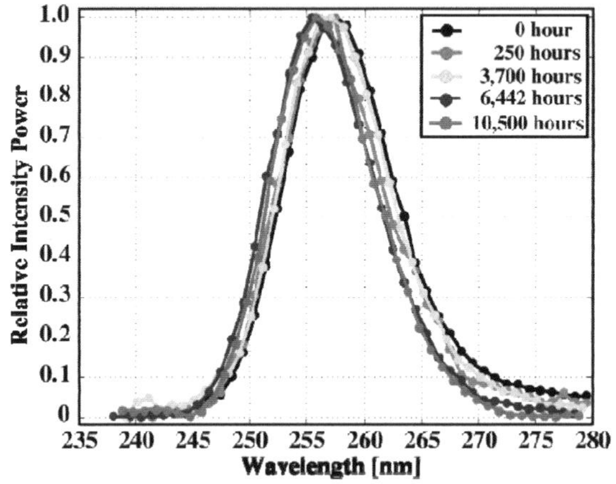

Figure 4. UV LED spectral measurements during the first 10,500 hours of operation.

Figure 4 shows the spectral measurement results from the beginning (0th hour), and at 250, 3700, 6442, and 10,500 hours of operation. In the first three measurements, the spectral center maintained its original peak wavelength position at ~257 nm within 0.5 nm. For the measurement at 6,442 and 10,500 hours, the central spectral peak shifted to 255-266 nm. The measurement is estimated as +/- 1 nm, due to the instrumentation spectral resolution, but mostly due to the mechanical rotational backlash of the grating element in the spectrometer. The small shift to shorter wavelength, if it is real, will not affect the charge management based on Au photoelectric effects.

2.4. Other laboratory uses of UV LEDs

In addition to these intentional lifetime tests, we have used several UV LEDs as UV light sources for other tests such as investigation of mirror coating degradation by UV light irradiation. In these usages, the UV LEDs were driven to produce ~10-15 times higher UV light output than that for charge management for several months. From these routine uses of UV LEDs, we have not observed appreciable UV output power degradation.

3. UV LED Radiation Hardness Measurement

3.1. Radiation hardness of the LED and particle detectors

Hardness against ionizing particle radiations is a necessary requirement for space-borne instruments. The components on the satellite must operate robustly or maintain its basic function in the presence of the high fluence particle radiation. There have been excellent studies on visible light and infrared (IR) LED radiation hardness by Johnston et al (see [6] and references therein) who compared performances of LEDs from several vendors. They used the 50-MeV protons from the 76" cyclotron Crocker Nuclear Laboratory at the University of California at Davis, and tested several LEDs emitting in the visible and near IR region. LEDs that were radiation hard for from high 10^{10} to 10^{11} protons/cm^2 were considered well suited for use in space flight. Particle detection in a large accelerator facility such as LHC has generated more intense interest in GaN detector. There have been hopes to use a GaN detector at 10^{15} neutrons/cm^2 fluence [7]. However, there has been no report for AlGaN UV LED radiation hardness measurement.

3.2. Experimental setup for radiation hardness measurement of UV LEDs

To determine the radiation hardness for the UV LEDs, we have used the same proton source as that in reference [6], i.e. the Crocker cyclotron at UV Davis. However in our experiments, most significant exposure was conducted with 63 MeV protons. Figure 5 shows the experimental setup. Two 255 nm UV LEDs were placed in the proton beam path formed by the cyclotron beam output aperture and by an aperture in a piece of thick aluminum block. Further, to reduce potentially serious radiation induced artefacts in the monitor photodiodes, the UV LEDs were obliquely oriented to allow the UV detectors to be placed outside of the primary proton beam. Further, to further reduce the radiation effects due to migrated particles from cascaded scatterings, we elected to use a SiC photodiode for detection of light from one UV LED. SiC has similar crystal structure as that of GaN, and has been considered radiation hard compared with Si or Ge materials. The selection of a SiC photodiode will practically enhance the radiation hardness of the UV LED charge management system. The second monitoring photodiode was a silicon detector (UV-20).

The UV LEDs were operating electrically during the proton radiation, to better simulate the space flight working conditions. We again used 10% duty cycle, 20-kHz modulated drive current for UV LEDs. The UV light emission from the UV LEDs was continuously monitored during the proton beam time. The average DC value of the photocurrent was recorded for the UV light emission measurement. The protons passed obliquely though the bottom of the UV LED TO-39 package before encountering the LED die. The average proton energy at the die is thus likely slightly less than the nominal 63.8 MeV. For the high flux measurement, the Crocker facility calibrated the proton fluence. For the low flux measurement, in which our experiment is a secondary package, the proton flux entering our apparatus was measured with a Faraday cup.

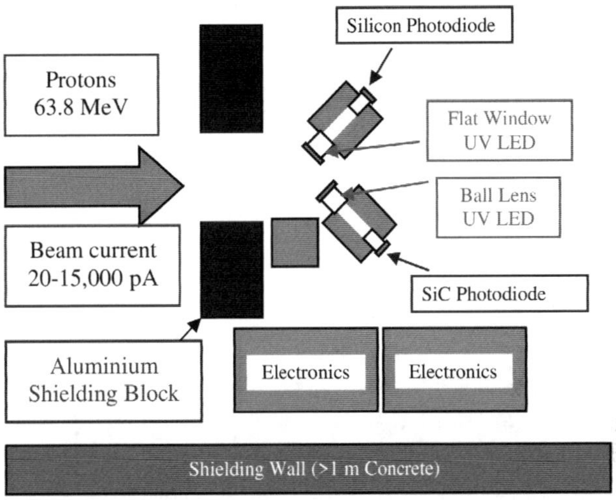

Figure 5. Experimental setup (top view) for UV LED proton radiation tests.

3.3. UV LED light output level during the proton irradiation

Figure 6 shows the measurement result for the UV LED light output level detected by the SiC detector, while the proton flux was accumulatively increased to 2×10^{12} protons/cm^2. SiC detector was used for detecting UV LED output level. We have four runs of the cyclotron. The first three were for purposes of the primary experiment. The first is a very low fluence calibration run to set the proton flux. In the second run, the proton fluence was increased to 1×10^{10} protons/cm^2. In the third run, the proton fluence was increased to 6.3×10^{10} protons/cm^2. The fourth run was a very high beam current (15,000 pA) dedicated to our UV LED experiment. We did not observe an appreciable UV LED light output drop during real time monitoring of the photodiodes for the UV LED and SiC detector pair. This experiment result demonstrates the extreme radiation hardness for the GaN/AlGaN UV LEDs.

There was a small photocurrent drop (~10%) from the silicon photodiode. It is suspected that the drop was due to radiation damage of the silicon photodiode by scattered protons through the mounting block of the photodiode. We are planning more tests for different UV LEDs and UV detector pairs.

Journal of Physics: Conference Series **154** (2009) 012028 doi:10.1088/1742-6596/154/1/012028

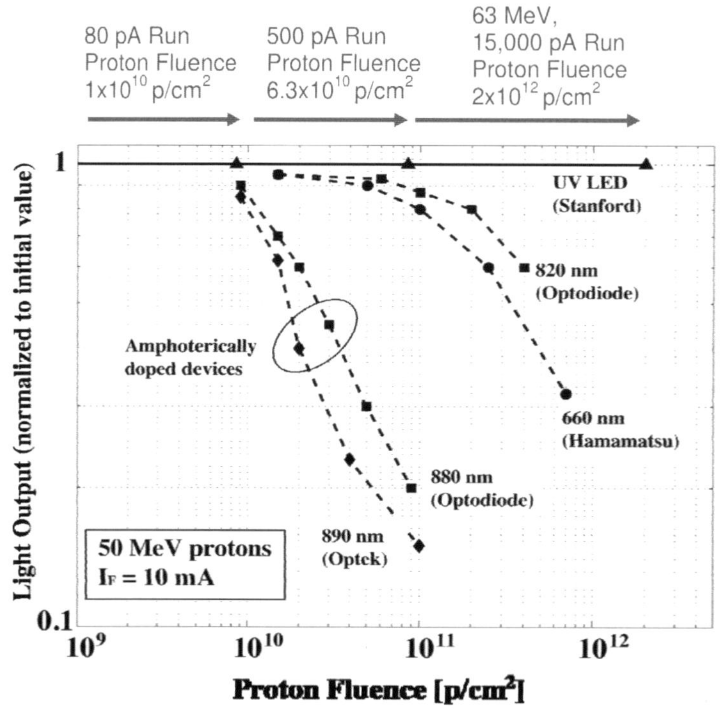

Figure 6. UV LED light output vs. proton fluence. The SiC photodiode was used. Data points other than UV LED are from reference [6]

3.4. UV LED emission spectrum measurements before and after the proton irradiation

Both of the UV LEDs were characterized before and after the proton irradiation. The latter was conducted when UV LEDs further cooled off from the high fluence proton activation. The same spectrometer described earlier was used for the spectral measurement. Figure 7 shows the measurement result. These measurements reveal no shift of either the center wavelength or the spectrum half-width in either device.

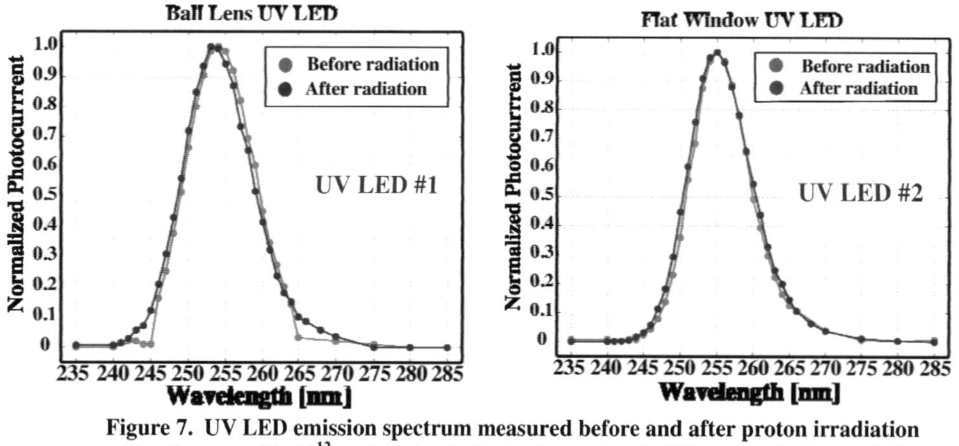

Figure 7. UV LED emission spectrum measured before and after proton irradiation with a fluence of $2{\times}10^{12}$ protons/cm2. Central wavelength 255 nm for both, no shift

4. Summary and conclusion

Our long term measurements show that optical power emitted by AlGaN UV LED is stable to within 3% over 15,000 hours in gaseous nitrogen. Equivalent observations of a similar device in high vacuum show essentially no effect in 5,000 hours. From a system engineering point of view, this result is especially encouraging given that we observe only the monitor photocurrent which lumps in all possible effects of reduced UV LED output, increased absorption in the ball lens or photodiode window, or reduced detection efficiency of the detector. Thus ~3% represents a worst case contribution from the UV LED itself.

Spectral measurements made periodically as the device is aged show convincingly that a critical device characteristic, the center emission wavelength, does not change to long wavelength as the device ages. The consistent center wavelength and spectral width ensure that there is no disproportionate drop of photons at the short wavelength end of the spectrum.

Two UV LEDs were irradiated with ~63MeV protons to an accumulated fluence of 2×10^{12} protons/cm^2, or ~100 years of radiation dose in the LISA orbit. Measurements of the light output and the spectrum show that there are no significant changes in UV LED light output. Spectral measurements before and after proton irradiation also reveal no significant shift in center wavelength or spectral width.

The reliable long term operations of AlGaN UV LED in either dry nitrogen or high vacuum, and under high flux radiation are critical steps for space qualification of UV LED as high reliability devices in the space environment. A system based on UV-emitting LED thus appears to be an attractive alternative to mercury discharge lamp for LISA proof mass charge control, as well as other precision space missions requiring minimized charge induced noise and damages.

Acknowledgement

This research was partially supported by NASA Grant NNX07AK65G, "Modular Gravitational Reference Sensor for Space Gravitational Wave Detection". We gratefully acknowledge Sensor Electronics Technology for providing UV LEDs. We are indebt to Mark McKelvey, Robert McMurray, and Paul Davis from NASA Ames Research Center for providing proton time. University of California Davis Croker Proton Facility staff operated the proton source and provided calibration.

References

[1] K.-X. Sun, B. Allard, S. Williams, S. Buchman, and R. L. Byer, "LED Deep UV Source for Charge Management," presented at Amaldi 6 Conferences on Gravitational Waves, June 2005, *Classical and Quantum Gravity*, 23(8):S141-S150, 2006.

[2] K.-X. Sun, S. Higuchi, A. Goh, B. Allard, D. Gill, S. Buchman, and R. L. Byer, "Spectral and Power Stability Tests of Deep UV LEDs for AC Charge Management," 6th LISA International Symposium, GSFC June 2006, AIP proceeding vol. 873, pp.215-219

[3] K.-X. Sun, S. Higuchi, B. Allard, D. Gill, S. Buchman, and R. L. Byer, "LIGO Test Mass Charging Mitigation Using Modulated LED Deep UV Light, " LIGO Science Collaboration (LSC), OWG & SWG Joint Meeting, Hanford, Washington, March 22, 2006, LIGO Documentation G050143-00-Z

[4] Stanford University Patent Application on Charge Management with Modulated UV Light Sources (K.-X. Sun, B. Allard, S. Buchman, and R. L. Byer)

[5] Osinski, "Short wavelength visible light emitters based on group III nitrides and their potential for space applications", SPIE Proc. CR-66, p. 92-120 (1997)

[6] A. H. Johnston and T. F. Miyahira, "Characterization of Proton Damage in Light-Emitting Diodes", IEEE Trans. Nuclear Science, p. 2500-2507, Vol. 47 (2000)

[7] J. Grant et al, "GaN as a radiation hard particle detector", Nuclear Instruments and Methods in Physics Research A vol. 576 p.60–65 (2000)

Journal of Physics: Conference Series **154** (2009) 012029 doi:10.1088/1742-6596/154/1/012029

Alternative Opto-Mechanical Architectures for the LISA Instrument

Dennis R Weise, Pierangelo Marenaci, Peter Weimer, Hans Reiner Schulte, Peter Gath, and Ulrich Johann

EADS Astrium GmbH, Claude-Dornier-Str., 88039 Friedrichshafen, Germany

E-mail: dennis.weise@astrium.eads.net

Abstract. As part of the on-going LISA Mission Formulation study under ESA contract, EADS Astrium has recently suggested and investigated a variety of novel LISA payload architectures utilizing so-called "In-Field Pointing" for accommodation of seasonal constellation dynamics. Here, the annual variation in the angle between the interferometer arms of roughly $\pm 1^{\circ}$ is compensated by steering the lines of sight of the individual telescopes with a small actuated mirror located in an intermediate pupil plane inside the telescopes. This introduces a certain flexibility for the overall payload configuration and allows for very compact designs. In particular, it enables a "single active proof mass" mode with a true cold redundancy between a nominal and a backup GRS system on board each spacecraft, and thus enhances mission robustness.

1. Introduction

A main task of the recent phase of the LISA Mission Formulation study conducted by EADS Astrium under ESA contract has been the investigation of principal alternatives for the realization of the LISA opto-mechanical payload architecture. Although a well-consolidated baseline design utilizing two active proof masses had been developed [1–3], the investigation of other approaches to realize the LISA instrument is considered essential in order ensure that the optimum configuration for the mission is found, with respect to performance, technical feasibility, complexity, risk, maturity, implementation programmatics, cost, etc.

One of the most promising alternatives to the original baseline configuration identified by EADS Astrium is a payload concept with only a single active proof mass per spacecraft, combined with the application of "In-Field Pointing" for the accommodation of constellation breathing. The underlying idea of this concept is that, from a conceptual point of view, a striking, intuitive, and elegant configuration for the LISA constellation would be one with a total of just three proof masses, located in full free fall at the corners of the constellation triangle, as schematically illustrated in Figure 1.

In-Field Pointing refers to one possibility for steering the active lines of sight of the two telescopes on board each spacecraft, which is necessary in order to follow the seasonal variation of the angle between the interferometer arms of the constellation triangle due to orbital mechanics. Beam steering is accomplished here within the field of view of each telescope by a small, actuated mirror, located in an intermediate pupil plane of the telescope. Since thus the telescopes can be rigidly attached to a common optical bench, In-Field Pointing integrates more coherently

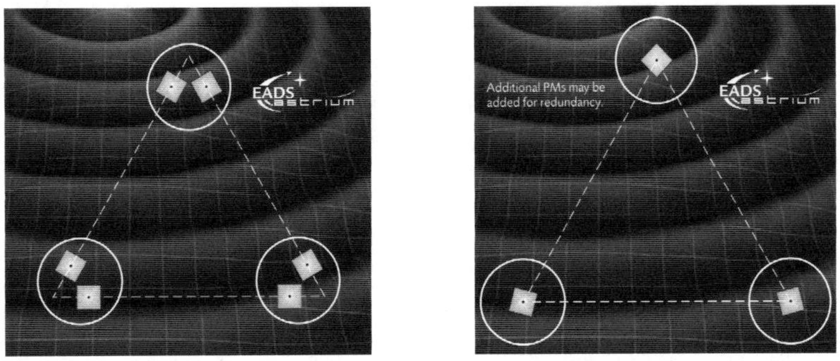

Figure 1. Historic "Two Active PM" configuration (left) and pure drag-free "Single Active PM" configuration (right).

with a Single Active Proof Mass architecture than the alternative of "Telescope Pointing". In Telescope Pointing, line of sight steering is accomplished by rotating the entire assembly of telescope, optical bench, and Gravitational Reference Sensor (GRS) as a whole, so that in principle each interferometer arm has to be delimited by its own proof mass.

While the technical realization of In-Field Pointing in compliance with the various performance requirements of the LISA instrument is extremely demanding, it still appears to offer a number of crucial benefits with respect to the original baseline configuration, when combined with a Single Active Proof Mass configuration:

- Since each spacecrafts actively operates only a single GRS system at time, a **full cold redundancy** to a potential second GRS system on board can be realized. This improves mission reliability and mitigates issues concerning GRS caging.
- Depending on the choice of proof mass geometry, a virtually **pure drag-free condition** can be realized, thereby reducing the criticality of electrostatic guidance and noise. Even if the heritage from the LISA Pathfinder mission is maximized by maintaining a cubical proof mass shape, only the rotational degrees of freedom require suspension and a simplification of the DFACS design should remain possible. In particular, detrimental stiffness cross coupling from suspension of the lateral degrees of freedom is avoided, so that an overall reduction of acceleration noise on the PM appears feasible.
- By enabling a "single-piece" opto-mechanical system with a **single optical bench**, In-Field Pointing avoids the need for the so-called backside fiber link, which is required in architectures using Telescope Pointing to establish an optical phase reference between the two separate optical benches on board each spacecraft. Its feasibility in terms of reciprocal pathlength-stability has not yet been experimentally demonstrated to date, and thus remains a potential show stopper for the current baseline design.
- Telescope Pointing requires the actuation of a large mass with extremely low torque noise, which may not be representatively testable under 1 g. In contrast, In-Field Pointing should ensure full on-ground testability of the beam pointing. Furthermore, a flexible harness to the payload core is avoided, so that potential impacts on the pointing performance due to harness cracking are mitigated. Launch locks for the pointing device are no longer required. Finally, continuous changes in self gravity due to breathing accommodation are minimized by the small size of the active components.
- In-Field Pointing introduces a high **flexibility** for the choice of the overall payload layout, and enables very compact and thermo-mechanically stable architectures. It thus provides the potential for savings in overall mass, volume, and power consumption with respect to the baseline configuration.

Journal of Physics: Conference Series **154** (2009) 012029 doi:10.1088/1742-6596/154/1/012029

To demonstrate that the above features can principally be materialized, a coherent payload architecture with a Single Active Proof Mass and In-Field Pointing has been developed, solving in particular the two main challenges of designing a suitable telescope providing the required wide field of view in the plane of pointing, and controlling the pathlength variations caused by active pointing to an acceptable level within the measurement band.

2. In-Field Pointing

Since the seasonal variation in the angle between the interferometer arms is about $60 \pm 0.8°$ [4], both telescopes on board each spacecraft require a field of view of at least $\pm 0.5°$ (± 9 mrad) within the constellation plane, if the actuation range is shared between them and margin is included. Orthogonal to the pointing plane, i. e. in the out-of-plane direction, the required field of view is given by the maximum acquisition range to be accommodated, which is on the order of $\pm 200\,\mu$rad.

Further constraints driving the design of the telescope have been the need to minimize stray light into the receive path, as well as maintaining ease of accommodation without compromising optical pahtlength stability. Both aspects are provided by the specialized off-axis system illustrated in Figure 2. It achieves an intrinsic avoidance of stray light through the absence of any obscuration in a fully reflective, two-stage design. The back optics is folded for straightforward interfacing with a vertical optical bench housing the main interferometry.

The design provides an intermediate pupil at a magnification of 10 for positioning of the "In-Field Pointing Mechanism" (IFPM) with a flat pointing mirror. An overall magnification of 100 compresses the 400 mm diameter virtual external pupil to an accessible internal pupil of 4 mm diameter. As shown in Figure 2, an excellent optical performance over the entire field of view is achieved. The design wavefront error is $< \lambda/100$ rms for any pointing angle, with only minimal variations over the field.

A conceptual design for the IFPM, developed in collaboration with TNO, is depicted in Figure 3. It employs elastic Haberland hinges embedded in a monolithic of a stable material (TiAlV) to cover the required tilt range of $\pm 2.5°$. The actuation is based on a walking piezo motor, which combines high resolution with a large travel range, compatible to the required dynamic range of about 10^7. This dynamic range is based on the allocation of a maximum of $1\,\text{nrad}/\sqrt{\text{Hz}}$ contribution to the external line of sight jitter of each telescope, corresponding to

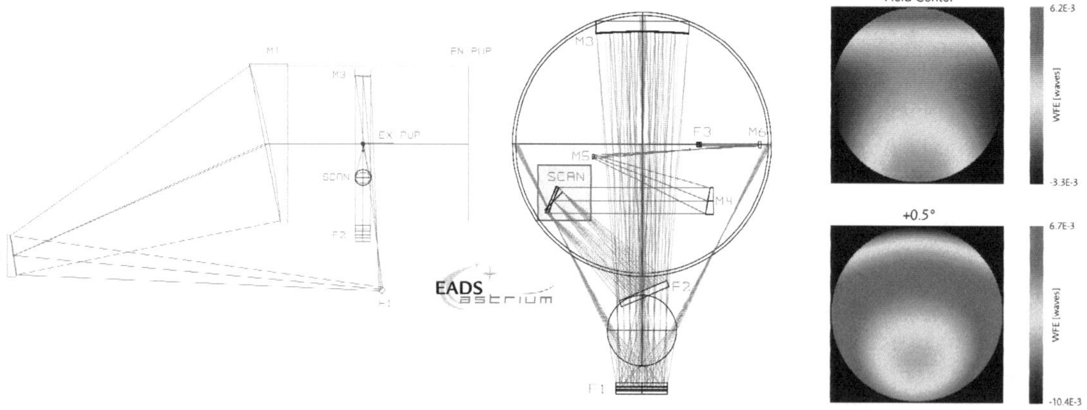

Figure 2. Optical design (left) and optical performance (right) of the specialized wide-field, off-axis telescope. It provides an intermediate pupil for positioning of the In-Field Pointing Mirror (red box).

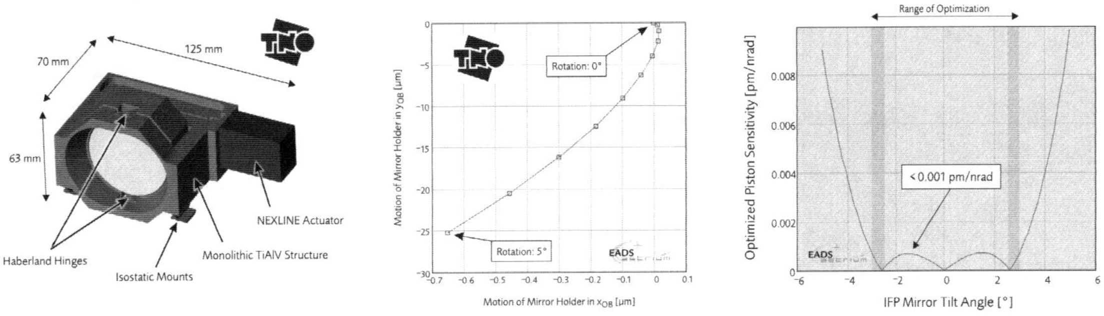

Figure 3. Conceptual design of the In-Field Pointing Mechanism by TNO, Delft (left). The middle plot shows the deterministic translation of the mirror surface due to hinge deformation. Nonetheless, the piston performance can be optimized to negligible sensitivity (right).

$10\,\mathrm{nrad}/\sqrt{\mathrm{Hz}}$ pointing jitter at the location of the IFPM (magnification 10).

Active pointing with the IFPM inside the telescope potentially introduces a new source of pathlength noise, as the system probes any dependence of the optical pathlength through the telescope on the pointing angle. While the nominal pointing adjustment on annual timescales in general will not produce spectral noise components within the measurement band of $10^{-4} - 1\,\mathrm{Hz}$, any coupling of pointing jitter to piston noise via this effect might be detrimental. Therefore, pathlength noise contributions driven by pointing jitter of the IFPM have been investigated in great detail, and in particular the following specific measures are taken to control these:

- The optical design of the telescope was tuned to minimize the sensitvity of the optical pathlength through the telescope to the pointing angle.
- Geometrical coupling of mirror tilt to piston is principally avoided by placing the IFPM in an intermediate telescope pupil and applying a Gimbal type mirror hinging.
- As illustrated in the middle panel of Figure 3, the mechanics of the Haberland hinge has been derived by FEM to determine the resulting piston motion of the IFPM mirror. By fine-tuning its position inside the holder, using only moderate adjustment tolerances of 10's of microns, the effective piston due to hinge mechanics can be compensated to negligible amounts (Figure 3, right panel).

While these measures allow to reduce pointing jitter to piston crosstalk due to the IFPM to levels that can well be accommodated in the overall pathlength noise budget, EADS Astrium has nonetheless suggested an ancillary metrology system to monitor pathlength noise generated within the telescope. The underlying idea of this system, known as Active Optical Truss, is to refer the phase of the TX wavefront leaving the telescope directly to the optical bench reference frame. Hence, the application of this system is in general not restricted to this specific noise source, and thus can in principle be of benefit to any LISA payload architecture. In consequence, it has meanwhile been incorporated also in the present baseline architecture [4].

3. Adding Redundant GRS Systems

Since the LISA metrology is intrinsically referenced to the center of mass of each proof mass in the constellation, the location of these fiducial points is also restricted by geometrical projection effects resulting from pointing jitter. Ideally, as sketched in Figure 4, the proof masses should be located on their respective line of sight. To be more exact, the optical point of reference each proof mass must be aligned to is the *phase center* of its associated telescope, i. e. the point about which the spacecraft can be rotated without producing a pathlength signal.

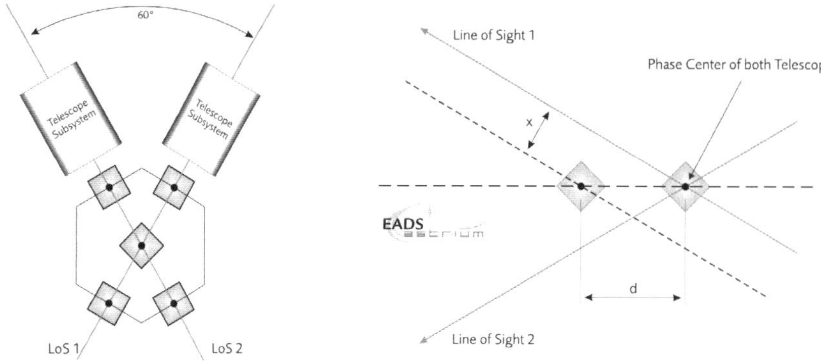

Figure 4. Ideally, the center of mass of each proof mass should be located exactly on its associated line of sight (left). Only minimal offsets x on the order of 2 mm are allowed due to geometric projection effects (right).

As soon as the center of mass is offset from the phase center (right panel of Figure 4), a lateral lever arm x is introduced, which again converts pointing jitter to piston noise. Since the effect is in principle deterministic, it can be calibrated, so that the applicable noise level is effectively that of the *pointing knowledge*. The pointing knowledge, i. e. the measured value for the relative pointing between RX and TX beam in each telescope, is established by Differential Wavefront Sensing on the science receiver [4], and estimated to be on the order of $0.5\,\mathrm{nrad}/\sqrt{\mathrm{Hz}}$. In consequence, the lateral lever arm x should not become larger than 2 mm to keep the associated noise contribution below about $1\,\mathrm{pm}/\sqrt{\mathrm{Hz}}$.

Given this small scale, adding a redundant GRS to a single active proof mass configuration in a geometry as depicted in the right panel of Figure 4 would unavoidably lead to increased measurement noise when switching to the redundant system. While such a graceful degradation of the measurement performance would probably be acceptable in a failure case, it nonetheless motivates measures to improve the pointing knowledge and thus increase the distance d beyond which the measurement noise is compromised.

Since the main limit to the above mentioned pointing knowledge is given by pointing jitter of the Point Ahead Angle Mechanism (PAAM), which is used to adjust the relative pointing between TX and RX beam, the immediate approach is to reduce or at least more accurately monitor its jitter. Such a PAAM metrology system can be realized using the same metrology principles as already applied for the inter- and intra-spacecraft interferometry (see [4]), and thus does not have significant impacts on the payload design and budgets. Like the Active Optical Truss, the robustness added by the PAAM metrology is thus beneficial in general, so that the latest baseline architecture also has inherited it.

4. Payload Architecture Variants

Based on the above considerations, a detailed opto-mechanical design for a payload architecture with a single active proof mass per spacecraft and the application of In-Field Pointing for accommodation of constellation breathing has been developed. It includes a single piece, vertical optical bench made from Zerodur, and a redundant GRS system in close vicinity to the nominal one, in line with the geometry shown in Figure 4. All payload components are supported isostatically from a low thermal expansion CFRP structure, which also provides the interface to the spacecraft.

The architecture is illustrated in Figure 5 together with all other payload variants that have meanwhile been established by EADS Astrium within the LISA Mission Formulation study.

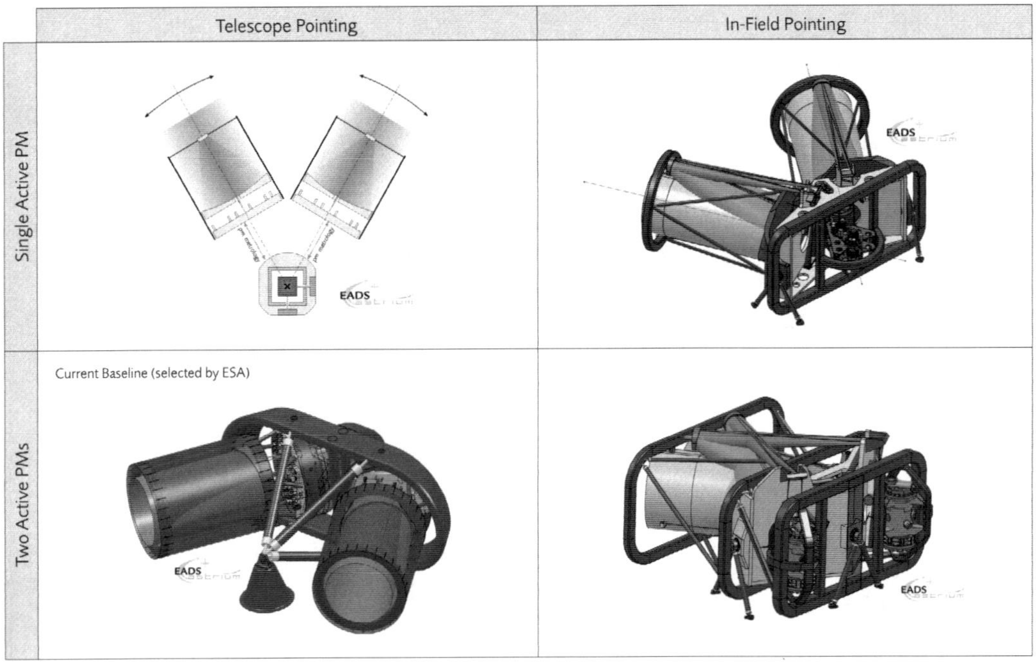

Figure 5. Overview of investigated payload architectures. For the combination of Telescope Pointing with a Single Active Proof Mass, only a basic concept has been conceived, confirming that it does not lead to a useful overall solution. The models are *not* to scale.

Notably, In-Field Pointing can also reasonably be combined with two active proof masses per spacecraft to yield an extremely compact payload, smaller in size than any other design so far. The baseline configuration currently selected by ESA however falls back to the more historic Telescope Pointing [4].

In conclusion, EADS Astrium has meanwhile explored the trade space of alternative payload architectures for the LISA instrument in great detail, and demonstrated that payload architectures utilizing only a single active proof masss per spacecraft are feasible, attractive, and do not show principal show stoppers.

Acknowledgments

This work was supported by the European Space Agency under ESA Contract No. 18756/04/NL/HB. We further acknowledge the on-going fruitful collaboration with TNO (Delft), in particular concerning the conceptual design of the IFPM, as well as the always enlightening discussions within the ESA/NASA study group and with the LISA International Science Team.

References

[1] Weise D, Braxmaier C, Gath P, Schulte H R and Johann U 2006 *Proceedings of the Sixth International Conference on Space Optics* ed Wilson A (ESTEC)

[2] Weise D, Braxmaier C, Kersten M, Holota W and Johann U 2006 *Proceedings of the 6th International LISA Symposium* ed Merkowitz S M and Livas J C (AIP) pp 389–394

[3] Gath P F, Johann U, Schulte H R, Weise D and Ayre M 2006 *Proceedings of the 6th International LISA Symposium* ed Merkowitz S M and Livas J C (AIP) pp 647–653

[4] Weise D, Marenaci P, Weimer P, Berger M, Schulte H R, Gath P and Johann U 2008 *Proceedings of the Seventh International Conference on Space Optics*

Journal of Physics: Conference Series **154** (2009) 012030 doi:10.1088/1742-6596/154/1/012030

A High Sensitivity Heterodyne Interferometer as a Possible Optical Readout for the LISA Gravitational Reference Sensor and its Application to Technology Verification

Martin Gohlke[1,2], Thilo Schuldt[2,3], Dennis Weise[1], Ulrich Johann[1], Achim Peters[2] and Claus Braxmaier[1,3]

[1] EADS Astrium, Claude-Dornier-Straße, 88039 Friedrichshafen, Germany
[2] Humboldt-Universität zu Berlin, Hausvogteiplatz 5-7, 10117 Berlin, Germany
[3] Hochschule für Technik, Wirtschaft & Gestaltung, Brauneggerstr. 55, 78462 Konstanz, Germany

E-mail: martin.gohlke@astrium.eads.net

Abstract. The space-based gravitational wave detector LISA (Laser Interferometer Space Antenna) utilizes a high performance position sensor in order to measure the translation and tilt of the free flying proof mass with respect to the optical bench. Depending on the LISA optical bench design, this position sensor must have up to pm/$\sqrt{\text{Hz}}$ sensitivity for the translation measurement and up to nrad/$\sqrt{\text{Hz}}$ sensitivity for the tilt measurement. We developed a heterodyne interferometer, combined with differential wavefront sensing, for the tilt measurement. The interferometer design exhibits maximum symmetry where measurement and reference arm have the same frequency and polarization and the same optical path-lengths. The interferometer can be set up free of polarizing optical components preventing possible problems with thermal dependencies not suitable for the space environment. We developed a mechanically highly stable and compact setup which is located in a vacuum chamber. We measured initial noise levels below 10 pm/$\sqrt{\text{Hz}}$ (longitudinal measurement) for frequencies above 10 mHz and below 20 nrad/$\sqrt{\text{Hz}}$ (tilt measurement) for frequencies above 1 mHz. This setup can also be used for other applications, for example the measurement of the coefficient of thermal expansion (CTE) of structural materials, such as carbon fiber reinforced plastic (CFRP).

1. Introduction

In the LISA concept, the detector should be sensitive enough to detect gravitational waves with induced strains of amplitude $h = \Delta l/l \sim 10^{-23}$. Therefore, it is necessary to measure the relative distance between the three spacecraft with an accuracy of a few picometers. But not only gravitational waves will change the distance. External disturbances like solar radiation pressure or solar wind will change the position of the spacecraft and thereby might affect the interferometer signals caused by gravitational waves. Therefore, a free-flying proof mass – as part of the gravitational reference sensor (GRS) – inside the satellite is taken as reference for a purely gravitational orbit. As the external disturbances only act on the surface of the satellite, the distance between proof mass and its housing (which is rigidly connected to the satellite) is changing. In case of a drag-free controlled satellite (drag-free attitude control system, DFACS),

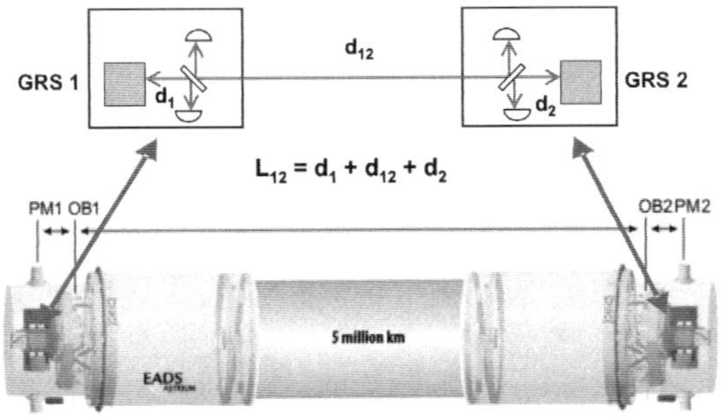

Figure 1. Schematic of the strap-down-architecture. The distance measurement is split into two local, from proof mass (PM) to the optical bench (OB), and one long distance measurement.

any change of the proof mass position is measured and the satellite is controlled in such a way that it is centered around the proof mass at any time, canceling all non-gravitational forces acting on the spacecraft. The proof mass also acts as the end mirror of the interferometer arms.

In the so-called strap-down-architecture (cf. Fig. 1 and [1]) each interferometric measurement of the distance between two satellites is split into three independent measurement: (i) the distance between the proof mass and the optical bench on the first satellite, (ii) the distance between the optical benches aboard the two satellites, and (iii) the distance between the optical bench on the second satellite and its proof mass. The proof mass position sensor is part of the so-called science interferometer and must fulfill requirements of $1\,\mathrm{pm}/\sqrt{\mathrm{Hz}}$ (for frequencies above 2.8 mHz with an f^{-2} relaxation down to $30\,\mu\mathrm{Hz}$) for the translation measurement and at least $20\,\mathrm{nrad}/\sqrt{\mathrm{Hz}}$ for the tilt measurement (for frequencies above 0.1 mHz with an f^{-1} relaxation down to $30\,\mu\mathrm{Hz}$, DFACS requirements).

These requirements necessitate an optical readout (ORO) of the proof mass position utilizing laser interferometry. Our interferometer concept represents a possible ORO.

2. Interferometer Setup

Our experimental setup is based on a highly symmetric design and represents a heterodyne interferometer with spatially separated beams [2–5]. Reference and measurement beams of the interferometer have the same frequency and polarization. Also, their optical pathlengths (especially inside optical components) are similar. Schematics of our interferometer setup are shown in Figs. 2 and 3, and a photograph in Fig. 4. The phase $\phi(t)$ between the signals on measurement and reference photodiode is proportional to the displacement Δl of the measurement mirror:

$$\phi(t) = \frac{4\pi n}{\lambda} \cdot \Delta l(t)\,, \tag{1}$$

where λ is the vacuum wavelength of the light and n the refractive index of the medium the light is traveling in.

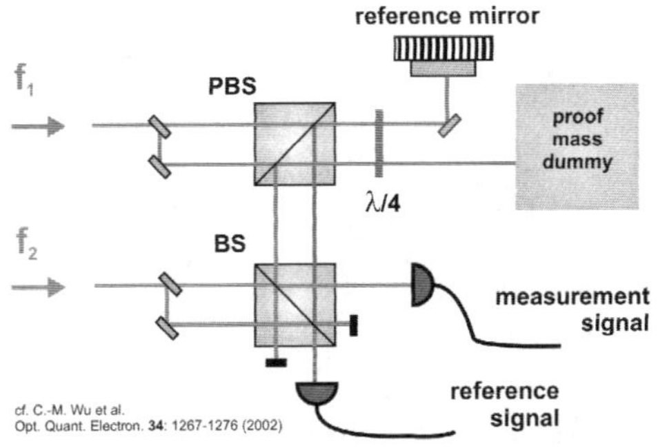

Figure 2. Schematic of the heterodyne interferometer (BS: beamsplitter; PBS: polarizing beamsplitter). The signal at the photodiodes are given by $I_{meas} \sim AB \cdot \cos(\Delta\omega t - \phi(t))$ and $I_{ref} \sim AB \cdot \cos(\Delta\omega t)$. Here, A and B denote the amplitudes of the laser, f_1 and f_2 the frequencies and $\Delta\omega = 2\pi|f_1 - f_2|$.

For in-quadrature measurement, the two signals

$$S_1 = \frac{1}{2}AB\cos\phi(t) \quad \text{and} \quad S_2 = \frac{1}{2}AB\sin\phi(t) \tag{2}$$

are generated, where the phase measurement is obtained by

$$\phi(t) = \tan^{-1}\frac{S_2}{S_1}\,. \tag{3}$$

In order to measure the tilt of the proof mass, the method of Differential Wavefront Sensing (DWS) [6] is implemented. Therefore, we use quadrant photodiodes and the phase difference between opposing halves is measured.

The interferometer setup can be subdivided into two parts: (i) the optical setup for frequency generation, and (ii) the interferometer board. The optics for frequency generation are placed on an optical table using standard optical components. The interferometer board is placed inside a vacuum chamber operated at pressures below 10^{-3} mbar. For mechanical stability and compactness, purpose-built optical component mounts with a beam-height of 2 cm are used.

2.1. Heterodyne Frequency Generation

An NPRO-type (non-planar ring-oscillator) Nd:YAG laser at a wavelength of 1064 nm is used as light source. Part of its output power is split and shifted in frequency by use of two acousto-optic modulators (AOMs). The AOMs are working at a frequency of 79.99 MHz and 80.00 MHz, respectively, resulting in a heterodyne frequency of 10 kHz. The two laser beams are fiber-coupled and sent to the interferometer inside the vacuum chamber via optical single mode fibers. The RF-signals driving the AOMs are generated by use of two phase-locked direct digital synthesizers (DDS).

2.2. Interferometer Setup

The 300mm × 440mm × 40mm interferometer board is made of cast aluminum which offers low internal stresses and therefore reduced long-term drifts of the material. The two frequencies are fiber-coupled to the interferometer board. Polarizers at the fiber inputs and outputs ensure a proper and clean polarization. The optical power of each beam is $\sim$ 2.5 mW at the fiber output, this is equivalent to 100μW on the proof mass. Both beams are split at a wedged glass plate, the front side reflections are used in the interferometer while the back side reflections are sent to two monitor photodiodes (PD1 and PD2 in Fig. 3). Their signals are taken for intensity stabilization of the laser beams after fiber outcoupling. The amplitudes of the RF signals driving the AOMs are used for actuation of the light intensities. The beams transmitted at the glass

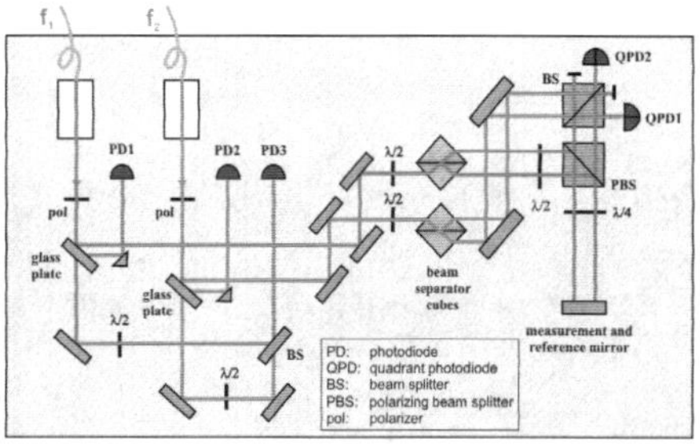

Figure 3. Schematic of the heterodyne interferometer.

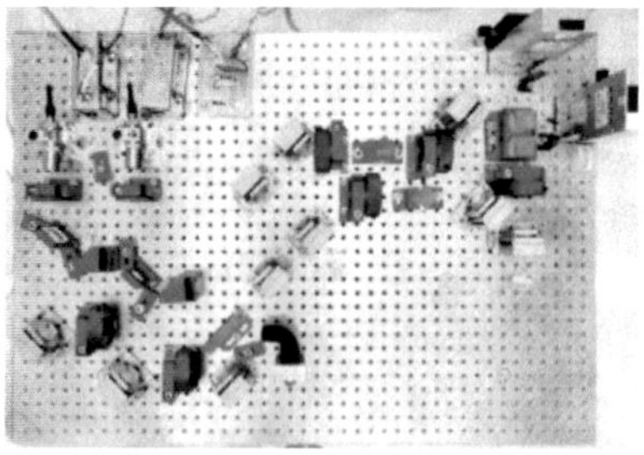

Figure 4. Picture of the interferometer inside the vacuum chamber.

plates are superimposed on a third photodiode (PD3 in Fig. 3). The generated heterodyne signal is phase-locked to a 10 kHz signal which is obtained by mixing the DDS output signals. The signal at the photodiode corresponds to an interferometer signal, where the splitting in measurement and reference beam takes place at the beamsplitter before the AOMs outside the vacuum chamber. Therefore, it contains all differential phase effects caused by the fibers and the AOMs. A piezoelectric actuation of one mirror in one arm of the interferometer outside the vacuum chamber is used for actuation of the feedback loop. The two beams used for the interferometer are both split into two parallel output beams. A symmetric beamsplitter is used providing the same optical path-lengths inside the component. Also, ghost reflections of the input beams are not reflected back in direction of the input beam.

The beams with frequency f_1 are representing measurement and reference beams of the interferometer, the beams with frequency f_2 are used for generating heterodyne signals at the photodiodes. The beams with frequency f_1 are first reflected by a polarizing beamsplitter (PBS) towards the measurement and reference mirror, which in our experiment is realized by one fixed mirror. After passing twice a quarter waveplate ($\sim \lambda/2$), both beams are transmitted at the PBS and superimposed with the beams with frequency f_2 at a (non-polarizing) beamsplitter (BS). Two quadrant photodiodes (QPD1 and QPD2 in Fig. 3) measure the heterodyne signals. Their sum signals are taken for translation measurement, and the signals of opposing halves for tilt measurement.

2.3. Digital Phase Measurement

The signals of the quadrant photodiode are first pre-amplified inside the vacuum chamber. The sum signals and the signals for differential wavefront sensing are generated and amplified outside the vacuum chamber by use of analog electronics. All signals are anti-aliasing filtered by use of a 6th order Bessel filter (corner frequency: 20 kHz). The signals are then simultaneously digitized by a National Instruments field programmable gate array (FPGA) computer board at a frequency of 160 kHz and with a 16 bit resolution. The FPGA board is programmed by LabVIEW. A digital phasemeter is implemented where the corresponding input signals to the FPGA board are multiplied and low-pass filtered. For in-quadrature measurement, the signals phase-shifted by 90° are generated. The data is reduced by a factor of 8000 and transferred to a LabVIEW host program which carries out the calculation as given in Equation (3). The program also monitors phasejumps by π and therefore offers a dynamic range not limited by $\lambda/2$ in mirror translation.

3. Measured Noise Performance

We performed a noise measurement where the measurement and the reference mirror are represented by the same fixed mirror (cf. Fig. 3). The power spectrum density (PSD) of the translation measurement is shown in Fig. 5, the PSD of the corresponding tilt measurement in Fig. 6. The peak near 3 Hz was caused by a missing synchronization of the FPGA clock and the DDS. It is removed in later measurements.

In translation measurements (cf. Fig. 5), noise level below $1\,\mathrm{nm}/\sqrt{\mathrm{Hz}}$ for frequencies above 0.1 mHz and below $10\,\mathrm{pm}/\sqrt{\mathrm{Hz}}$ for frequencies above 10 mHz were obtained. In tilt measurements (cf. Fig.6) the noise level is below $1\,\mu\mathrm{rad}/\sqrt{\mathrm{Hz}}$ for frequencies above 0.1 mHz and below $10\,\mathrm{nrad}/\sqrt{\mathrm{Hz}}$ for frequencies above 10 mHz.

4. Applications

The interferometer setup as described above is a highly sensitive measurement system which can be used to investigate other key aspects in the LISA context. For example:

- determine the coefficient of thermal expansion of ultra-stable structural materials, e.g. carbon fiber reinforced plastic (CFRP),

- testing the noise performance and linearity of various LISA mechanisms, e.g. the actuator of the in-field pointing mechanism as well as the mechanism itself, and
- investigate the quality of mirror surfaces in the pm-range.

Dilatometry
Based on the above presented setup we developed a highly sensitive dilatometer for characterizing the dimensional stability of ultra-stable materials. The principle of the dilatometer is shown in Fig.7. We are changing the temperature of the tube to be tested via radiative heating/cooling and measure the linear expansion with our interferometer with sub-nm resolution. The CTE is given as:

$$CTE = \frac{1}{\Delta T}\frac{\Delta L}{L}\ [K^{-1}]\,. \tag{4}$$

Measurement and reference mirrors are placed at the top and at the bottom inside a tube made by the material under investigation (cf. Fig. 7). The mirror mounts are specifically designed in order not to influence the CTE measurement. Therefore, the reflective surface of each mirror is in the same plane as the clamping points of the mirror mount inside the tube. We also developed a test tube support and a heating/cooling system. For measurement, a sine thermal cycling with a period up to several hours is applied to the device under test (cf. Fig. 8).

Due to limitations of the vacuum chamber and the interferometer setup, we are restricted to tubes with an inner diameter of 20 mm, an outer diameter of $\sim$ 27 mm and a length of 10 cm. The temperature of the device under test can be varied between 20° C and 60° C.

We first performed a measurement with a CFRP tube with a known CTE of $\sim -6.47 \cdot 10^{-7}\,\mathrm{K}^{-1}$. Our result with the interferometer test facility provided a CTE of $\sim -6.10 \cdot 10^{-7}\,\mathrm{K}^{-1}$. In the next step, we procured a new CFRP tube with a predicted CTE below $-2.5 \cdot 10^{-8}\,\mathrm{K}^{-1}$. We measured a CTE of $\sim -3.6 \cdot 10^{-7}\,\mathrm{K}^{-1}$, which does not agree with the theoretical value [7].

Presently, we and the company are analyzing the manufacturing process as the determination of the CTE includes uncertainties.

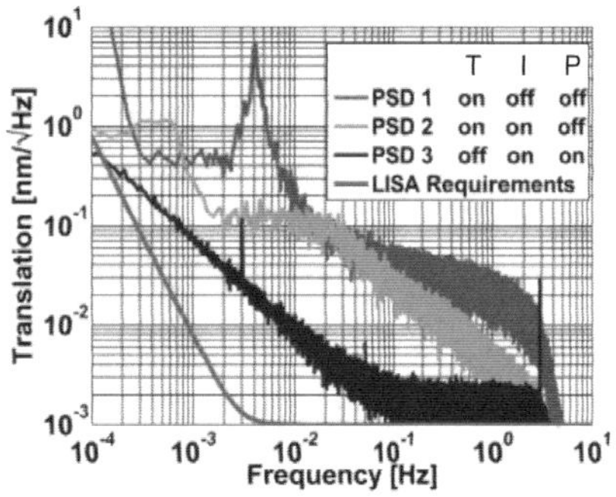

Figure 5. Power spectrum densities (PSD) of the translation noise measurements. The smooth line at the bottom represents the LISA requirements. At the beginning (upper line) with temperature stabilization (T) but without intensity- (I) and phase- (P) stabilization. The lower line represents the current noise performance.

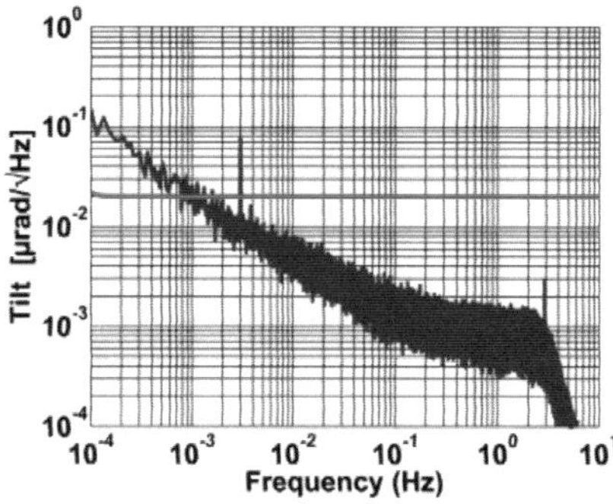

Figure 6. PSD of the tilt noise measurement (T: off, I: on, P: on). The smooth line represents the DFACS requirements. The roll off at 3 Hz is caused by a digital low pass filter in the data acquisition chain.

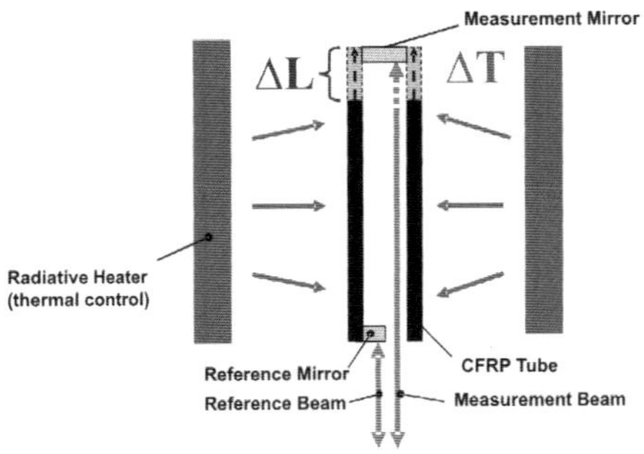

Figure 7. Sketch of our CTE measurement facility.

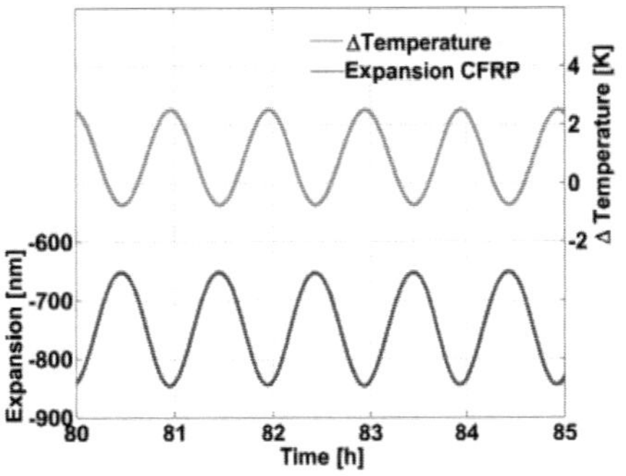

Figure 8. Typical time series of a CTE measurement run. The upper curve represents the cycled temperature of the CFRP probe, the lower curve the measured expansion.

5. Summary and Outlook

We developed a compact setup of a high sensitivity heterodyne interferometer. Noise levels below $10\,\mathrm{pm}/\sqrt{\mathrm{Hz}}$ in translation measurement and below $10\,\mathrm{nrad}/\sqrt{\mathrm{Hz}}$ in tilt measurement – both for frequencies above 10 mHz – were demonstrated in lab experiments. In a current project, the interferometer is adapted as dilatometer for measurement of the thermal expansion of carbon fiber reinforced plastics. Measurements were carried out validating the capability of our setup. In general, the test facility offers the possibility to characterize ultra-stable materials with a CTE down to $10^{-8}\,\mathrm{K}^{-1}$.

Current activities also include the development of a new quasi-monolithic setup using thermally highly stable glass ceramics as breadboard materials in the interferometer.

Acknowledgments

This work is partially supported by the German Aerospace Center (Deutsches Zentrum für Luft- und Raumfahrt e.V.) within the program LISA Performance Engineering (DLR contract number: 50OQ0701).

The authors thank the Albert-Einstein-Institute Hannover, Evgeny Kovalchuk and Klaus Palis from the Humboldt-University Berlin and Hans-Reiner Schulte from EADS Astrium Friedrichshafen for their support and the fruitful discussions.

References

[1] Gath P, Johann U, Schulte H R, Weise D and Ayre M 2006 *Laser Interferometer Space Antenna – 6th International LISA Symposium* (AIP conference proceedings) pp 647–653
[2] Schuldt T, Kraus H J, Weise D, Braxmaier C, Peters A and Johann U 2006 *Proceedings of the 6th International Conference on Space Optics (ICSO 2006)*
[3] Schuldt T, Gohlke M, Weise D, Johann U, Peters A and Braxmaier C 2007 *Proceedings of SPIE* (6717, Optomechatronic Sensors and Instrumentation III, 67160F)
[4] Schuldt T, Gohlke M, Weise D, Johann U, Peters A and Braxmaier C *International Journal of Optomechatronics* **1**
[5] Wu C, Lin S and Fu J *Opt. Quantum Electron. 12* **34** 1267–1276
[6] Morrison E, Meers B J, Robertson D I and Ward H *Appl. Opt. 22* **33**
[7] Cordero J, Heinrich T, Schuldt T, Gohlke M, Lucarelli S, Weise D, Johann U, Peters A and Braxmaier C 2008 *Proceeding of SPIE* (7018, Advanced optical and mechanical technologies in telescopes and instrumentation)

Characterization of Polarization Sensitive, High Efficiency Dielectric Gratings for Formation Flight Interferometry

Ke-Xun Sun, Patrick Lu, and Robert L. Byer

Ginzton Laboratory N-119, Stanford University, Stanford, CA94305-4088, USA

Jerald A. Britten, Hoang T. Nguyen, James D. Nissen, Cindy C. Larson, Michael D. Aasen, Thomas C. Carlson, and Curly R. Hoaglan

Lawrence Livermore National Laboratory, Livermore, CA 94550, USA

kxsun@stanford.edu

Abstract. Reflective diffraction gratings enable novel optical configurations that simplify and improve laser interferometers. We have proposed an all-reflective grating interferometer that can be used in LISA type interferometers for space gravitational wave detection [1]. One configuration requires a highly polarization sensitive grating. We report on characterizations of a grating made atop high reflective dielectric layers. Using a direct measurement method, the diffraction efficiency at the Littrow angle for s-polarization is measured as 97.3% and for p-polarization 4.2%, leading to a s/p polarization diffraction ratio of 23.2. The depolarization from s- to p-polarization is measured to be $\sim 1.7\times10^{-4}$, and from p- to s-polarization 1.8×10^{-4}. We derived a transfer matrix based on these measurements. Furthermore, we have developed a more accurate method for diffraction efficiency measurement using a grating cavity. These measurements are encouraging steps taken towards the requirements of an ideal grating interferometer.

1. Introduction

Grating interferometers have been proposed for ground-based gravitational wave detectors such as Laser Interferometer Gravitational Observatory (LIGO) [1]. Grating interferometers may also benefit Laser Interferometer Space Antenna (LISA) by both simplifying optical configurations and enhancing instrument performance [2].

LISA [3] and other formation flights for fundamental physics must make connected distance measurements from the gravitational reference sensor (GRS) center to the GRS housing and to other spacecrafts. In the Modular GRS (MGRS) design [2], we proposed using a double-sided reflective grating at the connecting points, so that both the GRS internal interferometry and external inter-spacecraft interferometry can be independently optimized. For the external interferometry, a highly polarization-sensitive, high-efficiency grating will be needed. Ideally, the diffraction efficiency should be near unity in the high efficiency polarization, but near zero in the low efficiency polarization. In addition, the depolarization, or the cross coupling of the two polarizations, should be as low as possible.

Such a grating can be difficult to find if we are limited to conventional metal gratings. However, the exciting development of multilayer dielectric (MLD) gratings provides a promising solution [4]. In this paper, we report on characterizations of two high-efficiency dielectric gratings designed and fabricated at Lawrence Livermore National Laboratory. Using a direct measurement method, the diffraction efficiency of one grating at the Littrow angle for s-polarization is measured to be 97.3% and for p-polarization 4.2%, leading to a s/p polarization diffraction ratio of 23.2. The depolarization from s- to p- was measured to be ~1.7×10^{-4}, and from p- to s- 1.8×10^{-4}. Further improvement of the grating characteristics are needed. However, these characteristics are already very promising for an initial build of the grating interferometer described in [2].

We further report measurement results using a grating cavity for a more accurate determination of diffraction efficiency and thermal wavefront distortion measurements made to characterize grating power handling.

2. Reflective Interferometry Using a Polarization Sensitive Grating

We first review the grating interferometer design [2] for the external laser interferometry for formation flight represented by LISA. Figure 1 shows the grating interferometer configuration. The grating is designed to be highly polarization sensitive. As shown in the example in Figure 1, the external laser beam (green-colored in drawing) with the majority of its polarization lying in the plane of incidence (p-polarization), will specularly reflect into the zeroth order and be transmitted to the remote spacecraft. The small portion of the beam that is in the orthogonal, or s-, polarization will diffract onto the detector, forming the local oscillator. The incoming beam from the remote spacecraft is polarized orthogonal to the outgoing beam. By proper design, the incoming beam is mostly diffracted onto the detector to mix with the local oscillator to produce the interference signal for the science measurement. In the grating-based MGRS, the external laser light does not illuminate the proof mass. Therefore, the power and wavelength of the external laser beam can be selected to optimize the measurement precision to the remote spacecraft. For example, for LISA a green laser beam would improve the sensitivity. For DECIGO (Deci-hertz Gravitational wave Observatory) and BBO (Big Bang Observer) the extreme sensitivity requirements could be met by using shorter wavelengths such as the 532 nm, 355 nm or 266 nm, all harmonics of the Nd:YAG laser at 1064 nm. This configuration essentially has the basic functions of external interferometry required by LISA, but has a significantly simplified structure. Also, we have shown that by proper selection of the grating period, the incoming and outgoing beams can have an angular separation equal to the "look ahead angle" [2].

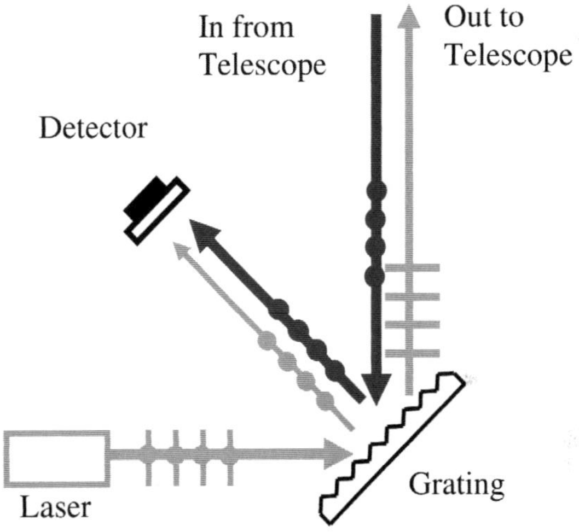

Figure 1: Optical configuration for laser interferometer using a polarization sensitive grating.

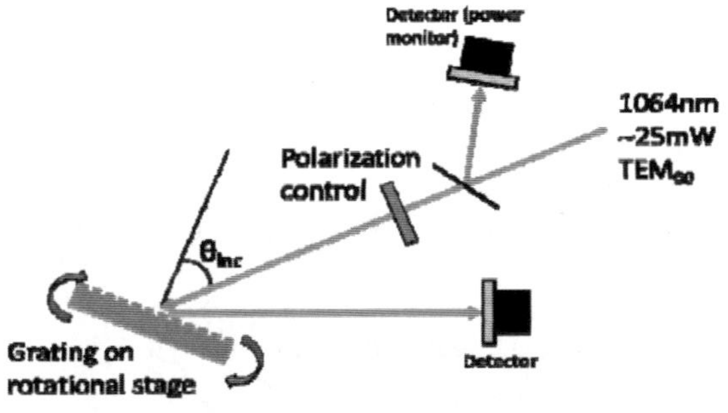

Figure 2: Experimental setup for measuring angle dependent polarization diffraction ratio.

3. Incident angle dependence of diffraction efficiency for each polarization

The design of a grating enjoys freedom to achieve the desired diffraction characteristics. The diffraction angles are determined by the grating period. The diffraction efficiency depends on the details of the grating shape, depth and design of the underlying stack [5], as does the polarization selectivity. The verification of strong polarization dependence of diffraction efficiency is a first step towards the polarization grating interferometer proposed in [2]. If verified, we can adjust the grating period and other parameters to tune the grating diffraction angle according to the requirement of the interferometer.

Our first measurements were therefore made to confirm the feasibility of using a grating to separate s- and p-polarized light. An all-dielectric diffraction grating consisting of a SiO_2/HfO_2 multilayer stack coated on a 100 mm diameter BK7 substrate at the University of Rochester Laboratory for Laser Energetics, with a 1740 line/mm grating patterned and etched into the top 575 nm-thick SiO_2 capping layer at LLNL, was used for this experiment.

The experimental setup is illustrated in Figure 2. A monochromatic 1064 nm, 25 mW laser beam from an NPRO laser was directed through a mode cleaner and a half-wave plate. The latter allowed for polarization rotation. This laser beam was directed at the grating mounted on a rotational stage. A detector was placed so that it captured light diffracting off the grating. Measurements were taken for a range of incident angles and results are plotted in Figure 3.

Figure 3(a) shows the -1 order diffraction efficiency for s-polarized light. The maximum efficiency of 97.3% was located near the Littrow angle 67.8°. Moreover the diffraction efficiency was maintained above 90% over a large angular range, from 61° to 77°. The large angular range of high efficiency diffraction makes a variety of interferometer arrangements possible. Figure 3(b) shows the -1 order diffraction efficiency for p-polarized light. The maximum diffraction efficiency was only ~4.2%, also located at the Littrow angle. The majority of the p-polarized light was specularly reflected into the zeroth order. Figure 3 (c) shows the ratio between s- and p- polarization diffraction efficiencies. The lowest ratio of ~23.2 was located at 67.8°, the Littrow angle, with higher ratios around both sides.

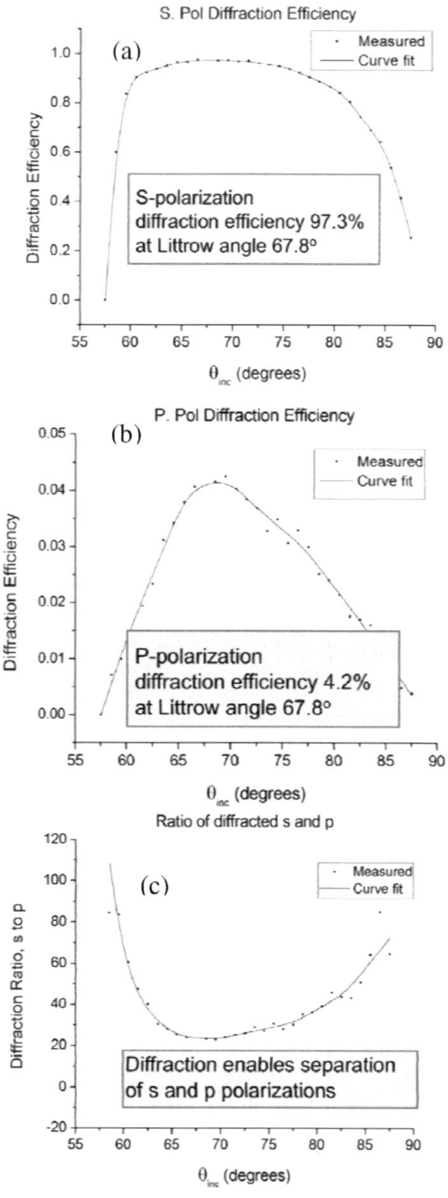

Figure 3: (a) Diffraction efficiency as a function of angle for s-polarized incident light. (b) Diffraction efficiency as a function of angle for p-polarized incident light. (c) Polarization-dependent diffraction efficiency ratio.

The grating diffracted significantly more for s-polarized light across all angles of incidence than that p-polarized light. This valuable property will help the implementation of the polarization grating interferometer.

4. Depolarization characterization

Ideally the grating should keep the polarization of incident light. Depolarization can cause undesirable mixing of the two polarizations, making noise enter the signal, and also resulting in loss of signal photons. Therefore, a grating with minimal depolarization is desired. Characterization of depolarization is necessary.

Figure 4 shows the experimental setup used to characterize the depolarization of the grating. These measurements were made near the Littrow angle. Light from an NPRO laser was diffracted off the grating into the -1 order and analyzed for s- and p- content, thereby determining the power transfer matrix of the grating diffraction. Most notably, only a very small amount of power from one polarization was diffracted into the other. Table 1 shows the measured data. Table 2 shows the calculated power transfer matrix.

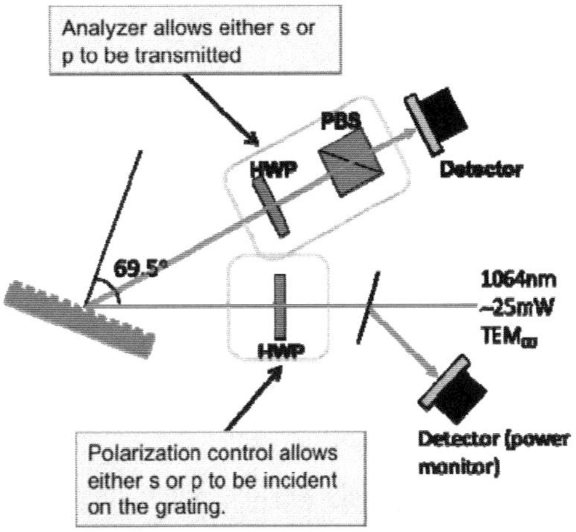

Figure 4: Experimental setup for depolarization measurements.

For s-polarized incident light, there was a transfer coefficient 1.67×10^{-4} for light diffracted into the p-polarization. For p-polarized light diffracting into s- the coefficient was 1.78×10^{-4}. These polarization extinction ratios rival those of transmissive polarizing beamsplitter cubes.

Table 1: Measured depolarization data

Incident Pol.	Incident Power (mW)	Exiting Component	Exiting Component Power (mW)	Transfer Coeff.
S	23.4	S	22.3	0.955
	23.4	P	3.90×10^{-3}	1.67×10^{-4}
P	22.9	S	4.07×10^{-3}	1.78×10^{-4}
	23.6	P	0.893	0.038

Table 2: Calculated power transfer matrix

		Output s	Output p	Polarization Extinction Ratio
Input	S	0.955	1.67×10^{-4}	5729
	p	1.78×10^{-4}	0.038	213

5. Cavity measurements for precise determination of diffraction efficiency at Littrow angle

Precise grating diffraction efficiency characterization is required for the calibration of interferometers. However, directly measuring the diffraction efficiency at the exact Littrow angle requires a beamsplitter to sample the returning beam and extra calibration steps, potentially adding uncertainty. We have developed a highly precise method of measuring diffraction efficiency at the Littrow angle consisting of a linear Fabry-Perot cavity that uses the -1 diffracted order of the grating as one of the end reflectors. The efficiency of the grating is calculated from measurements of cavity finesse. Note that cavities in this configuration also have implications for displacement sensing.

This experiment characterized a 1740 line/mm MLD grating on a 200x100 mm BK7 substrate, similar to the one earlier described but optimized with respect to grating height and linewidth for maximal diffraction efficiency at the Littrow angle. Our experimental setup is illustrated in Figure 4 (a). As in the previous section, light was emitted by a 1064nm NPRO and passed through an F=50 mode cleaner. This beam was modulated at 12 MHz in an electro-optic modulator (EOM) to provide sidebands for Pound-Drever-Hall locking of the mode cleaner and also to create a frequency reference signal used to measure cavity spectral characteristics. A piezo-mounted mirror allowed for scanning of the cavity length, and a photodiode placed after the cavity monitored the transmitted power. The mirror motion was independently measured by a Michelson interferometer, which makes it possible to calibrate the PZT element movement. Figure 4b shows a photograph of the 20 cm x 10 cm grating measurement setup.

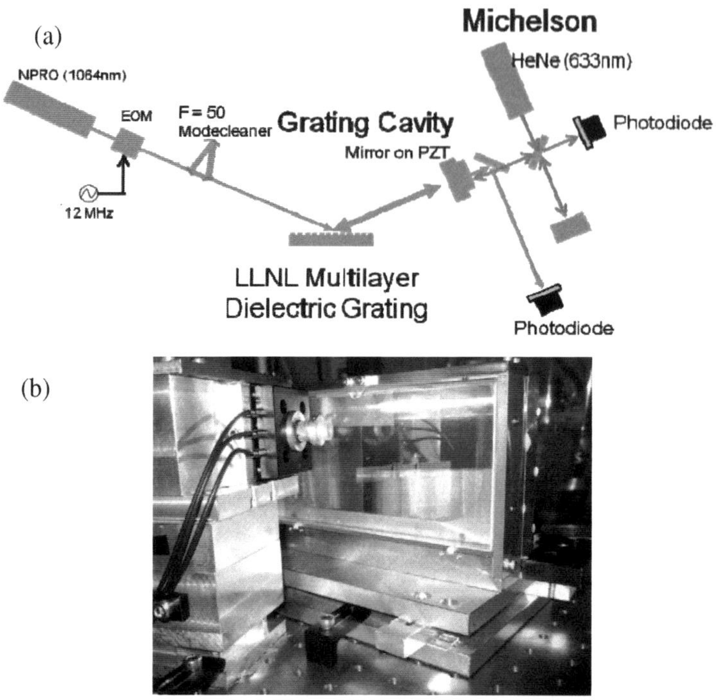

Figure 4: (a) Experimental setup for determination of grating efficiency at Littrow angle. (b) Photograph of the grating and Littrow cavity.

Figure 5 summarizes the cavity transmission while the piezo was scanned across a free spectral range of the cavity. We measured the finesse of a grid of points across the grating; in this example we present data gathered from the center point of the grating. In Figure 5(a), a full free spectral range of the cavity is shown, with both carrier and sideband peaks visible. The 12 MHz carrier-sideband and 24 MHz sideband-sideband beating components of the signal were digitally filtered out. Presenting the x-axis as the actuation distance is made possible by the Michelson interferometer, which allows for a mapping between scan time and distance traversed by the mirror. This calibration removes any nonlinearities in the piezo. In addition, by measuring the sideband spacing and using the known 12 MHz modulation frequency, it is straightforward to compute the free spectral range of the cavity and thus the cavity length. From these calculations, we arrived at a free spectral range (FSR) of 1.59 GHz, corresponding to a cavity length of 94.4 mm.

Figure 5(b) displays a zoomed view of one carrier peak and its sidebands. By performing a least squares fit on this plot to a model consisting of three summed Lorentzians, one can deduce the full-width-of-half-maximum (FWHM) of the central peak by comparing it to the sideband spacing. For this example, we have a FWHM of 1.97 MHz.

These two plots allow us to compute the finesse of the cavity formed between the grating and reference mirror to be 806. The reference mirror was independently characterized by taking two other mirrors and measuring the finesse of the cavities formed from each pairing of the mirrors in turn. The result was a reflectivity of 99.798%. With these two numbers, the diffraction efficiency of the grating is calculated to be 99.425%.

6. Grating power handling capability

As a test of the power handling capabilities of the dielectric grating, we illuminated a 1 mm patch of grating with 34W of 1064 nm light from a master oscillator power amplifier system. The illuminated spot was monitored with a Shack-Hartmann wavefront sensor, confirming that no significant wavefront distortions were present [6]. The power levels tested were significantly greater than the 2W required by LISA, indicating that similar gratings will be good candidates for BBO and DECIGO, which require ~100W of laser power.

7. Summary

A dielectric grating was measured for its efficiency across a range of incident angles for both polarizations, showing a polarization diffraction ratio of greater than 23.2 across all angles. We have also measured the depolarization of the grating, showing that less than 0.018% of incident light diffracts into the other polarization. In addition, a second grating was characterized for its diffraction efficiency at the Littrow angle. This was done using highly precise cavity measurements. Using this method, we have found grating diffraction efficiencies in excess of 99.4%, with five significant digits of precision. In light of these grating measurements, our proposed interferometer configuration shows future promise.

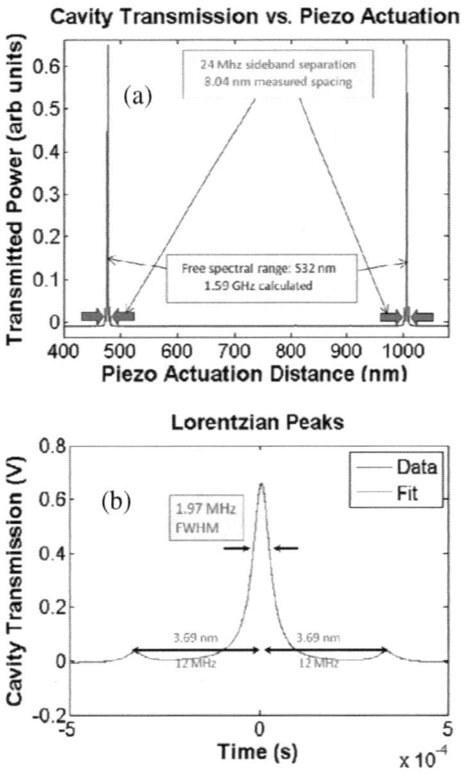

Figure 5: (a) Cavity transmission data during scan of cavity's length. (b) Zoomed view of the transmission peak that shows a cavity finesse of 806

Acknowledgements

This research is partially supported by NASA Beyond Einstein Science Foundation Grant NNX07AK65G "Modular Gravitational Reference Sensor for Space Gravitational Wave Detection", and by NSF grant for LIGO program.

References

[1] K.-X. Sun, R.L. Byer, "All-reflective Michelson, Sagnac, and Fabry–Perot Interferometers Based on Grating Beam Splitters," Optics Letters Vol. 23, pp 567-69 (1998)
[2] K.-X. Sun, G. Allen, S. Buchman, D. DeBra, R.L Byer, "Advanced Gravitational Reference Sensor for High Precision Space Interferometers," Class. Quantum Grav. Vol. 22, S287-S296, (2005)
[3] LISA Science Team "System and Technology Report", ESA (2000)
[4] M.D. Perry, R.D. Boyd, J.A. Britten, D. Decker, B.W. Shore, C. Shannon, E. Shults, and L. Li, "High Efficiency Multilayer Dielectric Diffraction Gratings" Optics Letters Vol. 20, 940-942 (1995)
[5] H. Wei and L. Li, "All-Dielectric Reflection Gratings: A Study of the Physical Mechanism for Achieving High Efficiency," Appl. Opt. Vol. 42, 6255-6260 (2003)
[6] P.P. Lu, K.-X. Sun, R. L. Byer. "Thermal Loading for Dielectric Gratings", manuscript in preparation

Journal of Physics: Conference Series **154** (2009) 012032 doi:10.1088/1742-6596/154/1/012032

Double Degree of Freedom Pendulum Facility for the study of Weak Forces

R Stanga[1,2], L Marconi[3,2], C Grimani[4,2], M Bassan[5,6], G Pucacco[5,6], E Reali[5,6], R Simonetti[6], N Finetti[7]

[1] Dipartimento di Astronomia e Scienza dello Spazio, Università degli Studi di Firenze, Florence, Italy.

[2] Istituto Nazionale di Fisica Nucleare, Florence, Italy.

[3] Dipartimento di Fisica, Università degli Studi di Firenze, Florence, Italy.

[4] Istituto di Fisica, Università degli studi di Urbino, Urbino, Italy.

[5] Dipartimento di Fisica, Università degli studi di Roma Tor Vergata, Roma, Italy.

[6] Istituto Nazionale di Fisica Nucleare, Roma Tor Vergata, Italy.

[7] Dipartimento di Fisica, Università degli Studi dell'Aquila, L'Aquila, Italy and Istituto Nazionale di Fisica Nucleare, Laboratori Nazionali del Gran Sasso, Gruppo Collegato dell'Aquila, Italy.

E-mail: marconi@fi.infn.it

Abstract. The LISA Test-Mass (TM) is sensitive to weak forces along all 6 Degrees of Freedom (DoFs). Extensive ground testing is required in order to evaluate the influence of crosstalks of various read-outs and actuators operating on different DoFs. To this purpose, and to better represent the flight conditions, we are developing a facility with 2 soft DoFs which consists of two stage roto-translational pendulum. This facility will measure the forces and stiffnesses simultaneously acting on the Test Mass along 2 different soft DoFs. The advantages with respect to a single DoF test bench are a more effective identification and debug of spurious effects between the TM and the capacitive position sensor surrounding it and the possibility to test actuation crosstalk with closed feedback loop. In particular, it allows us to measure the residual disturbance along one DoF when we close the control loop on the other one.

1. Introduction

The LISA requirement for the residual acceleration noise in the free-falling frame of the test-mass m_{TM} is [1]:

$$\frac{S_{fx}^{\ 1/2}}{m_{TM}} \leq 3\cdot 10^{-15}\left[1+\left(\frac{f}{3mHz}\right)^2\right]\frac{m}{s^2\sqrt{Hz}}$$

for $0.1mHz < f < 0.1Hz$. A drag-free operation mode is necessary, with minimal coupling between TM and spacecraft, so that the above requirements can be met.

A capacitive sensor system (GRS, *Gravitational Reference Sensor)* [2], is used to read out the position of the TM with respect to the spacecraft. This represents the error signal provided to the control loop that uses micro-thrusters as actuators to force the spacecraft to follow the TM.

Journal of Physics: Conference Series **154** (2009) 012032 doi:10.1088/1742-6596/154/1/012032

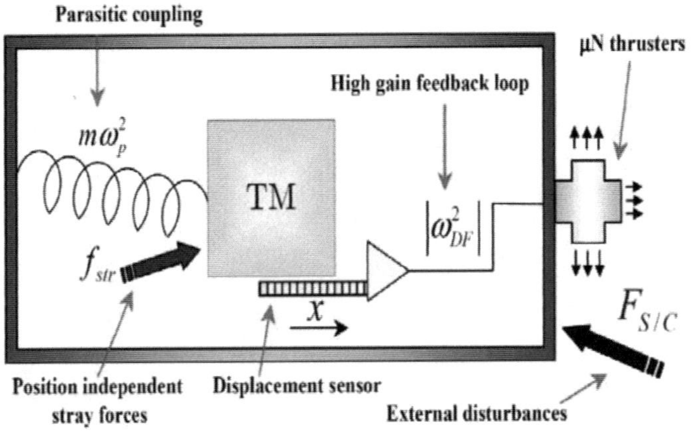

Figure 1. Referring to the diagram on the left, in the case of a single-mass/single-axis control loop, the residual closed-loop TM acceleration is:

$$a_n \approx \frac{f_{str}}{m_{TM}} + \omega_p^2 \left(x_n + \frac{F_{S/C}}{M\omega_{DF}^2} \right)$$

Ground tests are required to study the residual weak forces that may couple the TM to the GRS. The challenge for these tests is to cancel gravity, in order to simulate a free fall condition along as many DoFs as possible. Up to now, very good results were obtained with a torsion pendulum, that only allows to study one single DoF [1].

2. The Roto-Translational Pendulum

The geometry of the roto-translational pendulum [3], [4], [5] allows us to have 2 soft DoFs. The rotations around the two fibers are indicated by the independent angles are φ and (θ-φ) (figure 2). The φ rotation approximates a linear (translational) displacement x at the tip of the arm of the crossbar.

Table 1. Dimensions of the roto-translational pendulum.

	φ pendulum	**(θ-φ) pendulum**
Material	tungsten	Tungsten
length	90 *cm*	90 *cm*
diameter	100 μm	25 μm
momentum of inertia	0.02 kgm^2	3.38 10^{-6} kgm^2
frequency	1.5 *mHz*	2.4 *mHz*

The roto-translational pendulum is hosted in an Al5083 vacuum chamber 3*m* high, and 1*m* in diameter. Two remotely controlled motors allow us to rotate the φ fiber around its axis, and to raise or lower its suspension point. The central φ fiber holds a crossbar; at the tip of one of its arms we connect the (θ - φ) fiber, at the bottom of which the TM is hanging. To balance the crossbar, three more masses are rigidly fastened to the tips of the remaining crossbar arms. We minimized the quadrupole moment with an appropriate distribution of the balancing masses. The TM position is measured by capacitive sensors mounted in a cubic box (the GRS); the GRS can be remotely muved along 6 DoFs, in order to place the TM at its center with an accuracy of a few μm.

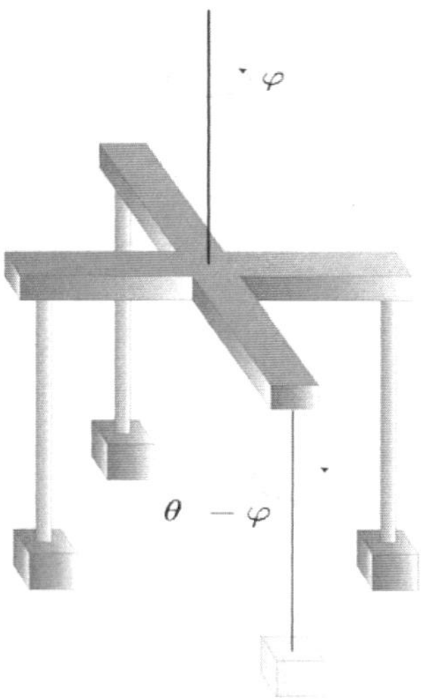

Figure 2. Schematic of the roto-translational pendulum.

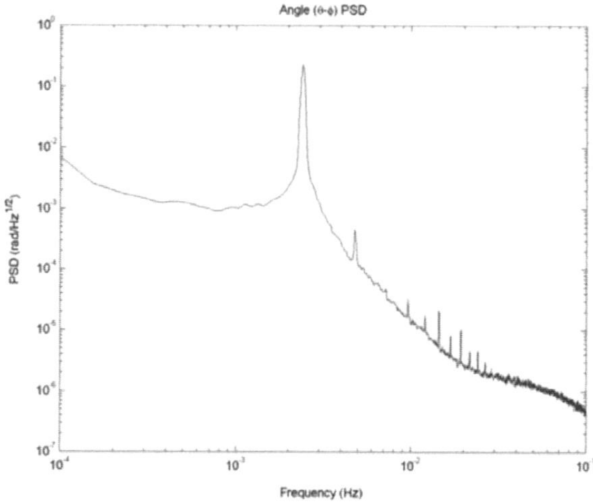

Figure 3. Measurement of the resonance frequency of the (θ - φ) pendulum.

With a scroll plus turbomolecular pump system we rapidly achieve a pressure of 10^{-5} *mbar*; after heating the vacuum chamber at about 80 $^{\circ}C$, we obtained a pressure of about 10^{-6} *mbar*.

Figure 4. View of the double pendulum and the GRS.

Figure 5. The GRS and the 6 DoF manipulator.

3. Read Out

The position read out of the TM is achieved with capacitive and optical devices. The optical system is used for calibration.

3.1. Capacitive Read Out

The capacitive sensing and feedback electronics were designed building on the experience of analogous devices previously developed for similar, single DoF applications [6], [7]. Each of two identical channels reads out a pair of capacitive sensors placed along the x axis of the GRS on opposite sides of the TM: the two pairs are placed parallel to each other, but at a different y position.
The signal is modulated at about 100kHz, and then read and demodulated with a lock-in amplifier. Sum and differences of these two channels outputs provide information about translation along x axis and rotation (θ-φ) as defined in Par.2, respectively. The capacitors are also used as electrostatic actuators, to close a feedback position loop on the TM.

Figure 6. *Electronics board of the capacitive read out.*

Figure 7. *The GRS and the TM.*

3.2. Optical Read Out

In addition to the capacitive readout, an optical readout scheme is also implemented, based on a commercial autocollimator (Elcomat Vario 300). The autocollimator can be mounted in two different positions, to measure either the angle φ, or the angle (θ-φ) using to two mirrors rigidly mounted on the TM and the crossbar, respectively.

Figure 8. *Vacuum chamber with the autocollimator.*

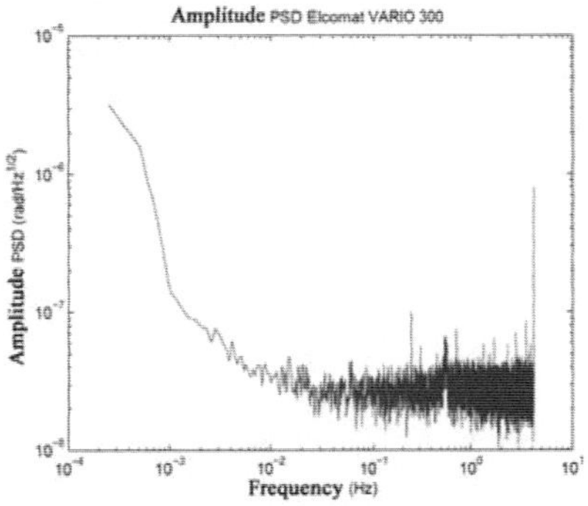

Figure 9. *Noise spectrum of the optical read out system: 50nrad/√Hz are basically in agreement with the instrument's performances.*

Journal of Physics: Conference Series **154** (2009) 012032 doi:10.1088/1742-6596/154/1/012032

4. Magnetic Noise Measurements

Fluctuations in the local magnetic field may couple to the magnetic moment of the roto-translational pendulum and act as weak forces between TM and its environment. We used three triaxial magnetic probes to measure the magnetic field and its gradient around the TM position, placing a μ-metal cylinder 0.4*mm* thick around the sensor, as a magnetic shield. We estimate the magnetic moment of the pendulum to be a few $10^{-9} Am^2$ on the basis of measurements [7] on the same test mass, and in a similar setting, and with this value we compute the expected torque noise generated by magnetic field fluctuations. As it is shown in figure 11, this noise is negligible, even when no magnetic shield is present (red curve).

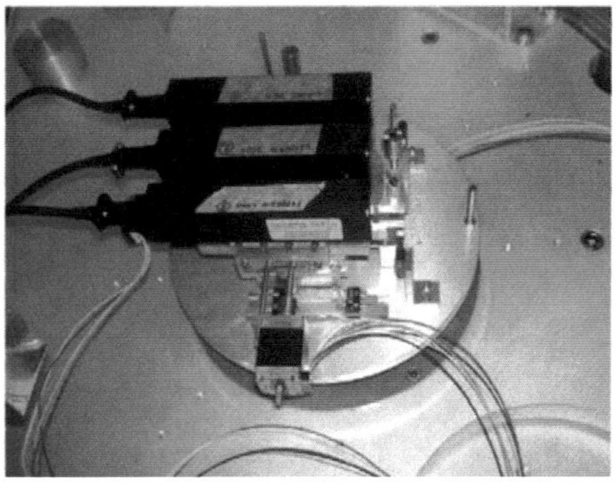

Figure 10. *The magnetic field sensors.*

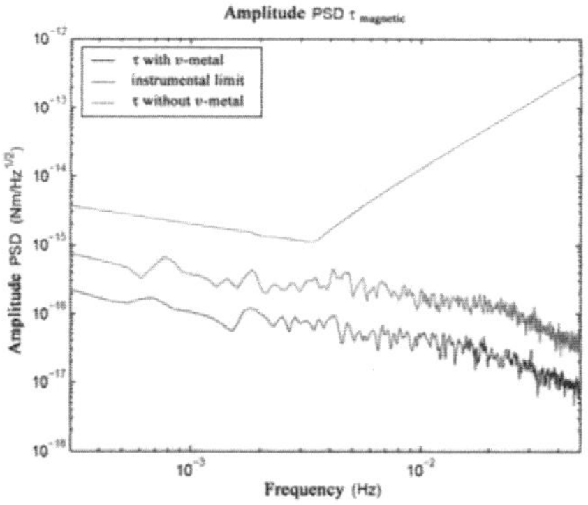

Figure 11. *Torque (τ) noise induced by the fluctuation of the magnetic field. The thermal noise limit is plotted in green color.*

5. Calibration and Qualification Measurements

We calibrated the displacement sensor on the bench; then we measured the readout noise with the two DoFs blocked, the TM being held in position by a shaft and the crossbar resting on a support. The sensitivity is better than 2 $nm/\sqrt{Hz}$ down to 50 mHz, and it is adequate to reach the thermal noise limit of the two pendulum stages; at lower frequencies the noise is higher, with a consistent contribution from the residual motion of the test mass.

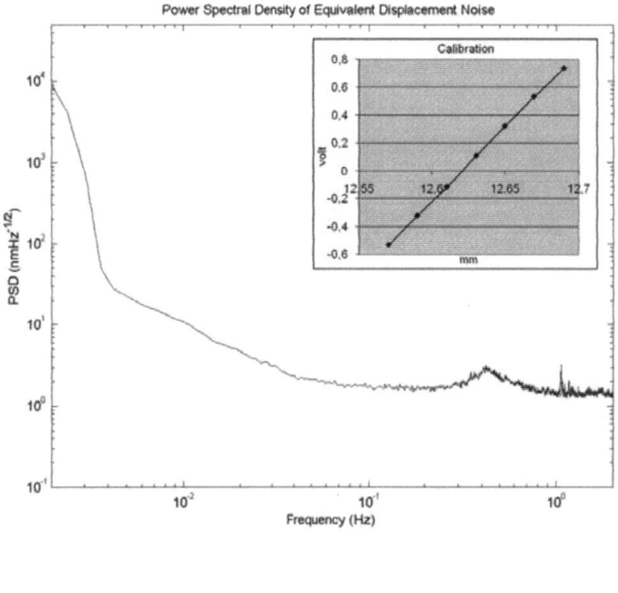

Figure 12. *The plot on the left shows the spectral density, expressed in displacement units (nm/√Hz) of the readout noise plus environmental noise. The shaft that holds the TM is not perfectly rigid, and the residual motion of the TM generates the broad peak around 0.4Hz The inset shows the calibration of the capacitive position sensor.*

6. Commissioning of the Roto-translational Pendulum

We integrated the sensor and the TM to the facility: the calibration of the capacitive readout is found consistent with the autocollimator read out of the position angle. Then we qualified the pendulum, first blocking the degree of freedom in translation, allowing for the TM to rotate around its axis inside the sensor. We measured for the rotation degree of freedom a frequency of $2.35mHz$, with the quality factor Q of about 750, at a pressure of $10^{-6}mbar$. As soon as a third electronic board will be delivered, we will be able to unblock the degree of freedom in translation.

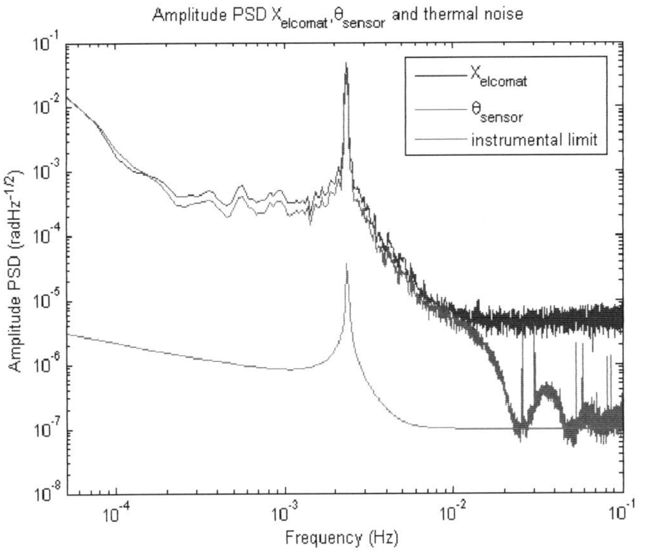

Figure 13. *PSD in angle measured for the degree of freedom in rotation. In blue, the optical autocollimator read out; in green the capacitive readout, measured with a time constant of 20s; in red the thermal noise and electronic readout noise.*

Figure 13 shows one of the first measurements of the angular PSD. Bringing the curves to the thermal noise level is one of the goals of the commissioning.

As we said, the capacitors are also used as electrostatic actuators, to close a feedback position loop on the TM. A Labview routine processes the position signal read from the electrostatic sensors, filters it to obtain a critical damping, and feedbacks a voltage modulated at $200Hz$ to the capacitors.

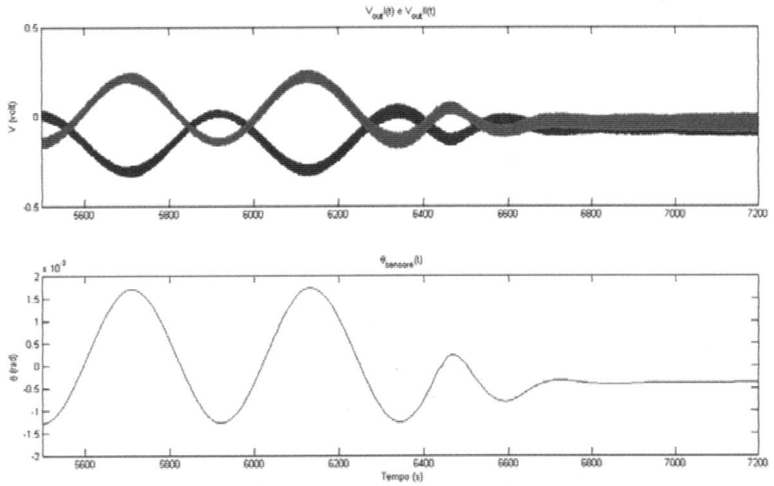

Figure 14. *Closing the feedback loop on rotation: in the top panel the voltage read with the capacitive device is plotted versus time; in the bottom panel, the angle is reported.*

7. Next Steps

The next steps of the activity on the 2DoFs pendulum will be:

- To improve the electronic noise figure below 50*mHz*.
- To achieve the noise specification for each degree of freedom (basically the thermal noise of the fibers that is in terms of force about 10^{-13} $N/\sqrt{Hz}$ at frequencies below the resonant frequency of the torsion pendulum).
- To start measurements with the 2DoFs simultaneously free and to apply a feedback control to either one or both of them.

8. References

[1] Vitale S et al., 2002 *Nuclear Physics B* **110** 210 and references therein
[2] Dolesi R et al., 2003 *Class. And Quant. Grav.* **20** S99
[3] Hoyle C D et al., 2003 *Proc. Marcel Grossman Meeting*
[4] Su Y et al., 994 *Phys. Rev. D* **50** 3614
[5] Stanga R et al, 2006 *J.Phys.: Conf. Ser.* **32** 180
[6] Hueller M et al., 2002 *Class. And Quant. Grav* **19** 1757
[7] Hueller M, 2003 *PhD degree thesis*

Improved Grating Angular Sensor for LISA and MGRS

Ke-Xun Sun, Patrick Lu, and Robert L. Byer

Ginzton Lab N-119, Stanford University, Stanford, CA 94305-4088, USA

kxsun@stanford.edu

Abstract. LISA requires high precision angular beam pointing and telescope steering. In this paper, we report recent results for an improved grating angular sensor. We have achieved better than 0.2 nrad/Hz$^{1/2}$ at 1 kHz with 14 mW of incident power, a factor of 5 improvement over our previously reported results. At 1 Hz we achieved 1-2 nrad/Hz$^{1/2}$. We realized these improvements by enclosing the grating angular sensor assembly in a vacuum chamber and mounting the optics components on a zerodur glass plate, thereby lowering the noise floor at low frequencies. Furthermore, by upgrading the electronics and thus the detector power handing capability, we also investigated sensitivity scaling versus incident laser power. The results will benefit the design of grating angular sensors.

1. Introduction

Spacecraft constellation missions, such as LISA, DECIGO, and BBO, require high precision angular pointing and steering, primarily for measuring relative position between spacecraft. LISA also needs to measure proof mass angular orientation. The precision required for the proof mass angular orientation measurement is ~100 nrad/Hz$^{1/2}$, and for telescope steering ~1 nrad/Hz$^{1/2}$. As a future technology for LISA type missions, the modular gravitational reference sensor (MGRS) will have similar angular sensing needs.

Grating angular sensors outperform traditional angular sensors based on simple specular reflection thanks to grating angular magnification and beam cross-section compression [1]. We have conducted a series of experiments and demonstrated the performance of grating angular sensors. In the first proof of principle experiment [1], we used an off-the-shelf, 1200 line/mm grating, and demonstrated 10 nrad/Hz$^{1/2}$ angular sensitivity over a mere 5 cm of working distance with 1 mW of incident power at 1 kHz.

In a second experiment [2], we improved angular sensitivity and demonstrated the robustness of the symmetric grating angular sensor against laser frequency noises. We custom-designed a 935 line/mm grating, and used it in conjunction with a 1064 nm Nd:YAG NPRO laser to form a symmetric grating angular sensor. With an input laser power of 4 mW, the angular sensitivity of the symmetric grating angular sensor at 1 kHz was observed to be ~1 nrad/Hz$^{1/2}$ [2]. In the presence of large frequency excursions of the laser, the angular sensor output remained constant [2].

Also, we developed three different methods of fabricating localized gratings on the surface of a proof mass [3]: electron-beam lithography, transfer-imprinting, and focused ion beam milling. We fabricated grating patterns with submicron features and line densities of 935 lines/mm using each of the methods. These fabrication techniques will enable patterning of the proof mass for precision angular sensing.

In addition, we demonstrated a dual range grating sensor, in which the first order diffraction is used to achieve high angular sensitivity, and the zeroth order is used to increase the dynamic range of the angular sensor [4].

In this paper, we report on recent progress in grating angular sensor performance. We improved the environmental shielding of the grating angular sensor. By mounting the optics components on a zerodur glass plate and enclosing the grating angular sensor assembly with a vacuum chamber, we lowered the noise floor at low frequencies. At 1 Hz, the noise floor was lowered to 1-2 nrad/Hz$^{1/2}$, which meets LISA requirement at this frequency. Furthermore, by upgrading the electronics and thus the detector power handing capability, we were able to investigate sensitivity scaling versus incident laser power. We have achieved better than 0.2 nrad/Hz$^{1/2}$ sensitivity at 1 kHz with 14 mW of incident laser power.

2. Angular measurements at lower frequencies

The LISA sensitivity band ranges from 30 μHz to 1 Hz, and the BBO sensitivity band from 0.1 Hz to 10 Hz. Therefore we have attempted to measure the grating sensor performance at low frequencies.

Figure 1 shows the improved experimental setup. For better isolation, we acquired a vacuum enclosure and associated turbo pumping system. The vacuum pressure in the chamber was first pumped down to 1×10^{-5} torr and then the pump was stopped to reduce vibration during data taking. Fine realignment was needed since the change of air pressure induced minute chamber deformations. The laser beam was directly coupled into the chamber via a window, and the beam path was enclosed in tubing to reduce air flow-induced beam walking. A zerodur glass plate was placed inside the vacuum chamber. The angular sensor was then mounted on an aluminum breadboard via the zerodur plate as the first rigid contact. The movement of the optical mounts thus largely followed the zerodur plate. Our intention was to take advantage of the low coefficient of thermal expansion (~5×10^{-8} 1/C°) of zerodur, thereby improving low frequency performance in the presence of normal environmental

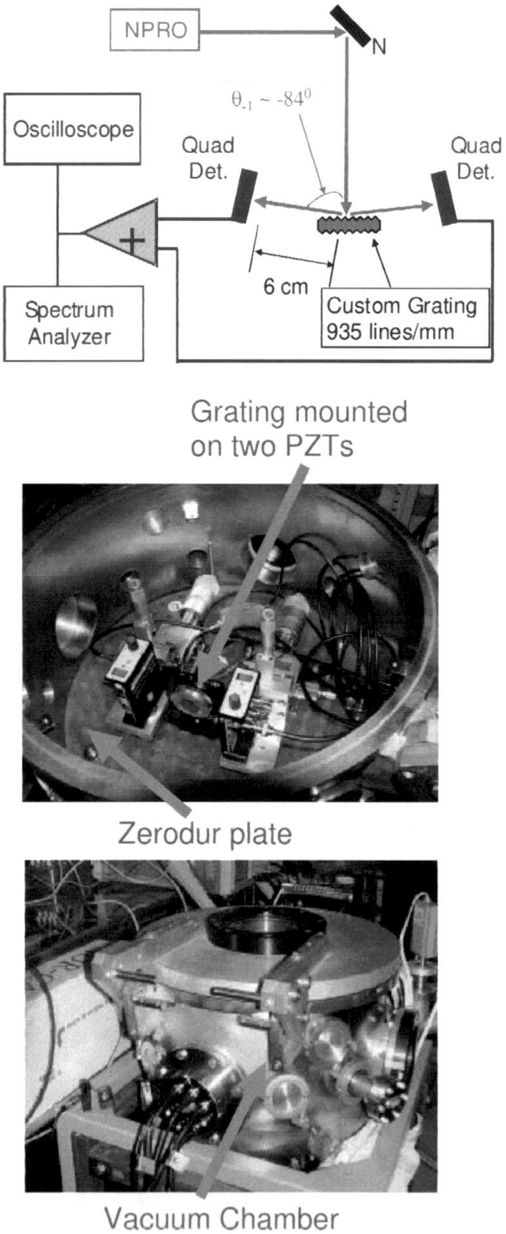

Figure 1. Improved experimental setup for the grating angular sensor.

temperature change.

Figure 2 (a) and (b) show the measurement results. Figure 2 (a) shows results from 0.1 to 10 Hz. For calibration, a 1 Hz, 0.5 μrad grating rotation was applied by differentially driving a pair of PZTs. We observed ~1-2 nrad/Hz$^{1/2}$ of angular sensitivity in the frequency range from 1 Hz to 10 Hz. Above 10 Hz the increased noise level was due to seismic noise in the lab. The noise floor at 0.1 Hz was measured to be ~5-10 nrad/Hz$^{1/2}$. From 0.4 to 0.1 Hz, the increase of noise approximately scaled as *1/f*. However, a 3-6 dB correction could be needed due to electronics coupling.

Figure 2 (b) shows the measurement result from 0.1 to 1000 Hz. A 10 Hz, 0.5 μrad rotation of the grating was applied as a calibration signal. From 0.1 to 10 Hz, similar behavior as that in Figure 2 (a) was shown. From 10 to 50 Hz, the noise floor was dominated by ground vibrations.

There is still much room for improvement in the measurement apparatus, depending on available resources. However, the current performance level has exceeded our expectations for the initial measurements at low frequencies.

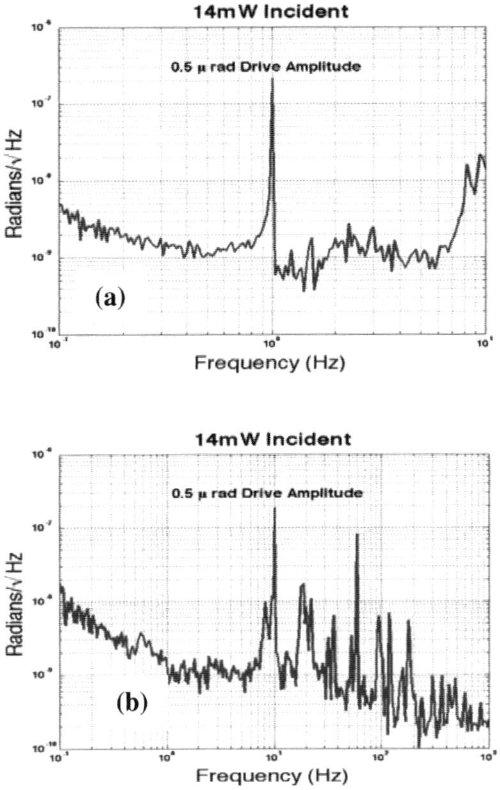

Figure 2. (a) Grating Sensitivity measurement for the grating angular sensor from 0.1 to 10 Hz. A 0.5 μrad, 1 Hz calibration angular motion was applied to the grating. (b) Sensitivity measurement for the grating angular sensor from 0.1 to 1000 Hz. A 0.5 μrad, 10 Hz calibration angular motion was applied to the grating. In both (a) and (b), the noise floor at 1 Hz is approximately 1~2 nrad/Hz$^{1/2}$. Also in both graphs, 10 Hz seismic vibrations show as strong noise sources. In (b) the 60 Hz line interference was also large. At low frequencies, the thermal and ground motion are dominant noise source for this particular experimental setup. A ~6dB correction to the 0.1 Hz measurement might be need to account for electronics circuit responses in the near dc region.

3. Angular sensitivity scaling versus incident laser power level

We have employed quad-photodiodes and direct detection for the grating angular sensors. Therefore the output signal is proportional to the incident laser power, limited by the saturation of photodiode and electronics circuitry. In the recent experiment, we improved the photodetector electronics of the grating angular sensor by optimizing gain and other parameters of the circuits. The photodetector could thus receive higher laser powers without saturation.

We investigated sensitivity scaling versus incident laser power. These measurements were taken around 1 kHz where environmental noises were lower, and the measured sensitivity represents the intrinsic optics and electronics performance of the grating angular sensor. The angular sensitivity was measured ~2 nrad/Hz$^{1/2}$ for 1 mW of incident laser power, ~0.8 nrad/Hz$^{1/2}$ for 2.4 mW, 0.6 nrad/Hz$^{1/2}$ for 4 mW, and 0.2 nrad/Hz$^{1/2}$ for 14 mW. The 0.2 nrad/Hz$^{1/2}$ sensitivity represents a factor of 5 improvement over our previously reported results. Figure 3 shows the spectrum from this measurement, including the 0.5 μrad rotation calibration signal at 1 kHz. The dependence of the noise floor, or the angular sensitivity, on input laser power is plotted in Figure 4. The linear fit indicates the dependence follows a power law with a exponent ~ -0.77, which could result from a combination of residual laser intensity noise and electronics noise.

Above 14 mW incident laser power, we observed the onset of detector saturation. However, the current photodetector can be further improved to handle higher incident laser power.

The laser power scaling of angular sensitivity is useful for selecting correct laser power level when designing systems for the grating angular sensor realistically. To illustrate this point, we plot power scaling of the signal-to-noise ratio (SNR) in Figure 5 for several signal strengths.

The top curve shows the SNR measured from our experiments using 0.5 μrad calibration signal. Under it are four extrapolated curves where the signal strengths are taken to be the proof mass angular sensing requirement of the LISA GRS (0.1 μrad/Hz$^{1/2}$), the further reduction of coupling noise (10 nrad/Hz$^{1/2}$), the telescope pointing requirement (1 nrad/Hz$^{1/2}$), and the noise floor we measured using 14 mW of incident power (0.2 nrad/Hz$^{1/2}$). As an example, to achieve 10 nrad/Hz$^{1/2}$ resolution sensing with an SNR~10, the input laser power should be ~1.2 mW.

4. Conclusion

A novel grating angular sensor that meets LISA requirements has been experimentally demonstrated. The sensitivity of 0.2 nrad/Hz$^{1/2}$ is better than the LISA gravitational reference sensor requirement by a factor of 500. This sensitivity level is measured at 1 kHz. However, it represents the intrinsic performance for the grating angular sensor. The first low frequency measurements reported here show that environmental noises dominate the noise floor. The highly sensitive performance of the grating angular sensor should be extended to lower frequency bands with proper environmental control.

The design of gratings and their fabrication on proof masses have many applications in precision spaceflight and industry. The grating angular sensor is expected to have better mechanical and thermal stability than a conventional angular

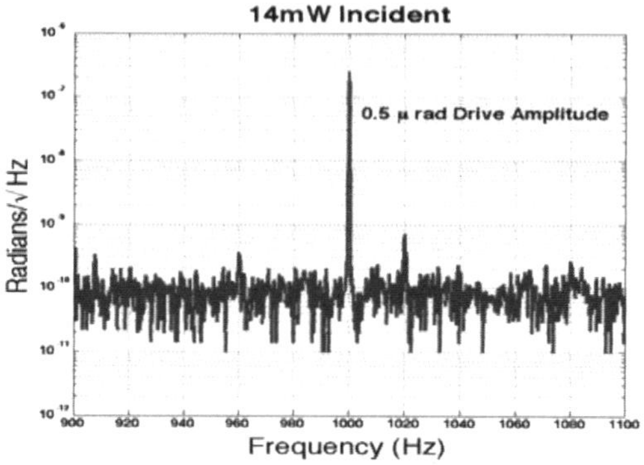

Figure 3: Input laser power was 14 mW. The noise floor was below 2x10^{-10} rad/Hz$^{1/2}$

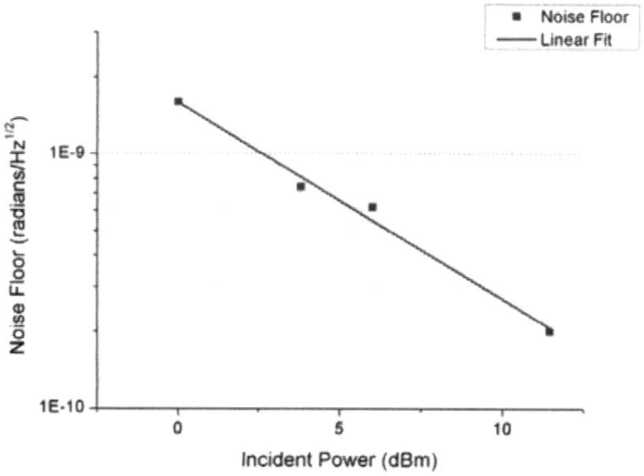

Figure 4. Sensor noise floor vs. incident laser power.

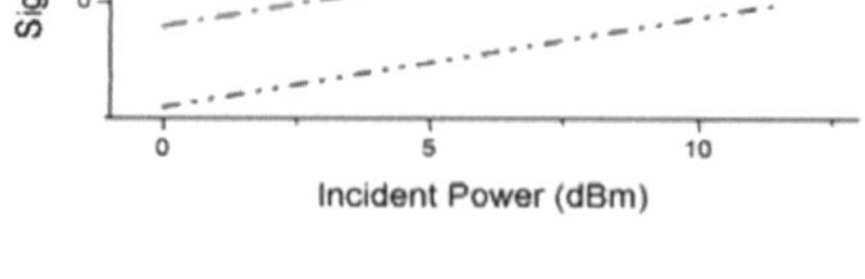

Figure 5: Plot of grating angular sensor SNR as a function of incident power for various assumed signals. The slope of the fit is 0.77.

sensor, and has broad implications for gravitational reference sensing, telescope pointing, and spacecraft flight control [5].

Acknowledgements

This research is partially supported by NASA Beyond Einstein Science Foundation Grant NNX07AK65G "Modular Gravitational Reference Sensor for Space Gravitational Wave Detection", and by the Jet Propulsion Laboratory DRDF program.

References

[1] K.-X. Sun, S. Buchman, and R. Byer, "Grating Angle Magnification Enhanced Angular and Integrated Sensors for LISA Applications", *J. Physics*; CS 32, 167 (2006).
[2] K.-X. Sun, P. Lu, and R. Byer, "A Robust, Symmetric Grating Angular Sensor for Space flights", AIP Proceedings, 6th LISA International Symposium, 2006.
[3] P. Lu, K.-X. Sun, and R. Byer, "Grating Fabrication for Gravitational Wave Interferometers and LISA GRS", AIP Proceedings, 6th LISA International Symposium, 2006
[4] A.J. Swank, K.-X. Sun, and D. B. DeBra, "Determining Gravitational Attraction by Mass Property Measurements", AIP Proceedings, 6th LISA International Symposium, 2006
[5] K.-X. Sun, G. Allen, S. Williams, S. Buchman, D. DeBra, and R.L. Byer, "Modular Gravitational Reference Sensor: Simplified Architecture to future LISA and BBO", *J. Physics*: CS, 32, 137-146 (2006).

Initial interferometric pre-investigations for LISA

E D Fitzsimons, J Bogenstahl, J Hough, C Killow, M Perreur-Lloyd, D I Robertson, S Rowan and H Ward

Institute for Gravitational Research, Department of Physics and Astronomy, University of Glasgow, Glasgow G12 8QQ, UK

E-mail: e.fitzsimons@physics.gla.ac.uk

Abstract. LISA, the Laser Interferometer Space Antenna, is a proposed ESA/NASA space based gravitational wave detector. In order to help meet the many technological challenges of LISA, the ESA precursor mission LISA Pathfinder (LPF) will test some of the key enabling technologies for LISA. LPF however will only go so far, and much work is needed to take LPF technology to a state suitable for LISA. One such area is the use of polarising Mach-Zehnder interferometers. We report on the design and initial construction of an experiment to test the use of such interferometric techniques, as well as suitable component mounting mechanisms.

1. Introduction

The LISA Pathfinder (LPF) mission goes some way to testing key technologies for LISA, but there are some elements of the LISA baseline design that LPF will not test. One of these is the use of polarising optics for beam steering. The interferometers used for probing the test masses on LPF are non-polarising Mach-Zehnder interferometers. In LISA it is planned to use a combination of non-polarising and polarising Mach-Zehnder interferometers, with the test mass read-out interferometer in particular being of the latter form [1]. Adopting this design requires that the ~pm stability of polarising optics in the mHz region be tested, and shown to be suitable for adoption in LISA.

2. Experimental concept

2.1. Hydroxide-catalysis bonding

In order to validate polarising interferometry as being suitable for use on LISA, an experiment must measure the stability of such arrangements in a LISA-like configuration. One possible solution is to use a low expansion base plate and hydroxide-catalysis bond components to it to form an interferometer [2]. The required stability of such systems has already been demonstrated, for example by the engineering model of the LPF Optical Bench Interferometer (OBI) [3]. Further more the LISA baseline design calls for hydroxide catalysis bonding as the preferred method of construction for the LISA optical benches [1].

2.2. Interferometer design

An interferometric test bench has been designed for this purpose, consisting of multiple Mach-Zehnder interferometers. There will be two non-polarising reference interferometers, one unequal arm length non-polarising interferometer to measure laser frequency noise [4], and one measurement interferometer into which can be placed polarising optics. The bench consists of a

300 mm square baseplate made of Ultra Low Expansion (ULE) glass ceramic, the top surface of which has been polished locally flat to $\lambda/10$ over any 30 mm length. Fused silica optics are then hydroxide catalysis bonded onto this surface, this is described in Section 4. Light is injected onto the bench through a monolithic fibre collimator as designed for the OBI [5]. Called the 'FIOS' (Fibre Injector Optical Subassembly), it consists of a small fused silica baseplate onto which the main components are attached. The low thermal expansion of the fused silica gives a stable lens-fibre separation, resulting in a stable beam. Two FIOSs are required, one for each of the beams, these will be bonded to either side of a mounting post which is bonded to the baseplate. The layout of the interferometer is shown in Figure 1. The polarising optics will be mounted on a separate small low thermal expansion baseplate (made of ZERODUR), that can be inserted in the measurement path, illustrated in the figure by a dashed box. This introduces a degree of versatility into the measurement capability and will allow a variety of different subassemblies to be tested, enabling, for example, comparison of different mounting techniques for polarising optics.

3. Construction plan

The test bench is being constructed using the same procedures and techniques as the OBI. In doing so there must be a carefully defined order of construction to ensure the alignment of the bench. Broadly, there are three different types of item to be attached to the baseplate, the mirrors and beam splitters, the FIOSs and the photodiodes. The mirrors and beam splitters can be further subdivided into two classes, critical and non-critical. Non-critical components are those that can be bonded down with larger tolerances, and any beam position errors arising from these larger tolerances can be compensated for with the positioning of the critical components. The critical components are M7 and M8 (the input beam steering mirrors for the FIOSs) and BS4, BS6, BS11 and BS12, the four beam recombiners. All other optical components (except the polarising components shown in the dashed box, which are treated separately) are non-critical. The procedure for construction is then: *(a)* template bonding of the non-critical components, *(b)* bonding of FIOS2, *(c)* precision alignment and bonding of M7, *(d)* bonding of FIOS1, *(e)* precision alignment and bonding of M8, *(f)* precision alingment and bonding of the four beam recombiners and *(g)* gluing of the six photodiodes, aligned to beam centres.

4. Construction of the test bench

4.1. Templating

4.1.1. Template design and manufacture The template used for construction step *(a)* is made of brass plate $\sim$ 4 mm thick. It sits on three ball bearings on the top surface of the bench, and locates against the baseplate on a further three ball bearings mounted in blocks under the template. For each component, there is a 'cut-out' in the template, with three ball bearings mounted in holes along two of the sides, one on the short side and two on the long side. In this way, a component placed into a cut out will sit against the three ball bearings. The template itself was manufactured using a 5-axis CNC mill, and after machining, was measured using a coordinate measuring machine (CMM). The positions of all the component locating ball bearings were measured, and the mean deviation of the ball bearings from the nominal positions was $\sim 30\,\mu$m.

4.1.2. Template bonding During template bonding, the baseplate, having been initially cleaned to specifications suitable for bonding (see [2]), was inclined to an angle of $\sim 5°$. The template is designed so that all the component locating ball bearings are to the same side. As a result, the inclination of the baseplate provides a small restoring force which holds the components in place against the ball bearings. The components were bonded in turn, having first been suitably cleaned like the baseplate, using $0.6\,\mu\mathrm{l\,cm^{-2}}$ of sodium silicate solution. The bonds were left

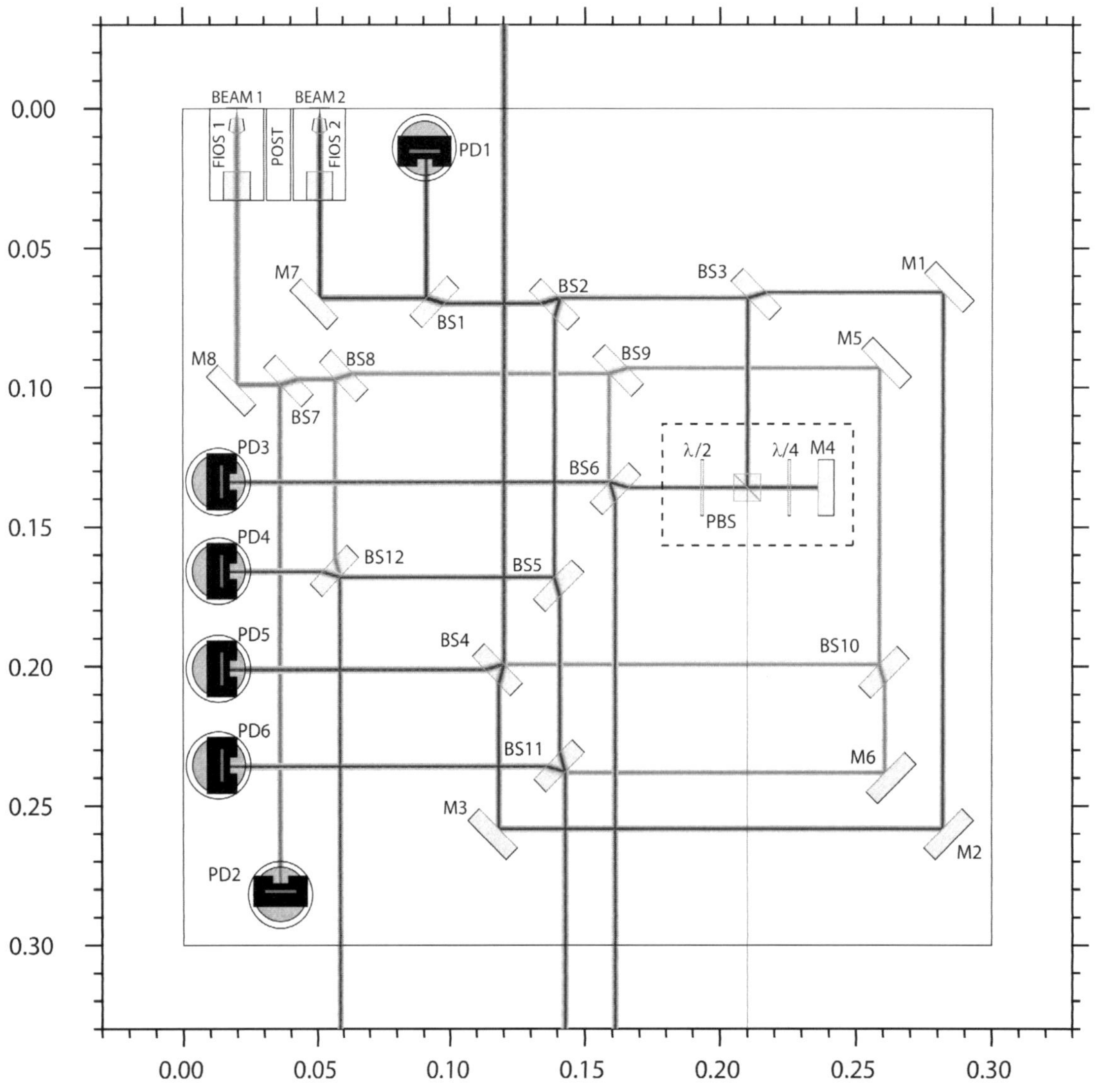

Figure 1. OptoCad [6] representation of the test bench; the polarising optics are shown bounded by a dashed box

to cure for 24 hours before the template was removed, using three screw risers built into the template which raised it above the height of the optics before it was lifted off. In addition to the non-critical mirrors and beam splitters, the FIOS mounting post was also bonded with the template. Figure 2 shows the bench after the templating procedure just prior to the removal of the template.

4.1.3. Component position accuracy After the bonds had thoroughly cured, the components were measured with the CMM to determine the accuracy of positioning. Since it is only the *relative* position of the components that is of interest (alignment to the sides or any other absolute reference frame is not required) the measured positions of the components was optimised by applying small rotations and translations to the measurement coordinate frame, and minimising the square sum of the difference between the altered measured positions and the nominal ideal positions. Translations of the components from the nominal position are largely irrelevant (given the known accuracy of the template), however angular deviations are of interest. Figure 3 shows the angular deviation from a perfect 45° of each component. All of the components fall within tolerances and these deviations can easily be accommodated by precision alignment of the critical components.

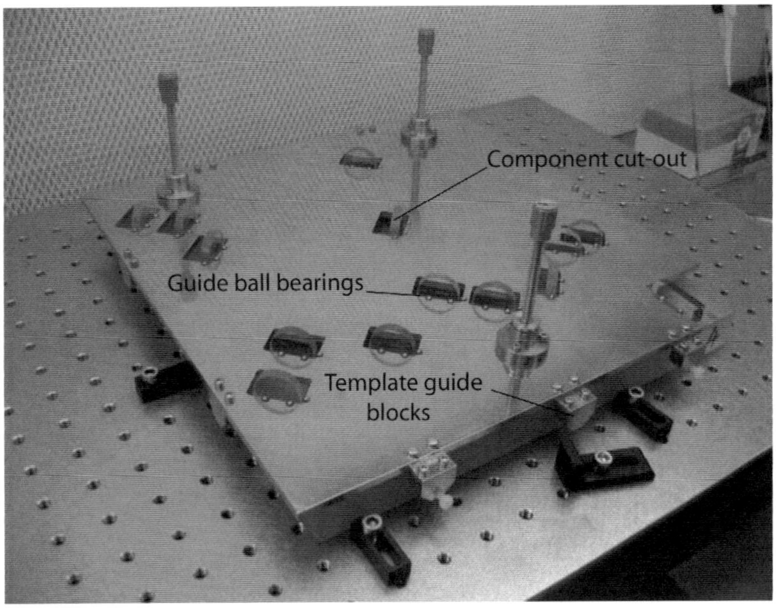

Figure 2. Photograph showing the bench after the template bonding procedure with the template still in place

4.2. FIOS bonding

The test bench is currently in the process of having FIOS2 bonded to the mounting post (step *(b)*). The bench has been incorporated into a mounting jig which holds the bench at a slight angle from vertical, so that the bonding surface of the mounting post is at a slight inclination from horizontal. This, as per the template bonding, is to provide a small restoring force against the actuators that will be used to position the FIOS. The FIOS will be floated on a buffer fluid (typically octane), this prevents optical contacting [7] of the two polished surfaces. The beam from the FIOS is monitored using a pair of quadrant photodiodes mounted in a frame. The position of the photodiode centres with respect to the frame is well known (having been previously calibrated using the CMM), thus by positioning the photodiodes such that the beam is centred on both of them, and measuring the frame with the CMM, the beam vector can be calculated. Since the ideal position of this beam is already known, it is possible to use an iterative alignment technique to position the FIOS in the correct place. The position of the FIOS is adjusted using the actuators which have a sub-micron resolution and contact the FIOS with ruby balls. Once optimum alignment is achieved, the FIOS is lifted off, bonding fluid applied and the FIOS relocated against the ruby balls. By rotating the bench through 180° the same technique can be used for FIOS1. In practice however, after FIOS2 is bonded the input mirror will then be bonded, using the precision actuators with the bench inclined. The beam from FIOS2 will then be fully profiled with respect to the baseplate, such that FIOS1 can be aligned

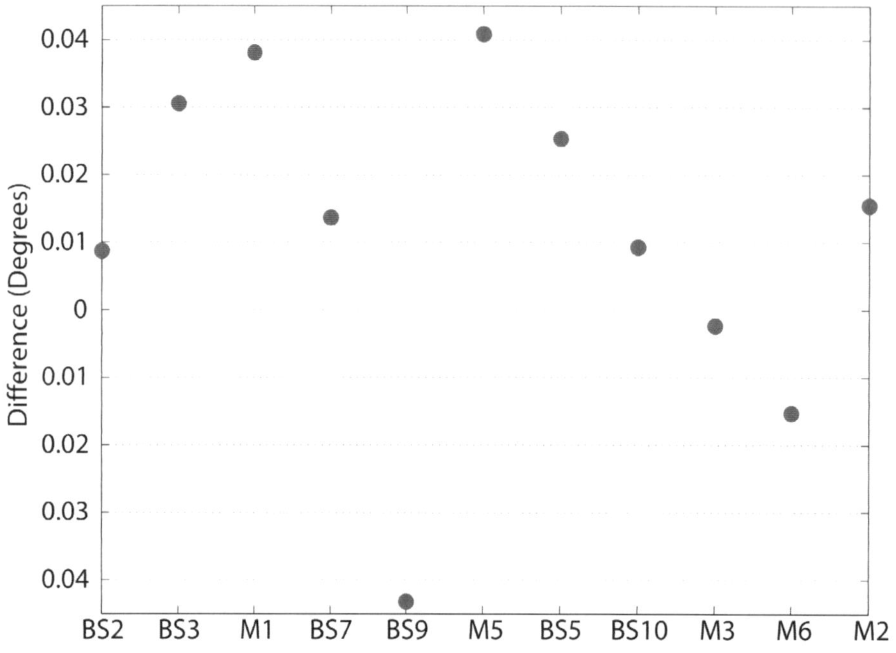

Figure 3. Measured angular deviations of templated components from nominal

to match FIOS2 as well as possible, both in height and vertical angle. After this the second input mirror can be precision bonded. The purpose of the input mirrors is to add a degree of freedom in the horizontal to allow the beams to be steered to the recombining points as well as possible.

4.3. Final construction steps

After the completion of steps *(b)* to *(e)*, the beam recombiners must be bonded. As before, the recombiner will be floated on a buffer fluid to allow adjustment of the component using the precision actuators. Beams will be injected from both FIOSs, and a off-bench photodiode used to measure the recombined beam. The contrast of the interference pattern will be optimised by adjusting the position of the recombiner. Once contrast is maximised, the component will be bonded in place. Only after the four recombiners are in position will the photodiodes be glued in place (step *(g)*), centered onto the recombined beams.

5. Test plan

Once construction is complete, the test bench will be integrated into a vacuum tank. A removable gantry incorporating precision actuators will be used to position the sub assembly containing the polarising optics into position. Laser light will be provided by a pair of Nd:YAG lasers operating at $\lambda = 1064\,\mathrm{nm}$. The lasers will be offset phase locked together, with the slave laser trailing the master by 79.99 MHz. The master laser will then be frequency shifted by an acousto-optic modulator (AOM) by 80 MHz. This will give a 10 kHz offset between the two beams, giving the heterodyne frequency for the experiment. These beams are injected onto the bench through single mode fibres, via a vacuum feedthrough, and onto the bench through the FIOSs. The two beams interfering in each of the interferometers will generate a 10 kHz heterodyne beat note at the photodiodes. By comparing the phase fluctuations of the beat notes of the reference and measurement interferometers, the stability of the measurement path can be established. Additionally, the measurement and reference interferometers all have approximately matching path length mis-matches (measurable by analysis of the heterodyne signals) [4]. This minimises coupling of laser frequency noise into the measurement. The frequency noise monitoring interferometer has deliberately offset arm lengths to exaggerate laser frequency noise and allow it to be compensated for, either in post-processing or by feeding back to the laser, see [4]. Using the replaceable sub assembly in the measurement path, the stability of polarising Mach-Zehnder interferometers can be verified, and a variety of mounting techniques for polarising optics can be tested in order to select the best mounting techniques for LISA.

6. Conclusion

We have presented the design, initial construction of, and future plans for, a hydroxide catalysis bonded test bench designed to test the concept of polarising Mach-Zehnder interferometers for use in LISA. Such investigations are vital in advancing the existing technology developed for LISA Pathfinder towards the more challenging LISA baseline. By applying LISA Pathfinder construction techniques (precision adjustable hydroxide catalysis bonding and monolithic fibre injectors in particular) we ensure the most representative test possible of these optical assemblies.

Referances

[1] D Weise et al. 2008 *See this volume*
[2] E J Elliffe, J Bogenstahl, A Deshpande, J Hough, C Killow, S Reid, D Robertson, S Rowan, H Ward and G Cagnoli 2005 *Class. Quantum Grav.* **22** S257-S267
[3] G Heinzel et al. 2005 *Class. Quantum Grav.* **22** S149-S154
[4] D Robertson, C Killow, H Ward, J Hough, G Heinzel, A Garcia, V Wand, U Johann and C Braxmaier 2005 *Class Quantum Grav.* **22** S155-S163
[5] D Robertson et al. 2008 *See this volume*
[6] OptoCAD by Roland Schilling. http://www.rzg.mpg.de/~ros/OptoCad, 2008

Journal of Physics: Conference Series **154** (2009) 012035 doi:10.1088/1742-6596/154/1/012035

Study of test-mass charging process in the LISA missions due to diffuse γ-rays

N Finetti[1], R Scrimaglio[1], C Grimani[2,3] and M Fabi[2]

[1] Dipartimento di Fisica dell'Università degli Studi dell'Aquila and Gruppo INFN dell'Aquila collegato ai Laboratori Nazionali del Gran Sasso, Via Vetoio, 67010 Coppito, L'Aquila, Italy.
[2] Istituto di Fisica dell'Università degli Studi di Urbino "Carlo Bo", Via S. Chiara, 27, 61029, Urbino (PU), Italy.
[3] Sezione INFN di Firenze, Via G. Sansone 1, 50019 Sesto Fiorentino, Firenze, Italy .

E-mail: noemi.finetti@aquila.infn.it

Abstract. Gravitational inertial sensors will be placed on board the Laser Interferometer Space Antenna (LISA) and aboard its precursor mission LISA Pathfinder (LISA-PF) in order to detect low frequency gravitational waves in space. Free-floating test-masses (Au_7Pt_3 cubes) will be housed in inertial sensors for detecting possible laser signal variations induced by gravitational waves. Charging of the LISA test-masses due to exposure of the spacecraft to cosmic radiation and energetic solar particles will affect operation of gravitational inertial sensors. In this paper we report on the role of diffuse γ-rays in charging the LISA and LISA-PF test-masses with respect to protons and helium nuclei. The diffuse γ-ray flux in the Galaxy has been interpolated taking into account the outcomes of recent calculations. A comparison with γ-ray observations gathered by different experiments (COMPTEL and EGRET, Milagro, Whipple, HEGRA, TIBET) has been carried out. Simulations of the test-mass charging process have been performed by means of the FLUKA2006.3b package. Monte Carlo simulations of the interaction of cosmic particles with the LISA spacecraft indicate that the diffuse γ-ray contribution to the average steady-state test-mass charging rate and to the single-sided power spectrum of the charge rate noise is marginal with respect to that due to galactic cosmic-rays.

1. Introduction

The Laser Interferometer Space Antenna, in its Earth-like orbit around the Sun, will be affected by space environment disturbance. Solar energetic particles and galactic cosmic-rays (GCRs) with energies greater than 100 $MeV(/n)$ will be able to penetrate the apparatus charging the free-floating test-masses (TMs) which interact electromagnetically with surrounding conducting surfaces of the inertial sensors (ISs). This charging process simulates the effects of gravitational waves and limits the sensitivity of LISA to gravitational wave detection [1–7]. The precursor mission LISA Pathfinder, operating at the Earth-Sun L1 Lagrange point, will be similarly affected by this radiation environment noise. For this reason an UV discharge system is provided for both LISA and LISA-PF. This work concerns the role of diffuse γ-rays (undetected by on board radiation monitors) in charging the LISA and LISA-PF test-masses. Monte Carlo (MC) simulations of the TM charging process have been performed by means of the FLUKA2006.3b package [8, 9] taking into account a simplified geometry of the apparatus [4]. The average steady-state charging rate and the single-sided power spectrum of the charge rate noise induced by diffuse γ-rays with respect to those due to GCR protons and helium nuclei are reported in this paper, while absolute values can be inferred from [1].

2. Diffuse galactic γ-ray spectrum

Observations of diffuse γ-ray flux in the Galaxy have been performed by different experiments as shown in Fig. 1: the COMPTEL and EGRET data are taken from [12,18]; the Milagro point is obtained from [13] with a conversion factor equal to 2.43 ± 0.55 as reported in [19]; other data points are upper limits set by Whipple [14], HEGRA [15,16] and TIBET [17]. EGRET observations reveal an excess above 1 GeV respect to the expected γ-ray spectrum calculated under the assumption that the locally observed cosmic-ray fluxes correspond to the GCR spectra. Recent calculation [10] have been performed assuming that GCR of energy below 100 TeV are accelerated by supernova remnant shock waves and that the shock compression ratio is supernova remnant age dependent. In [10] GCR proton and electron spectra have been obtained injecting the average source spectra (calculated from an ensemble of supernova remnants) in the Galaxy and using a 3-D convection-diffusion equation and the diffuse γ-ray spectrum has been produced by interaction of these GCRs with the interstellar medium and the interstellar radiation field. Results of model proposed in [10] are in agreement with the γ-ray experimental data at almost all energies except around 85 GeV (Fig. 1). We point out that averaged GCR spectra which reproduce the γ-ray experimental data so well result to be flatter than the GCR spectra observed locally (for a detailed discussion see [10]).

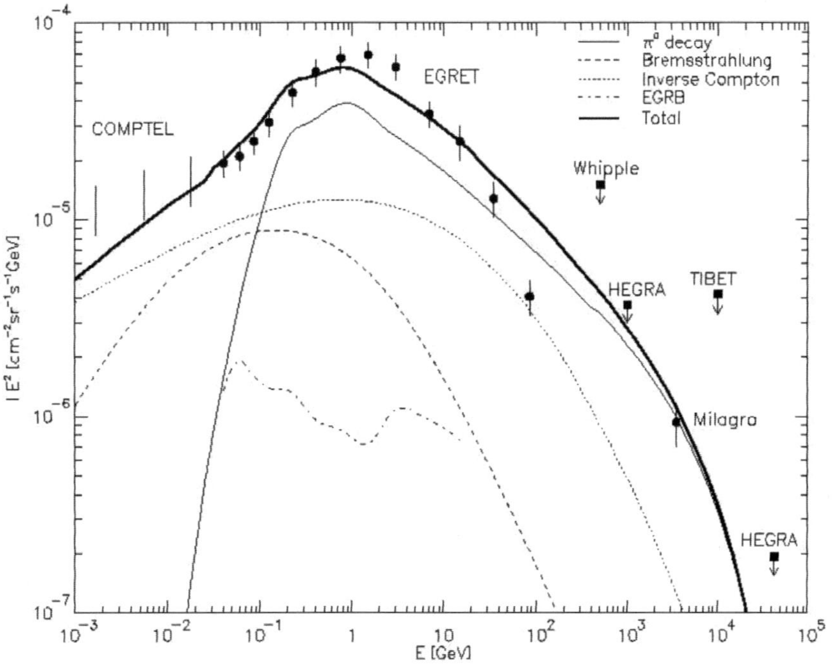

Figure 1. Averaged diffuse γ-ray spectrum in the Galaxy: results of calculation [10, 11] are compared with experimental data points (COMPTEL and EGRET [12,18], Milagro [13,19]) and upper limits (Whipple [14], HEGRA [15,16] and TIBET [17]). Different model components have been taken into account in the total γ-ray spectrum computation: π^0 decay, bremsstrahlung and inverse Compton [10] and extragalactic diffuse γ-ray background (EGRB) taken from [18].

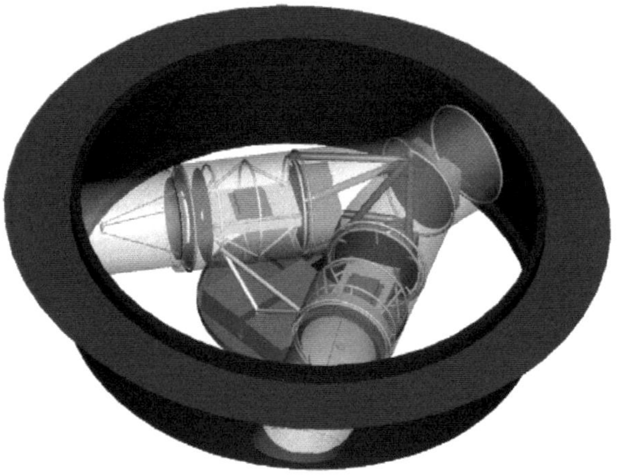

Figure 2. LISA carries identical science instruments on each of the three spacecraft. Like it is shown in this picture (from http://lisa.jpl.nasa.gov/), each spacecraft's instrument is made up of two optical assemblies and lasers, mounted on a disk-shaped radiator, housed inside a Y-shaped tubular structure that is supported by the spacecraft's cylindrical walls.

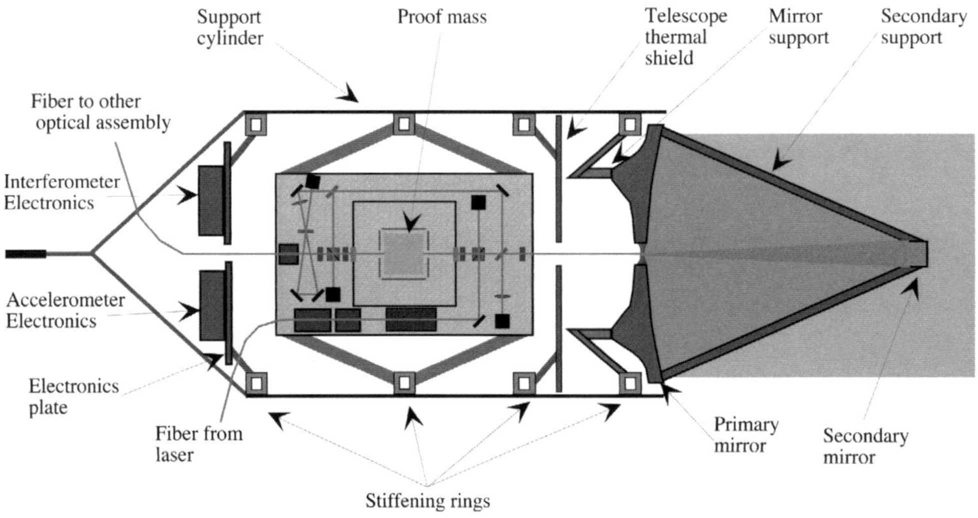

Figure 3. Optical assembly shown in detail (from http://lisa.jpl.nasa.gov/): here the TM is indicated as proof-mass.

3. FLUKA simulations

The LISA apparatus consists of an equilateral constellation of three spacecraft 5 million km apart. A gravitational wave will induce small variations in the laser signal as it runs along the space interposed between two free-floating TMs located aboard different spacecraft. In order to detect gravitational waves, each LISA module accommodates two interferometer telescopes mounted inside a Y-shaped payload structure (Fig. 2). Two free-floating TMs are housed in ISs located in optical benches (Fig. 3) mounted behind the telescopes. Each TM is a Au_7Pt_3 cube (46 mm×46 mm×46 mm) surrounded by electrodes lodged in a molybdenum housing and the assembly is accommodated in a titanium vacuum enclosure. This IS structure is the current design of the sensor aboard LISA-PF mission. In order to estimate the radiation environment noise, simulations of the LISA and LISA-PF TM charging process have been performed by means of the FLUKA2006.3b package [8, 9] taking into account a simplified geometry of the apparatus [4]. In carrying out MC simulation an isotropic primary flux has been generated uniformly on a spherical surface containing the whole apparatus and energies of primaries have been sampled randomly in the interval $[E_{\min}, E_{\max}]$ following the given distribution of the primary flux $F(E)$. The MC timelines have been normalized on the basis of energy differential flux definition, that is taking into consideration the fact that N_0 primaries are generated in a time interval $[0, t]$ with $t = N_0/(GF \cdot \int FdE)$, where GF is the geometrical factor ($GF = 4\pi^2 R^2$ for a spherical surface of radius R) and the flux integration is performed from $E_{\min}$ up to $E_{\max}$. For each primary (event-by-event analysis), the MC model tracks all particles penetrated or produced inside the IS calculating the difference between those entering and leaving the TM. Taking into account that IS materials are thick enough, a threshold of 100 keV has been imposed for secondary particle production and transport inside the IS [8, 9] .

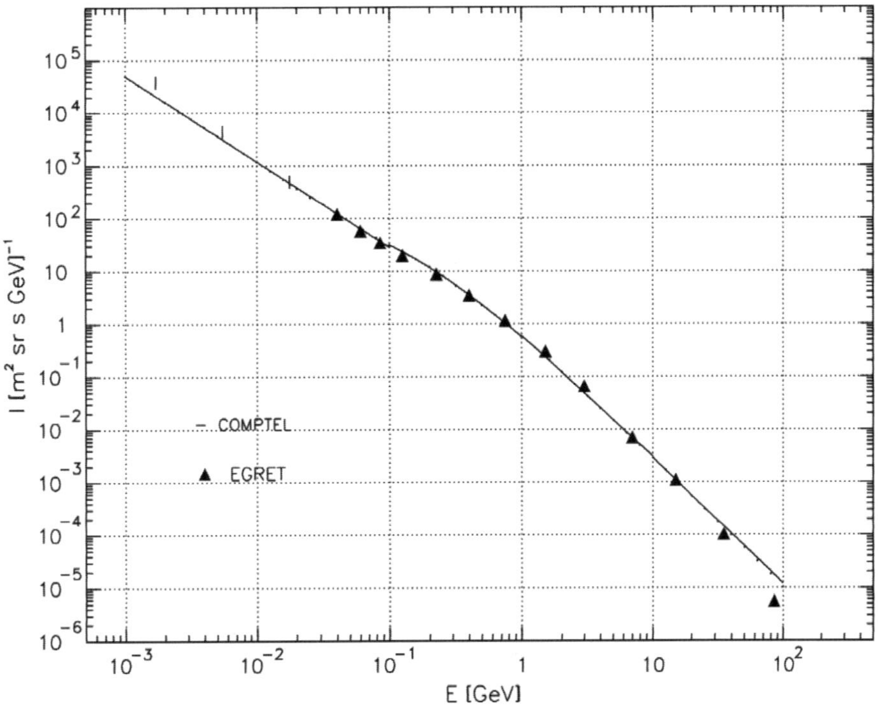

Figure 4. The diffuse γ-ray flux in the Galaxy in the energy interval $0.001 - 100$ GeV has been interpolated taking into account the outcomes of calculations reported in [10] (dots). In this figure fit functions (unbroken lines) and experimental data points [12, 18] are shown as well.

4. Simulation results

MC simulations have been performed for protons and ^{4}He nuclei (the two most abundant primary nuclei in GCRs) at the minimum of solar activity and for γ-rays. The adopted spectra for protons and ^{4}He nuclei have been chosen of the form $F = \alpha \cdot (E + \beta)^{-\delta} \cdot E^{\sigma}$ where E is the kinetic energy in $GeV(/n)$ ranging in the interval $0.1 - 100GeV(/n)$ and the differential flux F is in $m^{-2}sr^{-1}s^{-1}(GeV(/n))^{-1}$ [5]. From a fit of experimental data it results: $\alpha = 18000$, $\beta = 1.09$, $\delta = 3.66$, $\sigma = 0.87$ for protons and $\alpha = 850$, $\beta = 0.99$, $\delta = 3.10$, $\sigma = 0.35$ for ^{4}He nuclei [6, 7]. We point out that GCR protons and ^{4}He nuclei with energies below 100 $MeV(/n)$ are not able to penetrate the shielding overlying the TM. For γ-rays the energy E has been sampled randomly in the interval $0.001 - 100$ GeV following a distribution (Fig. 4) interpolated taking into account the outcomes of calculations reported in [10]. The corresponding γ-ray differential flux fit functions are reported in Tab. 1. During each run, the total exposure time was 600 s for protons and ^{4}He nuclei and 6000 s for γ-rays. We point out that only primaries with energies below 100 $GeV(/n)$ have been considered since events with energies above 100 $GeV(/n)$ are extremely rare for both GCRs and γ-rays.

Table 1. Fit functions of the γ-ray distribution used for MC simulations: the γ-ray energy E is in GeV and the γ-ray differential flux F is in $m^{-2}sr^{-1}s^{-1}GeV^{-1}$.

Energy interval	Fit function	Parameter values
0.001 GeV – 0.1 GeV	$F = \alpha{\cdot}E^{-\beta}$	$\alpha = 0.67730$; $\beta = 1.6214$
0.1 GeV – 1 GeV	$F = \alpha{\cdot}(E+\sigma)^{-\beta}$	$\alpha = 0.68316$; $\sigma = 0.085615$; $\beta = 2.2654$
1 GeV – 10 GeV	$F = \alpha{\cdot}E^{-\beta}$	$\alpha = 0.58984$; $\beta = 2.2810$
10 GeV – 100 GeV	$F = \alpha{\cdot}E^{-\beta}$	$\alpha = 0.70851$; $\beta = 2.3888$

In MC simulations, the average TM charging rate R has been calculated by using the following formula:

$$R = \sum_q (qek_q)$$

where e is the elementary charge equal to the absolute value of the electron charge, qe is the total charge generated in the TM for a single primary impinging on the apparatus (event-by-event analysis) and k_q is the rate of events which generate a total charge equal qe in the TM during the exposure to primary flux. Since the charging process causes stochastic fluctuations, the single-sided spectral density of the shot noise S_R associated with the charging current has been evaluated by considering the stochastic nature of the process (Poisson distribution)

$$S_R = \sqrt{\sum_q [2(qe)^2 k_q]}$$

We remind that taking into account time-dependent fluctuations of the total charge $Q(t)$ accumulated by the TM at the time t, at a given frequency f the single-sided spectral density of the noise due to charge fluctuations $S_Q(f)$ is given by

$$S_Q(f) = \frac{S_R}{2\pi f}$$

From MC simulations a negative charging rate induced by diffuse γ-rays has been predicted while a positive charging rate has been obtained for both GCR protons and ^{4}He nuclei. The negative charging rate for γ-rays is due to the fact that the number of secondary electrons stopped inside the TM is higher than that of secondary positrons, while for GCR protons and ^{4}He nuclei the charging process is mainly from direct primary stopping in the TM. Moreover, from MC simulations the average steady-state charging rate induced by diffuse γ-rays R_γ with respect to that due to GCR protons R_p and that caused by ^{4}He nuclei R_α resulted $R_\gamma/R_p = -0.5{\cdot}10^{-3}$ and $R_\gamma/R_\alpha = -3.4{\cdot}10^{-3}$ (the minus signs in the two charging rate ratios are due to the negative charging induced by diffuse γ-rays with respect to the positive one due to GCRs) and for the single-sided power spectrum of the charge rate noise related to diffuse γ-rays S_{R_γ} with respect to that caused by GCR protons S_{R_p} and that due to ^{4}He nuclei S_{R_α} we obtained $S_{R_\gamma}/S_{R_p} = 7.5{\cdot}10^{-2}$ and $S_{R_\gamma}/S_{R_\alpha} = 27.6{\cdot}10^{-2}$.

5. Conclusions

MC simulations of the interaction of cosmic particles with the LISA spacecraft indicate that the contribution to the average steady-state TM charging rate due to diffuse γ-rays is marginal with respect to that caused by GCRs. Moreover, in MC simulations the negative charging rate for γ-rays results from the fact that the number of secondary electrons stopped inside the TM is higher than that of secondary positrons, while for GCR protons and ^{4}He nuclei the charging process is mainly from direct primary stopping in the TM (positive charging rate).

Acknowledgments

Authors are very grateful to Dr. Vincenzo Acconcia (IT System Manager; Università degli Studi di Urbino) for providing his valuable technical support.

References

[1] Araújo H M et al. 2004 arXiv:astro-ph/0405522
[2] Vocca H et al. 2005 *Classical and Quantum Gravity* **22** S319
[3] Grimani C et al. 2005 *Classical and Quantum Gravity* **22** S327
[4] Vocca, H et al. 2004 *Classical and Quantum Gravity* **21** S665
[5] Grimani C et al. 2004 *Classical and Quantum Gravity* **21** S629
[6] Grimani C et al. 2007 *Proc. 30th ICRC in Merida* **SH 3.2.326**
[7] Grimani C. et al. 2007 poster for the 7th Edoardo Amaldi Conference on Gravitational Waves in Sydney.
[8] Fassò A, Ferrari A, Ranft J. and Sala P R, CERN-2005-10 (2005), INFN/TC_05/11, SLAC-R-773
[9] Fassò, A et al 2003 Computing in High Energy and Nuclear Physics 2003 Conference (CHEP2003), La Jolla, CA, USA, March 24-28, 2003, (paper MOMT005), eConf C0303241 (2003), arXiv:hep-ph/0306267
[10] Satyendra T 2006 arXiv:astro-ph/0603599
[11] Satyendra T 2006 private communication
[12] Strong A W, Moskalenko I V and Reimer O 2004a *ApJ* **613** 962
[13] Sinnis G 2005 *Proc. 29th ICRC in Pune* **4** 81
[14] LeBohec S et al. 2000 arXiv:astro-ph/0003265
[15] Aharonian F A et al. 2001, *A&A* **375** 1008
[16] Horns D and Schmele D 1999 *Proc. 26th ICRC in Salt Lake City* **OG 3.2.24**
[17] Amenomori M et al. 1997 *Proc. 25th ICRC in Durban* **3** 117
[18] Strong A W, Moskalenko I V and Reimer O 2004b arXiv:astro-ph/0405441
[19] Erlykin A D and Wolfendale A W 2005 *Proc. 29th ICRC in Pune* **4** 47

Electrostatic-control performance measurement of the inertial sensor with a torsion pendulum

H B Tu, Y Z Bai, Z B Zhou[1] and J Luo

School of Physics, Huazhong University of Science and Technology, Wuhan 430074, P.R. China

E-mail: zhouzb@mail.hust.edu.cn

Abstract. An electrostatic-controlled torsion pendulum was used to simulate the operation of the inertial sensor in flight. The twist motion of the proof mass was monitored by a capacitance transducer and was controlled by an electrostatic actuator. The influences of the parasitic coupling of the capacitance transducer, the magnetic field, and the translation motion coupling were measured. The torque noise of the controlled pendulum came to about 6×10^{-13} N m Hz$^{-1/2}$ from 2 mHz to 0.1 Hz, mainly limited by the translation-rotation coupling and the back action of the capacitance sensor, which could be suppressed by extending the gap of the capacitance sensor.

1. Introduction

High precision inertial sensors play a paramount role in gravitational experiments in space [1-3]. These inertial sensors can be classified as two operation modes. In one mode the sensors act as a free-falling reference to guide the micro-Newton thrusters to compensate the non-gravitational forces received by the spacecraft, namely the drag-free control, such as those in the LISA and GOCE missions. In the other mode the sensors act as an accelerometer to measure the accelerations of the spacecraft or the expected forces, such as those in the CHAMP, GRACE, and MicroSCOPE missions. In both modes the proof masses (PMs) should be isolated from any residual disturbances. For LISA mission, the spurious accelerations of the PM should be smaller than 3×10^{-15} m s^{-2} Hz$^{-1/2}$ from 0.1 mHz to 0.1 Hz, which is a tremendous challenge for ground-based measurement.

For high precision inertial sensors, the performance verification on ground is limited by 1 *g* Earth's gravity acceleration. One scheme is to use high voltage to levitate the PM, and ONERA succeed to develop the STAR and SuperSTAR accelerometers using this scheme. However, the experimental level was limited to 10^{-9} m s^{-2} Hz$^{-1/2}$ due to the very stiff coupling of the high electric field and seismic noises [4]. Another scheme is to suspend the PM by a thin fiber to act as a torsion pendulum or balance. A torque resolution better than 10^{-14} N m Hz$^{-1/2}$ has been achieved by several groups with torsion balance or pendulum [5-7]. University of Trento had made plentiful progress to investigate the residual disturbances of the PMs for LISA and LISA Pathfinder with the torsion pendulum [7-10].

The torsion pendulum can be used not only to investigate the residual disturbances of the PM based on its high sensitivity to the weak force, but also to simulate the operation of the inertial sensor in flight based on the weak coupling of the fiber [11, 12]. For the second application, the gap between the electrodes and the PM is in general smaller than 1 mm, and the negative stiffness k_e due to the capacitance transducer is much larger than that of the fiber k_f. It means that the torsion pendulum is unstable if without the feedback control. It is beneficial to estimate the performance of the electrostatic actuator for controlling the non-sensitive motions of the PM in the LISA mission, and to investigate the cross coupling between the non-sensitive and sensitive motions of the PM and the drag-free control on ground. An electrostatic-controlled torsion pendulum was developed to investigate the performance of the inertial sensor in our laboratory, and the progress was outlined.

2. Experimental setup

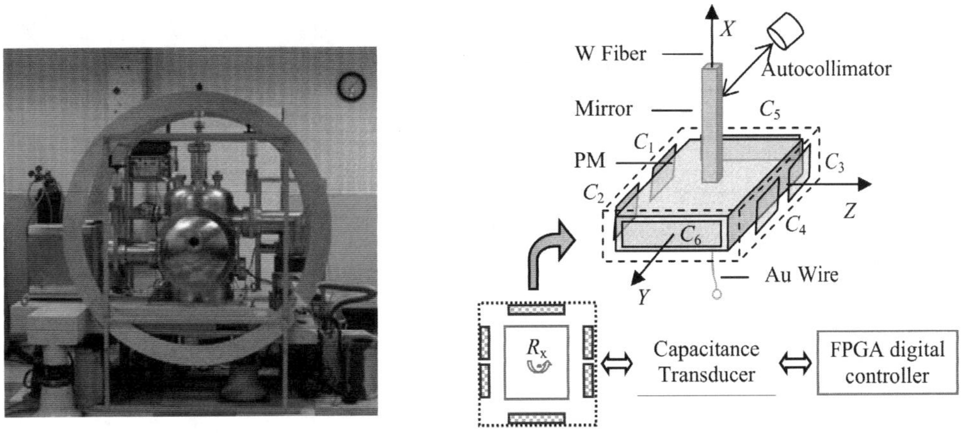

Figure 1 Left: Photo of the experimental setup; Right: Schematic diagram of the electrostatic-controlled torsion pendulum. It consists of a sensor head, capacitance transducer and a digital controller. Six electrodes are used to measure both the translation and the twist motions of the PM, and control the twist motion.

The schematic diagram of the electrostatic-controlled torsion pendulum is shown in figure 1. A titanium cuboid PM of about 4 cm × 4 cm × 1 cm combined with a mirror was suspended by a 25 μm tungsten fiber with a length of about 40 cm. A 5 μm Au wire with a length of about 4 cm was used to discharge the PM, and its elastic coefficient was measured as 4.2×10^{-11} N m rad^{-1} by comparing the free oscillating periods of the torsion pendulum with and without the discharge wire, which was about 300 times smaller than that of the suspension tungsten fiber. The total mass of the pendulum was about 78 g and the moment of inertia was about 1.995×10^{-5} kg m^2. The free oscillation period T_0 of the pendulum with the discharge wire was (248.4 ± 0.9) s, and the quality factor Q was about 2800 under the vacuum of 2×10^{-5} Pa. The total elastic coefficient k_f was $(1.28 \pm 0.01) \times 10^{-8}$ N m rad^{-1}.

The electrode frame was made of Au-coated glass-ceramics, which was mounted on a six degree-of-freedom manipulator. Three pairs of electrodes were used not only to measure the Y, Z translations and twist of the PM, but also to control the twist of the PM and investigate the twist-translation coupling. The equilibrium capacitance gaps d_0 were about 150 μm. C_1~C_4 and C_5~C_6 were about 4.9 pF and 13.3 pF, respectively. A capacitance transducer circuitry based on the AC capacitance and transformer bridge [11, 12], combined with a digital control loop based on a field-programmable-gate-array (FPGA), was developed. The noise of the capacitance transducer was 2×10^{-6} pF Hz$^{-1/2}$ from 10 mHz to 1 Hz, and was dominated by f^{-1} noise at the frequency range below 10 mHz.

Four coils were used to investigate the effects of the magnetic field. A pair of large Helmholtz coils with diameters of about 1.6 m was symmetrically set outside the vacuum chamber along the Y axis, as shown on the left in figure 1, while two small coils were mounted near the electrodes inside the chamber along the Z axis. Each small coil with a diameter of about 27 mm and turn number of 130 was made of 0.4 mm enameled copper. In order to distinguish the torque from the susceptibility and residual dipole moment of the PM, both of the small coils were symmetrically departed 20 mm along the Y axis and 61.5 mm along the Z axis away from the center of the PM, as shown in figure 2. Four heaters were set close to the center of each electrode along the Z axis to investigate the effects of the temperature, and four temperature sensors were employed to measure temperature variations. The resistance of each heater was about 5 Ω. An aluminum shield box with a thickness of about 4 mm was used to partly shield the PM from the disturbances of residual gas and heat flow.

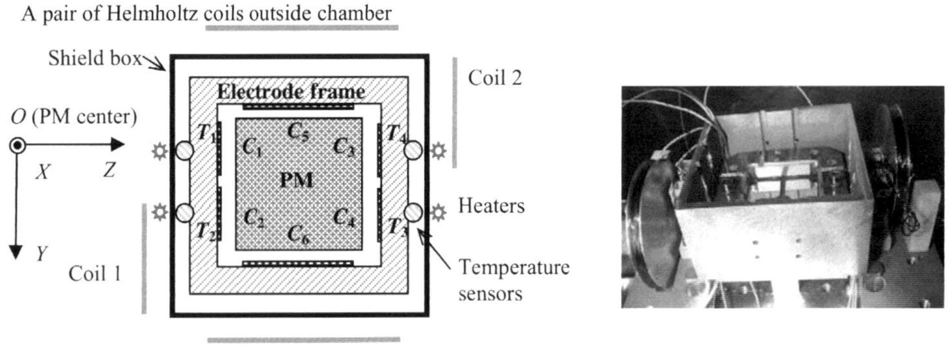

Figure 2 Left: Schematic diagram of the residual disturbances experiment. Right: Photo of the layout in chamber.

3. Current Experimental Results

3.1. *Torque sensitivity calibration*

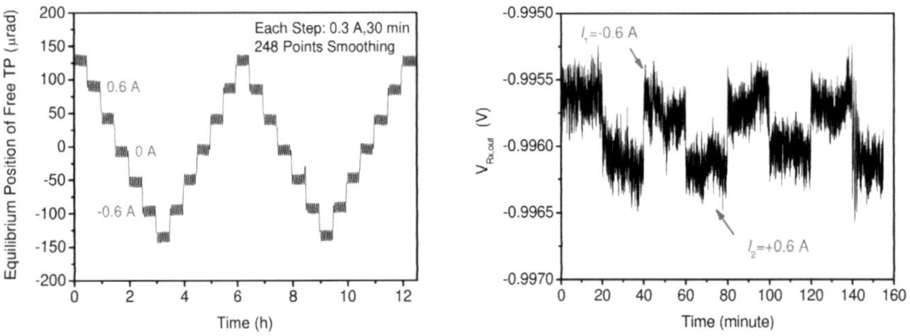

Figure 3 Left: Position variation of the torsion pendulum in free oscillation mode. Right: Feedback voltage variation in electrostatic-controlled mode.

A torque originating from the interaction of the remnant dipole moment $\boldsymbol{M}_r$ of the PM with an applied magnetic field $\boldsymbol{B}$ induced by the Helmholtz coils can be employed to calibrate the torque sensitivity of the electrostatic actuators. The magnetic torque is firstly measured based on the elastic torque of the suspending fiber in the case of the free oscillation pendulum, where the electrode frame was dropped below the PM by driving the manipulator, and then the sensitivity of the electrostatic actuator can be determined by the same magnetic torque under the control mode. In the free oscillation mode, when the currents through the Helmholtz coils varied from -0.6 A to 0.6 A, the twist variation of the pendulum was (180.6 ± 5.1) μrad as shown on the left in figure 3, which was simultaneously monitored by an autocollimator with a resolution of 0.1 μrad $\mathrm{Hz}^{-1/2}$. In the control mode, the feedback voltage variation was (-0.40 ± 0.07) mV as the same (± 0.6 A) currents were alternately applied through the coils, as shown on the right in figure 3. Therefore, the torque sensitivity could be determined as $S_{\mathrm{Nm}} = (1.73 \pm 0.31) \times 10^8\,\mathrm{V\,(N\,m)^{-1}}$, which was in good agreement with the theoretical estimation given by $d_0 \big/ \left(V_b \sum_{i=1}^{4} C_i \right)$, here V_b is a 5 V DC bias voltage for linearizing the electrostatic actuators. An AC V_b would be considered to apply to the PM in order to suppress the charge coupling in the further experiments. The twist angle synchronously recorded by the autocollimator was less than

1 μrad, which contributed an uncertainty of 10^{-14} N m. Furthermore, the magnetic field variation, which generated by the Helmholtz coils from -0.6 A to 0.6 A, was directly measured as about -101.4 μT along *Y* axis in the space of the PM by a magnetometer with a precision of 10 nT. Then the *Z* component $\boldsymbol{M}_{rz}$ of the residual dipole moment of the PM could be estimated as $(-2.3 \pm 0.4) \times 10^{-8}$ A m^2 based on the coupling $\Delta\boldsymbol{\tau} = \boldsymbol{M}_r \times \boldsymbol{B}$.

3.2. The parasitic stiffness measurement

The parasitic stiffness k_e of the capacitance transducer is a very important parameter for determining the parameters of the control loop and estimating the back-action torque noise of the pendulum. For the twist of the torsion pendulum R_x, it can be expressed as

$$k_e \approx -\frac{(3b^2 + 3ad_0 + h^2)V_T^2}{12d_0^2}\sum_{i=1}^{6} C_i \quad \text{with } V_T^2 = V_{Rx,out}^2 + V_b^2 + V_p^2/2 \tag{1}$$

where *a* is the side length of the PM, *b* is the distance between the adjacent electrodes, *h* is the width of the electrode, $V_{Rx,out}$ and V_p, respectively, are the control voltage and a pumping voltage with a frequency of 100 kHz and an amplitude of 4 V applied to the PM. The twist position of the PM relative to the frame was periodically set by the FPGA unit. Thanks to the electrostatic control, the PM was controlled to fellow up with the setting position. The electrostatic-controlled torque is equal to the torque induced by k_e if neglecting the effect of the suspension fiber, which is given by

$$\Delta V / S_{Nm} = k_e \times \Delta\theta. \tag{2}$$

The error signal variation of each step was 0.01V, which corresponded to the twist angle was (12.4 ± 0.2) μrad, and the corresponding feedback control voltage variation ΔV was (0.012 ± 0.002) V, as shown in figure 4. Therefore, k_e could be obtained as $(5.6 \pm 1.4) \times 10^{-6}$ N m rad^{-1}, which agreed with the theoretical value given by equation (1) and $k_e / k_f \sim 438$ was deduced.

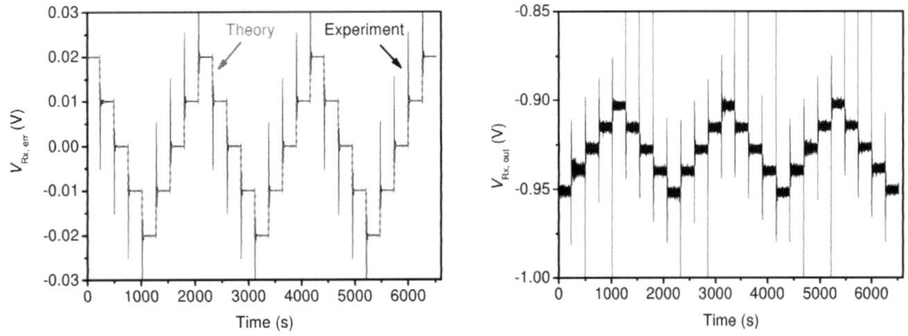

Figure 4 Left: The dash and solid lines respectively represent the expected and experimental error signals. Right: Corresponding control voltage.

3.3. Magnetic and thermal disturbance measurement

If assuming the susceptibility χ and the residual dipole moment $\boldsymbol{M}_r$ are uniform for the whole PM, the torque contributed by the magnetic field of the two coils inside the chamber can be written as

$$\boldsymbol{\tau}_m = \boldsymbol{M}_r \times \boldsymbol{B} + \int_V \boldsymbol{r} \times \nabla\left[(\boldsymbol{M}_r/V + \chi\boldsymbol{B}/2\mu_0)\cdot\boldsymbol{B}\right]dv, \tag{3}$$

where $\boldsymbol{B} = \boldsymbol{B}_1 + \boldsymbol{B}_2$ is the composite magnetic induction, *V* is the volume of the PM, and $\boldsymbol{r}$ is the displacement vector of dv. Thanks to the symmetrical layout of the two small magnetic coils, $\boldsymbol{M}_r$ will contribute null torque to the pendulum when the exciting currents in the two coils are same in amplitude but contrary in direction, which allows independent measuring the influence of χ.

When the current directions of the two coils were contrary and synchronously changed between 0 A and 0.6 A (typical gradient of magnetic induction was 7×10^{-3} T m^{-1} at the center of the PM), the feedback voltage variation ΔV was (-1.9 ± 0.3) mV and the corresponding torque was 1.1×10^{-11} N m. Based on the numerical calculation of equation (3) with the parameters of the coils, the susceptibility of the PM was measured about $(4.2 \pm 0.8) \times 10^{-4}$. When the currents directions of the two coils were uniform (typical magnetic induction was 2×10^{-4} T at the center of the PM), the feedback voltage variation ΔV was (1.1 ± 0.2) mV and the coupling torque was -6.4×10^{-12} N m. Solving equation (3) combining with the values of χ and $\boldsymbol{M}_{rz}$, the Y component of $\boldsymbol{M}_r$ was obtained as $\boldsymbol{M}_{ry} = (9.5 \pm 1.8) \times 10^{-8}$ A m^2.

The pair of heaters near T_1, T_3 and another pair of heaters near T_2, T_4 were alternately heated to periodically produce a significant temperature difference $T_d = (T_1 - T_2 + T_3 - T_4) / 2$ between the electrodes. The temperature modulation experiment showed that the control voltage variation was less than 0.3 mV when the temperature difference T_d came to 0.55 ^oC, it means that the temperature coupling factor of the electrostatic torsion pendulum was below 3×10^{-12} N m K^{-1}, which was limited by the sensitivity of the electrostatic-controlled pendulum. The upper-limit temperature coupling factor obtained from the preliminary experiments is about 15 times of those of the radiometer and radiation effects predicted [13, 14], which would be carefully investigated in the further experiments.

3.4. Coupling measurement between translation and twist motions

Moving the electrode frame along Z axis in the range of $\pm$ 45 μm by periodically driving the manipulator, as shown on the left in figure 5, the variation of the feedback voltage ΔV reflected the coupling came from the translation was shown in the right picture of figure 5. The result showed that the coupling intensity was various as $(0.5 \sim 5) \times 10^{-11}$ N m μm^{-1} at different displacement of the translation. A similar result was obtained by the coupling measurement along Y axis translation. Near the capacitance equilibrium position, the coupling coefficient was about 2.5×10^{-11} N m μm^{-1}. Main coupling was estimated from the imbalance of the parasitic capacitances derived from conductive wires, which contributed a nonzero angle $\Delta\theta$ between the PM and the frame, and then produced a coupling torque as $k_e \times \Delta\theta$. The coupling from the presence of the stray DC potentials on the electrodes and the spurious motion of the manipulator stage were investigated in the further experiments.

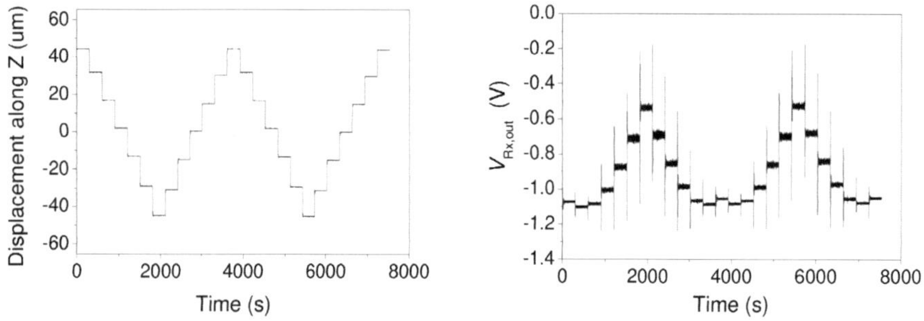

Figure 5 Left: Displacement of the electrode frame along Z axis. Right: Variation of the control voltage.

3.5. Current torque resolution of the controlled pendulum

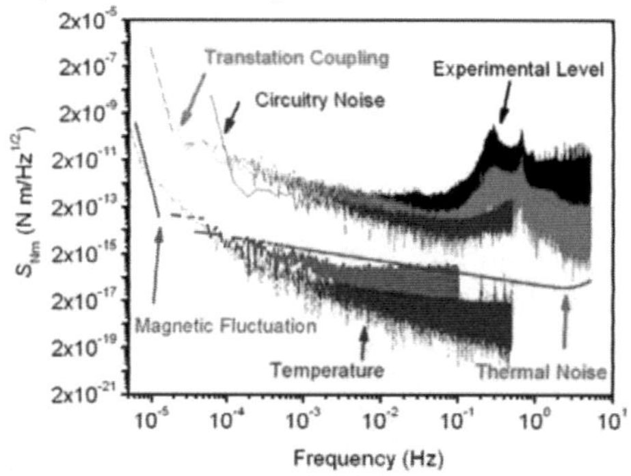

Figure 6 Torque noise of the controlled pendulum.

Feedback voltage as well as the temperature and magnetic field fluctuations of the experimental environment was recorded for a long time. The equivalent torque noise of the controlled pendulum came to 6×10^{-13} N m Hz$^{-1/2}$ in the frequency range from 2 mHz to 0.1 Hz, as shown in figure 6, which was mainly limited by the coupling disturbance from the translations seismic noise. The limit due to the capacitance transducer was 3 times lower than the experimental level.

The fluctuations of the temperature and magnetic field of the experimental environment were lower than 0.03 °C / day and 5 μT / day, respectively. For the ground-based testing of the inertial sensor, the limits of the temperature and magnetic field fluctuations as well as the thermal noise were much lower than the present experimental level.

4. Discussions

If the capacitance gap was extended to 4 mm as LISA Pathfinder, and the DC bias voltage V_b down to 0.1 V, the parasitic stiffness of the capacitance transducer will sharply decrease by a factor of about 8×10^4, which will significantly suppress the translation-rotation coupling and the back action due to the capacitance transducer. If so, the translation-rotation coupling and the back action torque disturbances would be less than the thermal noise of the pendulum of 2×10^{-15} N m Hz$^{-1/2}$ at 1 mHz. For requirement of performance measurement of the PM for LISA, high-*Q* torsion balances should be developed, for example using fused silica fiber suspension [15].

A torsion balance has been constructed to test the multi-degree-of-freedom control of the PM, and also investigate the cross coupling. In addition, a dual pendulum is being considered to investigate the drag-free control on ground, where one PM should be controlled to track to another PM. It is useful for designing and verifying the drag-free control of LISA and LISA Pathfinder [16].

Acknowledgments: This work was supported by the National Natural Science Foundation of China under grants 10675047.

References

[1] Touboul P, Foulon B, Willemenot 1998 *Acta Astronutica* 45 605
[2] Mester J, Torii R, Worden P, Lockerbie N, Vitale S and Everitt C W F 2001 *Class. Quantum Grav.* **18** 2475
[3] Touboul P, Foulon B, Rodrigues M, Marque J P 2004 *Aerosp. Sci. Technol.* 8 431
[4] Willemenot W and Touboul P 2000 *Rev. Sci. Instrum.* **7** 302
[5] Schlamminger S, Choi K Y, Wagner T A, Gundlach J H and Adelberger E G 2008 *Phys. Rev. Lett.* **100** 041101
[6] Tu L C, Guan S G, Luo J, Shao C G and Liu L X 2007 *Phys. Rev. Lett.* **98** 201101
[7] Carbone L, Ciani G, Dolesi R, Hueller M, Tombolato D, Vitale S, Veber W J and Cavalleri A 2007 *Phys. Rev. D* **75** 042001
[8] Hueller M, Cavalleri A, Dolesi R, Vitale S and Weber W J 2002 *Class. Quantum Grav.* **19** 1757
[9] Carbone L, Cavalleri A, Dolesi R, Hoyle C D, Hueller M, Vitale S and Weber W J 2003 *Phys. Rev. Lett.* **91** 151101
[10] Carbone L, Cavalleri A, Dolesi R, Hoyle C D, Hueller M, Vitale S and Weber W J 2004 *Class. Quantum Grav.* **21** S611
[11] Zhou Z B, Gao S W and Luo J 2005 *Class. Quantum Grav.* **22** S537
[12] Zhou Z B, Qu S B, Tu H B, Bai Y Z, Wu S C, Wan Q Y an Luo J 2008 *Int. J. Mod. Phys. D.* **17** 985
[13] Hueller M, Armano M, Carbone L, Cavalleri A, Dolesi R, Hoyle C D, Vitale S and Weber W J 2005 *Class. Quantum Grav.* **22** S521
[14] Pollack S E, Schlamminger S and Gundlach J H 2006 *AIP Conference Proceedings* **873** 158
[15] Nicolodi D, *An improved torsion pendulum for on-ground verfication of the LISA gravitational reference sensor*, poster in 7th LISA conference, 16-20 June, 2008, Barcelona
[16] Private discussion with Gerhard Heinzel

7th International LISA Symposium IOP Publishing
Journal of Physics: Conference Series **154** (2009) 012037 doi:10.1088/1742-6596/154/1/012037

Design of a Highly Stable and Uniform Thermal Test Facility for MGRS Development

Sei Higuchi, Ke-Xun Sun, Daniel B. DeBra, Saps Buchman, Robert L. Byer

Hansen Experimental Physics Laboratory, Stanford University, 452 Lomita Mall, Stanford, CA 94305, USA

E-mail: seihiguchi@stanford.edu

Abstract. We have designed combined passive and active thermal control system to achieve sub microkelvin temperature stability and uniformity over an optics bench size enclosure, which has an analogous structure to the LISA spacecraft. For the passive control, we have constructed a new thermal enclosure that has a multilayer structure with alternative conducting and insulating layers, which enables the temperature uniformity and ease the burden of the active control. The thermal enclosure becomes an important test facility for Modular Gravitational Reference Sensor (MGRS) development. For the active control, we have developed a model predictive control (MPC) algorithm, which will regulate temperature variations of the proof-mass (PM) down to sub-microkelvin over the LISA science band. The LISA mission requires extremely tight temperature control, which is as low as 30 $\mu K/\sqrt{Hz}$ over 0.1 mHz to 1 Hz. Both temporal stability and spatial uniformity in temperature must be achieved. Optical path length variations on optical bench must be kept below 40 $pm/\sqrt{Hz}$ over 0.1 mHz to 1 Hz. Temperature gradient across the proof mass housing also must be controlled to reduce differential thermal pressure. Thermal disturbances due to, for example, solar radiation and heat generation from electronics, are expected to be significant disturbance source to the LISA sensitivity requirements. The MGRS will alleviate the thermal requirement due to its wider gap between the proof-mass and the housing wall. However, a thermally stable and uniform environment is highly desirable to achieve more precise science measurement for future space science missions.

1. Introduction

This research focuses on developing a thermal test facility for modular gravitational reference sensor (MGRS) ground verification testing. It aims to provide thermal stability better than 30 $\mu K/\sqrt{Hz}$ over 0.1 mHz to 1 Hz by taking advantage of active control. Its extension is suitable for in-flight thermal control for the LISA spacecraft to compensate solar radiation. Limited thermal mass of the LISA spacecraft and relatively low frequency range in which LISA is interested, the spacecraft calls for active compensation mechanism to satisfy the thermal requirements. In addition to stability, thermal gradient across the proof mass housing needs to be minimized to avoid differential radiation pressure. Thus, spatial uniformity is also important to fully ensure LISA's strain sensitivity. The graphical interpretation of the thermal stability goal is shown in Figure 1. Spectral temperature variations is drawn in the figure. The primarily goal of control system is to push down the curve below the LISA thermal requirement region by both passive insulation and active control.

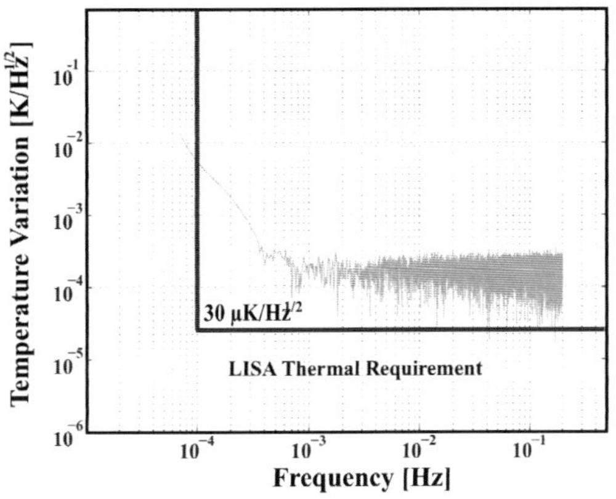

Figure 1. Graphical interpretation of thermal requirement. We aim to push the spectral curve below the thermal requirement by passive insulation and active control.

Thermal disturbances including solar radiation, and temperature gradients across the proof mass housing is expected to be significant disturbance source to the LISA sensitivity requirements. Even a small temperature gradient can produce distortions in the housing structure, which results in a change in the mass attraction force [7]. Temperature fluctuations also cause radiometer effects with residual gas pressure. Lastly, the fluctuations in the temperature gradient across the proof-mass housing will cause the differential thermal radiation pressure that adds extra acceleration to proof mass. P. L. Bender[1] claims that the passive thermal isolation clearly will become worse at frequencies substantially below 0.1 mHz and suggests the need of active temperature control system as well as larger gaps around the proof-mass. the current nominal gap size is 2 mm[8] while the MGRS gap size is an order of a few cm [4, 5, 6].

The LISA spacecraft must reject non-gravitational-wave forces, since the gravitational waves reaching to the vicinity of the Earth have a very small amplitude. Minimizing the non-gravitaional-wave accelerations, which come from both the environment and the spacecraft themselves, is the task of MGRS for drag-free control. The main environmental disturbances to LISA are the forces from the Sun: fluctuations in solar radiation pressure and pressure from the solar wind. The job of the spacecraft is to shield an internal proof mass that floats freely, not attached to the spacecraft, from external disturbances.

2. Design of highly stable and uniform thermal chamber

Figure 2 shows the overview of the experimental system. We have a thermal enclosure specially designed for ground testing, which simulates the LISA spacecraft flying in the deep space. Heating pads and temperature sensors are attached around the enclosure. They are connected to a computer through GPIB interface. Python-based software controls experiment and also logs data.

The left of Figure 3 demonstrates a schematic of the thermal chamber. The top part corresponds to the spacecraft's solar panel module that are made of aluminum honeycomb, polymide and carbon fiber reinforce plastic (CFRP). However, both polymide and CFRP are not available at reasonable prices, they are replaced by polystyrene form and acrylic sheets. They have similar thermal properties. On the very top of the solar module part, six heating pads are mounted over a 2-mm-thick copper sheet so that we can simulate solar radiation heat

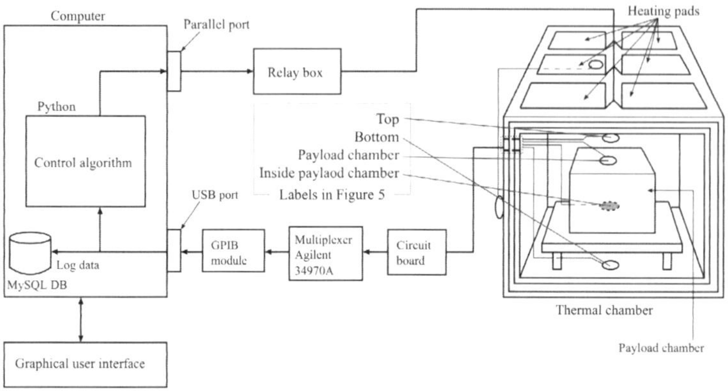

Figure 2. Overview of the thermal experimental system

input to spacecraft. The heating pads are covered with air bubble sheet to reduce wasted heat dissipation to air. These heating pads are powerful enough to simulate the solar constant, 1350 W/m^2. All heating pads are connected to an external mechanical relay box and controlled via the computer. The top view of the thermal chamber and a door are shown on the right.

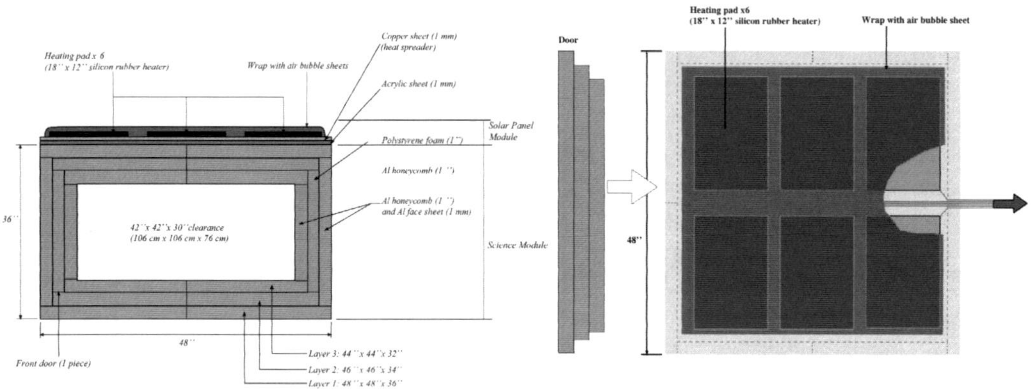

Figure 3. Schematic of the new thermal chamber: side view (left) and top view (right).

The bottom part is associated with the science module of the LISA spacecraft, which we plant to place optics system, MGRS test-object, and also conduct active control. In the space environment, because there is not convective heat transfer, thermal coupling will be weaker than the lab environment. However, we anticipate the metal structure won't have enough temperature attenuation capability particularly at the lower frequency range. Therefore, we decided to design a multi-layer structure blended with aluminum honeycomb and polystyrene form. The very outer layer, which is made of 1-inch thick aluminum honeycomb and 2-mm thick aluminum face sheets, provides first order temperature uniformity particularly along the face sheets. Honeycomb structure makes the whole system significantly lighter and also works as a thermal insulation. Then the polystyrene-form layer attenuates temperature fluctuations. The third layer completes a sealed control volume as a heat spreader that equalize temperature profile around the inner space of the chamber. The whole chamber is placed on a 8-feet by 4-feet optics table that simulates the cold plate of the spacecraft. Finally, the entire experimental system is placed within a clear plastic thermal tent, primarily to cut down on air drafts. A small chamber is placed inside the new chamber that simulates the Y-tube telescope of the LISA spacecraft. It is referred as a payload chamber. The completed chamber appears in Figure 4.

Figure 4. New thermal chamber constructed at Stanford University

3. Dynamic characteristics against disturbance temperature input

The Figure 5 demonstrates the dynamic characteristics of the new thermal chamber against the disturbance input. The disturbance sinusoidal input is applied by the heating pads at various frequencies. Output temperature is measured at four locations indicated in Figure 2. The experimental data is fitted as a second or third-order LTI model with a time-delay element. The sensitivity inside the payload chamber is 2.2×10^{-3} at 0.1 mHz. Table 1 shows the zero-frequency gain, time-delays and time constants (T.C.). New multi-layer structure chamber enables to achieve lower sensitivity to disturbance input at the frequency range of LISA's interest.

Table 1. Key parameters of the dynamic characteristics

Parts	Zero frequency gain (K/K)	Time-delay (sec)	T.C. 1 (sec)	T.C. 2 (sec)	T.C. 3 (sec)
Top	1.9585×10^{-1}	7.6422×10^{-5}	2.5047×10^{3}	6.2663×10^{3}	none
Bottom	6.5523×10^{-2}	8.2531×10^{3}	1.4059×10^{4}	1.4096×10^{4}	none
Cu chamber	9.9816×10^{-2}	1.7701×10^{3}	7.0631×10^{3}	2.9948×10^{4}	none
Inside	9.0222×10^{-2}	1.3148×10^{4}	8.4859×10^{3}	1.3148×10^{4}	2.4117×10^{5}

4. Results: active control simulation and experimental spectral stability performance of the new chamber

Figure 6 demonstrates a 12-day simulation of the proof-mass temperature being controlled with the MPC algorithm applied the LTI model obtained previously. The control input is also shown. The simulation accounts for 50 $\mu K/\sqrt{Hz}$ of sensor noise added at the plant output. The temperature is maintained approximately ±20 μK at steady-state. During the simulation, the disturbance temperature was modeled as a combination of sinusoidal functions,

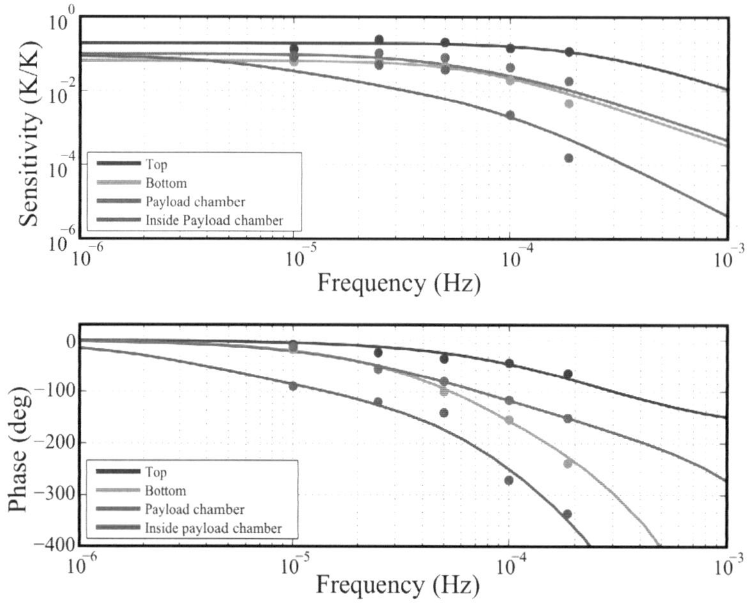

Figure 5. Bode plot: solid line fitted LTI model. Dots represent experimental data points.

$d(t) = \sum_{i=1}^{3} A_i \sin(\omega_i t + \phi_i)$ and Gaussian random noise. The controller did not have any information of the disturbance's parameters (frequency, phase, and amplitude) at $t = 0$. The controller adaptively estimates the unknown parameters of disturbance.

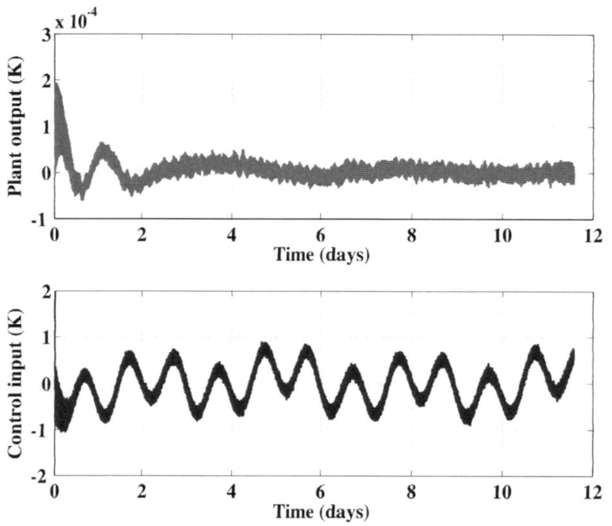

Figure 6. Simulation results of active control

Figure 7 compares the root mean squared spectral density (amplitude spectral density, ASD) of temperature variations of the proof-mass, the new thermal chamber, and the old thermal chamber. The proof-mass data is simulation shown in Figure 6 while thermal chamber data is obtained through experiment. The spectral stability of the new chamber has been improved nearly by an order of magnitude at 1 mHz over the old chamber. Finally, the expected thermal stability will be below 30 $\mu K/\sqrt{Hz}$ above 0.1 mHz with the aid of active control.

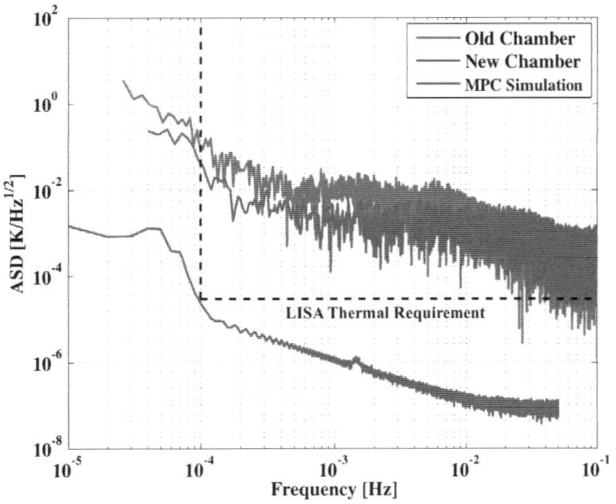

Figure 7. Spectral thermal stability of the new and old thermal chamber (experiment) and active control result (simulation). The below of dashed lines indicates the LISA thermal requirement: 30 $\mu K/Hz^{1/2}$.

5. Conclusion

The experimental performance of the new multi-layer thermal chamber and the active control simulation results have demonstrated the potential of satisfying the LISA thermal stability requirement. Continuing development will achieve more precision.

6. Acknowledgments

This research was supported by NASA, NNX07AK65G, Modular Gravitational Reference Sensor for Space Gravitational Wave Detection, and Department of Aeronautics & Astronautics, Stanford University. Authors also would like to acknowledge Mark McKelvey, Fred Witteborn at NASA Ames Research Center, and Hansen Experimental Physics Laboratory at Stanford University.

References

[1] Bender P 2003 *Class. Quantum Grav.* **20** S301
[2] Buckman S et al 2004 *5th International LISA Symposium, ESTEC*
[3] Sun K, Allen G, Buchman S, DeBra D and Byer R 2005 *Class. Quantum Grav.* **22** 10 S287
[4] Sun K, Allen G, Williams S, Buchman S, DeBra D, Byer R 2006 *Journal of Physics CS* **32** 137
[5] Sun K et al 2006 *AIP Conference Proceedings***873** 515
[6] Sun K, Johann U, DeBra D, Buchman S, Byer R 2007 *Journal of Physics: Conference Series* **60** 272
[7] Swank A 2006 *Class. Quantum Grav.***23** 3437
[8] Astrium 2006 *LISA Final Technical Report***LI-RP-DS-009**

Journal of Physics: Conference Series **154** (2009) 012038 doi:10.1088/1742-6596/154/1/012038

Gravitational disturbances in drag-free spacecraft

Antonio Pulido Patón[1*], Clive C Speake[2], Christian Trenkel[3], Sachie Shiomi[4] and Wei-Tou Ni[1]

[1]Center for Gravitation and Cosmology, Purple Mountain Observatory, Chinese Academy of Sciences, Beijing West Road 2, Nanjing 210008, P. R. China.
[2]School of Physics and Astronomy, University of Birmingham, Edgbaston, Birmingham B15 2TT, United Kingdom.
[3]Astrium UK, Gunnels Wood Road, Stevenage, Hertfordshire, SG1 2AS, United Kingdom.
[4]Space Geodesy Laboratory, Department of Civil Engineering, National Chiao Tung University, Hsinchu, Taiwan, 300, ROC.

E-mail: *antonio@pmo.ac.cn

Abstract. The performance of drag-free spacecraft will be ultimately limited by the uncompensated gravitational field and gravity field gradient of the spacecraft structure and payload. We identify sources of disturbance of gravitational origin and we explore a new concept for determining gravitational fields in the vicinity of the proof mass. The benefits of these diagnostic tools for future missions, like ASTROD, are discussed.

1. Introduction

Geodesic motion is at the core of numerous space-borne experimental tests of Gravity and Fundamental Physics. Tests of the Equivalence Principle and the Gravitational Inverse Square Law, the identification of the nature of the Pioneer Anomaly, measurements of the Gravitational Constant and its spatial-temporal variations, tests of relativistic gravity, i.e., light deflection and light retardation, geodesic deviation and Lense-Thirring effect, and the detection and measurement of gravitational waves, are within the most fundamental experimental searches in Gravitation theory. To perform such tests, proof masses in free fall are generally required. The technique to place a proof mass in geodesic/free-fall motion is known as drag-free technology.

Currently the state-of-the-art technology in drag-free performance presumes that we can achieve an acceleration noise in free-fall motion with an accuracy of about 3×10^{-14} m s^{-2} Hz$^{-1/2}$, in the measurement bandwidth between 1 mHz and 30 mHz. This is the acceleration noise performance expected for the LTP (LISA Technology Package) in the mission LISA Pathfinder [1]. LISA Pathfinder is the technology precursor of the LISA (Laser Interferometer Space Antenna) mission [2]. On the other hand, other mission concepts like ASTROD (Astrodynamical Space Tests of Relativity using Optical Devices) [3] aims at improving drag-free performance compared to LISA at frequencies below the LISA MBW, i.e., up to micro-hertz frequencies, while the Japanese gravitational wave antenna DECIGO (DECi-hertz Interferometer Gravitational wave Observer) [4] requires drag-free performance at deci-hertz frequencies.

It is worth noticing that the verification of an anomalous acceleration towards the Sun, the Pioneer Anomaly, requires DC drag-free performance. Also in the case of the ASTROD mission concept, the measurement of the relativistic parameter β, and the determination of the asteroids masses require long integration times and therefore DC or quasi-DC drag-free performance is beneficial. The idea of designing a space capsule with a near zero gravitational force acting on the test bodies had previously been considered within the project SEE (Satellite Energy Exchange). The SEE project [5] aims to measure the gravitational constant and its time variation by measuring the gravitational energy exchange between two bodies.

In this paper, we consider the self-gravity effects in inertial sensors. The presence of residual local gravitational fields is an obstacle for further improvement of drag-free performance. Local gravity gradients couple the proof mass to the spacecraft contributing to the acceleration noise because of spacecraft jittering and structural distortion. Local gravity fields are also a key requirement in order to achieve high drag-free performance at DC. In what follows we discuss new ideas about plausible gravitational diagnostic tools to actively tune drag-free performance in orbit.

2. Gravitational disturbances

In the current state-of-the-art drag-free technology, the coupling of the proof mass and the spacecraft is mainly due to residual electrostatic and local gravitational gradients. Electrostatic couplings are caused by proximity of metallic surfaces to the proof mass. Nevertheless electrostatic sensors, commonly used in drag-free satellites, require close gaps to achieve high sensitivity. Electrostatic sensing is also intrinsically unstable, and, therefore, we face the problems related to the need for low-frequency force feedback for ultra low-frequency drag-free operation. To solve these problems, optical sensors have been considered for implementation in the next-generation drag-free spacecraft [6, 7]. Given that scenario, drag-free performance will then be limited by the spacecraft self-gravitational field. To suppress gravitational noise we must identify its origin and, if possible, we have to introduce gravitational diagnostic tools and compensation techniques. Here is a list of possible sources of gravitational noise:

- Thermal and other distortions of the spacecraft and payload. Thermal-gravitational effects contribute as a fluctuating force noise [8] and they are likely to shift the local gravitational field value. There could also be some distortions due to, for example, tidal effects [9].
- Gravitational force and gradients due to micrometeorites passing nearby the spacecraft [10].
- Changes of the local gravitational field due to asymmetrical thruster propellant consumption.
- Changes of the local gravitational field due to active thermal control techniques (i.e. movable radiation panels, changing phase materials for absorption and rejection of heat, etc).
- Changes of the local gravitational field due to the need to move payload constituents (i.e. communicating antennae, moving telescopes for laser beam pointing, etc).
- Gravitational noise due to density inhomogeneities of the proof mass and defects in the constructing materials. Density inhomogeneities in the proof mass induce, to first order, residual proof mass dipole and quadrupole moments. These moments are coupled to the spacecraft gravitational gradients. In drag-free spacecraft, the proof mass is fixed in position and orientation, and ideally located in a minimum of the local gravitational potential. In that case, noise caused by density inhomogeneities is minimized.

In the case of ASTROD, some of the spacecraft will be exposed to a significant change of solar irradiance flux due to variations of spacecraft-Sun relative distance. Given that scenario, the spacecraft are likely to require active thermal control techniques. In the ASTROD constellation, the line of sight between one of the spacecraft and the distant spacecraft varies along the orbit. That will affect the spacecraft self-gravity due to the need to move the telescopes for laser beam pointing. In a preliminary design, dummy telescopes are considered to compensate this change of gravitational field at the proof mass location. The fact that spacecraft self-gravity could change

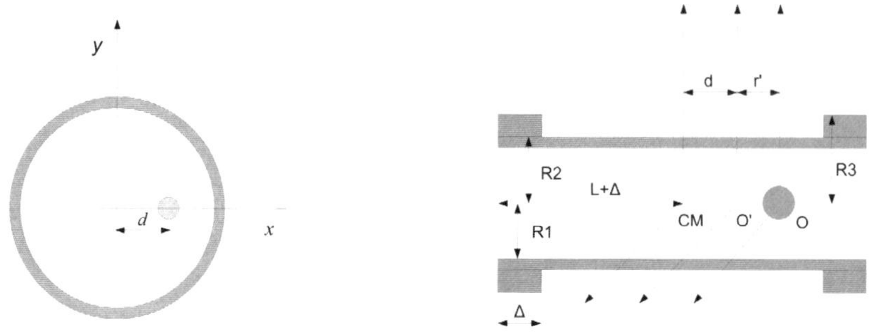

Figure 1. Gravity gradiometer concept for gravitational diagnostic in drag-free spacecraft. The inner proof mass of the gradiometer is in free fall motion and it is inserted inside an outer mass shell. The gravitational field inside the spherical outer mass shell vanishes whatever the position the proof mass inside the shell (left). Alternative gravity gradiometer configuration with a cylindrical-type outer mass shell geometry (right).

along the orbit is an important motivation to explore gravitational diagnostic techniques. In what follows, we describe a gravity gradiometer to measure "in situ" the local gravitational field at the free-falling proof mass location.

3. Gravity diagnostic tools

A standard drag-free spacecraft, which employs only one proof mass, can not monitor the spacecraft self-gravitational field gradients. The idea of a gravity gradiometer for an inertial sensor is shown in figure 1. The proof mass in drag-free motion lies inside an outer mass shell. The outer mass shell must fulfill two requirements: (a) it should provide a second reference mass to measure gravitational gradients; and (b) the outer mass shell should not contribute to the local gravitational field acting on the proof mass. To achieve the former, the center of mass of the inner and outer masses have to be shifted with respect to each other. By suppressing the acceleration experienced by the inner mass due to "internal" gravity in the gradiometer, the spacecraft gravitational gradients are monitored by measuring the differential acceleration (or displacement) of the inner mass relative to the outer shell. With (b) we ensure that the gravitational gradient measured is due to the surrounding spacecraft. We also need to eliminate coupling to gradients and higher orders of the outer shell gravitational potential below the gravity gradient noise requirement. Gravitational gradiometers with outer spherical and cylindrical shells are depicted in figure 1. An obvious appealing property of the spherical geometry (figure 1 (left)) is that the local gravitational field is zero whatever the proof mass position inside the shell.

3.1. A gravity gradiometer for self-gravity diagnostic

Due to manufacturing and engineering difficulties in constructing a spherical shell, we explore other non-spherical geometries with similar properties. We consider the cylindrical-type outer shell geometry shown in figure 1 (right). To start the discussion it is useful to think of the proof mass acceleration as due to a multipole expansion around the origin. The proof mass acceleration along an axis, in terms of multipole moments, is given by [11],

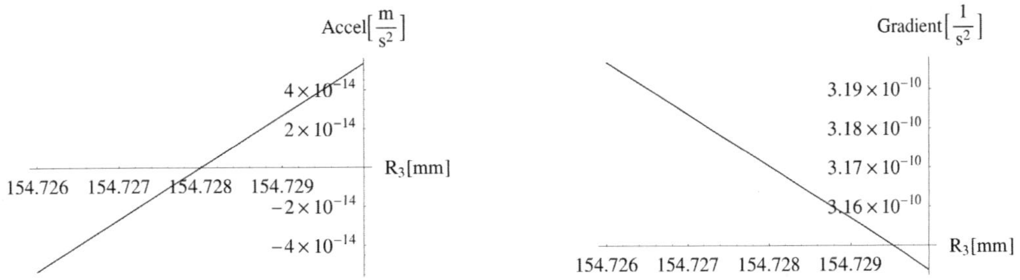

Figure 2. The inner mass acceleration due to the outer shell is zero for a value of $R_3 \simeq 154.728$ mm (left). The residual gravity gradient is of the order of 3×10^{-10} s^{-2} (right). The geometry parameters are chosen to be L=200, R_1=100, R_2=120, Δ=80 and d=20 milimeters.

$$a_z = \frac{4\pi G}{M_{tm}} \sum_{l=1}^{\infty} \sum_{m=-l}^{+l} \sqrt{\frac{(l-m)(l+m)}{(2l-1)(2l+1)}}\, q_{l-1,m} Q_{lm} \tag{1}$$

where M_{tm} is the mass of the inner test body, q_{lm} are the inner multipole moments of the test mass and Q_{lm} are the outer multipole moments of the outer shell. By inspection of equation (1), we conclude that the axial field, a_z, can be made small by eliminating as many multipole terms in the expansion as possible. For the purposes of this work, it is more convenient to use the following expansion of the gravitational potential for a cylindrically symmetric outer mass shell. If we restrict ourselves to displacements of a point test mass along the axis of symmetry we find that (see Appendix A)

$$V(z) = -G(4\pi)^{1/2} q_{00} \sum_{L=0}^{\infty} \sqrt{\frac{4\pi}{2L+1}}\, Q_{L0} z^L \tag{2}$$

where z defines the displacement of the proof mass with respect to the operating position. It should be noticed (see figure 1) that the operating position of the free-falling proof mass is shifted by a distance d with respect to the center of mass of the outer shell. In equation (2) the outer multipoles, Q_{LM}, are calculated with respect to the reference frame with origin at O' (see figure 1 (right) and appendix A for explanation). It is also interesting to note that in general the outer multipoles of a body can be calculated about an arbitrary origin following [12]. Equation (2) shows that the gravitational gradient is proportional to outer shell multipole Q_{20}. As mentioned above, the gravitational field inside the outer spherical shell geometry vanishes whatever the position of the inner proof mass inside the shell. If we consider an infinitely long hollow cylinder the gravitational field is also zero whatever the proof mass position inside the shell. This is not the case when we consider a hollow cylinder of finite length. However we can design a "stepped" cylindrical geometry (see figure 1 (right)) for which the local acceleration, proportional to Q_{10}, can be made zero given the right parameter values. A drawback of that geometry is that the inner mass will be in unstable equilibrium on the cylinder axis. Stabilizing forces should then be applied passively or actively. As a comment, here we are trying to design a hollow cylinder which behaves gravitationally like a spherical shell in the inside region by eliminating its outer multipoles, Q_{LM}. That can be compared with, for example, the test masses design strategy for the STEP (Satellite Test of the Equivalence Principle) mission [13]. The geometry of the STEP test masses is chosen to eliminate the inner multipoles, q_{lm}, in order to construct monopole test masses.

For a "sufficiently" far away spacecraft structure and payload, the masses configuring the gravity gradiometer can be considered as point masses. In that approximation, we can estimate the sensitivity to spacecraft gravity gradients, K_{gg}, by the ratio of the residual differential acceleration between the inner and outer mass, due to their internal gravitational interaction, and the center of mass separation, d, i.e. $\Delta a_{res} \simeq K_{gg}d$. Figure 2 (left) shows a particular set of geometrical parameters for which we have a zero local acceleration. However because of machining inaccuracies and inaccuracy in center of mass knowledge and positioning, the acceleration has an uncertainty of the order of 10^{-13} m s^{-2}. We have considered those inaccuracies to be at the micrometer level. With this particular choice of parameters the expected sensitivity to "external" gravity gradients is about 5×10^{-12} s^{-2}. The residual gravitational gradient due to the outer shell is about 3×10^{-10} s^{-2} (see figure 2 (right)). These numbers are sufficiently below the allocated magnitude of the uncertainty in $K_{gg} \sim 3 \times 10^{-8}$ s^{-2} for ASTROD I [14].

In a practical scheme we envisage being able to modulate the spacecraft gravity field acting on the gradiometer by periodically changing the position of the spacecraft using the thrusters. Active correction and nulling of the asymmetry in the mass distribution could be accomplished using trim masses.

4. Conclusions

In the next generation of drag-free spacecraft the ultimate obstacle to improve the quality of geodesic motion will be the residual self-gravity field at the location of the free-falling proof mass. Gravitational diagnostic tools on ground and/or in orbit will be of demand for future drag-free spacecraft. We have emphasized that a standard drag-free spacecraft with one proof mass only, is unable to measure the spacecraft self-gravitational field gradient. Measurement of the gravity gradient may serve to help distinguish between gravitational and non-gravitational interactions between the drag free sensor and the spacecraft. We have proposed a gravity gradiometer with two masses, one of which is the free-falling proof mass, to measure "in situ" the self-gravitational field at the proof mass location. The relative motion of the inner and outer masses will be most likely measured interferometrically.

Acknowledgments

One of the authors (APP) has been supported by CAS Research Fellowship for International Young Researchers, grant no. 20070309. We (APP and WTN) would like to thank the National Natural Science Foundation of China (grant nos. 10705051, 10778710) for supporting this work.

Appendix A. Multipole expansion of the gravitational potential

We describe the gravitational interaction between an inner test mass and an outer mass shell (see figure 1) by means of a spherical harmonic expansion. In the general case, the gravitational potential energy between two rigid bodies in terms of multipole moments can be written as

$$V = -4\pi G \sum_{L=0}^{\infty} \sum_{M=-L}^{+L} \frac{1}{2L+1} \tilde{q}_{LM} Q_{LM} \qquad \text{(A.1)}$$

where $\tilde{q}_{LM}(Q_{LM})$ are the inner(outer) multipoles defined by

$$\tilde{q}_{LM} = \int_{v_t} \rho_t r'^L Y^*_{LM}(\Theta', \Phi') d^3x' \quad \text{and} \quad Q_{LM} = \int_{v_s} \rho_s r^{-(L+1)} Y_{LM}(\Theta, \Phi) d^3x \qquad \text{(A.2)}$$

and (r', Θ', Φ'), (r, Θ, Φ) refer to spherical coordinates of the inner and outer test masses. The density distributions, ρ_t and ρ_s, are assumed to be uniform. The multipoles are calculated in the same reference frame, in our case with origin at O' according to figures 1 (right) and A1.

By performing a translation of the inner multipole in equation (A.1), we can infer how the gravitational potential energy varies around O'. Following [15] we express the inner multipoles

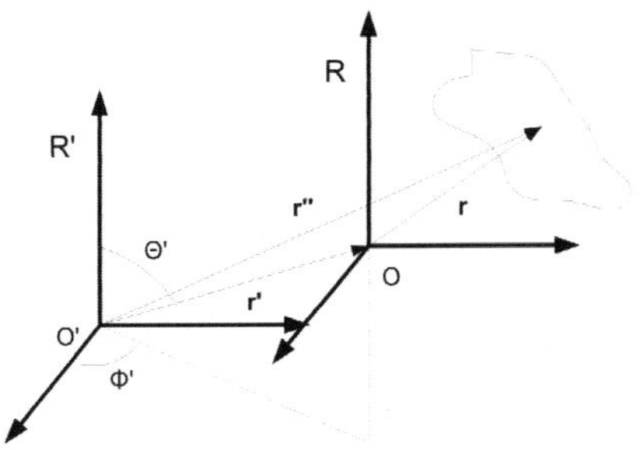

Figure A1. Reference frames to calculate the gravitational potential multipole expansion.

$\tilde{q}_{LM}$, calculated with respect to reference frame R' with origin at O', in terms of the inner multipoles q_{lm}, calculated with respect to the reference frame R with origin at O (see figure A1) as

$$\tilde{q}_{LM} = \sum_{l,l'=0}^{L} \sum_{m,m'} \sqrt{\frac{4\pi(2L+1)!}{(2l+1)!(2l'+1)!}} r'^{l'} \delta_{L,l+l'} C(l',m',l,m,L,M) Y^*_{l'm'}(\hat{r}') q_{lm} \tag{A.3}$$

By substituting (A.3) into (A.1), the potential energy can be written in terms of the translation parameter, r', and the direction (Θ', Φ') of the translation giving

$$V(r',\Theta',\Phi') = -4\pi G \sum_{L=0}^{\infty} \sum_{M=-L}^{+L} \frac{1}{2L+1} Q_{LM}$$
$$\sum_{l,l'=0}^{L} \sum_{m,m'} \sqrt{\frac{4\pi(2L+1)!}{(2l'+1)!(2l+1)!}} C(l',m',l,m,L,M) \delta_{L,l+l'} r'^{l'} Y^*_{l',m'}(\Theta',\Phi') q_{lm} \tag{A.4}$$

For the case of an inner monopole mass performing a displacement along the symmetry axis of a cylindrical shell, the equation (A.4) transforms into equation (2).

References

[1] McNamara P *et al* 2008 *Class. Quantum Grav.* **25** 114034
[2] Bell T E 2008 *Nature* **452** 7183 18
[3] Ni W T 2008 *Int. J. Mod. Phys.* D **17**(7) 921
[4] Kawamura S *et al* 2006 *Class. Quantum Grav.* **23** S125
[5] Sanders A J *et al* 1999 Meas. Sci. Technol. **10** 514
[6] Xu X and Ni W T 2003 *Adv. Space Res.* **32**(7) 1443
[7] Speake C C and Aston S M 2005 *Class. Quantum Grav.* **22** S269
[8] Stebbins R T *et al* 2004 *Class. Quantum Grav.* **21** S653
[9] Scheithauer S *et al* 2006 *Class. Quantum Grav.* **23** 7273
[10] Purdue P and Larson S L 2007 *Class. Quantum Grav.* **24** 5869
[11] Shiomi S *et al* 2001 *Class. Quantum Grav.* **18** 2533
[12] Trenkel C and Speake C C 1999 *Phys. Rev.* D **60** 107501-2
[13] Mester J 2001 *Class. Quantum Grav.* **18** 2475
[14] Shiomi S and Ni W T 2006 *Class. Quantum Grav.* **23** 4415
[15] D'Urso C and Adelberger E G 1997 *Phys. Rev.* D **55** 7970

Journal of Physics: Conference Series **154** (2009) 012039 doi:10.1088/1742-6596/154/1/012039

Determination of the neutron star mass-radii relation using narrow-band gravitational wave detector

C.H. Lenzi[1], M. Malheiro[1], R. M. Marinho[1], G. F. Marranghello[3] and C. Providência[2]

[1]Departamento de Física, Instituto Tecnológico de Aeronaútica, São José dos Campos/SP, Brazil
[2]Centro de Física Computacional, Departamento de Física, Universidade de Coimbra, Coimbra, Portugal
[3]Universidade Federal do Pampa, Bagé/RS, Brazil

E-mail: chlenzi@ita.br

Abstract. The direct detection of gravitational waves will provide valuable astrophysical information about many celestial objects. The most promising sources of gravitational waves are neutron stars and black holes. These objects emit waves in a very wide spectrum of frequencies determined by their quasi-normal modes oscillations. In this work we are concerned with the information we can extract from f and p_I-modes when a candidate leaves its signature in the resonant mass detectors ALLEGRO, EXPLORER, NAUTILUS, MiniGrail and SCHENBERG. Using the empirical equations, that relate the gravitational wave frequency and damping time with the mass and radii of the source, we have calculated the radii of the stars for a given interval of masses M in the range of frequencies that include the bandwidth of all resonant mass detectors. With these values we obtain diagrams of mass-radii for different frequencies that have allowed to determine the better candidates to future detection taking in account the compactness of the source. Finally, to determine which are the models of compact stars that emit gravitational waves in the frequency band of the mass resonant detectors, we compare the mass-radii diagrams obtained by different neutron stars sequences from several relativistic hadronic equations of state (GM1, GM3, TM1, NL3) and quark matter equations of state (NJL, MTI bag model). We verify that quark stars obtained from MIT bag model with bag constant equal to 170 MeV and quark matter in color-superconductivity phase are the best candidates for mass resonant detectors.

1. Introduction

The detection of gravitational waves (GWs) will have many implications in Physics and Astrophysics. Besides the confirmation of the general relativity theory, it will allow the investigation of several astrophysical phenomena, such as the existence of black holes and the mass and abundance of neutron stars, thus opening new scientific frontiers. The most promising sources for detection of GWs are neutron stars and black holes. These objects emit waves in a very wide spectrum of frequencies determined by their quasi-normal modes oscillations [1].

With the goal of analyzing the possibility of a future detection of the quasi-normal modes of compact stars by resonant mass detectors (RMDs), we focalize our attention in the region of the spectrum in the range at 0.8-3.4 kHz, which is the operation region of the antennas: ALLEGRO, EXPLORER, NAUTILUS, AURIGA [2], SCHENBERG, and MiniGrail. In particular we will work with the frequency band of the spherical detectors SCHENBERG and MiniGrail (2.8-3.4 kHz) [3, 4].

The SCHENBERG detector, which is installed at the Physics Institute of the University of Sao Paulo (at Sao Paulo city, Brazil), it is the second spherical detector ever built in the world and the first equipped with a set of parametric transducers. It has undergone its first test run in September 8, 2006, with three transducers operational. Recent information on the present status of this detector can be found in reference [5]. It is worth stressing also that among all known GW detectors, the spherical ones are the only ones capable to determine the direction of the incoming wave [6, 7].

In this paper we present an extension of the work made by Marranghello [1], where the authors show results for a restricted band of frequency that include the SCHENBERG and MiniGrail bandwidth. We analyze the case of a possible future detection made by all resonant antennas and compare the mass and radius range obtained from the frequency bands of the GWs modes to the mass and radius calculated with several relativistic equations of state (EoS) models.

In section 2 we introduce the f and p_I-mode, in section 3 we show the mass-radii diagrams for the f and p_I modes and compare them with the ones obtained with different relativistic EoS models, in section 4 we introduce the damping time of the f-mode and its mass-radii diagram and finally, in section 5, we make the last considerations.

2. The quasi-normal modes: f and p_I-modes

The neutron stars have a rich spectrum of frequencies because the fluid perturbation oscillates in many different modes. From the GW point of view the most important quasi-normal modes are the fundamental mode of the fluid oscillation (f-mode), the first pressure mode (p_I-mode), the first GW mode (w_I-mode) [8] and the r-modes that, under certain circunstances, can be an important source of GWs [9].

In this work we concentrate in the f and p_I-modes. The fundamental mode can be described by the density distribution inside the star, while the p-mode is the pressure restoration force. In reference [10], the authors have obtained an empirical formulae for the frequencies of these two modes as a function of the mass and radius using a wide sample of equations of state:

$$\nu_f = (0.79 \pm 0.09) + (33 \pm 2)\sqrt{\frac{M}{R^3}}, \tag{1}$$

$$\nu_p = \frac{1}{M}\left[(-1.5 \pm 0.8) + (79 \pm 4)\frac{M}{R}\right], \tag{2}$$

where the mass and the radii are given in km (remember that $M_\odot \approx 1.477$ km), while ν_f and ν_p are given in kHz. Using the empirical relations (1) and (2), we have calculated the radii R of the stars for a given interval of masses M in the range of frequencies that include the bandwidth (0.8-3.4 kHz) of all RMDs in operation. In Table (1) we can see the resonant frequencies of these detectors.

Through these relations we have obtained diagrams for p_I and f-modes that relate GW frequency with masses and radii of the sources. These diagrams allow us to determine the better candidates for a future detection by resonant antennas from the compactness of the star. We can see in figure (1) and (2) the f and p_I-mode's mass-radii diagrams where the different colors scale identify the different frequencies.

Table 1. Frequency band of the RMDs in operantion in the world.

Antena	Location	Freq.(Hz)	Type
ALLEGRO	Baton Rouge	890-920	Bar
EXPLORER	CERN	895-920	Bar
NAUTILUS	Frascai	905-925	Bar
AURIGA	Legnaro	850-930	Bar
SCHENBERG	São Paulo	3100-3300	Spherical
MiniGrail	Leiden	2800-3000	Spherical

3. Comparison of the mass-radius diagrams with the ones obtained by relativistic EoS models

To determine which relativistic models of compact stars emit GW in the frequency bands of the RMDs we compare the diagrams of the relations (1) and (2) with somesome mass/radius neutron star sequences obtained by different relativistic models that generate several equations of state for hadronic matter such as models NP, NPH, NPHQ with and without isovector-scalar δ [11], namely

- the models with parameters set GM1, GM3, NL3, TM1 [12, 13]

and some for quark strange matter as

- Nambu-Jona-Lasínio model (NJL) [14], color-flavor locked phase (CFL) [15] and the MIT bag model with different values of the bag constant [16] and hybrid star EoS.

Relativistic hadronic models have been widely used in order to describe nuclear matter, finite nuclei, stellar matter properties, and recently in the high temperature regime produced in heavy ion collisions [17]. Many variations of the well known quantum hadrodynamic model [18] have been developed and used along the last decades. Some of them rely on density dependent couplings between the baryons and the mesons [19–25] while others use constant couplings [26–28]. Still another possibility of including density dependence on the lagrangian density is through derivative couplings among mesons and baryons [29–31] or the coupling of the mediator mesons among themselves [32–34]. The relativistic model couplings are adjusted in order to fit expected nuclei properties such as binding energy, saturation density, compressibility and energy symmetry at saturation density, particle energy levels, etc. These same relativistic models are extrapolated to higher densities as in stellar matter and the results obtained for the neutron star masses and radii are quite good in comparison with the astronomical observations. In the case of bare quark stars, the strange matter inside the star is usually describe by the MIT bag model, a Fermi gas of free quarks with a vacuum energy known as the bag constant, or by chiral models like the Nambu-Jona-Lasínio (NJL) model that has a dynamical chiral symmetry breaking mechanism that originates mass for the quarks. Recently, the possibility that quarks can be paired at high densities and be in a color superconductive phase has originated new quark matter equations of state, that depending on the pairing interaction, can be quite stiff and produce large stars masses and radii [35–37]. The main feature of quark stars, since they are bound by the strong force and not by gravity, is that they are more compact and have smaller mass to radius ratio than a neutron star. As we will see, it is this fact that strange stars can have small radii, that explains the high frequency GWs modes produced by this type of stars.

We can see in the diagrams (1) and (2) that the frequency band of the RMDs are on the dark region, where it is expected that GWs generate from less compact neutron stars in both diagrams. This fact shows the impossibility of a future detection, by cylindrical antennas, of relativistic neutron star candidates emitting gravitational wave on p_I or f-mode. However, we can see on the diagrams that the spherical detectors bandwidth, MiniGrail and Schenberg, have some candidates near their resonant frequencies. The most probable source would correspond to a very compact object with radius smaller than 10 km. The models that fulfill this condition are models of strange quark stars, as preview in [1]. This fact is confirmed when we compare the compact star sequence generated from the MIT bag model (with bag constant $B^{1/4} = 170$ MeV), NJL model and CFL of quark matter. On the other hand the p_I-mode would only be expected to come from less compact neutron stars.

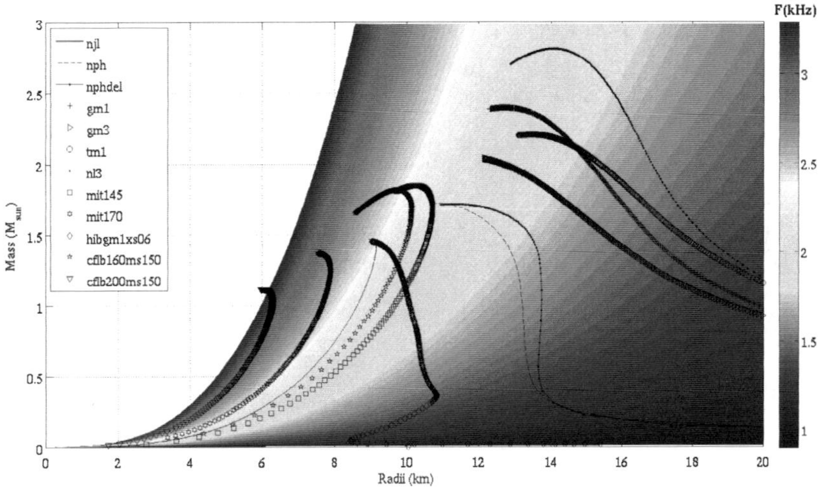

Figure 1. Mass-radius diagram of the empirical relation 1 abtained by Benhar *et al* for f_I-mode. We compare with some models of relativistic EoS. The different colors identify the different frequencies.

4. The damping time

How can we distinguish the f-mode in a putative detection? And how to determine the mass and radius of the star? The damping time is the response for these questions [1]. In [10] the authors obtained an empirical relation for the f-mode damping time as function of the radius and mass, described by:

$$\tau_f = \frac{R^4}{cM^3}\left[(8.7 \pm 0.2)\cdot 10^{-2} + (-0.271 \pm 0.009)\frac{M}{R}\right]^{-1}. \tag{3}$$

Even though the RMDs can not determine the damping time properties with small errors, we use the empirical relation (3) to calculate the damping time given the intervals of radius and mass (R, M) obtained with relation (1). We can get a new mass-radius diagram, but doing a distinction in the damping time. We compare the diagram with models of quark stars, CFL and MIT bag model with bag constant equal to 170 MeV. Results obtained can be seen in figure (3). Through these results we can estimate the masses and radii of the stars solving the inverse problem, as show in [9].

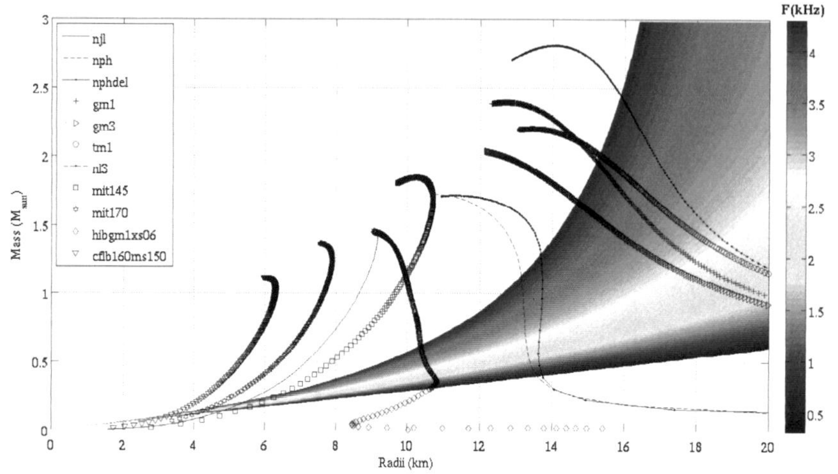

Figure 2. Mass-radius diagram of the empirical relations 2 abtained by Benhar *et al* for p_I-mode. We compare with some models of relativistic EoS. The different colors identify the different frequencies.

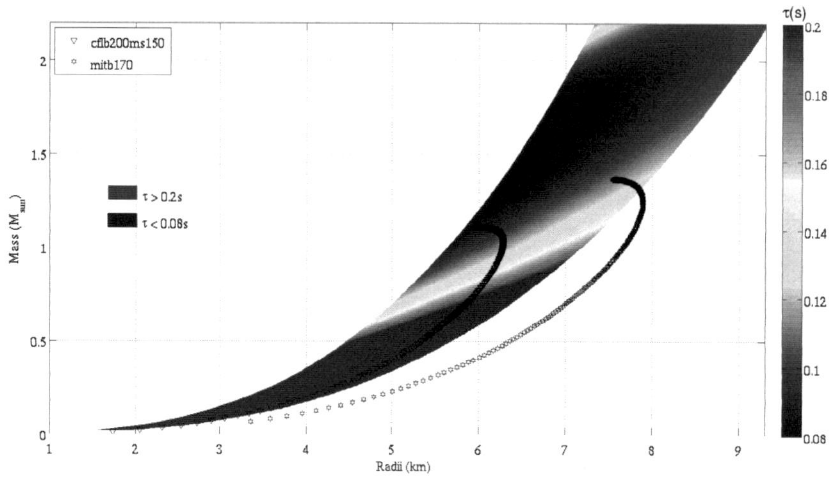

Figure 3. Mass-radius diagram of the empirical relations 2 abtained by Benhar *et al* for damping time of the f_I-mode. We compare the diagram with models that discribe quark stars, CFL of the quark matter with bag constant 200 MeV and gap 100 MeV and MIT bag model with bag constant equal to 170 MeV. The different colors identify the different damping time.

5. Summary

RMDs bandwidth are on the spectrum regions with a few (or none) neutron star candidates emiting GWs through their f and p_I-modes. However, the spherical detectors are on a region where the f-modes of very compact objects can be detected. All sequences of neutron stars described by quark matter models are on the region near to MiniGrail and Schenberg bandwidth, but the MIT bag model with bag constant equal to 170 MeV and CFL of the quark matter constant 200 MeV and the gap 100 MeV are the best candidates for these detectors, as we can see in figures (1) and (3). On the other hand the detection of the f and p_I-modes of neutron stars by bar detectors is unlikely, because their bandwidth is located in low frequencies.

Acknowledgments
This work was partially supported by FEDER and FCT (Portugal) under the project PTDC/FP/64707/2006. The CHL, MM, RMM thank the financial support given by CAPES through the fellowship 2071/07-0 and the international cooperation program Capes-Grices between Brazil-Portugal.

References

[1] Marranghello G F and de Araujo J C N 2006 *Class. Quatum Grav.* **23** 6345
[2] Astone P and *et al* 2007 *Phys. Rev.* D **76** 102001
[3] Gotard L and *et al* 2007 *Phys. Rev.* D **76** 102005
[4] Costa C and *et al* 2008 *Class. Quantum Grav.* **25** 184002
[5] Aguiar O D and *et al* 2008 *Class. Quantum Grav.* **40** 183
[6] Lenzi C H and *et al* 2008 *Phys. Rev.* D **78** 62005
[7] Lenzi C H and *et al* 2008 *Gen. Reltiv. Grav.* **40** 183
[8] Kokkotas K and Schutz B 1992 *MNRAS* **255** 119
[9] Kokkotas K D, Apostolatos T A and Andersson N 1999 *Astrophys. J.* **516** 307
[10] Benhar O, Ferreri V and Gualtieri L 2004 *Phys. Rev.* D **70** 124015
[11] Menezes D P and Providência C 2004 *Phys. Rev.* C **70** 58801
[12] Menezes D P, Melrose D B, Providência C and Wu K 2006 *Phys. Rev.* C **73** 25806
[13] Menezes D P and Providência C 2003 *Phys. Rev.* C **68** 35804
[14] Ruivo C, Souza C and Providência C 1999 *Nuc. Phys.* 59
[15] Alford M and Reddy S 2003 *Phys. Rev. D* **67** 074024
[16] Chodos A, Jaffe R L, Thorne C B and Weisskopf V F 1974 *Phys. Rev. D* **9** 3471
[17] Menezes D P, Providencia C, Chiapparini M, Bracco M E, Delfino A and Malheiro M 2007 *Phys. Rev.* C **76** 064902
[18] Serot B and Walecka J 1986 *Advances in Nuclear Physics* (Plenum-Press) p 16
[19] Lenske H and Fuchs C 1995 *Phys. Lett.* 355
[20] Fuchs C, Lenske H and Wolter H H 1995 *Phys. Rev.* C **52** 3043
[21] Typel S and Wolter H H 1999 *Nucl. Phys. A* 331
[22] Brown G E and Rho M 1991 *Phys. Rev. Lett.* **66** 2720
[23] Gaitanos T, Toro M D, Typel S, Baran V, Fuchs C, Greco V and Wolter H H 2004 *Nucl. Phys.* 24
[24] Niksic T, Vretenar D, Finelli P and Ring P 2002 *Phys. Rev.* C **66** 024306
[25] Vretenar D, Niksic T and Ring P 2003 *Phys. Rev.* C **68** 024310
[26] Lalazissis G A, Konig J and Ring P 1997 *Phys. Rev.* C **55** 540
[27] Sumiyoshi K, Kuwabara H and Toki H 1995 *Nucl. Phys.* 725
[28] Glendenning N K 2000 *Compact Stars* (New-York: Springer-Verlag)
[29] Delfino A, Coelho C T and Malheiro M 1995 *Phys. Rev.* C **51** 2188
[30] Delfino A, Coelho C T and Malheiro M 1995 *Phys. Lett* 361
[31] Chiapparini M, Delfino A, Malheiro M and Gattone A 1997 *Phys.* 41
[32] Horowitz C J and Piekarewicz J 2001 *Phys. Rev.* C **64** 062802R
[33] Bunta J K and Gmuca S 2003 *Phys. Rev.* C **68** 054318
[34] Bunta J K and Gmuca S 2004 *Phys. Rev.* C **70** 054309
[35] Weber F, Meixner M, Negreiros R P and Malheiro M 2007 *Int. J. of Mod. Phys* 1165
[36] Malheiro M, Linares L, Fiolhais M and Taurines A 2007 *Nucl. Phys.* 562c
[37] Malheiro M, Linares L, Malheiro M, Fiolhais M and Taurines A 2006 *Braz. J. of Phys.* **36** 1391

Journal of Physics: Conference Series **154** (2009) 012040 doi:10.1088/1742-6596/154/1/012040

DECIGO : The Japanese space gravitational wave antenna

Shuichi Sato[1], Seiji Kawamura[2], Masaki Ando[3], Takashi Nakamura[4], Kimio Tsubono[3], Akito Araya[5], Ikkoh Funaki[6], Kunihito Ioka[7], Nobuyuki Kanda[8], Shigenori Moriwaki[9], Mitsuru Musha[10], Kazuhiro Nakazawa[3], Kenji Numata[11], Shin-ichiro Sakai[6], Naoki Seto[2], Takeshi Takashima[6], Takahiro Tanaka[12], Kazuhiro Agatsuma[3], Koh-suke Aoyanagi[27], Koji Arai[2], Hideki Asada[13], Yoichi Aso[14], Takeshi Chiba[15], Toshikazu Ebisuzaki[44], Yumiko Ejiri[16], Motohiro Enoki[45], Yoshiharu Eriguchi[35], Masa-Katsu Fujimoto[2], Ryuichi Fujita[46], Mitsuhiro Fukushima[2], Toshifumi Futamase[43], Katsuhiko Ganzu[4], Tomohiro Harada[47], Tatsuaki Hashimoto[6], Kazuhiro Hayama[48], Wataru Hikida[38], Yoshiaki Himemoto[49], Hisashi Hirabayashi[17], Takashi Hiramatsu[18], Feng-Lei Hong[19], Hideyuki Horisawa[20], Mizuhiko Hosokawa[21], Kiyotomo Ichiki[3], Takeshi Ikegami[19], Kaiki T. Inoue[22], Koji Ishidoshiro[3], Hideki Ishihara[8], Takehiko Ishikawa[6], Hideharu Ishizaki[2], Hiroyuki Ito[21], Yousuke Itoh[50], Nobuki Kawashima[22], Fumiko Kawazoe[23], Naoko Kishimoto[6], Kenta Kiuchi[27], Shiho Kobayashi[24], Kazunori Kohri[25], Hiroyuki Koizumi[6], Yasufumi Kojima[51], Keiko Kokeyama[16], Wataru Kokuyama[3], Kei Kotake[2], Yoshihide Kozai[26], Hideaki Kudoh[3], Hiroo Kunimori[21], Hitoshi Kuninaka[6], Kazuaki Kuroda[18], Kei-ichi Maeda[27], Hideo Matsuhara[6], Yasushi Mino[14], Osamu Miyakawa[14], Shinji Miyoki[18], Mutsuko Y. Morimoto[6], Tomoko Morioka [3], Toshiyuki Morisawa[4], Shinji Mukohyama[39], Shigeo Nagano[21], Isao Naito[57], Kouji Nakamura[2], Hiroyuki Nakano[52], Kenichi Nakao[8], Shinichi Nakasuka[31], Yoshinori Nakayama[29], Erina Nishida[16], Kazutaka Nishiyama[6], Atsushi Nishizawa[30], Yoshito Niwa[30], Taiga Noumi[31], Yoshiyuki Obuchi[2], Masatake Ohashi[18], Naoko Ohishi[2], Masashi Ohkawa[32], Norio Okada[2], Kouji Onozato[3], Kenichi Oohara[32], Norichika Sago[33], Motoyuki Saijo[34], Masaaki Sakagami[30], Shihori Sakata[2], Misao Sasaki[12], Takashi Sato[32], Masaru Shibata[35], Hisaaki Shinkai[53], Kentaro Somiya[14], Hajime Sotani[36], Naoshi Sugiyama[37], Yudai Suwa[3], Rieko Suzuki[16], Hideyuki Tagoshi[38], Fuminobu Takahashi[39], Kakeru Takahashi[3], Keitaro Takahashi[12], Ryutaro Takahashi[2], Ryuichi Takahashi[37], Tadayuki Takahashi[6], Hirotaka Takahashi[54], Takamori Akiteru[5], Tadashi Takano[40], Keisuke Taniguchi[50], Atsushi Taruya[3], Hiroyuki Tashiro[4], Yasuo Torii[2], Morio Toyoshima[21], Shinji Tsujikawa[41], Yoshiki Tsunesada[55], Akitoshi Ueda[2], Ken-ichi Ueda[10], Masayoshi Utashima[56], Yaka Wakabayashi[16], Hiroshi Yamakawa[42], Kazuhiro Yamamoto[23], Toshitaka Yamazaki[2], Jun'ichi Yokoyama[3],

Chul-Moon Yoo[8], Shijun Yoshida[43], Taizoh Yoshino[28],

[1]Faculty of Engineering, Hosei University, kajinocho, Tokyo 184-8584, Japan, [2]National Astronomical Observatory Japan, Osawa 2-21-1, Tokyo 181-8588, Japan, [3]Department of Physics, The University of Tokyo, Tokyo 113-0033, Japan, [4]Department of Physics, Kyoto University, Kyoto 606-8502, Japan, [5]Earthquake Research Institute, The University of Tokyo, Tokyo 113-0032, Japan, [6]Institute of Space and Astronautical Science, Japan Aerospace Exploration Agency (JAXA), Kanagawa 229-8510, Japan, [7]Institute of Particle and Nuclear Studies, High Energy Accelerator Research Organization (KEK), Ibaraki 305-0801, Japan, [8]Department of Physics, Osaka City University, Osaka 558-8585, Japan, [9]Department of Advanced Materials Science, The University of Tokyo, Chiba 277-8561, Japan, [10]Institute for Laser Science, The University of Electro-Communications, Tokyo 182-8585, Japan, [11]NASA Goddard Space Flight Center, Code 663, 8800 Greenbelt Rd., Greenbelt, MD20771, USA, [12]Yukawa Institute for Theoretical Physics, Kyoto University, Kyoto 606-8502, Japan, [13]Department of Earth and Environmental Sciences, Hirosaki University, Aomori 036-8560, Japan, [14]California Institute of Technology, 1200 E. California Blvd. MC 18-34, Pasadena, CA 91125, USA, [15]Nihon University, Setagaya, Tokyo 156-8550, Japan, [16]Ochanomizu University, 2-1-1, Tokyo 112-0012, Japan, [17]Space Educations Center, Japan Aerospace Exploration Agency (JAXA), Kanagawa 229-8510, Japan, [18]Institute for Cosmic Ray Research, The University of Tokyo, Chiba 277-8582, Japan, [19]National Institute of Advanced Industrial Science and Technology (AIST), Ibaraki 305-8563, Japan, [20]Department of Aeronautics and Astronautics, Tokai University, Kanagawa 259-1292, Japan, [21]National Institute of Information and Communications Technology (NICT), Tokyo 184-8795, Japan, [22]School of Science and Engineering, Kinki University, Osaka 577-8502, Japan, [23]Max-Planck-Institute for Gravitational Physics (Albert-Einstein-Institute), Callinstr. 38 D-30167 Hannover, Germany, [24]Astrophysics Research Institute, Liverpool John Moores University, Twelve Quays House, Egerton Wharf, Birkenhead L41 1LD, UK, [26]Gunma Astronomical Observatory, Agatsuma-gun, Gunma 377-0702, Japan, [27]Department of Physics, Waseda University, Tokyo, 169-8555, Japan, [28]Nakamura-minami Nerima, Tokyo 176-0025, Japan, [29]Department of Aerospace Engineering, National Defense Academy, Yokosuka 239-8686, Japan, [30]Faculty of Intergrated Human Studies, Kyoto University, Kyoto 606-8501, Japan, [31]Department of Aeronautics and Astroautics, The University of Tokyo, Tokyo 113-8656, Japan, [32]Niigata University, Niigata 950-2181, Japan, [33]Highfield, Southampton SO17 1BJ, United Kingdom, [34]Department of Physics, Rikkyo University, Tokyo 171-8501, Japan, [35]Department of Earth Science and Astronomy, the University of Tokyo, Tokyo 153-8902, Japan, [36]Theoretical Astrophysics,Institute for Astronomy and Astrophysics, Eberhard Karls University of Tuebingen, Auf der Morgenstelle 10, 72076 Tuebingen, Germany, [37]Nagoya University, Graduate School of Science, Aichi 464-8601, Japan, [38]Department of Earth and Space Science, Osaka University, Osaka 560-0043, Japan, [39]Institute for Physics and Mathematics of the Universe (IPMU), The University of Tokyo, Chiba 277-8568, Japan, [40]Department of Electronics and Computer Science, Nihon University, Funabashi 274-8501 Japan, [41]Department of Physics, Tokyo University of Science, Tokyo, 162-8601, Japan, [42]Research Institute for Sustainable Humanosphere, Kyoto University, Kyoto 611-0011, Japan [43]Astronomical Institute, Tohoku University, Sendai 980-8578, Japan, [44]RIKEN, 2-1 Hirosawa, Wako 351-0198, Japan, [45]Faculty of Business Administration, Tokyo Keizai University, Tokyo 185-8502, Japan, [46]Theoretical Physics,Raman Research Institute, Sir C.V.Raman Avenue, Sadashivanagar P.O., Bangalore 560 080, India, [47]Department of Physics, Rikkyo University, Tokyo 171-8501, Japan, [48]University of Texas, 80 Fort Brown, Brownsville 78520, Texas, U.S.A., [49]Center for Educational Assistance, Shibaura Institute of Technology, Saitama 337-8570, Japan, [50]Department of Physics, University of Wisconsin-Milwaukee, 1900 East Kenwood Blvd. Milwaukee, WI 53211, USA, [51]Graduate School of Science, Hiroshima University, Hiroshima 739-8526, Japan, [52]Rochester Institute of Technology, 78 Lomb Memorial Drive, Rochester, NY 14623, USA, [53]Dept of Information Systems, Osaka Institute of Technology, Hirakata 573-0196, Japan, [54]Department of Management and Information Systems Science, Nagaoka University of Technology, Niigata 940-2188, Japan, [55]Graduate School of Science and Engineering / Physics, Tokyo Institute of Technology, Tokyo 152-8550, Japan, [56]Tsukuba Space Center, Japan Aerospace Exploration Agency (JAXA), Ibaraki 305-8505, Japan, [57]Numakage, Saitama-shi, Saitama 336-0027 Japan,

Abstract. DECi-hertz Interferometer Gravitational wave Observatory (DECIGO) is the planned Japanese space gravitational wave antenna, aiming to detect gravitational waves from astrophysically and cosmologically significant sources mainly between 0.1 Hz and 10 Hz and thus to open a new window for gravitational wave astronomy and for the universe. DECIGO will consist of three drag-free spacecraft, 1000 km apart from each other, whose relative displacements are measured by a differential Fabry-Perot interferometer. We plan to launch DECIGO in middle of 2020s, after sequence of two precursor satellite missions, DECIGO pathfinder and Pre-DECIGO, for technology demonstration required to realize DECIGO and hopefully for detection of gravitational waves from our galaxy or nearby galaxies.

1. Introduction

DECIGO (DECi-hertz Interferometer Gravitational wave Observatory) is the planned Japanese space gravitational wave antenna mission [1][2, 3, 4]. DECIGO is targeting to observe gravitational waves from astrophysically and cosmologically significant sources mainly between 0.1 Hz and 10 Hz, thus, to open a new window of observation for gravitational wave astronomy, and also for the universe.

The scope of DECIGO is to bridge (Fig. 1) the frequency gap between LISA [5] band and terrestrial detectors band such as advanced LIGO and LCGT [6]. The major advantage of DECIGO specializing in this frequency band is that the expected confusion limiting noise level caused by irresolvable gravitational wave signals from many compact binaries, such as white dwarf binaries in our Galaxy, is quite low above 0.1 Hz [7], therefore there is a potentially extremely deep window in this band.

Thus, as DECIGO will have sensitivity in the frequency range between LISA and terrestrial detectors band, DECIGO can serve as a follow-up for LISA by observing inspiraling sources that have moved above the LISA band, or as a predictor for terrestrial detectors by observing inspiraling sources that have not yet moved into the terrestrial detectors band.

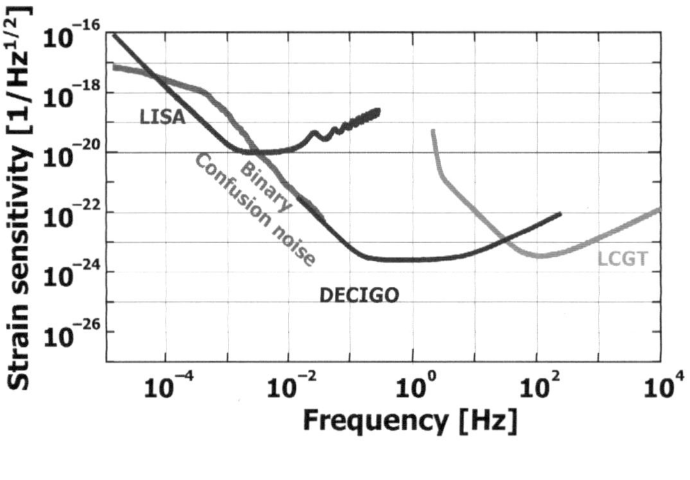

Figure 1. DECIGO design sensitivity with LISA and LCGT (on behalf of terrestrial detectors).

2. Pre-conceptual design

The pre-conceptual design of DECIGO consists of three drag-free spacecraft which keep triangular configuration with formation flying technique. The separation of each spacecraft (proof mass) is designed to be 1,000 km, whose relative displacements are measured by a differential Fabry-Perot (FP) interferometer (Fig. 2). The laser source is supposed to be frequency-doubled Nd:YAG laser with $\lambda = 532$ nm yielding output power of 10 W. The mass of the mirror is 100 kg with 1 m diameter, with low-loss High-reflectivity coatings, which enables the finesse of FP cavity to reach 10 with green light. Three sets of such interferometers sharing the mirrors as arm cavities comprise one cluster of DECIGO. As shown in Fig. 3, four clusters of DECIGO, located separately in the heliocentric orbit with two of them nearly at the same position, form the constellation DECIGO.

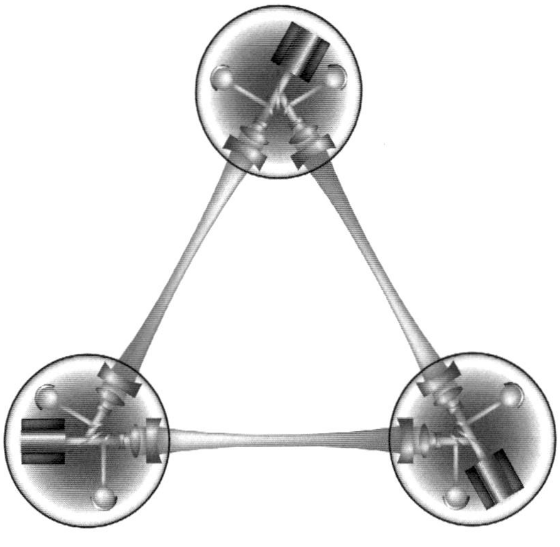

Figure 2. The pre-conceptual design of DECIGO. Three drag-free spacecraft keep 1000 km triangular configuration with formation flying technique. Each spacecraft will have light source and two proof masses.

The advantage of FP configuration is clearly that it can utilize much power of light for a better shot-noise-limited sensitivity than the transponder-type configuration (e.g. LISA). On the other hand, the FP configuration requires very accurate formation flying: the FP configuration requires the distance between two mirrors, thus, the distance between two spacecraft to be constant during continuous operations. This is a major difference between DECIGO and a transponder-type configuration, where the spacecraft are freely falling according to their local gravitational field.

3. Sensitivity goal and science

The target sensitivity of DECIGO, as shown in Fig. 4, is supposed to be limited by quantum noise in all frequency band: by the radiation pressure noise below 0.15 Hz, and by the shot noise above 0.15 Hz. In order to reach this sensitivity, all the practical noise should be suppressed well below this level. This imposes more stringent requirements than LISA for some subsystems of DECIGO, especially in the acceleration noise and frequency noise, therefore rigorous investigations are supposed to be indispensable for attainment of design sensitivity. Nonetheless, full success of DECIGO is expected to extract fruitful sciences.

- **Characterization of dark energy:** DECIGO will have enough sensitivity to detect gravitational waves coming from neutron star binaries at z=1 for five years prior to coalescences. Within this observable volume, about 50,000 neutron star binaries are expected to coalesce every year [8]. In addition to the physics of the neutron star, with precise analysis resolving gravitational wave signals coming from a number of binaries, it is possible to determine the acceleration of the expansion of the universe [1]. The constellation DECIGO is expected to have an angular resolution of about 1 arcsec, therefore, there is a chance to identify the host galaxies of each binary system. Thus, the acceleration of the expansion of the universe can also be measured by determining their red shifts optically [9], which will lead to better characterization of dark energy.

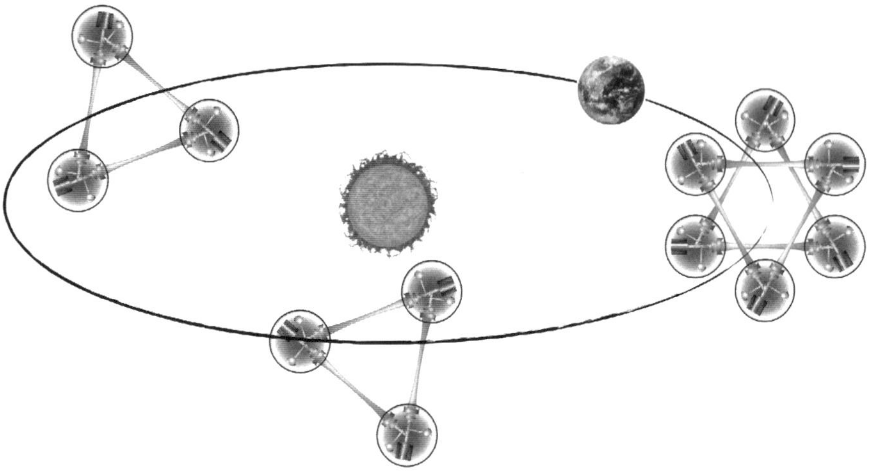

Figure 3. Constellation DECIGO in its orbit, which will have four clusters of DECIGO in total.

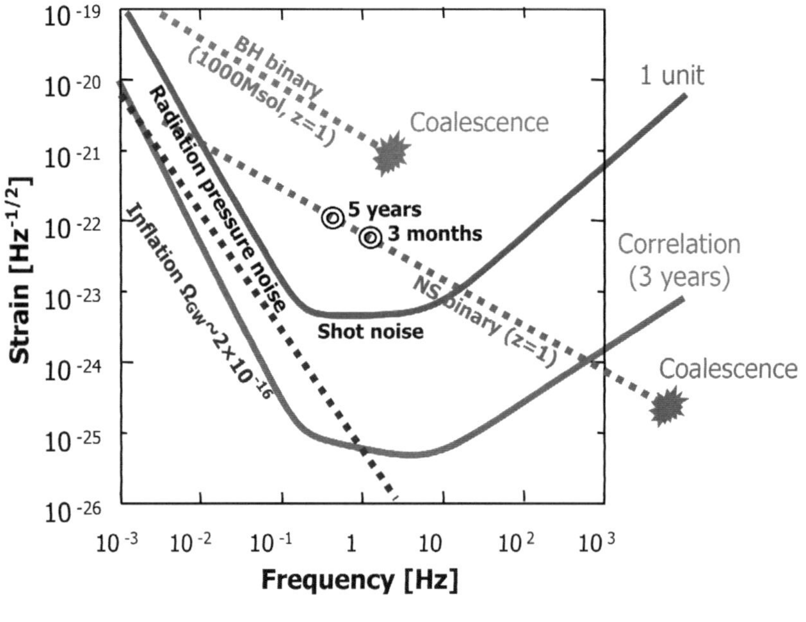

Figure 4. DECIGO design sensitivity (for 1unit and for 3 years correlation) and expected gravitational wave souces.

- **Formation mechanism of supermassive black holes:** DECIGO can detect gravitational waves coming from coalescences of intermediate-mass black hole binaries with an extremely high fidelity. For example the coalescences of black hole binaries of 1,000 solar masses at $z = 1$ give a signal to noise ratio of 6,000. This will make it possible to collect numerous data about the relationship between the mass of the black holes and the frequency of the coalescences, which will reveal the formation mechanism of supermassive black holes

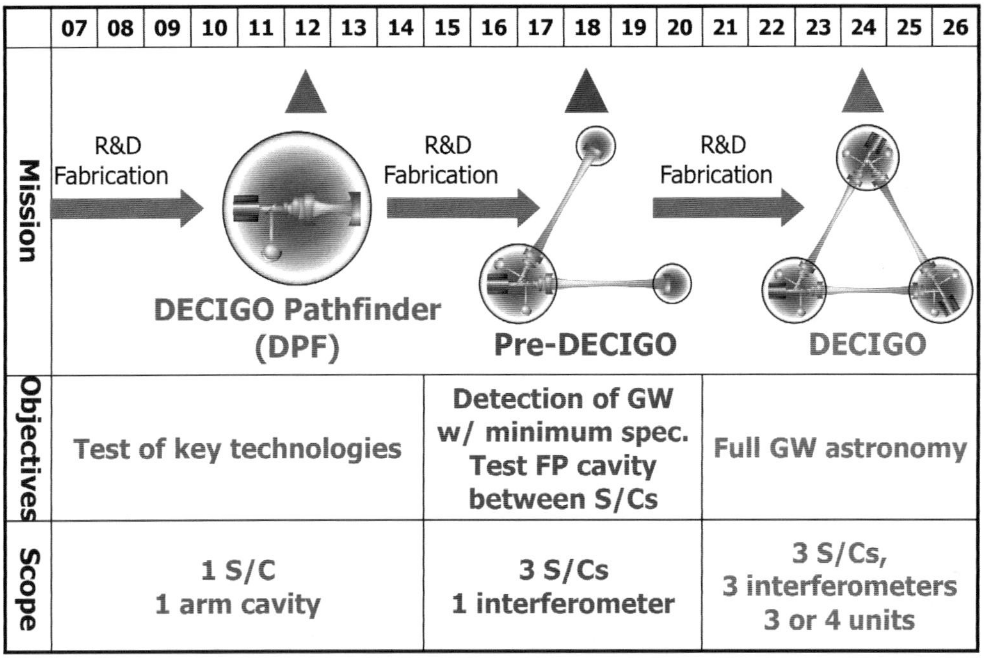

Figure 5. Roadmap of DECIGO. Pre-conceptual design of Pre-DECIGO is shown in addition to DPF and DECIGO.

in the center of galaxies.

- **Verification and characterization of inflation:** With correlation analysis of the data from the two clusters of DECIGO at nearly same location for three years, DECIGO will capable to detect stochastic background gravitational waves corresponding to $\Omega_{\rm GW} = 2 \times 10^{-16}$. According to the standard inflation model, it is expected that we could detect gravitational waves produced at the inflation period of the universe with DECIGO. This could be an extremely significant science driver for DECIGO because gravitational waves are the only means which make it possible to directly observe the inflation of the universe.

4. Roadmap

DECIGO is expected to be launched in the middle of 2020s (Fig. 5), before that, we plan to launch two precursor satellites: DECIGO pathfinder (DPF) [?] and pre-DECIGO(See Fig. 5). Major objective of these missions is a demonstration of key technologies for DECIGO just as LISA pathfinder [12] does for LISA, in addition, we also hope we can extract some scientific achievements with limited equipments allowed for these satellites in phases.

DPF tests the key technologies for DECIGO such as drag-free control of the spacecraft, stabilized laser system in space, precision laser metrology in space and test mass lock mechanism. At the same time, as DPF will have gentle sensitivity to the gravitational waves, it is expected that DPF will put some upper limit to the gravitational waves from the sources around center of our galaxy.

The technical objectives of Pre-DECIGO are demonstration of accurate formation flying, precision laser metrology with long baseline FP cavity and drag-free control for multiple spacecraft. Pre-DECIGO will have 100 km-scale FP cavity, therefore, it is supposed to have reasonable sensitivity to detect gravitational waves with minimum specifications. We hope that it will be launched around 2018.

Finally DECIGO is supposed to be launched around 2024 to open a new window of observation for gravitational wave astronomy.

5. DECIGO Pathfinder

DPF [10, 11] will employ a small-sized drag-free spacecraft that contains two freely falling proof masses, whose relative displacement is measured with a Fabry-Perot interferometer. A short Fabry-Perot cavity with finesse of 100 is illuminated by the frequency-stabilized Nd:YAG laser light yielding output powe rof 100 mW. The proof masses are clamped tightly for the launch and released gently in orbit. DPF is supposed to be delivered in the geocentric sun-synchronous orbit with an altitude of 500km. DPF will have strain sensitivity of $\sim 10^{-15}$ around the frequency band of 0.1-1 Hz.

The primary objective of DPF is to demonstrate key technologies for DECIGO such as drag-free control system, FP cavity precision metrology system in orbit, frequency-stabilized laser in orbit, and the clamp release mechanism. In addition, the scientific objective of DPF is to detect rather unlikely events of intermediate-mass black hole ($10^3 - 10^4 M_{\rm sol}$) inspirals in our galaxy; it is possible to detect such events with the aimed sensitivity of DPF.

Recently, DPF was identified as one of the candidate missions for the small satellite mission series which had been initiated by the Japanese space agency, JAXA/ISAS. This program is to launch at least 3 small satellites in upcoming 5 years using standard bus systems, whose scope is to reduce the cost of missions significantly compared with the conventional missions, and thus to increase a chance to go to space for a variety of fields. DPF is now selected as one of the potential mission candidates for the second or third missions, so DPF will be launched in 2012 (second mission) in the best and earliest case.

6. Conclusions

The future Japanese space gravitational wave antenna, DECIGO, is expected to detect gravitational waves from various kinds of sources and thus to open a new window of observation for gravitational wave astronomy. We have started serious R&D for DPF as one of the candidate missions for the small-spacecraft mission series to demonstrate the technologies required to realize DECIGO.

References

[1] Seto N, Kawamura S and Nakamura T 2001 Possibility of direct measurement of the acceleration of the universe using 0.1 Hz band laser interferometer gravitational wave antenna in space *Phys. Rev. Lett.* **87** 221103
[2] Kawamura S et al 2006 The Japanese Space Gravitational Wave Antenna - DECIGO *Class. Quantum Grav.* **23** S125
[3] Kawamura S et al 2008 The Japanese Space Gravitational Wave Antenna - DECIGO *Journ. of Phys.: Conf. Ser.* **120** 032004
[4] Kawamura S et al 2008 The Japanese Space Gravitational Wave Antenna; DECIGO *Journ. of Phys.: Conf. Ser.* **122** 012006
[5] LISA: System and Technology Study Report, ESA document ESA-SCI (2000)
[6] Kuroda K et al 2002 Japanese large-scale interferometers *Class. Quantum Grav.* **19** 1237
[7] Farmer A J and Phinney E S 2003 The gravitational wave background from cosmological compact binaries *Mon. Not. R. Astron. Soc.* **346** 1197
[8] Cutler C and Harms J 2006 Big Bang Observer and the neutron-star-binary subtraction problem *Phys. Rev. D* **73** 042001
[9] Schutz B F 1986 Determining the Hubble constant from gravitational wave observations *Nature* **323** 310
[10] Ando M et al 2008 DECIGO pathfinder *Journ. of Phys.: Conf. Ser.* **120** 032005
[11] Ando M et al 2008 DECIGO pathfinder *in this volume*
[12] Anza S et al 2005 The LTP experiment on the LISA Pathfinder mission Class. Quantum Grav. 22 S125-S138

Journal of Physics: Conference Series **154** (2009) 012041 doi:10.1088/1742-6596/154/1/012041

Constraining effective quantum gravity with LISA

Nicolás Yunes[1,2] and Lee Samuel Finn[1]

[1] Department of Physics, The Pennsylvania State University, University Park, PA 16802, USA.
[2] Department of Physics, Princeton University, Princeton, NJ 08544, USA.

E-mail: LSFinn@psu.edu, nyunes@princeton.edu

Abstract.
All modern routes leading to a quantum theory of gravity – i.e., perturbative quantum gravitational one-loop exact correction to the global chiral current in the standard model, string theory, and perhaps even loop quantum gravity – require supplementing the Einstein-Hilbert action with a parity-violating Chern-Simons term. Such a term leads to amplitude-birefringent gravitational wave propagation: i.e., one (circular) polarization state amplified with propagation while the other is attenuated. The proposed Laser Interferometer Space Antenna (LISA) is capable of observing gravitational wave sources at cosmological distances, suggesting the possibility that LISA observations may place a strong bound on this manifestation of quantum gravity. Here we report on a calculation of the effect that spacetime amplitude birefringence has on the signal LISA is capable of observing from inspiraling supermassive black hole binaries at large redshift. We find that the birefringence manifests itself in the observations as an anomalous precession of the binary's orbital angular momentum as it evolves toward coalescence, whose magnitude depends on the integrated history of the Chern-Simons coupling over the worldline of radiation wavefront. We estimate that LISA could place bounds on Chern-Simons modified gravity that are several orders of magnitude stronger than the present Solar System constraints, thus providing a probe of the quantum structure of spacetime.

1. Motivation

Gravitational interactions described in General Relativity (GR) are symmetric under the discrete transformations of charge conjugation, parity- and time-reversal. All current routes to a quantum theory of gravity require the introduction of a parity-violating term to the Einstein-Hilbert action, which will only be apparent in the theory's gravitational sector. Experimental confirmation of gravitational symmetry breaking can serve as a guide to understand how the marriage of GR and quantum mechanics will take place. The observational details may even provide reason to choose among different approaches: e.g., different types of string theory or loop quantum gravity. Here we report on detailed calculations [1] of the effect this parity-violating correction has on gravitational wave propagation and how those effects manifest themselves in observations that the proposed Laser Interferometer Space Antenna (LISA) [2, 3] will make possible.

All modern approaches to a quantum gravitational theory of gravity lead to triangle-like anomalies, which must be canceled for mathematical consistency [4]. Cancellation of these anomalies leads to a Chern-Simons (CS) parity-violating contribution to the Einstein-Hilbert action. In four-dimensional compactifications of both perturbative and non-perturbative string theory, the inclusion of such a CS term is inescapable to cancel the Green-Schwarz anomaly and to satisfy duality symmetries in the presence of Ramond-Ramond scalars [6, 7]. Parity-violation may also be present in loop quantum gravity if one requires invariance under large gauge transformations of the Ashtekar variables [5], but the details of such a mechanism have not yet been worked out in detail. The effective quantum gravitational model that consists of the addition of such a CS term to the action goes by the name of CS modified gravity.

Gravitational experiments are the only route to test the CS extension, since this effective theory does not couple to photons. Solar system ephemerides can be used to place weak constraints on the magnitude of the CS coupling [8, 9, 10], but gravitational wave observations have the potential to place much stronger bounds. This is because the CS parity-violation affects GW propagation by inducing an *amplitude birefringence* [11, 7, 12]: i.e. a polarization and wavelength-dependent amplification or attenuation of the GW amplitude, whose amplitude increases with distance propagated. The proposed Laser Interferometer Space Antenna [2, 3] — a space-based gravitational wave detector — is capable of observing inspiraling supermassive black holes as deep as redshift 30 and over a period long enough that the radiation wavelength evolves over two orders of magnitude. The combination of propagation over cosmological distances and change of polarization amplitudes with signal wavelength offers the possibility of placing a strong bound on the magnitude of the CS coupling constant [1].

This paper is organized as follows: Section 2 describes the basics of CS modified gravity and its effect on GW propagation; Section 3 explains the observational implications of such an effect for GWs emitted by binary systems in the inspiral phase; Section 4 presents a rough approximation of the bound that LISA could place on the CS coupling parameter given a GW detection; Section 5 concludes and points to future research. In the remainder of this paper, we shall use the conventions of Misner, Thorne and Wheeler [13], described in more detail in [1].

2. Chern-Simons modified gravity

The CS modification consists of enhancing the Einstein-Hilbert action through the addition of the product of a Pontryagin density term and a CS coupling function $\theta(x^\alpha)$ [11]. This effective theory yields the modified field equations

$$G_{\alpha\beta} + C_{\alpha\beta} = 8\pi T_{\alpha\beta}, \tag{1}$$

where G_{ab} is the Einstein tensor, T_{ab} is the stress-energy tensor of the matter fields and C_{ab} is a Cotton-like tensor, given by [11]

$$C^{\alpha\beta} = -\left[\theta_{,\epsilon}\epsilon^{\epsilon\gamma\delta(\alpha}\nabla_\gamma R^{\beta)}{}_\delta - \nabla_\delta\theta_{,\gamma}\,\epsilon^{\mu\nu\gamma(\alpha}R^{\beta)\delta}{}_{\mu\nu}\right], \tag{2}$$

θ is a function of spacetime coordinates, $\epsilon^{\alpha\beta\gamma\delta}$ is the Levi-Civita tensor density, and $R_{\mu\nu}$ is the Ricci tensor. Parenthesis in the superscript stand for symmetrization, commas in a subscript stand for partial differentiation and the ∇_μ operator stands for covariant differentiation.

For gravitational wave perturbations propagating in a Friedmann-Robertson-Walker (FRW) background within CS modified gravity, the line element may be written as

$$ds^2 = a^2(\eta)\left[-d\eta^2 + (\delta_{ij} + h_{ij})\,d\chi^i d\chi^j\right] \tag{3}$$

where η is conformal time, χ^i are comoving spatial coordinates, δ_{ij} is the Euclidean metric, $a(\eta)$ is the scale factor, and h_{ij} is a transverse-traceless (TT) metric perturbation. Focusing attention on a plane-wave solutions in co-moving coordinates we introduce

$$h_{lm}(\eta, \chi^j) = \frac{A_{lm}}{a(\eta)}e^{-i\left[\phi(\eta) - \kappa n_k \chi^k\right]}, \tag{4}$$

where the amplitude $\mathcal{A}_{lm}$, the unit vector in the direction of wave propagation n_k and the conformal wavenumber $\kappa > 0$ are all constant. Resolving the amplitude into left- and right-handed polarizations simplifies the dispersion relation for the gravitational wave phase, which is found to satisfy

$$i\phi''_{\rm R,L} + \left(\phi'_{\rm R,L}\right)^2 + \mathcal{H}' + \mathcal{H}^2 - \kappa^2 = \frac{i\lambda_{\rm R,L}\left(\theta'' - 2\mathcal{H}\theta'\right)\left(\phi' - i\mathcal{H}\right)\kappa/a^2}{\left(1 - \lambda_{\rm R,L}\kappa\theta'/a^2\right)} \tag{5}$$

where primes denote differentiation with respect to conformal time, $\mathcal{H} = a'/a$ is the conformal Hubble parameter and $\lambda_{\rm R,L} = +1$ (or -1) for right- (or left-) polarized GWs.

The evolution equation for the phase clearly indicates the characteristic signature of CS modified gravity. The left-hand side of Eq. (5) is the familiar GR dispersion relation for wave propagation in FRW spacetimes, while the entirety of the CS correction appears on the right-hand side. This correction is a purely imaginary source term whose sign depends on the radiation wave polarization state and whose magnitude depends on the wavenumber and in particular on derivatives of the CS scalar. A critical aspect of this correction is that it will lead to a phase correction that depends on the entire integrated history of the CS coupling.

Treating the CS contribution to the dispersion relation as small, we can solve Eq. (5) perturbatively. Assuming that θ' and θ' evolve on cosmological timescales, and that the present-day values of θ''_0/a_0^2 and $\mathcal{H}_0\theta'_0/a_0^2$ are small, the accumulated CS correction to the plane wave phase in propagating from conformal time η_i to the present is

$$\Delta\phi_{1({\rm R,L})} = i\lambda_{\rm R,L}\frac{k_0}{H_0}\xi(z), \tag{6}$$

where we have defined the redshift-dependent CS form-factor $\xi(z)$ via

$$\xi(z) = \frac{H_0^2}{2}\int_0^z dz\,(1+z)^{5/2}\left[\frac{7}{2}\frac{d\theta}{dz} + (1+z)\frac{d^2\theta}{dz^2}\right], \tag{7}$$

and where $k_0 = a_0\kappa_0$ is very much greater than the Hubble constant $H_0 = \dot{a}/a$, with overhead dots standing for time differentiation and z for redshift.

The correction to the phase is purely imaginary with a sign that depends on the circular polarization state of the waves. Thus, one circular polarization state will be amplified and the other attenuated. The correction to the phase also depends linearly on the wavenumber – it is stronger for shorter wavelengths than for longer wavelengths. Finally, through Eq. (7), its magnitude also depends on the integrated history of the CS coupling constant between the redshift of the source and the present. In the next section we will take advantage of all these effects to identify the signature of CS parity violation in the gravitational wave signal from inspiraling supermassive black hole binaries as observed by LISA.

3. Observational implications

The proposed LISA gravitational wave detector will be sensitive to gravitational waves in the $(3 \times 10^{-5}, 100)$ mHz band, with peak sensitivity in the $(1, 10)$ mHz band. Radiation from inspiraling black hole binaries with components of mass on order $10^6\,{\rm M}_\odot$ will radiate in this band during the final year before coalescence. The radiation itself will be nearly monochromatic, with its instantaneous radiation frequency increasing adiabatically over two orders of magnitude over this last year. As designed, LISA will be capable of detecting sources like these at redshifts up to approximately 30.

Black hole binaries in the early inspiral phase will radiate gravitational waves which in CS gravity become

$$h_{\rm R,L} = h^{GR}_{\rm R,L}\exp\left[\lambda_{\rm R,L}\frac{k_0}{H_0}\xi(z)\right] \sim h^{GR}_{\rm R,L}\left[1 + \lambda_{\rm R,L}\frac{k_0(t)}{H_0}\xi(z)\right] \tag{8}$$

Journal of Physics: Conference Series **154** (2009) 012041 doi:10.1088/1742-6596/154/1/012041

where $h^{GR}_{\mathrm{R,L}}$ is the GR prediction for the right/left-polarized gravitational waveform. One can clearly see that the CS correction exponentially enhances or suppresses the gravitational wave amplitude depending on the polarization of the signal. Moreover, since the correction depends linearly on wavenumber, it also depends on time in a polynomial fashion.

Focus attention on the ratio of the radiation amplitudes in the right and left circular polarization states:

$$\frac{h_{\mathrm{R}}}{h_{\mathrm{L}}} = \frac{h^{GR}_{\mathrm{R}}}{h^{GR}_{\mathrm{L}}} \exp\left[\frac{2k_0(t)\xi(z)}{H_0}\right] = \frac{1+x}{1-x} \tag{9}$$

where $k_0(t)$ is the observed time-dependent radiation wavenumber (i.e., $f_0(t)/2\pi c$). In the absence of the CS correction the exponential term in ξ is unity and the ratio defines the quantity x, which is, in fact, equal to $\cos\iota$, where ι is the direction angle between the line-of-sight to the binary system and the binary system's orbital angular momentum.

In the presence of the CS correction the relationship between the observable x, defined by equation (10), and the actual inclination angle ι becomes

$$x = \frac{\sinh\left(\frac{k_0(t)\xi(z)}{H_0}\right) + \cosh\left(\frac{k_0(t)\xi(z)}{H_0}\right)\hat{\chi}}{\cosh\left(\frac{k_0(t)\xi(z)}{H_0}\right) + \sinh\left(\frac{k_0(t)\xi(z)}{H_0}\right)\hat{\chi}} \sim \cos\iota + \frac{k_0(t)\xi(z)}{H_0}\sin^2\iota + \mathcal{O}\left(\xi^2\right). \tag{10}$$

If the binary components have significant spins then spin-orbit coupling will lead to a time-dependence of $\cos\iota$, and thus x. The effect of the CS coupling is to introduce *apparent* evolution of $\cos\iota$, which tracks the radiation wavelength in a well-determined way. The time-dependence of this contribution is unique to the effects of the CS contribution, and thus, distinguishable from the physical precession of the binary system. In the next subsection we estimate the magnitude of the CS correction that LISA may be able to observe.

4. Constraining quantum gravity with LISA

Given a GW detection with LISA from a supermassive black hole binary at large redshift, how well can we measure the CS correction? The answer to this question may be obtained by performing a full-co-variance matrix analysis that accounts for all eighteen parameters that characterize the binary system: eleven intrinsic ones (two component masses, two component spins and their orientation, the orbital phase, reference time when the phase, spins and eccentricity are measured), six extrinsic ones (two angles that define the orbital plane orientation, the source distance and angular location, the inclination angle), and one CS-related parameter ξ. This analysis would have to account for higher harmonics that arise due to higher-order contributions to the GW amplitude [14], as well as non-vanishing eccentricity. Performing such a study is of course a formidable task that has not yet been carried out.

We have made a series of plausible approximations to arrive at a crude estimate of the bound that LISA observations of an inspiraling binary black hole system could place on a CS correction to GR. First, we assume that we have two gravitational wave detectors, each of which observes a distinct polarization state of the radiation incident from the distant binary. Let the noise power spectral density in both detectors be white with magnitude S_0 across the entire, relevant band. Second, classify the eighteen parameters that characterize the binary into those that only determine the GW amplitude $\mu_i(t)$ (the distance, the inclination angle, and the CS parameter ξ), and those that affect only or principally the real part of the signal phase $\psi_i(t)$ (the orbital phase, the sky location, the binary period, and other parameters that describe the spins and angular momentum). Third, assume there is no correlation between these two sets of parameters, in which case the Fisher matrix becomes block-diagonal. In this case the inverse Fisher matrix is also block diagonal, with each block equal to the inverse of the corresponding block of the Fisher matrix. Finally, focus attention on the parameter block that includes the CS parameter ξ.

For concreteness, let us now assume that there has been a GW detection from a non-precessing binary black hole system with known component masses $M = 10^6 M_\odot (1+z)^{-1}$, seen plane on ($\hat{\chi}_0 = 0$) at a redshift of $z = 15$. Such a system will produce a GW that, during the final year of observation, sweeps through the LISA band from a radiation wavelength of $c(10^{-4}\mathrm{Hz})^{-1}$ to $c(10^{-2}\mathrm{Hz})^{-1}$. Setting aside the antenna pattern functions, our approximate Fisher analysis suggests that the accuracy to which we could measure the CS parameter is given by

$$\delta\xi = \sqrt{\left(\mathbf{\Gamma}^{-1}\right)_{\xi\xi}} = 1.7 \times 10^{-20} \left(\frac{\sqrt{S_0}}{10^{-20}\,\mathrm{Hz}^{-1/2}}\right) \left(1 + z - \sqrt{1+z}\right), \tag{11}$$

which corresponds to a "1-sigma" upper bound on ξ of order 10^{-19}. Such a result compares favorably, with Solar System experiments, which are orders of magnitude weaker and can constrain ξ (and thus θ) only locally.

Theory does not yet offer any real constraints on ξ: i.e., its magnitude depends strongly on the theoretical framework that prescribes the function θ. For example, in perturbative string theory, the theoretically expected magnitude of θ is on the order of the Planck scale. In the non-perturbative regime, however, Ramond-Ramond charges will source the CS correction, leading to a significantly larger value [17]. Even in perturbative frameworks, there are some models where ξ could be enhanced, such as for late-time vanishing string couplings [18, 19, 20, 21, 22, 23, 24, 25, 26, 27, 28]. An experimental measurement or bound on the magnitude of the CS coupling may thus select among alternative theoretical models for the quantum theory of gravity.

5. Conclusions

Parity violation is a defining characteristic of modern theoretical models for quantum gravity. One manifestation of this parity violation is a spacetime amplitude birefringence for gravitational (but not electromagnetic) wave propagation. LISA observations of the gravitational waves from cosmologically distant inspiraling black hole binary systems will show the effects of this spacetime birefringence as an apparent evolution of the orbital angular momentum of the binary relative to the detector line-of-sight. This evolution has a unique time dependence that allows it to be distinguished from physical precession of the binary, which may also lead to an evolving inclination angle. We estimate that LISA observations are sufficiently sensitive to bound the parameter space of modern, theoretical models for quantum gravity and, thus, probe the fundamental nature of the universe.

Acknowledgments

This manuscript is based on [1], which was written in collaboration with Stephon Alexander. We would also like to thank Ben Owen and Frans Pretorius for helpful discussions and comments. N. Y. acknowledges the support of the National Science Foundation (NSF) via grant PHY-0745779. LSF acknowledges the support of NSF awards PHY 06-53462 and PHY 05-55615, and NASA award NNG05GF71G. Lastly, we acknowledge the support of the Center for Gravitational Wave Physics, which is funded by the National Science Foundation under Cooperative Agreement PHY 01-14375.

References

[1] Alexander S, Finn L S and Yunes N 2008 *Phys. Rev. D* **78** 066005 (*Preprint* 0712.2542)

[2] Bender P *et al.* 1998 Laser Interferometer Space Antenna for the Detection and Observation of Gravitational Waves: An International Project in the Field of Fundamental Physics in Space LISA Pre-Phase A Report Max-Planck-Institut für Quantenoptik Garching mPQ 233

[3] Danzmann K and Rüdiger A 2003 *Class. Quantum Grav.* **20** S1–S9
[4] Alvarez-Gaume L and Witten E 1984 *Nucl. Phys.* **B234** 269
[5] Ashtekar A, Balachandran A P and Jo S 1989 *Int. J. Mod. Phys.* **A4** 1493
[6] Polchinski J Cambridge, UK: Univ. Pr. (1998) 531 p
[7] Alexander S H S and Gates S James J 2006 *JCAP* **0606** 018 (*Preprint* hep-th/0409014)
[8] Alexander S and Yunes N 2007 *Phys. Rev. Lett.* **99** 241101 (*Preprint* hep-th/0703265)
[9] Alexander S and Yunes N 2007 *Phys. Rev.* **D75** 124022 (*Preprint* 0704.0299)
[10] Smith T L, Erickcek A L, Caldwell R R and Kamionkowski M 2008 *Phys. Rev.* **D77** 024015 (*Preprint* 0708.0001)
[11] Jackiw R and Pi S Y 2003 *Phys. Rev.* **D68** 104012 (*Preprint* gr-qc/0308071)
[12] Alexander S and Martin J 2005 *Phys. Rev.* **D71** 063526 (*Preprint* hep-th/0410230)
[13] Misner C W, Thorne K S and Wheeler J A 1973 *Gravitation* (San Francisco: W. H. Freeman & Co.)
[14] Arun K G, Iyer B R, Sathyaprakash B S, Sinha S and Broeck C V D 2007 *Phys. Rev.* **D76** 104016 (*Preprint* 0707.3920)
[15] Finn L S 1992 *Phys. Rev. D* **46** 5236–5249
[16] Finn L S 1993 (*Preprint* gr-qc/9403021)
[17] Svrcek P and Witten E 2006 *JHEP* **06** 051 (*Preprint* hep-th/0605206)
[18] Brandenberger R H and Vafa C 1989 *Nucl. Phys.* **B316** 391
[19] Tseytlin A A and Vafa C 1992 *Nucl. Phys.* **B372** 443–466 (*Preprint* hep-th/9109048)
[20] Nayeri A, Brandenberger R H and Vafa C 2006 *Phys. Rev. Lett.* **97** 021302 (*Preprint* hep-th/0511140)
[21] Sun C Y and Zhang D H 2006 (*Preprint* hep-th/0611101)
[22] Wesley D H, Steinhardt P J and Turok N 2005 *Phys. Rev.* **D72** 063513 (*Preprint* hep-th/0502108)
[23] Alexander S, Brandenberger R H and Easson D 2000 *Phys. Rev.* **D62** 103509 (*Preprint* hep-th/0005212)
[24] Brandenberger R, Easson D A and Kimberly D 2002 *Nucl. Phys.* **B623** 421–436 (*Preprint* hep-th/0109165)
[25] Battefeld T and Watson S 2006 *Rev. Mod. Phys.* **78** 435–454 (*Preprint* hep-th/0510022)
[26] Brandenberger R H, Nayeri A, Patil S P and Vafa C 2006 (*Preprint* hep-th/0608121)
[27] Brandenberger R 2007 (*Preprint* hep-th/0702001)
[28] Brax P, van de Bruck C and Davis A C 2004 *Rept. Prog. Phys.* **67** 2183–2232 (*Preprint* hep-th/0404011)

Journal of Physics: Conference Series **154** (2009) 012042 doi:10.1088/1742-6596/154/1/012042

Perspectives for Testing Quantum Aspects of Gravity using LISA

O D Aguiar[1], K H C Castello-Branco[2], O D Miranda[1], J C N de Araujo[1] and E Abdalla[3]

[1]Instituto Nacional de Pesquisas Espaciais, São José dos Campos, Brazil

[2]Universidade de São Paulo, São Carlos, SP, Brazil

[3]Universidade de São Paulo, São Paulo, SP, Brazil

odylio@das.inpe.br, karlucio@ursa.ifsc.usp.br, oswaldo@das.inpe.br, jcarlos@das.inpe.br, and eabdalla@fma.if.usp.br

Abstract. LISA should be able to detect the gravitational waves from the QNM ringdown of supermassive black holes in the 10^5 - 10^8 solar mass range. On the other hand, it is reasonable to think that any quantum theory of gravitation should impose the quantization of the energy levels of these QNM. Here we discuss the possibility of distinguishing quantum aspects of gravity using LISA to observe QNM overtones of highly excited supermassive black holes.

PACS numbers: 04.80.Nn, 95.55.Ym

1. Introduction

Gravitational waves can not escape from within the event horizon of a black hole. However, outgoing waves which originate at the spacetime outside the horizon are possible and are the so called quasi-normal modes (QNMs). There is an extensive amount of literature on QNMs [1, 2, 3, 4, 5, 6], which include good reviews [7, 8].

QNMs of perturbed black black holes are the characteristic damped oscillations which display the signatures of these objects. It is expected that they will provide the definitive observational evidence on the existence of black holes. In fact, since QNMs depends uniquely on black hole's mass and angular momentum, they will give us the possibility to determine these parameters by performing black hole gravitational wave spectroscopy. Besides, it has been speculated that QNMs might provide not only the classical signature of black holes but also give some information about their quantum aspects [9]. This opens an interesting possibility to testing quantum aspects of gravity by observing QNMs from existing astrophysical black holes. LISA and other laser interferometer antenna projects in space, such as BBO and DECIGO, seem to be the best choices for observing gravitational waves coming from these QNMs, because the expected signal-to-noise ratio can be enormous in many possible scenarios.

Based on Bohr's correspondence principle, it is possible to think of the high overtones of the ringing frequencies of a black hole as being equally spaced [10]. On the other hand the lowest QNM frequency might be related to the mean irreducible mass associated with the quantum ergosphere [11].

Therefore the quantum aspects of gravity might be imprinted either on the asymptotic behavior of the high overtones (on their relative frequency separation) or on the wavelength of the fundamental ones (their absolute frequencies), or on both.

QNMs excited by coalescence of supermassive black holes (SMBHs) from galaxy mergers or from newly formed SMBHs, among other astrophysical events, are very important sources for LISA [12].

LISA should be able to detect the gravitational waves from the QNM ringdown of supermassive black holes in the 10^5 - 10^8 solar mass range throughout the observable Universe [13], as well as having the potential to perform no-hair tests [14].

Here we analyze the perspectives for testing quantum aspects of gravity using LISA.

2. Quasinormal Modes of Supermassive BHs and Possible Astrophysical Scenarios of Excitation

From theoretical calculations it appears that there is a high probability that most energy is emitted in the fundamental mode of the quadrupole (l=2) [15], leaving the other overtones of the quadrupole and the other multipoles with a much smaller portion of the total energy emitted. The physical explanation of this might be related to the time scale of the free falling ("plunge") of bodies from the innermost stable circular orbit (~ $6GM/c^3$) into a BH and the last stable orbit period (~ $2\pi \cdot 6GM/c^3$), which are both around the period of the fundamental mode of the quadrupole (~ $2.7 \cdot 6GM/c^3$). In other words, the time scales of most of the physical processes that excite the BH are closely tuned to the fundamental (n=0, l=2) QNM. This makes the amount of energy that goes to the other modes very sensitive to the details of the astrophysical event that produces excitation. For a "point test particle" of mass m falling radially into a Schwarzschild black hole of mass $M >> m$, Davis et al. 1971 found the distribution of the total energy to the multipoles l=2, l=3, l=4, l=5, l=6 to be about 0.879, 0.105, 0.0134, 0.00191, 0.000268, respectively.

Expressions for the frequencies of QNMs with very large imaginary parts can be found in Anderson 93 [16], Nollert 93 [17], Liu 95 [18], Nollert 99 [7], Kokkotas 99 [8], and Abdalla 07 [19].

Two important ingredients that should be taken into account are the rotation of BHs [20, 21] and the accretion of matter (dust shells and thick accretion disks) onto the black hole [22, 23, 24, 25].

We should analyze how they will affect the precision of the QNM frequency measurement and, therefore, on our capability of testing quantum aspects of gravity with these measurements.

QNMs of SMBHs can be strongly excited and emit gravitational waves under a finite number of possibilities. We may consider the following:
- the formation of a SMBH from a protostar;
- the coalescence of two smaller SMBHs;
- the capture of stars by a SMBH;
- the falling of large amounts of baryonic matter (rocks, dust, and gas) into the SMBH.

It is possible that there are other exotic possibilities (maybe involving cosmic strings or dark matter), but we will restrict our analysis to the above set.

After a quick inspection of the above list, we can conclude that almost all major possible sources of excitation of the QNMs of SMBHs have time scales related to the free falling time ("plunge") from the innermost stable circular orbit (from ~ 3 RSch = RISCO) and the last stable orbit periods. The exception might be the formation of the SMBH from a supermassive very low metalicity protostar and the cases close to equal-mass SMBHs.

3. Analysis and Discussion Considering Future Observations from LISA

Using numerical relativity simulations of non-spinning binary black holes mergers Berti et al. 2006 and 2007 [13, 14] analyzed the problem of detecting ringdown waveforms and of estimating the source parameters, showing that LISA has the potential to perform no-hair tests of general relativity. They computed the expected signal-to-noise ratio for ringdown events, the relative parameter estimation accuracy, and the resolvability of different modes. They also discussed the extent to which uncertainties on physical parameters, such as the black hole spin and the energy emitted in each mode, will affect the ability of performing black hole spectroscopy. Ioka and Nakano 2007 [26] also studied the problem of higher perturbative order of QNMs in binary BH mergers. They found that the second-order QNMs (l=4) have frequencies twice those of the first-order ones (l=2) and the GW amplitude is up to ~10% of that of the first order one, in agreement with the previous findings of Davis et al. 1971 [15]. They also compared these characteristic GW amplitude curves (first-, second-, and with third-order) with the sensitivity curves of LISA and Ultimate DECIGO.

How feasible is it testing quantum aspects of gravity using LISA and others laser interferometer projects in space?

One really promising thing when we talk about these detectors is: they will measure gravitational waves from astrophysical events with huge signal to noise ratios, especially when they come from SMBHs. Some of these events will be seen at the border of the observable universe.

In order to perform our analysis we are going to choose one event with high signal to noise ratio and, likewise, with smaller errors for the determination of the astrophysical parameters such as rotation (spin). Our SMBH, therefore, should have its QNMs in the highest sensitivity band of LISA, namely from ~ 2 to 20 mHz, which is in agreement with the results found by Berti et al. 2006 [13]. Our SMBH will have 3.7 x 10^6 solar masses, which is the mass of the putative BH at the center of our Milky Way. For scenarios at high redshifts the total mass should go down from the 3.7 x 10^6 solar mass value by a factor of (z + 1).

In Figure 1 we plotted the characteristic GW amplitude (empirical) curves for the fundamental (n=0) and the seven first excited overtones plus the 70th excited overtone for the first-order (l=2, quadrupole) QNM of a Schawzschild BH for three possible astrophysical scenarios. The excited overtones 8th to 69th were omitted in order to avoid overloading the graph with curves. The LISA sensitivity curve plotted is for bursts and S/N ~ 5. This means that the fundamental and the first 70 overtones for l=2 are detectable with this chosen sensitivity threshold. A Kerr BH would have its QNMs shifted in frequency to the right, to higher frequencies.

We assumed for this calculation the parameters listed in Table 1. The total amplitude emitted in the form of GWs follows the known expression [27, 28, 29]:

$$h_{eff} = 2\times10^{-21}\left(\frac{\varepsilon}{0.01}\right)\left(\frac{d}{10Mpc}\right)^{-1}\left(\frac{\mu}{M_{solar}}\right)$$, where ε was taken as 0.01.

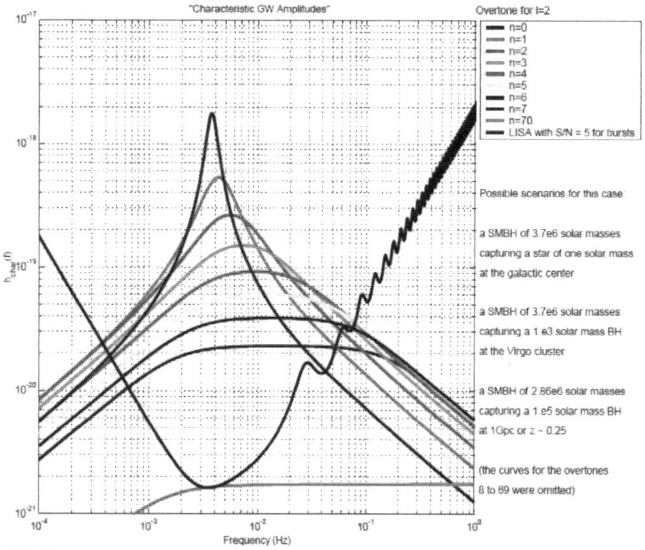

n	Re(σ)	Im(σ)	|σ|	M	f = ω/2π	Q
0	0.37367	-0.08896	0.38411	3.7E+06	3.70793 E-03	4.2004
1	0.34671	-0.27391	0.44185	3.7E+06	4.26530 E-03	1.2658
2	0.30105	-0.47828	0.56514	3.7E+06	5.45541 E-03	0.62944
3	0.25150	-0.70515	0.74866	3.7E+06	7.22695 E-03	0.35666
4	0.20751	-0.94684	0.96931	3.7E+06	9.35697 E-03	0.21916
5	0.16930	-1.19561	1.20754	3.7E+06	1.16566 E-02	0.14160
6	0.13325	-1.44791	1.45403	3.7E+06	1.40360 E-02	0.092029
7	0.09282	-1.70384	1.70637	3.7E+06	1.64719 E-02	0.054477
8	0.00000	-2.00324	2.00324	3.7E+06	1.93377 E-02	0.00000
9	0.06326	-2.30264	2.30351	3.7E+06	2.22363 E-02	0.027473
10	0.07655	-2.56083	2.56197	3.7E+06	2.47313 E-02	0.029893
15	0.08859	-3.82511	3.82614	3.7E+06	3.69345 E-02	0.023160
20	0.08739	-5.08147	5.08222	3.7E+06	4.90597 E-02	0.017198
25	0.08497	-6.33563	6.33620	3.7E+06	6.11646 E-02	0.013411
30	0.08260	-7.58868	7.58913	3.7E+06	7.32594 E-02	0.010885
40	0.08182	-10	10.16	3.7E+06	9.81097 E-02	0.0081
70	0.07259	-18	17.65	3.7E+06	1.70418 E-01	0.0041
120	0.06580	-30	30.15	3.7E+06	2.91017 E-01	0.0022
160	0.06285	-40	40.14	3.7E+06	3.87520 E-01	0.0016
200	0.06084	-50	50.14	3.7E+06	4.84033 E-01	0.0012
1000	0.05138	-250	250.13	3.7E+06	2.41458 E+00	0.000205
1001	0.05138	-250	250.38	3.7E+06	2.41700 E+00	0.000205

Figure 1. Characteristic GW amplitude (empirical) curves for the fundamental (n=0) and the seven first excited overtones plus the 70th excited overtone for the first-order (l=2, quadrupole) QNM of a Schawzschild BH for three possible astrophysical scenarios assuming general relativity as the underlying theory of gravity. We kept the mass falling into the SMBH at least 10 times smaller, otherwise the distribution of energy among the QNMs would be different from the one assumed. On the right a table gives the set of parameters for the l=2 QNM overtones of a 3.7 x 10^6 solar mass Schwarzschild black hole.

The partition of energies among the multimodes were assumed to follow the one proposed by Davis et al. 1971 [15], and the partition among the overtones were assumed to be proportional to the ratios $Q_n^2/\Sigma Q_n^2$, where $Q_n = \pi f_n t_n$ is the quality factor of the QNM overtone n, and ΣQ_n^2 is the sum of all (infinite) Q_n^2. This partition is assumed valid only in the case the astrophysical events that excite the QNMs have their Fourier peak below the fundamental mode (n=0) frequency. This is not the case of equal-mass coalescences, for example, but it is when $m_{captured} << M_{SMBH}$. The characteristic GW amplitude curves plotted in Figure 1 are empirical, but are in agreement with the shape of the curves found by Ioka and Nakano 2007 [26].

The results might be off from a rigorous calculation, but they are satisfactory for the point we wish to make. It is clear from this graph that the fundamental overtone, when emitted, masquerades the shape of the other overtones, making it difficult for one to determine their nominal frequencies and Qs. Even using special techniques as the ones mentioned by Berti et al. 2006 and 2007 [13, 14], it will almost be impossible to determine frequencies of very high overtones with the required precision for measuring the frequency spacing. The high overtones form a kind of single "flat" background signal. Perhaps only the fundamental and the five first excited overtones will be determined with any satisfactory precision.

The situation might be a little bit better in the cases of close to equal-mass coalescences. Medium overtones (n ~ 10) might be more excited compared to the ones with the partition of energy assumed above, but still this does not help the very high overtones (n > 100) to be measured with precision.

If one wants to find quantum aspects of gravity from SMBH QNMs measured using LISA, one has probably to find them in the absolute frequencies and spacing among the fundamental and first overtones.

It is puzzling to note that the angular frequency (ω) of the fundamental QNM for l=2 is only 20% off from the interesting expression:

$$\omega\,(l=2;n=0) \sim \frac{\hbar}{\pi L_P^2 M}$$

, where πL_P^2 is the area of a sphere with a diameter equal to the Planck length, and M is the mass of the star;

which can be rewritten as:

$$\omega\,(l=2;n=0) \sim \frac{M_P}{\pi T_P M}$$

, where M_P is the Planck mass is T_P is the Planck time.

4. Hawking Radiation

There is one curious coincidence connecting the Hawking radiation, SMBHs, and the laser interferometer space antennas such as LISA, BBO, and DECIGO. Even though it is strange to talk about black body radiation for "thermal" temperatures of 10^{-11} K to 10^{-15} K, these are the temperatures where SMBHs in the 10^4 – 10^8 solar mass range will peak in the sensitive band (10^{-4} - 1 Hz) of LISA, BBO and DECIGO for the Hawking gravitational radiation. The flux emitted in this band, however, is negligible and, so is the correspondent h. This is a pity, because the Hawking radiation on gravitational waves would be an interesting tool for probing quantum aspects of gravity. Only for very low mass black holes might the flux (of high frequency gravitational waves) be measurable some day.

5. Conclusion

Three astrophysical events involving Schawzschild BHs were chosen as examples that would produce the same set of characteristic GW amplitude curves for LISA, with high signal to noise ratio. The curves were calculated empirically, assuming a partition of energies among the multimodes and among the overtones of the quadrupolar mode and that general relativity is the underlying theory of gravity (could other theories give significantly different results?). The signal to noise ratio was strong enough to keep up to the 70th excited overtone within the reach of the burst sensitivity curve for LISA with S/N ~ 5.

From these curves it was apparent that the fundamental overtone, when emitted, masquerades the shape of the other overtones, making it difficult for one to determine their nominal frequencies and Qs. Even using special techniques it is unlikely that the frequencies of very high overtones can be determined with the required precision for measuring the frequency spacing.

From the measurements of SMBH QNM using LISA, it is likely that only the fundamental and first overtones will be available for one to look for quantum aspects of gravity. Perhaps the frequencies of the fundamental QNMs themselves are directly related to quantum principles.

6. Acknowledgments

We thank Kostas Kokkotas, Emanuele Berti, Curt Cutler, Stefano Vitale, and Stephen Merkowitz for useful information and interesting discussions. We also thank the referee for bringing some points to our attention. This work has been supported by FAPESP (grant No. 2006/56041-3) and MCT/INPE.

References

[1] Fröman, N *et al* 1992 *Phys. Rev.* **D45** 2609
[2] Andersson, N, Araujo, M E and Schutz, B F 1993 *Class. Quantum Grav.* **10** 735
[3] Andersson, N, Araujo, M E and Schutz, B F 1993 *Class. Quantum Grav.* **10** 757
[4] Andersson, N, Araujo, M E and Schutz, B F 1994 *Phys. Rev.* **D49** 2703
[5] Abdalla, E, Chirenti, C B M H, Saa, A 2007 *JHEP* 0710:086
[6] Abdalla, E, Chirenti, C B M H, Saa, A 2006 *Phys. Rev.* **D74** 084029
[7] Nollert, H-P 1999 *Class. Quantum Grav.* **16** R159
[8] Kokkotas, K "Quase-Normal Modes of Stars and Black Holes", arXiv:gr-qc/9909058
[9] Maggiore, M 2008 *Phys. Rev. Lett.* **100**, 141301
[10] Hod, S 1998 *Phys. Rev. Lett.* **81** 4293
[11] York Jr, J W 1983 *Phys. Rev.* **D28** 2929
[12] Aguiar, O D *et al* 1998 "Some Specific Sources in our Galaxy for the Laser Interferometer Space Antenna (LISA)", Proc. of the 2nd International Lisa Symposium on the Detection and Observation of Gravitational Waves in Space, Pasadena, CA, USA, Ed. William M. Folkner, AIP Conference Proceedings 456, Woodbury, NY
[13] Berti, E, Cardoso, V and Will, C M 2006 *Phys. Rev.* **D73**, 064030
[14] Berti, E, Cardoso, J, Cardoso, V and Cavaglià, M, 2007 *Phys. Rev.* **D76** 104044
[15] Davis, M *et al* 1971 *Phys. Rev. Lett.* **27** 1466
[16] Andersson, N 1993 *Class. Quantum Grav.* **10**, L61
[17] Nollert, H.-P 1993 *Phys. Rev.* **D47** 5253
[18] Liu, H 1995 *Class. Quantum Grav.* **12** 543
[19] Abdalla, E and Giugno, D 2007 *Brazilian J. Phys.* **37** 450
[20] Dorband, E N *et al* 2006 *Phys. Rev.* **D74** 084028
[21] Berti, E and Cardoso, V 2006 *Phys. Rev.* **D74** 104020
[22] Leung, P T *et al* 1997 *Phys. Rev. Lett.* **78** 2894
[23] Leung, P T *et al* 1999 *Phys. Rev.* **D59** 0440034
[24] Papadopoulos, P and Font, J 2001 *Phys. Rev.* **D63** 044016
[25] Nagar *et al* 2007 *Phys. Rev.* **D75** 044016
[26] Ioka, K and Nakano, H 2007 *Phys. Rev.* **D76** 061503(R)
[27] Thorne, K S 1987 "Gravitational Radiation", In: "300 years of gravitation", ed. Stephen Hawking & Werner Israel, Cambridge University Press, Cambridge, USA, 330
[28] Gültekin, K, Coleman, M and Hamilton, D P 2006 *Ap. J.* **640** 156
[29] Kokkotas, K, 2007 "GWs from Binary Systems", Lecture 4, Mini-Course "Generation mechanisms of gravitational waves", In: INPE Advanced School II - Compact Objects, http://www.das.inpe.br/school/2007/lectures.htm

Journal of Physics: Conference Series **154** (2009) 012043 doi:10.1088/1742-6596/154/1/012043

Gravitational-wave memory revisited: memory from the merger and recoil of binary black holes

Marc Favata

Kavli Institute for Theoretical Physics, University of California, Santa Barbara, CA 93106-4030

E-mail: `favata@kitp.ucsb.edu`

Abstract. Gravitational-wave memory refers to the permanent displacement of the test masses in an idealized (freely-falling) gravitational-wave interferometer. Inspiraling binaries produce a particularly interesting form of memory—the Christodoulou memory. Although it originates from nonlinear interactions at 2.5 post-Newtonian order, the Christodoulou memory affects the gravitational-wave amplitude at leading (Newtonian) order. Previous calculations have computed this non-oscillatory amplitude correction during the inspiral phase of binary coalescence. Using an "effective-one-body" description calibrated with the results of numerical relativity simulations, the evolution of the memory during the inspiral, merger, and ringdown phases, as well as the memory's final saturation value, are calculated. Using this model for the memory, the prospects for its detection are examined, particularly for supermassive black hole binary coalescences that LISA will detect with high signal-to-noise ratios. Coalescing binary black holes also experience center-of-mass recoil due to the anisotropic emission of gravitational radiation. These recoils can manifest themselves in the gravitational-wave signal in the form of a "linear" memory and a Doppler shift of the quasi-normal-mode frequencies. The prospects for observing these effects are also discussed.

1. Introduction and motivation

In the standard picture of the gravitational-wave (GW) signal from a coalescing binary, the oscillating amplitude starts small at early times, grows to some peak amplitude, and than decays back to its zero-value at late times. For example, the dominant ($l = m = 2$) mode of the GW polarizations from coalescing binary black holes (BBHs) follows this standard picture. However, some sources exhibit a gravitational-wave *memory* in which the values of the GW polarization amplitudes differ at late and early times:

$$\Delta h^{\rm mem}_{+,\times} = \lim_{t\to+\infty} h_{+,\times}(t) - \lim_{t\to-\infty} h_{+,\times}(t), \tag{1}$$

where t is the observer's time. In an ideal GW detector (eg., a ring of freely-falling test masses), the memory causes a permanent displacement that persists after the GW has passed.

Gravitational-wave memory comes in two types: The *linear memory* has been known since the 1970's ([1] and references therein) and arises from sources that produce a net change in the time-derivatives of their *source-multipole-moments.* A simple example of a source with linear memory is a binary on a hyperbolic orbit (gravitational two-body scattering) [2]. To see how memory arises in this system, consider the leading-order transverse-traceless (TT) GW field,

$h_{jk}^{\rm TT} \propto \ddot{\mathcal{I}}_{jk}^{\rm TT}/R$, where R is the distance to the observer, $\mathcal{I}_{jk} = \eta M[x_i x_j]^{\rm STF}$ is the source quadrupole moment, M is the total mass, η is the reduced-mass ratio, x_j is the relative binary separation vector, and STF means to take the symmetric-trace-free part. Using the Newtonian equations of motion, one can easily see that at large separations (before and after scattering) $\ddot{x}_j \to 0$, but $\dot{x}_j$ is a constant for unbound systems and has a different direction at late and early times. This leads to a memory $\Delta h_{jk}^{\rm TT} \propto \eta M \Delta[\dot{x}_j \dot{x}_k]^{\rm TT}/R$.

Other systems with linear memory are those whose components change from being bound to unbound (or vice versa). This includes binaries whose members are captured, disrupted, or undergo mass loss. Gravitational-waves with linear memory have been studied in the context of supernova explosions (see [3] for a review and references), mass loss due to neutrino emission [4], and gamma-ray-burst jets [5]. A general formula for the linear memory is given by [6, 1]:

$$\Delta h_{jk}^{\rm TT} = \Delta \sum_{A=1}^{N} \frac{4M_A}{R\sqrt{1-v_A^2}} \left[\frac{v_A^j v_A^k}{1 - \boldsymbol{v}_A \cdot \boldsymbol{N}} \right]^{\rm TT}, \tag{2}$$

where A is an index that labels the N masses M_A with velocities $\boldsymbol{v}_A$ that are unbound in their initial or final states (or both), and $\boldsymbol{N}$ is a unit vector that points from the source to the observer.

The *nonlinear memory* was discovered independently by Payne [7], Blanchet and Damour [8], and Christodoulou [9]. It is often referred to as the "Christodoulou memory." The nonlinear memory arises from a change in the *radiative multipole moments* that is sourced by the energy-flux of the radiated GWs. One can heuristically understand the origin of the nonlinear memory as follows: Consider the relaxed Einstein field equations (EFE) in harmonic gauge: $\Box \bar{h}^{\alpha\beta} = -16\pi\tau^{\alpha\beta}$, where $\bar{h}^{\alpha\beta}$ is the gravitational field tensor and $\tau^{\alpha\beta}$ depends on the matter stress-energy tensor, the Landau-Lifshitz pseudotensor $t_{\rm LL}^{\alpha\beta}$, and other terms quadratic in $\bar{h}^{\alpha\beta}$ [10]. Of the many nonlinear terms in $t_{\rm LL}^{\alpha\beta}$, there is a particular piece that is proportional to the stress-energy tensor for GWs: $T_{jk}^{\rm gw} = \frac{1}{R^2}\frac{dE^{\rm gw}}{dt d\Omega} n_j n_k$, where $\frac{dE^{\rm gw}}{dt d\Omega}$ is the GW energy flux and n_j is a unit radial vector. When applying the standard Green's function to the right-hand-side of the relaxed EFE, this piece yields the following correction term to the GW field [11]:

$$\delta h_{jk}^{\rm TT} = \frac{4}{R} \int_{-\infty}^{T_R} dt' \left[\int \frac{dE^{\rm gw}}{dt' d\Omega'} \frac{n'_j n'_k}{(1 - \boldsymbol{n}' \cdot \boldsymbol{N})} d\Omega' \right]^{\rm TT}, \tag{3}$$

where T_R is the retarded time. This shows that part of the distant GW field is sourced by the loss of GW energy. Thorne [6] has shown that the nonlinear memory [Eq. (3)] can be described in terms of the linear memory [Eq. (2)] if the unbound objects in the system are taken to be the individual radiated gravitons with energies $E_A = M_A/(1 - v_A^2)^{1/2}$ and velocities $v_A^j = c\, n_A'^j$.

The Christodoulou memory is a particularly interesting manifestation of the nonlinearity of general relativity. Although it arises from multipolar interactions beginning at the 2.5-post-Newtonian (PN) order [12], the nonlinear memory affects the GW amplitude at leading (Newtonian) order. Like GW tails the memory depends on the entire past-history of the source; but unlike most other nonlinear PN effects, the memory is non-oscillatory and causes a slowly-growing shift in the + polarization[1].

Here we focus on the memory from merging BBHs. While the oscillatory pieces of the GW polarizations during the inspiral are known to 3PN order [13], the Christodoulou memory has, until recently, only been calculated to leading-(Newtonian)-order[2] [11, 12]. However a proper

[1] For circular orbits $h_\times$ is unaffected by the Christodoulou memory. However, during the inspiral there is a DC contribution to $h_\times$ at 2.5PN order from nonlinear corrections to the radiative current octupole moment [12].

[2] Blanchet et. al [13] have shown that the 0.5PN correction to the memory is exactly zero for quasi-circular orbits. The 3PN corrections to the memory pieces of the inspiral waveform are reported in Ref. [14].

determination of the memory's detectability requires not only knowing how it slowly accumulates during the inspiral, but also how the memory rapidly grows and saturates to its final value during the merger and ringdown. Earlier estimates of the memory's detectability have either made crude order-of-magnitude estimates [6] or considered only the memory during the inspiral [15].

In principle numerical relativity (NR) simulations can compute the memory at all stages of the coalescence. In practice extracting the memory from these simulations faces several difficulties: NR simulations can best resolve the dominant $l = m = 2$ spin-weighted spherical-harmonic mode of the Ψ_4 curvature scalar; but for circular orbits the memory is present only in the $m = 0$ modes, which are smaller by 5PN orders (or about four orders-of-magnitude) during the late inspiral. The memory is also sensitive to the two integration constants that must be determined when computing the metric perturbations from the curvature perturbations. For simulations that compute the metric perturbations directly, the $m = 0$ modes enter at the same PN order as the $l = m = 2$ mode, but are still numerically smaller by nearly three orders of magnitude. The sensitive dependence of the memory to the binary's past history means that large errors can result unless the simulations start with large initial binary separations. See Ref. [14] for further discussion. In the absence of results from NR, the purpose of this work is to estimate the evolution and saturation of the memory during the merger and ringdown phases.

2. Calculating the memory from binary black hole coalescence

The leading-order PN multipole moment expansion of the oscillatory and memory pieces of the GW polarizations can be expressed as [16]:

$$h_{+}^{(0)} - ih_{\times}^{(0)} \approx \frac{1}{8R}\sqrt{\frac{5}{2\pi}}\left[(1+\cos\Theta)^2 e^{2i\Phi} I_{22}^{(2)} + (1-\cos\Theta)^2 e^{-2i\Phi} I_{2-2}^{(2)}\right], \quad \text{and} \tag{4}$$

$$h_{+}^{\text{mem}} \approx \frac{\eta M h^{\text{mem}}}{384\pi R}\sin^2\Theta(17+\cos^2\Theta), \quad h^{\text{mem}} \equiv \frac{1}{\eta M}\int_{-\infty}^{T_R} |I_{22}^{(3)}(\tau)|^2 d\tau \approx \frac{16\pi}{\eta}\left(\frac{\Delta E_{\text{rad}}}{M}\right). \tag{5}$$

Here (Θ, Φ) are the direction to the observer and $I_{2\pm 2}^{(n)}$ is the nth-time-derivative of the spherical harmonic coefficient of the source mass-quadrupole moment[3].

To model the evolution of the source-quadrupole moment we restrict ourselves to non-spinning, circularized BBHs and follow the "effective-one-body" (EOB) approach (see [17] and references therein), calibrated to the results of NR simulations. Before using the full EOB formalism, it is instructive to first consider a bare-bones version of EOB called the *minimal waveform model* (MWM). The MWM is a simple, analytic model for the inspiral, merger, and ringdown that qualitatively captures some of the important physics while minimizing complexity. It consists of modeling the multipole moments during the inspiral by their leading-order PN expressions and then matching to a sum of quasi-normal modes (QNMs). During the inspiral the source quadrupole-moment derivatives are

$$I_{2\pm 2}^{(q),\text{insp}} = 2\sqrt{\frac{2\pi}{5}}\eta M r^2 (\mp 2i\omega)^q e^{\mp 2i\varphi}. \tag{6}$$

Here $\omega \equiv \dot{\varphi} = (M/r^3)^{1/2}$ is the orbital frequency, $r = r_m(1 - T/\tau_{\text{rr}})^{1/4}$ is the orbital separation, φ is the orbital phase, $\tau_{\text{rr}} = (5/256)(M/\eta)(r_m/M)^4$, $T = t - t_m$, and r_m is the orbital separation at the "matching time" t_m. For $t > t_m$ the quadrupole-moment derivatives are modeled as a sum of ringdown QNMs:

$$I_{2\pm 2}^{(2+p),\text{ring}} = \sum_{n=0}^{n_{\max}} (-\sigma_{22n})^p A_{22n} e^{-\sigma_{22n} T}, \tag{7}$$

[3] These "scalar" moments are related to the more familiar STF-source-quadrupole moment $\mathcal{I}_{ij}$ via $I_{2m} = (16\pi\sqrt{3}/15)\mathcal{I}_{ij}\mathcal{Y}_{ij}^{2m\,*}$, where the $\mathcal{Y}_{ij}^{2m}$ are related to the ordinary spherical harmonics by $Y^{2m} = \mathcal{Y}_{ij}^{2m} n^i n^j$ [10].

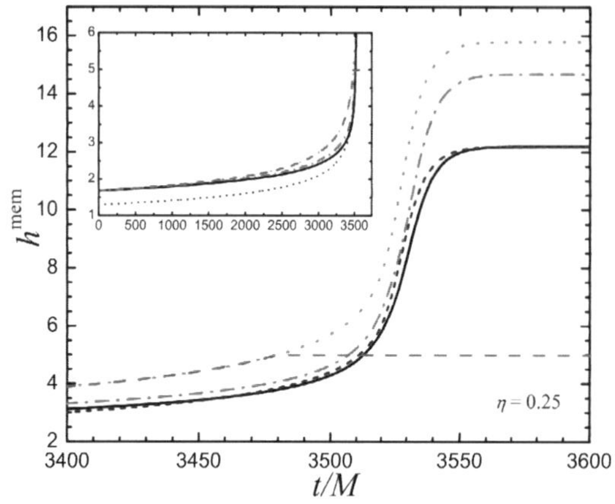

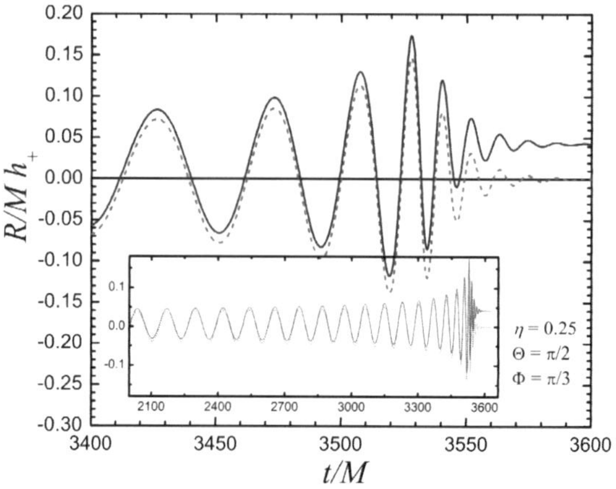

Figure 1. The left plot shows the evolution and saturation of the memory $h^{\rm mem}$ near the merger time. The solid (black) line uses the full-EOB formalism calibrated to numerical relativity simulations as in Ref. [20]. The dashed-dotted (green) line uses this same formalism but without the EOB amplitude correction factors $f_{22}^{\rm NQC} F_{22}$. The dotted (cyan) curve is the minimal-waveform model. The short-dashed (blue) curve is the minimal-waveform model multiplied by a "fudge" factor ≈ 0.77 so that it matches the full-EOB curve at late times. The red (dashed) curve is similar to the treatment in [15]: the inspiral memory is truncated at an orbital separation $r = 5M$. Note that the PN/EOB corrections tend to reduce the memory's magnitude. The right plot shows the h_+ polarization with memory (sold/blue) and without (dashed/red). This is computed using the full-EOB model. The inset plots show the early-time evolution. Both plots are for an equal-mass binary with the matching to the ringdown signal at $t_m/M \approx 3522$.

where $\sigma_{lmn} = i\omega_{lmn} + \tau_{lmn}^{-1}$, with QNM angular frequencies ω_{lmn} and damping times τ_{lmn} given in Ref. [18]. These QNMs depend on the final mass M_f and the dimensionless spin parameter a_f of the BH merger remnant and are determined by NR simulations [eg., the fits in Eqs. (C5)-(C6) of [19]]. The coefficients A_{22n} are determined by matching Eqs. (6)-(7) at $t = t_m$ for $2 \le (q, p+2) \le n_{\rm max} + 2$. Substituting these relations into $h^{\rm mem}$ yields a simple expression for the memory's evolution:

$$h^{\rm mem}_{\rm MWM} = \frac{8\pi M}{r(T)} H(-T) + \left\{ \frac{8\pi M}{r_m} + \frac{1}{\eta M} \sum_{n,n'=0}^{n_{\rm max}} \frac{\sigma_{22n}\sigma^*_{22n'}}{\sigma_{22n} + \sigma^*_{22n'}} A_{22n} A^*_{22n'} \left[1 - e^{-(\sigma_{22n} + \sigma^*_{22n'})T} \right] \right\} H(T), \tag{8}$$

where $H(T)$ is the Heaviside function. Choosing $n_{\rm max} = 2$ and $r_m = 3M$ (corresponding to the light-ring of a Schwarzschild BH), yields a final saturation value of the memory $\Delta h^{\rm mem}_{\rm MWM} \approx 16$.

In addition to the MWM, we can model the evolution of $I_{2\pm2}$ using the EOB approach of Ref. [20], where the freely adjustable EOB parameters are determined by fitting to simulations from the Jena and Caltech/Cornell NR groups. The main differences between the MWM and the full-EOB calculation are: (i) the EOB equations of motion are solved to determine the binary separation and orbital frequency during the inspiral, (ii) PN correction factors to the amplitude of $I_{2\pm2}^{(q),{\rm insp}}$ are included as in Ref. [20], and (iii) we match to 5 QNMs at 5 points near the EOB-deformed light-ring for $q = p + 2 = 2$. The results of the MWM and full-EOB model are presented in Figure 1. Further details are presented in Ref. [16].

These results can be used to compute the signal-to-noise ratio (SNR) for the memory signal [16]. For initial LIGO the memory will be undetectable, but advanced LIGO may have a slim chance of detecting the memory from nearby BBH mergers (SNR ≈ 8 for a $50M_\odot/50M_\odot$ binary at 20 Mpc). LISA has good prospects for detecting the memory from supermassive BBH mergers at large redshifts: for example a $10^5M_\odot/10^5M_\odot$ binary at redshift $z = 2$ will have a SNR ≈ 9.

3. Signature of radiation recoil in the gravitational-wave signal

Numerical relativity (NR) calculations of gravitational radiation recoil have found that BHs can receive kicks as large as ~ 175 km/s for nonspinning BHs [21], ~ 450 km/s for equal-mass BHs with spins anti-aligned along the orbital angular momentum [22], and ~ 4000km/s for the "super-kick" configuration (spins anti-aligned in the orbital plane) [23]. These moderate to large recoils can have interesting electromagnetic signatures [24, 25, 26, 27, 28, 29, 30, 31], and can also leave signatures in the GW signal. This GW signature could confirm the nature of the recoil (anisotropic GW emission vs. 3-body recoil) and determine the kick magnitude and direction.

During the inspiral the center-of-mass recoil is a negligible high-PN-order effect; but if the masses and spins are accurately determined from the inspiral waves, the final kick after the merger can be inferred from NR simulations. Following the merger the recoil grows large enough to leave potentially detectable effects on the GW signal. One of these effects is an additional contribution to the GW memory. This "kick memory" arises from Thorne's memory formula [Eq. (2)] for a system of N unbound masses:

$$\Delta h_{jk}^{\rm TT} = \sum_{A=1}^{N-1} \frac{4E_A^{\rm graviton}}{R} \left[\frac{n_A^j n_A^k}{1 - \boldsymbol{n}_A \cdot \boldsymbol{N}}\right]^{\rm TT} + \frac{4M_f}{R\sqrt{1-V_{\rm kick}^2}} \left[\frac{V_{\rm kick}^j V_{\rm kick}^k}{1 - \boldsymbol{V}_{\rm kick} \cdot \boldsymbol{N}}\right]^{\rm TT}. \tag{9}$$

Here the sum in the first term is over the $N-1$ individual gravitons radiated throughout the entire coalescence; this is the Christodoulou memory [$\delta h_{jk}^{\rm TT}$; Eq. (3)]. The second term is the memory ($\Delta h_{jk}^{\rm TT,kick}$) from the remaining particle in the system—the kicked BH with final mass M_f and recoil velocity $V_j^{\rm kick}$. Since the Christodoulou memory roughly scales with the radiated energy, $\Delta h^{\rm Chris.} \sim 4\Delta E_{\rm rad}/R$, we can see that the ratio of the two memories scales like

$$\frac{\Delta h^{\rm kick}}{\Delta h^{\rm Chris.}} \sim \frac{V_{\rm kick}^2}{\Delta E_{\rm rad}/M} \sim 3\times 10^{-3} \left(\frac{V^{\rm kick}}{3000\ {\rm km/s}}\right)^2 \left(\frac{3\%}{\Delta E_{\rm rad}/M}\right). \tag{10}$$

Since plausible SNRs for the Christodoulou memory are < 100, it seems unlikely that the "kick memory" will be easily detected.

Prospects are better for detecting a Doppler shift of the QNM oscillations. Measuring this Doppler shift relies on accurately measuring the masses and spins from the inspiral, and using NR simulations to determine the rest-frame QNMs. Comparison with the observed QNMs then determines the line-of-sight (LOS) recoil velocity, $V_{\rm LOS} = \boldsymbol{V}_{\rm kick} \cdot \boldsymbol{N}$. The accuracy with which $V_{\rm LOS}$ can be determined then depends on how accurately the QNMs can be measured. Using Eq. (7.2) of Ref. [18], we can roughly estimate the relative error in the LOS recoil: $\sigma_{V_{\rm LOS}}/V_{\rm LOS} \sim 2/(\rho V_{\rm LOS})$, where ρ is the SNR from the QNM signal. For large SNRs and moderately large kicks, this implies a relative error of a few tens of percent.

4. Conclusions

While previous works have computed the memory only during the inspiral, here the memory is computed through the merger and ringdown. The calculation has used a simple, fully-analytic model for the coalescence as well as a previously developed effective-one-body (EOB) approach calibrated to numerical simulations. Prospects for detecting the memory are poor for initial and advanced LIGO, but are promising for supermassive BH mergers that LISA will detect with high signal-to-noise ratios ($\gtrsim 1000$). Binaries that recoil after merger also show a linear memory, as well as a Doppler shift of the quasi-normal-mode frequencies. These effects will be difficult to detect except in strong sources with moderately large kicks.

Although challenging, it will be important for numerical relativity simulations to test the semi-analytic computations of the memory discussed here. Future Mock LISA Data Challenges [32] should also consider including waveforms with memory in their data sets. This will provide a better understanding of LISA's ability to measure this interesting effect.

Acknowledgments

This research was supported in part by the National Science Foundation under Grant No. PHY05-51164. I am grateful to several participants of the 7th LISA Symposium for helpful discussions concerning this work.

References

[1] Braginskii V B and Thorne K S 1987 *Nature (London)* **327** 123–125
[2] Turner M 1977 *Astrophys. J.* **216** 610–619
[3] Ott C D 2008 (*Preprint* arXiv:0809.0695[astro-ph])
[4] Epstein R 1978 *Astrophys. J.* **223** 1037–1045
[5] Sago N, Ioka K, Nakamura T and Yamazaki R 2004 *Phys. Rev. D* **70** 104012 (*Preprint* arXiv:gr-qc/0405067)
[6] Thorne K S 1992 *Phys. Rev. D* **45** 520–524
[7] Payne P N 1983 *Phys. Rev. D* **28** 1894–1897
[8] Blanchet L and Damour T 1992 *Phys. Rev. D* **46** 4304–4319
[9] Christodoulou D 1991 *Phys. Rev. Lett.* **67** 1486–1489
[10] Thorne K S 1980 *Rev. Mod. Phys.* **52** 299
[11] Wiseman A G and Will C M 1991 *Phys. Rev. D* **44** R2945
[12] Arun K G, Blanchet L, Iyer B R and Qusailah M S S 2004 *Class. Quantum Grav.* **21** 3771–3801 **22**, 3115 (2005) (*Preprint* arXiv:gr-qc/0404085v4)
[13] Blanchet L, Faye G, Iyer B R and Sinha S 2008 *Class. Quantum Grav.* **25** 165003 (*Preprint* arXiv:0802.1249v2[gr-qc])
[14] Favata M 2008 (*Preprint* arXiv:0812.0069v1[gr-qc])
[15] Kennefick D 1994 *Phys. Rev. D* **50** 3587–3595
[16] Favata M Nonlinear gravitational-wave memory from binary black hole mergers (in preparation)
[17] Damour T 2008 *International Journal of Modern Physics A* **23** 1130–1148 (*Preprint* arXiv:0802.4047v1[gr-qc])
[18] Berti E, Cardoso V and Will C M 2006 *Phys. Rev. D* **73** 064030 (*Preprint* arXiv:gr-qc/0512160)
[19] Baker J G, Boggs W D, Centrella J, Kelly B J, McWilliams S T and van Meter J R 2008 *Phys. Rev. D* **78** 044046 (*Preprint* arXiv:0805.1428v2[gr-qc])
[20] Damour T, Nagar A, Hannam M, Husa S and Brugmann B 2008 *Phys. Rev. D* **78** 044039 (*Preprint* arXiv:0803.3162v2[gr-qc])
[21] González J A, Sperhake U, Brügmann B, Hannam M and Husa S 2007 *Physical Review Letters* **98** 091101 (*Preprint* arXiv:gr-qc/0610154)
[22] Pollney D, Reisswig C, Rezzolla L, Szilágyi B, Ansorg M, Deris B, Diener P, Dorband E N, Koppitz M, Nagar A and Schnetter E 2007 *Phys. Rev. D* **76** 124002 (*Preprint* arXiv:0707.2559v1[gr-qc])
[23] Campanelli M, Lousto C O, Zlochower Y and Merritt D 2007 *Phys. Rev. Lett.* **98** 231102 (*Preprint* arXiv:gr-qc/0702133)
[24] Merritt D, Milosavljević M, Favata M, Hughes S A and Holz D E 2004 *Astrophys. J. Letters* **607** L9–L12 (*Preprint* arXiv:astro-ph/0402057)
[25] Loeb A 2007 *Physical Review Letters* **99** 041103 (*Preprint* arXiv:astro-ph/0703722)
[26] Volonteri M and Madau P 2008 *Astrophys. J. Letters* **687** L57–L60 (*Preprint* arXiv:0809.4007v1[astro-ph)
[27] Devecchi B, Dotti M, Rasia E, Volonteri M and Colpi M 2008 (*Preprint* arXiv:0805.2609v1[astro-ph])
[28] Mohayaee R, Colin J and Silk J 2008 *Astrophys. J. Letters* **674** L21–L24 (*Preprint* arXiv:0709.3321v4[astro-ph])
[29] Schnittman J D and Krolik J H 2008 *Astrophys. J.* **684** 835–844 (*Preprint* arXiv:0802.3556v2[astro-ph])
[30] Shields G A and Bonning E W 2008 *Astrophys. J.* **682** 758–766 (*Preprint* arXiv:0802.3873v1[astro-ph])
[31] Haiman Z, Kocsis B, Menou K, Lippai Z and Frei Z 2008 (*Preprint* arXiv:0811.1920v1[astro-ph])
[32] Arnaud K A, Babak S, Baker J G, Benacquista M J, Cornish N J, Cutler C, Finn L S, Larson S L, Littenberg T, Porter E K, Vallisneri M, Vecchio A, Vinet J Y and Data Challenge Task Force T M L 2007 *Class. Quantum Grav.* **24** 551 (*Preprint* arXiv:gr-qc/0701170)

Bayesian model comparison in gravitational wave data analysis

Massimo Tinto[1] **and Richard Umstätter**[1]

[1] Jet Propulsion Laboratory, California Institute of Technology, Pasadena, CA 91109

E-mail: `Massimo.Tinto@jpl.nasa.gov`

Abstract. We estimate the probability of detecting a gravitational wave signal from coalescing compact binaries in simulated data from a ground-based interferometer detector of gravitational radiation using Bayesian model selection. The simulated waveform of the chirp signal is assumed to be a spin-less Post-Newtonian (PN) waveform of a given expansion order, while the searching template is taken to be either of the same PN family as the simulated signal or one level below its PN expansion order. Within the Bayesian framework we estimate the detection probabilities and the statistical uncertainties due to noise as a function of the signal-to-noise ratio (SNR), and the posterior distributions of the parameters characterizing the chirp signal. Our analysis indicates that the detection probabilities are not compromised when simplified models are used, while the accuracies in the determination of the parameters can be significantly worsened.

1. Introduction

Ground-based interferometer detectors of gravitational waves have reached their design sensitivity, and have started to search [1, 2, 3, 4], in the kilohertz frequency band, for gravitational waves emitted by astrophysical sources such as spinning neutron stars, supernovae, and coalescing binary systems. Coalescing binary systems in particular are expected to be the first to be detected and studied. These signals have a unique signature that enables them to be extracted from wide-band data by matched filtering [5]. This signature is their accelerating sweep upwards in frequency as the binary orbit decays because of energy loss due to the emission of gravitational radiation. Coalescing binaries have a potential advantage over other sources in signal-to-noise ratio (SNR) by a factor that depends on the square-root of the ratio between the corresponding number of cycles in the wave trains [6].

The effectiveness of the matched filtering procedure relies on the assumption of exactly knowing the analytic form of the signal (possibly) present in the data. Recent breakthroughs in numerical relativity [7, 8, 9] have started to provide a complete description of the radiation emitted during the inspiral, merger, and ring-down phases of generic binary black hole merger scenarios. Although the ability of obtaining numerically all the templates needed in a data analysis search (perhaps hundred of thousands of them) might be practically impossible, in principle we should be able to compare these numerically derived waveforms against various analytic templates obtained under different approximating assumptions. Work in this direction has already started to appear in the literature [10, 11, 12] within the so called "frequentist framework", in which estimates in the reduction in SNRs and inaccuracies of the determination of the parameters characterizing the signal, due to the use of approximated waveforms, have been derived. Depending on the magnitude of these degradations one can decide whether to use these approximated waveforms as templates in a data analysis search.

Since it can be argued that contiguous PN approximations should well characterize the differences between the "true signal" present in the data and the highest-order PN approximation, we have performed such a comparison within the Bayesian framework. An analogous, frequentist analysis has recently been performed by Cutler and Vallisneri [13] for the case of super massive black holes binaries observed by LISA (the Laser Interferometer Space Antenna). Their approach relied on the use of the Fisher-Information matrix, which is known to give good results in the case of large (hundreds to thousands) SNRs. In the case of ground-based interferometers instead, since the expected SNRs will be probably smaller than 10, a parameter estimation error analysis based on the Fisher-Information matrix would lead to erroneous results [14].

In a recent paper [15] we have estimated the loss in probability of detection (i.e. loss of *evidence* of a signal to be present in the data) as a function of the SNR when the true signal present in the data is a spin-less waveform represented by a given PN approximating order and the search model is one PN order level below it. Our analysis was purposely limited to only two separate cases (i.e. PN3.5 vs. PN3.0 and PN2.0 vs. PN1.5) in order to cover the region of the PN approximations that have already, or are in the process of, being used in the analysis of the data collected by presently operated ground-based interferometers.

Our model comparison relies on a Bayesian Markov Chain Monte Carlo (MCMC) technique, as MCMC methods have successfully been applied to a large number of problems involving parameter estimations [16] in experimental data sets. In our analysis the chirp signals (the one present in the data and the one used as the model) are characterized by five parameters: the two masses of the system, m_1 and m_2, their time to coalescence t_C, the coalescence phase ϕ_C, and their distance r from Earth.

This paper is organized as follows. In Section 2 we provide a brief summary of the Bayesian framework and its implementation in our problem. After deriving the expressions of the likelihood function and the priors for the parameters searched for, in Sec. 3 we give a brief description of the Markov-Chain Monte Carlo sampling technique adopted for calculating the posterior distributions, and summarize the results of our numerical analysis. We find that (i) the difference in detection probability when using a simplified model rather than the true one is negligibly small in comparison to the uncertainties associated with different noise realizations, while (ii) the posterior credibility regions of the parameters reveal offsets from the true parameter values that can be much larger than the statistical uncertainty.

2. The Bayesian formulation

The Bayesian full probability model for our problem, which involves the comparison between the two possibilities of having either a signal and noise or just noise in the data, requires the determination of the likelihoods, the prior distributions for the parameters associated with the models, and the resulting posterior distributions. Let us consider a data stream $d(t) = s_t(\boldsymbol{\theta}; t) + n(t)$ containing the instrumental noise $n(t)$ and a chirp signal $s_t(\boldsymbol{\theta}; t)$ that we will regard as the "*true*" signal. Here, $\boldsymbol{\theta}$ is the vector representing all the parameters associated with the signal, and the noise is assumed to be a stationary Gaussian random process of zero mean. In the Fourier domain the observed data can equivalently be written as $\tilde{d}(\boldsymbol{\theta}; f) = \tilde{s}_t(\boldsymbol{\theta}; f) + \tilde{n}(f)$ (where tilde denotes the Fourier transform operation) and we will refer to this expression as model $\mathcal{M}_t$.

In what follows we will assume the "*true*" signal $\tilde{s}_t(\boldsymbol{\theta}; f)$ to be the gravitational wave emitted by a coalescing binary system and represented by a spin-less Post-Newtonian approximation in phase and Newtonian in amplitude for which $\boldsymbol{\theta} = \{m_1, m_2, r, t_C, \phi_C\}^T$. Here m_1 and m_2 are the masses of the rotating objects, t_C is the coalescence time, r the absolute distance to the binary system, and ϕ_C the phase of the signal at coalescence.

We will then describe the detection and estimation of the parameters of the "true" signal by relying on a spin-less lower-order Post-Newtonian waveform, $\tilde{s}_s(\boldsymbol{\theta}; f)$. This simpler model will be referred to as model $\mathcal{M}_s$.

The derivation of the detection probability implies a comparison between model $\mathcal{M}_s$ and the null-model, which postulates mere noise $\tilde{n}(f)$ within the data. This model will be referred to as model $\mathcal{M}_n$.

Since we have assumed the distribution of the random process associated with the noise of the detector to be Gaussian of zero-mean, it follows that the likelihood function is proportional to the exponential of the integral of the squared and normalized residuals between $\tilde{d}(\boldsymbol{\theta_t}; f)$ and signal template $\tilde{s}(\boldsymbol{\theta}; f)$ over the frequency band of interest $[f_L, f_U]$. We define the data $\tilde{d}(f) := \tilde{d}(\boldsymbol{\theta_t}; f)$ to be modeled by $\mathcal{M}_t$ with its "*true*" parameter vector $\boldsymbol{\theta_t}$. The comparison with the simplified model $\mathcal{M}_s$, results in the following expression for the likelihood function

$$p(\tilde{d}|\mathcal{M}_s, \boldsymbol{\theta}) \propto \exp\left(-2\int_{f_L}^{f_U} \frac{|\tilde{d}(f) - \tilde{s}_s(f, \boldsymbol{\theta})|^2}{S_n(f)} \mathrm{d}f\right) , \tag{1}$$

where $S_n(f)$ is the one-sided power spectral density of the noise.

By substituting $\tilde{d}(f) = \tilde{s}_t(f, \boldsymbol{\theta_t}) + \tilde{n}(f)$ into Eq. 1, the likelihood function becomes

$$p(\tilde{d}|\mathcal{M}_s, \boldsymbol{\theta}) \propto \exp\left(-2\int_{f_L}^{f_U} \frac{|\tilde{s}_t(f, \boldsymbol{\theta_t}) + \tilde{n}(f) - \tilde{s}_s(f, \boldsymbol{\theta})|^2}{S_n(f)} \mathrm{d}f\right) . \tag{2}$$

In analogy to Eq. 2, under model $\mathcal{M}_n$ (with no parameters) the likelihood assumes the following form

$$p(\tilde{d}|\mathcal{M}_n) \propto \exp\left(-2\int_{f_L}^{f_U} \frac{|\tilde{s}_t(f, \boldsymbol{\theta_t}) + \tilde{n}(f)|^2}{S_n(f)} \mathrm{d}f\right) . \tag{3}$$

For comparison reasons, in the case of using the "*true*" model the likelihood function becomes

$$p(\tilde{d}|\mathcal{M}_t, \boldsymbol{\theta}) \propto \exp\left(-2\int_{f_L}^{f_U} \frac{|\tilde{s}_t(f, \boldsymbol{\theta_t}) + \tilde{n}(f) - \tilde{s}_t(f, \boldsymbol{\theta})|^2}{S_n(f)} \mathrm{d}f\right) , \tag{4}$$

which will then give information about the impairment in detection when using model $\mathcal{M}_s$ instead of $\mathcal{M}_t$. Note, that the investigation of this question requires to do the two different model comparisons separately, i.e. $\mathcal{M}_n$ vs. $\mathcal{M}_s$ and $\mathcal{M}_n$ vs. $\mathcal{M}_t$ [21].

The next step needed for completing our Bayesian full probability model is the identification of suitable prior distributions for the five parameters characterizing the chirp signal. The derivation of suitable priors has been discussed in [15] and more extensively in [17, 18], and we refer the reader to those references for details. Note that the selection of the appropriate priors $p(\boldsymbol{\theta})$ bears significant influence on the evidence of a signal presence within noise, since the prior identifies the size of the parameter space which the evidence is based on. In short, the masses are assumed to be uniformly distributed over a specified range, $[m_{min}, m_{max}]$, and the prior distribution for the distance is chosen to be a cumulative distribution of having systems out to a distance x smaller than r, $P(x < r)$, proportional to the cube of the distance, x^3. In order to obtain a proper prior distribution that does not diverge once integrated to infinity, it is down-weighted by including an exponential decaying and includes the assumption of a uniform distribution for the masses. The sigmoid function that serves as the down-weighing term, is chosen in such way that a $(2-2)M_\odot$ inspiral system is assumed to be detectable with probabilities 0.1 and 0.9 out to distances 95 Mpc and 90 Mpc respectively [17, 18]. This choice depends on the assumed noise model which roughly coincides with the noise sensitivity curve of the LIGO-I detector [23].

As far as the time to coalescence is concerned, we have assumed it to be uniformly distributed over a time interval of 1 second centered around the value identified by the masses of the binary and the lower frequency cut-off of the detector [24]. This search range for the time to coalescence t_C is larger than that used in [17] because when using the simplified model the posterior peak can be offset from the true value by more than the posterior width. Finally, we have chosen the phase of the signal at coalescence, ϕ_C, to be uniformly distributed over the interval $[0, 2\pi]$, i.e. $p(\phi_C) = \mathrm{I}_{[0,2\pi[}(\phi_C)/(2\pi)$.

The choice of priors is different when it comes to the analysis of model $\mathcal{M}_n$. Since this model postulates mere noise, there are no parameters entering the likelihood, which is therefore a constant.

The final remaining step in defining the Bayesian procedure is to assign prior probabilities to the models themselves. Since we have no a priori knowledge, the unbiased choice is to assume equal probability for each. However, one can show that for SNRs equal to about 7 or larger the effects of assigning different prior detection probabilities are negligible. See the appendix in [15] for a detailed analysis of this point.

By applying Bayes' theorem using the likelihoods and priors defined above, we then derive the multidimensional posterior probability distribution for the model and its parameters

$$p(i, \boldsymbol{\theta}_i|\tilde{d}) = \frac{\left\{ \begin{array}{ll} p(\mathcal{M}_n) \cdot p(\tilde{d}|\mathcal{M}_n) & \text{if } i = 0 \\ p(\mathcal{M}_s, \boldsymbol{\theta}) \cdot p(\tilde{d}|\mathcal{M}_s, \boldsymbol{\theta}) & \text{if } i = 1 \end{array} \right\}}{\int p(\mathcal{M}_s, \boldsymbol{\theta}) \cdot p(\tilde{d}|\mathcal{M}_s, \boldsymbol{\theta}) \mathrm{d}\boldsymbol{\theta} + p(\mathcal{M}_n) \cdot p(\tilde{d}|\mathcal{M}_n)}, \quad (5)$$

where $i \in \{0, 1\}$ corresponds to the two models $\{\mathcal{M}_n, \mathcal{M}_s\}$. In the same way, it is possible to derive the posterior for the comparison of $\mathcal{M}_n$ vs. $\mathcal{M}_t$. A Bayesian analysis naturally justifies Occam's Razor [19, 20] due to the penalization of unreasonably complex models by integrating over the parameter space resulting in the preference for a simpler model.

3. Bayesian model selection and posterior probabilities

There exist various techniques for tackling the multi-dimensional problem of estimating the posterior distributions. We have implemented a relatively new procedure, called Reversible Jump MCMC (RJMCMC) technique [25, 26], which simultaneously addresses the problems of model selection and parameter estimation. The RJMCMC is combined with traditional fixed dimension MCMC techniques that sample from the parameters of the current model. See [15] for details.

Since our simulations were computationally time-intensive, we have created data sets from "*true*" wave forms of three hypothetical binary inspiral systems whose masses were selected in order to get a fair representation of the entire mass-parameter space. We also considered two scenarios where the "*true*" wave form was either of PN 2.0 or PN 3.5 order. The detection of each scenario was attempted by either a PN 1.5 or a PN 3.0 wave form respectively [15].

The distance of each binary system was varied in order to obtain different signal-to-noise ratios. The noise realizations were drawn in such a way that they corresponded to the approximated expression of the one-sided power spectral density of initial LIGO [27]

$$S_n(f) = \frac{S_0}{5} \left[\left(\frac{f_0}{f} \right)^4 + 2 \left[1 + \left(\frac{f}{f_0} \right)^2 \right] \right] \quad (6)$$

with $S_0 = 8.0 \times 10^{-46}$ Hz^{-1} being the minimum noise of the detector and $f_0 = 175$ Hz the frequency at which the sensitivity of the detector reaches its maximum. The noise samples were generated directly in the Fourier domain by first simulating white noise samples, and then by scaling their amplitudes according to the required noise spectrum $S_h(f)$. The simulated data

Journal of Physics: Conference Series **154** (2009) 012044 doi:10.1088/1742-6596/154/1/012044

were sampled at 4096 Hz for a duration of about 24 s, and they were produced by embedding the different signals into noise samples that where generated in the Fourier domain.

The integration bandwidth for the likelihood was chosen from 12 Hz up to the frequency of the last stable orbit or 600 Hz, whichever is the smaller. Since the SNR is negligible above 600 Hz, we fixed this to be the upper frequency cut-off. For the binary system ($m_1 = 1.8M_\odot, m_2 = 0.6M_\odot$) the frequency at the last stable orbit is equal to 1832.2 Hz, which gives an integration limit of 600 Hz. This translates in 14355 complex samples that contribute to the posterior distribution. In the case of the binary system with ($m_1 = 45.0M_\odot, m_2 = 0.52M_\odot$) the last stable orbit is at 96.6 Hz, resulting in 2065 complex samples involved in the determination of the likelihood function. Finally, for the high mass binary system ($m_1 = 45.0M_\odot, m_2 = 30M_\odot$) in our set of systems the last stable orbit is at 58.6 Hz, implying now only 1138 complex samples over which the likelihood is calculated.

After we conducted the MCMC simulations we derived the posterior detection probabilities for the competing models from the MCMC outputs regarding the three example binary systems, the four different model comparisons, and the different sets of SNRs. For each binary system we computed the posterior probabilities for the considered scenarios and contrasted the probability of detection based on a lower order PN expansion against the one using the true wave form. The results are shown in figures (3,4,5) of reference [15]. The main conclusion of our analysis is that the probability of detection is not impeded by using a simplified model for detecting wave forms of higher PN order in the low-SNR regime. The Bayesian approach provides the means to gain insight into the variation of the detection probability over different noise realizations. The difference between the posterior detection probabilities corresponding to the true and the simplified model is very small as compared to its variance over different noise realizations. It is worth mentioning that analyses performed within the frequentist framework [27, 28, 12] and aimed at comparing the detectability of a signal by using a simplified wave form were focused entirely on estimating the resulting loss of SNR. The Bayesian model comparison presented here has the inherent ability to estimate probabilities and their uncertainties due to noise, providing much more insights into this issue.

The impact of the use of lower PN order wave forms on the bias of posterior distributions of the parameters was then estimated by calculating the posterior distributions of the parameters characterizing the analyzed binary systems. We found that when the model matches the signal present in the data the posteriors cover well the true parameter values. However, when the searching model is different from that of the signal present in the data the posterior distributions display a marked displacement of the mass parameters from their "true values" that can be as large as several of their standard deviations. Very striking is also the error in the time to coalescence. The masses and time to coalescence are obviously the parameters subject to biases when using a simplified model. This is physically understandable since these three parameters define the phase of the signal.

4. Acknowledgments

This research was performed at the Jet Propulsion Laboratory, California Institute of Technology, under contract with the National Aeronautics and Space Administration.

References

[1] The LIGO project home page. http://www.ligo.caltech.edu
[2] The Virgo project home page. http://www.virgo.infn.it/
[3] The GEO-600 project home page. http://www.geo600.uni-hannover.de/
[4] The TAMA project home page. http://tamago.mtk.nao.ac.jp/
[5] Helström C W, 1968 *Statistical Theory of Signal Detection*, Pergamon Press, London (UK).
[6] Thorne K S 1987, Gravitational radiation. In *Three Hundred Years of Gravitation*, pages 330–458. Cambridge University Press. Hawking S W and Israel W, eds.

[7] Pretorius F 2005 Evolution of binary black-hole spacetimes. *Physical Review Letters*, **95**, 121101

[8] Campanelli M, Lousto C O, Marronetti P and Y. Zlochower 2006, Accurate evolutions of orbiting black-hole binaries without excision. *Physical Review Letters*, **96** 111102.

[9] Baker J G, Centrella J, Choi D, Koppitz M and van Meter J, 2006. Gravitational-wave extraction from an inspiraling configuration of merging black holes. *Physical Review Letters*, **96**, 111102.

[10] Buonanno A, Cook G B and Pretorius F, 2007. Inspiral, merger, and ring-down of equal-mass black-hole binaries. *Physical Review D*, **75**, 124018.

[11] Baker J G, van Meter J R, McWilliams S T, Centrella J and Kelly B J, 2007. Consistency of post-Newtonian waveforms with numerical relativity. *Physical Review Letters*, **99** 181101.

[12] Pan Y, Buonanno A, Baker J G, Centrella J, Kelly J B, McWilliams S T, Pretorius F, and van Meter J R, 2008. A data-analysis driven comparison of analytic and numerical coalescing binary waveforms: nonspinning case. *Phys. Rev. D*, **77**, 024014.

[13] Cutler C and Vallisneri M, 2007. Lisa detections of massive black hole inspirals: parameter extraction errors due to inaccurate template waveforms. *Phys. Rev. D*, **76**, 104018.

[14] Vallisneri M, 2008. Use and abuse of the fisher information matrix in the assessment of gravitational-wave parameter-estimation prospects. *Phys. Rev. D*, **77**, 042001

[15] Umstätter R and Tinto M, 2008. Bayesian comparison of Post-Newtonian approximations of gravitational wave chirp signals. *Physical Review D*, **77**, 082002.

[16] Gilks W R, Richardson S, Spiegelhalter P J, and Green D J, editors, 1996. *Markov chain Monte Carlo in practice*. Chapman and Hall, London (UK).

[17] Röver C, Meyer R and Christensen N, 2006. Bayesian inference on compact binary inspiral gravitational radiation signals in interferometric data. *Classical and Quantum Gravity*, **23** (15), 4895–4906.

[18] Röver C, Meyer R and Christensen N, 2007. Coherent Bayesian inference on compact binary inspirals using a network of interferometric gravitational wave detectors. *Physical Review D*, **75**(6), 062004

[19] Jaynes E T, 2003. *Probability Theory: The Logic of Science*. Cambridge University Press, Cambridge (UK).

[20] Loredo T J, 1992. *Statistical challanges in modern astronomy*, pages 275–297. Springer-Verlag, New York.

[21] Gregory P C, 2005. *Bayesian Logical Data Analysis for the Physical Sciences*. Cambridge University Press, Cambridge (UK).

[22] Sandage A, 2001. *Encyclopedia of Astronomy and Astrophysics*. Institute of Physics Publishing, Bristol (UK).

[23] Sigg D, 2004. Commissioning of LIGO detectors. *Classical and Quantum Gravity*, **21**, S40915.

[24] Pai A, Dhurandhar S V and Bose S, 2001. Data-analysis strategy for detecting gravitational-wave signals from inspiraling compact binaries with a network of laser-interferometric detectors. *Physical Review D*, **64**, 042004.

[25] Green P J, 1995. Reversible jump Markov chain Monte Carlo computation and Bayesian model determination. *Biometrika*, **82**(4), 711–732

[26] Green P J, Hjort N L and Richardson S editors. 2003. *Highly Structured Stochastic Systems*. Oxford University Press

[27] Chronopoulos A E and Apostolatos T A, 2001. Less accurate but more efficient family of search templates for detection of gravitational waves from inspiraling compact binaries. *Physical Review D*, **64**, 042003.

[28] Canitrot P, 2001. Systematic errors for matched filtering of gravitational waves from inspiraling compact binaries. *Physical Review D*, **63**, 082005.

Grids for efficient all sky search of white dwarf binaries in Mock LISA Data Challenge

A Błaut[1], A Królak[2] and M Piętka[3]

[1] Institute of Theoretical Physics, University of Wrocław, Wrocław, Poland
[2] Institute of Mathematics, Polish Academy of Sciences, Warszawa, Poland
[3] Faculty of Physics, Unversity of Białystok, Białystok, Poland

E-mail: ablaut@ift.uni.wroc.pl, krolak@impan.gov.pl, map@alpha.uwb.edu.pl

Abstract. We present construction of the 3 and 4 dimensional grids in the parameter space for all sky-search of gravitational wave signals from white dwarf binaries with LISA data. The 3 dimensional grid is for search of frequency and the sky position of the source and the 4 dimensional grid includes the spin down parameter. The grid solves the covering problem in the parameter space with the constraint that nodes of the grid coincide with Fourier frequencies (multiples of the inverse of the observation time).This allows the use of the fast Fourier transform (FFT) in the evaluation of the optimal statistic and greatly speeds up the search.

1. Introduction

Matched filtering detection of weak, quasi-monochromatic gravitational wave signals relies on an efficient placement of the templates in the bank. Optimally it should minimize the number of templates for a given allowed loss in the signal-to-noise ratio due to a signal offset. More specifically one considers the space of parameters of the modeled signal together with, generally non-flat, metric provided e.g. by the reduced Fisher matrix. Here we consider flat parameter space for all sky search for monochromatic and nearly monochromatic signals. We construct the grid of templates which is to be restricted in the frequency direction in order to speed the search by making use of FFT.

The paper is organized as follows. In section 2 we sketch a search method that enables fast evaluation of the detection statistic for gravitational wave signals from nearly monochromatic sources by using the FFT algorithm. In section 3 we define the covering problem with constraint. In section 4 we present two constructions of nearly optimal lattices in any number of dimensions that in a good approximation satisfy the constraint. Results are briefly illustrated in section 5 for the case of 3 and 4 dimensions. In Appendix we briefly summarize some basic definitions from the theory of lattices.

2. Grid on the parameter space

The response of the space-based detector LISA to a gravitational wave can be written [1, 2] as a linear combination of 4 functions $h^{(k)}$ depending on time. Each function $h^{(k)}$ depends on parameters ξ^{μ} called *intrinsic* parameters and the constant coefficients $a^{(k)}$ called *extrinsic* parameters:

$$h(t; a^{(k)}, \xi^\mu) = \sum_{k=1}^{4} a^{(k)} h^{(k)}(t; \xi^\mu). \tag{1}$$

In the maximum-likelihood search method the estimators $(\hat{a}^{(k)}, \hat{\xi}^\mu)$ are found by maximizing the log likelihood function with respect to parameters $(a^{(k)}, \xi^\mu)$. By solving explicitly the maximum likelihood equation for the extrinsic parameters one can define the so called $\mathcal{F}$-statistic and reduce the search to the intrinsic parameters space. For a signal from white-dwarf binaries buried in a stationary Gaussian noise and for intrinsic parameters defined by $(\xi^1, \xi^2, \xi^3, \xi^4) = (\omega, \dot{\omega}, \beta, \lambda)$, where ω is the angular frequency of the gravitational wave, $\dot{\omega}$ is the time derivative of ω, β and λ are the latitude and the longitude of the source, the detection statistic $\mathcal{F}$ consists of integrals of the form

$$\int_0^{T_o} y(t) m(t; \omega, \beta, \lambda) \exp\left[i\phi_{mod}(t; \omega, \dot{\omega}, \beta, \lambda) \exp i\omega t\right] dt, \tag{2}$$

where $y(t)$ are the noisy data and m, ϕ_{mod} are amplitude and phase modulation functions. T_o is the observation time. As the modulation functions m and ϕ_{mod} depend on the frequency ω we cannot apply the FFT algorithm directly in calculation of the $\mathcal{F}$-statistic. In order to do that we first analyze the data in a narrow band over which the slowly varying modulation function m is evaluated at the mid frequency of the band and, second, we introduce a linear parametrization of the phase

$$\phi_{mod} = \frac{1}{2}\dot{\omega}t^2 + A\cos\Omega t + B\sin\Omega t,$$

where frequency is absorbed in the new parameters A and B,

$$\begin{aligned} A &= \omega R \cos\beta \cos(\lambda - \eta_0) \\ B &= \omega R \cos\beta \sin(\lambda - \eta_0), \end{aligned}$$

where η_0 is the initial position of the LISA constellation on the orbit around the sun, $R = 1\text{AU}$ and $\Omega = 2\pi/\text{yr}$. This results in the following linear phase model

$$h(t) = A_0 \cos(\omega t + \frac{1}{2}\dot{\omega}t^2 + A\cos\Omega t + B\sin\Omega t + \phi_0) \tag{3}$$

with extrinsic (A_0, ϕ_0) and intrinsic $(\omega, \dot{\omega}, A, B)$ parameters. The reduced Fisher matrix [2] $\tilde{\Gamma}$ of the model has constant coefficients. We want to distribute the points of the grid in such a way that the distance defined by $\tilde{\Gamma}$ from any point of the parameter space to the nearest node of the grid is not larger than some fixed value r.

3. Covering problem with constraints

The problem of constructing the grid in the parameter space is equivalent to the problem of covering d-dimensional space with equal overlapping spheres of a given radius. The optimal covering would have minimal possible thickness (see [3] and appendix). The thinnest *lattice* coverings are known in dimensions up to 5. In what follows we consider only lattice coverings.

In our search scheme the calculation of the detection statistic $\mathcal{F}$ can be interpreted as a Fourier transform and computed efficiently by fast Fourier transform algorithm. For this reason the nodes of the grid must coincide with Fourier frequencies: $\Delta\omega, 2\Delta\omega, 3\Delta\omega, \ldots$ for some fixed frequency resolution $\Delta\omega$. This imposes a condition that one of the lattice basis vectors has a fixed length

$$|\vec{l}| = \sqrt{\tilde{\Gamma}\left[(\Delta\omega, 0, \ldots, 0), (\Delta\omega, 0, \ldots, 0)\right]}$$

and forbids an immediate use of the general results of the theory of lattice coverings. Instead one can formulate the *covering problem with constraint*: to find the thinnest lattice covering of the d-dimensional space with spheres of radius r and one of the basis vectors of the lattice having fixed length $|\vec{l}|$. As far as we know the general solution to the problem is not known. We present two constructions of lattices that in a good approximation satisfy the constraints. They can be viewed as possible solutions to the problem and may serve as starting point for alternative constructions. We illustrate the procedures in 3 and 4 dimensions.

4. Constructions of the grid

Let a vector $\vec{v}_0$ define the frequency resolution. We search for a lattice $\Lambda(w')$ with lattice basis w' satisfying two constraints that can be expressed as $w' = \left\{\vec{v}_0, \vec{w}'_1, \ldots, \vec{w}'_{d-1}\right\}$ and $R(\Lambda(w')) = r$, where $R(\Lambda)$ denotes the covering radius of a lattice Λ.

4.1. Procedure I

As an initial approximation for the lattice we take $\Lambda(w)$, $w = \{\vec{v}_0, \vec{w}_1, \ldots, \vec{w}_{d-1}\}$, with an arbitrary orthonormal $(d-1)$-dimensional basis $\{\vec{w}_1, \ldots, \vec{w}_{d-1}\}$. The construction of $\Lambda(w')$ is based on the following algorithm **AI**.

Algorithm AI
Input: $w = \{\vec{v}_0, \vec{w}_1, \ldots, \vec{w}_{d-1}\}$.
Output: $w' = \left\{\vec{v}_0, \vec{w}'_1, \ldots, \vec{w}'_{d-1}\right\}$, with covering radius $R(\Lambda(w')) \approx r$.
AI1. $w' = w$.
AI2. Repeat
AI3. $w' = \left\{\vec{v}_0, c\vec{w}'_1, \ldots, c\vec{w}'_{d-1}\right\}$, where $c = \frac{r}{R(\Lambda(w'))}$
AI4. Until $R(\Lambda(w')) \approx r$.
AI5. Return w'.

We notice that the orientation of the constraint vector $\vec{v}_0$ has no effect on the optimal constrained lattice, it is only the length of $\vec{v}_0$ and assumed value of covering radius that matters (and more precisely: only their ratio because the overall scale can be taken arbitrarily).

To construct an optimal constrained lattice one repeats the algorithm **AI** $[\{\vec{v}_0, \vec{w}_1, \ldots, \vec{w}_{d-1}\}]$ with $\vec{v}_0$ having different orientations with respect to $\{\vec{w}_1, \ldots, \vec{w}_{d-1}\}$ and chooses w' having the smallest thickness.

4.2. Procedure II

In the Procedure II one tries to find an optimal constrained lattice starting with a thinnest unconstrained lattice in d-dimensions. The idea is to shrink the optimal lattice as little as possible such that one of the basis vectors of the resulting lattice coincides with the constraint vector $\vec{v}_0$.

We notice first that for a given lattice Λ there always exists the shortest lattice vector $\vec{l}$ satisfying $|\vec{v}_0| \leq \vec{l}$ that we denote by $\vec{l}(\Lambda)$. We define the following algorithm **AII**.

Algorithm AII
Input: Lattice Λ; vector $\vec{v}_0$.
Output: Lattice Λ'; $\vec{l}(\Lambda') = \vec{v}_0$.
AII1. Find $\vec{l}(\Lambda)$.
AII2. Contract Λ along $\vec{l}(\Lambda)$ to obtain Λ_c with $|\vec{l}(\Lambda_c)| = |\vec{v}_0|$.
AII3. Rotate Λ_c to obtain a lattice Λ_{rc} with $\vec{l}(\Lambda_{rc}) = \vec{v}_0$.
AII5. Return Λ'.

The contraction of the initial lattice Λ is "minimal" in the sense that the thickness of the resulting lattice Λ' satisfies $\Theta(\Lambda) \leq \Theta(\Lambda') \leq \frac{|\vec{l}(\Lambda)|}{|\vec{v}_0|}\Theta(\Lambda)$.

For optimal initial lattices the algorithm **AII** defines the following function $f_{\vec{v}_0} : \mathbf{R} \to \mathbf{R}$: for $x = R(\Lambda)$, $f_{\vec{v}_0}(x) = R(\Lambda')$. For a given $r \in \mathbf{R}$ we denote by r_i the value of x for which the function $|f_{\vec{v}_0}(x) - r|$ reaches its minimum. The optimal constrained lattice is obtained by application of the algorithm **AII** to an optimal (unconstrained) lattice Λ with covering radius $R(\Lambda) = r_i$.

In dimensions $d = 2, 3, 4, 5$ as the initial lattices one takes the so called Voronoi's principal lattice of the first type, A_d^* [3] but the procedure can be generalized to any number of dimensions by taking as the input the best known lattice covering in a given dimension [3],[4].

5. Results

Figure 1 show results of the application of the two procedures to the model (3) in 3 and 4 dimensions for different covering radii. The resolution vector is $\vec{v}_0 = (2\pi, 0, \ldots)$ and the observation time is 2 years.

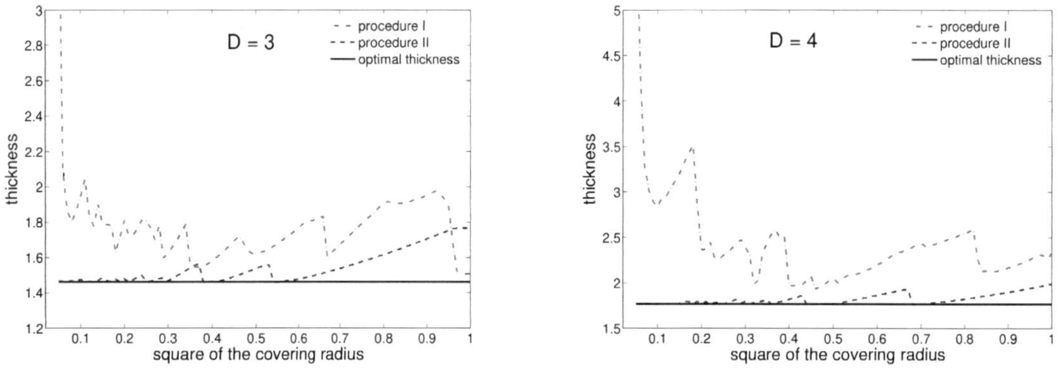

Figure 1. Thickness of the lattice in 3 and 4 dimensions.

Appendix

In general for any discrete set of points $\mathcal{S} = \{\vec{s}_1, \vec{s}_2, \ldots\}$ in $\mathbf{R}^n$ the *covering radius* R of $\mathcal{S}$ is defined as the least upper bound for any point of $\mathbf{R}^n$ to the closest point $\vec{s}_i$:

$$R(\mathcal{S}) = \sup_{\vec{x} \in \mathbf{R}^n} \inf_{\vec{s} \in \mathcal{S}} |\vec{x} - \vec{s}|.$$

Then spheres of equal radius r centered at the points $\vec{s}_i$ will cover $\mathbf{R}^n$ only if $r \geq R$.

A *lattice* Λ is a discrete subset of $\mathbf{R}^n$. Any lattice has a *basis* $\left\{\vec{l}_1, ..., \vec{l}_n\right\}$ of linearly independent vectors such that the lattice is the set of all linear combinations of $\vec{l}_i$'s with integer coefficients:

$$\Lambda = \left\{ \sum_{i=1}^{n} c_i \vec{l}_i : \quad c_i \in \mathbf{Z}, \qquad i = 1, 2, \ldots, n \right\}. \tag{A.1}$$

A lattice basis is not unique, in dimensions $d > 1$ there are infinitely many of them, but all the bases have the same number of elements called the *dimension* of the lattice.

The parallelotope consisting of points $c_1\vec{l}_1 + \ldots + c_n\vec{l}_1$ with $0 \leq c_i < 1$ is a *fundamental parallelotope* and is an example of an *elementary cell*, that is the building block containing one lattice point which tiles the whole $\mathbf{R}^n$ by translations of lattice vectors. There are infinitely many elementary cells but the volume of each elementary cell is unique for a given lattice Λ.

The *Voronoi cell* around any point $\vec{v}$ of Λ is the set of vectors $\vec{x}$ of $\mathbf{R}^n$ which are closer to $\vec{v}$ than to any other lattice vector:

$$V(\vec{v}) = \{\vec{x} : \quad |\vec{x} - \vec{w}| \geq |\vec{x} - \vec{v}| \quad \text{for all} \quad \vec{w} \in \Lambda\} . \tag{A.2}$$

All Voronoi cells of a given lattice are congruent convex polytopes and are another examples of elementary cells sometimes referred to as Wigner-Seitz cells or Brillouin zones.

For the lattice Λ having Voronoi cells congruent to polytope $V(\vec{v})$, where $\vec{v}$ is any of the lattice points, the covering radius is the circumradius $R(\Lambda)$ of $V(\vec{v})$ i.e. the largest distance between $\vec{v}$ and the vertices of $V(\vec{v})$.

The thickness of the lattice covering is given by:

$$\Theta(\Lambda) = \frac{\text{volume of } d\text{-dimensional sphere of radius } R(\Lambda)}{\text{volume of the elementary cell of } \Lambda} \tag{A.3}$$

References

[1] Królak A, Tinto M, Vallisneri M 2004 *Phys. Rev.* D **70**, 022003

[2] Jaranowski P, Królak A 2008 Gravitational-wave data analysis. Formalism and simple applications: the Gaussian case *Living Rev. Relativity*, **8**, 3 (cited on 6 June 2008) http://www.livingreviews.org/lrr-2005-3

[3] Conway J H, Sloane N J A 1993 *Sphere Packings, Lattices ang groups* (Springer-Verlag, New York)

[4] Schurmann A, Vallentin F http://www.math.uni-magdeburg.de/lattice_geometry

Journal of Physics: Conference Series **154** (2009) 012046 doi:10.1088/1742-6596/154/1/012046

Simulation of cosmological stochastic background in LISA

E J Buis[1], S Oemrawsingh[2] and G Vacanti[1]

[1]cosine Science & Computing BV, Niels Bohrweg 11, 2333 CA Leiden, The Netherlands
[2]University of California, Santa Barbara, CA, USA

E-mail: ebuis@cosine.nl

Abstract. A pilot study has been carried out to simulate stochastic gravitational wave background originating from first order phase transitions in the early universe. The space based gravitational wave detector LISA will be operational in the range of 10^{-4} to 0.1 Hz and could be sensitive to the red shifted gravitational waves from cosmological origin. In this study we have modeled the signals from first order phase transitions and we compared the signal both with the expected instrumental noise and realistic simulated foreground signals, originating from the white dwarf population in our galaxy.

1. Introduction

One of the most striking observables from the early universe that has been measured up to now, is most probably the cosmic microwave background radiation (CMB). The CMB provides us with an astonishing picture of the early universe at the period of photon decoupling. The Planck mission, which is scheduled for launch next year will provide an even more accurate measurement of the CMB. With its launch foreseen in the next decade the LISA mission [1] may offer the possibility of measuring the state of the universe in an even earlier phase. It is postulated that first order phase transitions may give rise to violent events which could be a source of gravitational waves [2, 3].

In this paper we describe a pilot study to simulate gravitational waves originating from a primordial plasma. A description of the source is given in the next section along with the expressions for the expected spectra. In section 3 we describe the foreground, while we discuss the Monte Carlo simulations in 4. We conclude in section 5.

2. Gravitational waves from phase transitions

In many cosmological models the state of the universe is described by an effective potential of a scalar field. The potential is symmetric with a single minimum. With the expansion of the universe and the temperature drop, the scalar potential changes shape and reaches a point where it exhibits a barrier between two degenerate minima. At this point, first order phase transitions may occur, when quantum tunneling occurs through the barrier from the high energy state (the so-called "false" vacuum) to a lower level (the "true" vacuum). In contrast to second order phase transitions (when no barrier is tunneled), a first order phase transition leads to bubbles of true vacuum, which expand and merge until the complete plasma is in the broken symmetry state. Collisions between these bubbles are a source of gravitational waves and give rise to subsequent turbulence in the primordial plasma. Besides bubble collisions, the turbulence of the plasma is an additional—and in most cases a stronger—source of gravitational waves.

The expected spectrum from the stochastic background is well described in the literature [4, 5]. Three important parameters that completely cover the characteristics of the gravitational wave spectrum are the rate of variation of the bubble nucleation rate, called β, and the ratio of the latent heat liberated at the phase transition to the energy density in the high energy phase, which is called α. The third parameter is the temperature of the plasma (T) and sets the scale of the symmetry breaking.

The intensity of stochastic background signals are conventionally expressed in a dimensionless quantity Ω (or Ωh^2 to define a parameter which is independent of the actual Hubble expansion rate [6]). The spectra of the gravitational waves is rather featureless and is characterized by the maximum amplitude and the frequency at which the maximum occurs. At lower frequencies the amplitude depends on the frequency f as $f^{2.8}$, and at higher frequencies as $f^{-1.8}$. The dependence of the amplitude frequency is related to the dynamics of plasma and the bubbles, which is detailed in [7, 8, 9].

The maximum intensity of the waves due to bubble collisions is given by:

$$\Omega_{\text{coll}} h^2(f_{\text{coll}}) \simeq 1.1 \times 10^{-6} \kappa^2 \left[\frac{H_*}{\beta}\right]^2 \left[\frac{\alpha}{1+\alpha}\right]^2 \left[\frac{v_b^3}{0.24+v_b^3}\right] \left[\frac{100}{g_*}\right]^{1/3} \tag{1}$$

while the peak frequency is given by

$$f_{\text{coll}} \simeq 5.2 \times 10^{-3} \text{mHz} \left[\frac{\beta}{H_*}\right] \left[\frac{T_*}{100 \text{ GeV}}\right] \left[\frac{g_*}{100}\right]^{1/6} \tag{2}$$

The exact value of α and β (as well as the fluid velocities v_b and u_s and the function κ) depend on the details of the symmetry breaking potential, but typical values for β/H_* (hereafter we used the dimensionless parameter β/H_* instead of β) range up to a few 10^3 [10]. Likewise, α is found in the range 0–1. The parameter g_* is the number of relativistic degrees of freedom at T_*. The number of degrees of freedom is model dependent and is in principle a free parameter in our study. However since the dependence of $\Omega_{\text{coll}} h^2$ on g_* is only $g_*^{1/3}$ we fix the parameter to the Standard Model value of 100 at or above the electroweak temperature. Note that some of the parameters are labeled $(*)$ to denote their values at the transition temperature T_*.

Gravitational waves from bubble collisions are accompanied by another and possibly a more dominant source of gravitational waves, i.e. the turbulence of the primordial plasma. The maximum amplitude and the peak frequency of this source is described in a similar way as the gravitational waves originating from bubble collisions:

$$\Omega_{\text{turb}} h^2(f_{\text{turb}}) \simeq 1.4 \times 10^{-4} u_s^5 v_b^2 \left[\frac{H_*}{\beta}\right]^2 \left[\frac{100}{g_*}\right]^{1/3} \tag{3}$$

while the peak frequency is given by;

$$f_{\text{turb}} \simeq 3.4 \times 10^{-3} \text{mHz} \frac{u_s}{v_b} \left[\frac{\beta}{H_*}\right] \left[\frac{T_*}{100 \text{ GeV}}\right] \left[\frac{g_*}{100}\right]^{1/6} \tag{4}$$

Here, the amplitude at lower frequencies to the frequency with f^2 and at higher frequencies $f^{-7/2}$.

Note that the red-shift of the signal to the present epoch has been included in the description of the signal intensity and frequencies.

3. Foreground and noise

The stochastic background signal as discussed in the previous section is expected to be weak in comparison with the foreground signals and noise experienced in the spacecraft. A study of the stochastic background signal requires therefore detailed knowledge on the instrumental noise and foreground. The foreground signals are dominated by the white dwarfs binaries and are described in many papers [11, 12]. White dwarf binaries are expected to be numerous. They will give rise to individual waveforms, but the addition of many binaries will add up to a stochastic foreground. This foreground signal has a broad bandwidth, ranging from 0.1 down to 10^{-4} or 10^{-5} Hz. Individual binaries may be resolved or may stand out in form of peaks on top of a otherwise broad foreground signal. Because the white dwarf binaries are located in the galactic plane, the foreground is relatively high (compared to the instrumental noise), but due to LISA's orbit around the Sun, the intensity will fluctuates with time with an annual periodicity. This periodicity may prove to be a handle to discriminate the foreground, since the stochastic background from cosmological origin is expected to be an isotropic signal.

4. Simulations and results

Since the existing science simulators for the LISA missions, such as the 'LISA simulator' [13] or 'Synthetic LISA' [14], only allow for point sources to be generated, simulation of an isotropic stochastic background cannot be done yet. A way out is to mimic an isotropic signal as the superposition of many isotropically distributed point sources. This solution is however relatively CPU time consuming and we therefore use the analytic description of the response of LISA [15] and convolve this with the expected stochastic background signal. An advantage of this method is that the generation of a signal spectrum is relatively fast, which allows us to generate the signals for the various input parameters. Given the number of input parameters a fast scan of the parameter space is advantageous. A drawback of the method is that no time trace is generated, which means that filtering algorithms cannot be tested on the signal of interest. A second

The annual averaged LISA response is shown in figure 1. Shown here is the frequency dependent response for the time delay interferometry parameter X. Details of the time delay interferometry (TDI) as used for the LISA mission can be found in [16].

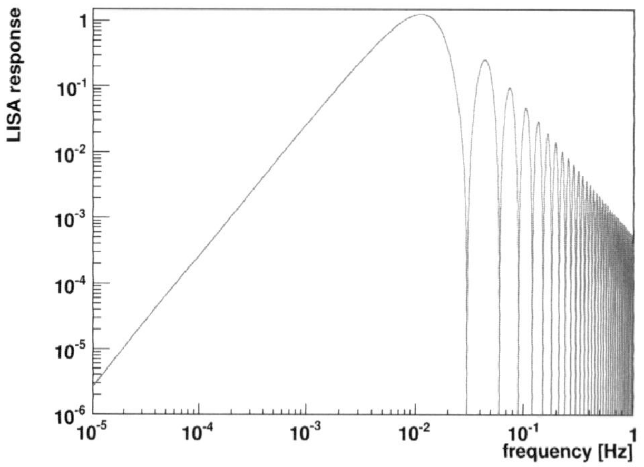

Figure 1. Annual averaged response of the LISA mission for the TDI variable X. The analytical expression for the response is taken from [15].

Journal of Physics: Conference Series **154** (2009) 012046 doi:10.1088/1742-6596/154/1/012046

In figure 2 we show the stochastic background spectrum for several possible values of the three free parameters as discussed in section 2.

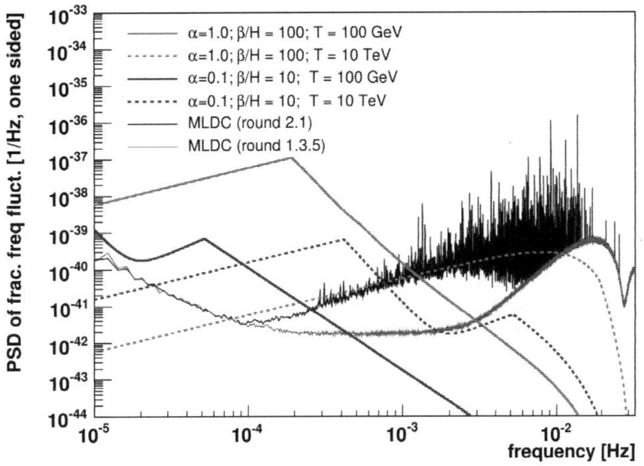

Figure 2. The fractional frequency fluctuation spectrum of the TDI variable X for various signals in the LISA mission. The colored lines indicate several possible realizations of the cosmological stochastic background. The foreground signals from the white dwarf binaries are in black, while the grey line represents the instrumental noise.

The noise and white dwarf binaries foreground are taken from the Mock LISA Data Challenge (MLDC) [17]. This is a project that provides training and blind data sets containing combinations of realistic simulated LISA noise and the signal from typical gravitational wave sources. Participants of this project have to perform data analysis to extract the signals from the noise, and correctly identify the gravitational wave properties. We have used two data as representatives for noise and foreground. The first data set contains consists the LISA instrumental noise (MLDC round 1.3), while the second data set (MLDC round 2.1) consists of both instrumental noise and monochromatic signals from about 30 million white dwarf binaries. Both data sets are obtained in the form of a time trace of 2^{22} data points. Since the cadence of the time trace of the data is 15 seconds, the highest frequency in the spectrum is 0.033 Hz. In figure 2 the frequency spectra of the MLDC are shown together with the expected signal from the stochastic background. The MLDC data sets are transformed using the tools provided by the Synthetic LISA software package to obtain a frequency spectrum.

From figure 2 it is clear that the stochastic background spectra differ widely for the various input parameters and that the sensitivity for these signals varies with free parameters in the phase transition models. A complete simulation of one point in the parameter space involves a number of steps. The simulation is started by choosing a point the parameters space. In this study we varied the three parameters: α, β/H_* and the temperature T_*. The resulting frequency spectrum for each point in parameter space is then convoluted with the response of LISA and added to the data from the LISA Mock Data Challenge. The spectra with and without signal added are compared using a Kolmogorov-Smirnov test [18] for a large range of input parameters. Using this test has the advantage that the spectra can be compared without having to assume an analytical expression for the expected distribution. For each point in the parameter space the Kolmogorov-Smirnov test returned the probability that the null hypothesis (i.e., that the two spectra are the same) is false. Scanning the parameter space in this way provides the knowledge on the details of the phase transition for which the LISA would be sensitive. In figure 3 we show the excluded values of β/H_* as a function of the temperature T_* for two values of α. Two exclusion plots are shown: in figure 3 (a) stochastic background is compared with instrumental noise only, while in (b) the signal of interest is compared with the white dwarf binaries. In figure

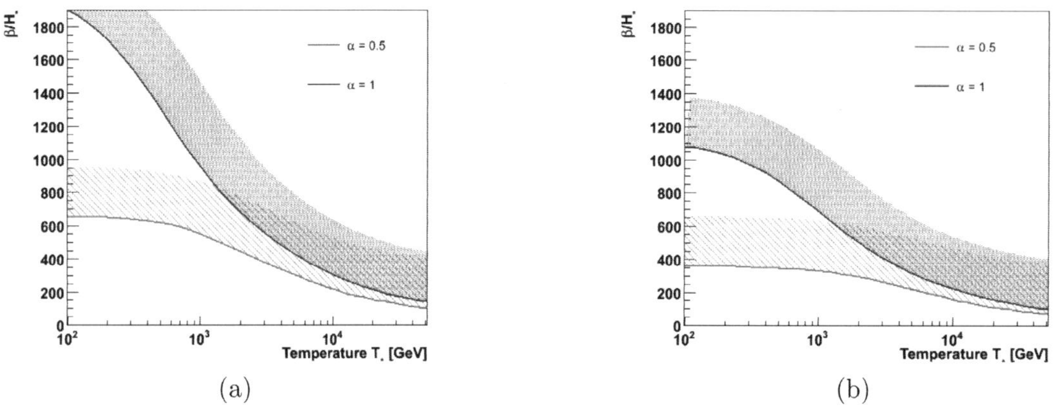

Figure 3. Excluded values for β_*/H as function of the temperature T_* at the phase transition for two values of α In (a) the spectrum from stochastic background is compared to LISA noise only, while in (b) the white dwarf foreground is added. The excluded region is on the shaded side of the solid lines.

4 we show similar exclusion plots, but here we show the excluded values for α as a function of the temperature T_*. From both figure 3 and 4 we see that the LISA mission is most sensitive to lower values for β_*/H and higher values of α, i.e. for a high nucleation rate and a large amount of released latent heat during the phase transition. Moreover we see that the presence of white dwarf binaries foreground largely affects the sensitivity. To increase the LISA's potential to discover gravitational waves from cosmological origin it is therefore important to remove these foreground signals from the data.

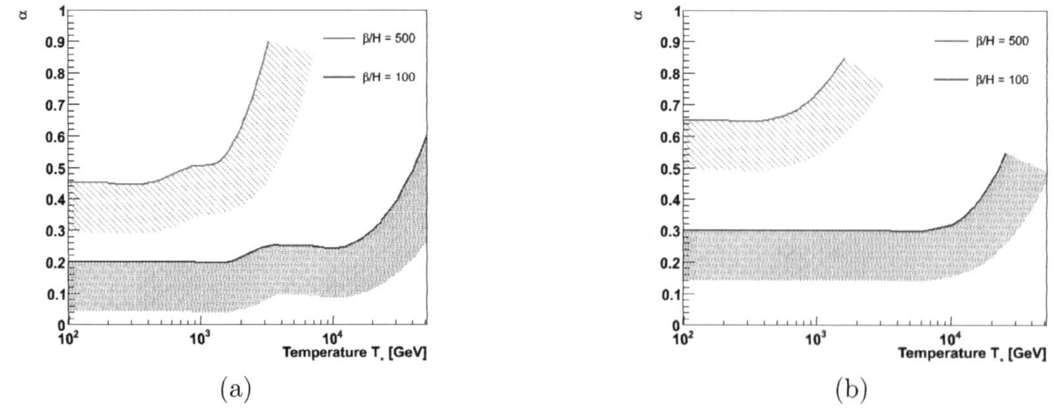

Figure 4. Excluded values for α as function of the temperature T_* at the phase transition for two values of $beta_*/H$. In (a) the spectrum from stochastic background is compared to LISA noise only, while in (b) the white dwarf foreground is added. The excluded region is on the shaded side of the solid lines.

5. Conclusion

We have studied and simulated the stochastic background from first order phase transitions in the early universe. Using the analytical expression for the response of LISA we have been able to compare the signal of interest to a realistic simulated dataset of noise sources and signals from astrophysical sources. Data sets that included the noise and foreground, were taken from the Mock LISA Data Challenge. To get a first estimation of LISA's sensitivity to the gravitational waves from cosmological origin, the statistical significance has been determined between the spectra containing the noise sources and foreground signals with and without the signal of interest.

Based on the study we arrive at the following conclusions. The stochastic background signal from cosmological origin may be observed with LISA if the released latent heat is strong enough (i.e. large value for α) in combination with a large bubble nucleation rate (i.e. low value for β_*/H). Moreover we see that LISA's sensitivity is the largest for the temperatures at which the phase transition occurs, i.e. below a few TeV. At higher temperatures the spectrum from the stochastic background waves shift to higher frequencies and out of the sensitive region of LISA. Due to a cadence of the time trace in the data under study, the maximum frequency in our study was 0.033 Hz. A similar study with a smaller value for cadence could provide a better determination of the sensitivity at higher frequencies.

Finally we have shown that the suppression of the white dwarf foreground substantially increases the discovery potential of the mission for gravity-waves from symmetry breaking in the early universe, in particular if the symmetry breaking occurs at an energy scale between 100 GeV and 1 TeV.

References

[1] Danzmann K 1997 *Classical and Quantum Gravity* **14** 1399
[2] Turner M S and Wilczek F 1990 *Phys. Rev. Lett.* **65** 3080
[3] Witten E 1984 *Phys. Rev. D* **30** 272–285
[4] Kamionkowski M, Kosowsky A and Turner M S 1994 *Phys. Rev. D* **49** 2837
[5] Grojean C and Servant G 2007 *Phys. Rev. D* **75** 043507
[6] Maggiore M 2000 *Physics Reports* **331** 283
[7] Nicolis A 2004 *Classical and Quantum Gravity* **21** L27
[8] Kosowsky A, Turner M S and Watkins R 1992 *Phys. Rev. Lett.* **69** 2026–2029
[9] Kosowsky A, Turner M S and Watkins R 1992 *Phys. Rev. D* **45** 4514–4535
[10] Apreda R, Maggiore M, Nicolis A and Riotto A 2002 *Nuclear Physics B* **631** 342–368
[11] Hils D, Bender P L and Webbink R F 1990 *Astrophys. J.* **360** 75
[12] Nelemans G, Yungelson L R and Portegies Zwart S F 2001 *Astron. Astrophys.* **375** 890
[13] Cornish N, Rubbo L and Poujade O The LISA simulator http://www.physics.montana.edu/LISA/
[14] Vallisneri M 2005 *Phys. Rev. D* **71** 022001
[15] Larson S L, Hellings R W and Hiscock W A 2002 *Phys. Rev. D* **66** 062001
[16] Armstrong J W, Estabrook F B and Tinto M 1999 *Astrophys. J.* **527** 814
[17] Mock LISA data challenge http://astrogravs.nasa.gov/docs/mldc/
[18] Press W H, Teukolsky S A, Vetterling W T and Flannery B P 1992 *Numerical recipes in C. The art of scientific computing* (Cambridge: University Press, —c1992, 2nd ed.)

Time-delay interferometry and the relativistic treatment of LISA optical links

S V Dhurandhar
IUCAA, Postbag 4, Ganeshkind, Pune - 411 007, India.
E-mail: sanjeev@iucaa.ernet.in

Abstract.
In order to attain the requisite sensitivity for LISA - a joint space mission of the ESA and NASA- the laser frequency noise must be suppressed below the secondary noises such as the optical path noise, acceleration noise etc. By combining six appropriately time-delayed data streams containing fractional Doppler shifts - a technique called time delay interferometry (TDI) - the laser frequency noise may be adequately suppressed. Here we investigate the problem of TDI in the general case of unequal up-down links and also include the effect of the Earth on the spacecraft and the optical links. We show that there are symmetries in the physics which can be successfully used to simplify the algebra of the TDI. We finally give the example of the first generation modified Sagnac observable in which the laser frequency noise is suppressed because of the symmetries.

1. Introduction

LISA - Laser Interferometric Space Antenna - is a proposed mission which will use coherent laser beams exchanged between three identical spacecraft forming a giant (almost) equilateral triangle of side 5×10^6 kilometres to observe and detect low frequency cosmic GW [1].

Laser frequency noise dominates the other secondary noises, such as optical path noise, acceleration noise by 7 or 8 orders of magnitude, and must be removed if LISA is to achieve the required sensitivity of $h \sim 10^{-22}$, where h is the metric perturbation caused by a gravitational wave. In LISA, six data streams arise from the exchange of laser beams between the three spacecraft approximately 5 million km apart. These six streams produce redundancy in the data which can be used to suppress the laser frequency noise by the technique called time-delay interferometry (TDI) in which the six data streams are combined with appropriate time-delays [2]. A mathematical foundation for the TDI problem for LISA was given in [3], by showing that the data combinations constituted an algebraic structure; the data combinations cancelling laser frequency noise formed the *module of syzygies* over the polynomial ring of time-delay operators.

We extend the work of [3] in which we had a commutative polynomial ring over the three time-delay operators. Because in the general case, the up-down links are unequal and because of the flexing of the arms, we have a non-commutative ring in six indeterminates. We show in this paper that the algebraic structure is still a module over this non-commutative polynomial ring. Secondly, for the LISA model that we consider, which has been optimised to reduce flexing, we have found symmetries which could be used to simplify the algebra. We discuss these symmetries and their implications; the symmetries can be used to construct a smaller ring - the original non-commutative ring is quotiented by the ideal generated by commutators which capture the symmetries. For the model of LISA we use the orbits given in [4] in which the orbits have been computed in the Sun and Earth's field (we have ignored the effects of Jupiter and other planets). We show that optimising the spacecraft orbits for minimal flexing is conducive to suppressing the laser frequency noise. For some TDI observables their modified first generation form may prove adequate; in this context we discuss the Sagnac observable. These computations will be useful in the development of a LISA simulators, the LISACode for instance [5].

2. The flexing of LISA's arms in the Sun's and Earth's field

The orbits of the spacecraft are chosen as follows [4]: first, we choose a set of Keplerian orbits in the Sun's field only which give a minimum flexing of the arms - the peak to peak variation in armlengths is the least $\sim$ 48000 km, see [6]. In the second step, on these base orbits we linearly add the perturbations due to the Earth's gravitational field. This choice is so, because we believe, that the optimum solution which yields the minimum flexing of the arms in the combined field of the Sun and Earth, lies close to the optimum solution already found in the field of the Sun only. We briefly describe the orbits below: we choose the Sun as the origin with Cartesian coordinates $\{X, Y, Z\}$ as follows: the ecliptic plane is the $X - Y$ plane and we consider a circular reference orbit of radius $R = 1$ A. U. centred at the Sun. Let $\delta_0 = 5\alpha/8$ where $\alpha = L_0/2R$ and $L_0 \sim 5,000,000$ km represents the nominal distance between two spacecraft of the LISA configuration. We choose the tilt of the plane of the LISA triangle to be $\delta = \pi/3 + \delta_0$ which has been shown in [6] to yield minimum flexing of the arms. We choose spacecraft 1 to be at its lowest point (maximum negative Z) at $t = 0$. This means that at this point, $Y = 0$ and $X \simeq R(1 - e)$. The orbit of spacecraft 1 is an ellipse with inclination angle $\sim \alpha$ with the ecliptic, eccentricity $e \sim \alpha/\sqrt{3}$. This fixes the orbit of spacecraft 1. The orbits of of spacecraft 2 and 3 are obtained by rotating the orbit of spacecraft 1 about the Z-axis by the angles 120° and 240° respectively, their phases are adjusted so that they maintain a distance of $\sim L_0$ between each other. To the first order in the eccentricity, the spacecraft form a rigid equilateral triangle of side L_0. It is only at the second order in the α (or eccentricity) that the flexing of the arms occurs.

The Earth's field is now included perturbatively via the CW framework [7]. The CW frame is chosen as follows: We take the reference particle to be orbiting in a circle of radius R with constant Keplerian angular velocity Ω. Then the transformation to the CW frame $\{x, y, z\}$ from the barycentric frame $\{X, Y, Z\}$ is given by,

$$x = X\cos\Omega t + Y\sin\Omega t - R, \; y = -X\cos\Omega t + Y\sin\Omega t, \; z = Z. \tag{1}$$

Consistent with the Keplerian orbits described above and to the first order in eccentricity, for the kth spacecraft, we have the following position coordinates in the CW frame:

$$x_k = -\frac{1}{2}\rho_0 \cos\phi_k,\ y_k = \rho_0 \sin\phi_k,\ z_k = -\frac{\sqrt{3}}{2}\rho_0 \cos\phi_k, \tag{2}$$

where $\rho_0 = L_0/\sqrt{3}$ and $\phi_k = \Omega t - 2\pi(k-1)/3 - \phi_0$.

LISA follows the Earth 20° behind. We consider the model where the centre of the Earth leads the origin of the CW frame by 20° - thus in our model, the 'Earth' or the centre of force representing the Earth, follows the circular reference orbit of radius 1 A. U. Also the Earth is at a fixed position vector $\mathbf{r}_\oplus = (x_\oplus, y_\oplus, z_\oplus)$ in the CW frame. We find that $x_\oplus = -R(1-\cos 20°) \sim -9\times 10^6$ km, $y_\oplus = R\sin 20° \sim 5.13\times 10^7$ km and $z_\oplus = 0$. We introduce the small parameter ϵ in terms of the quantity $\omega_\oplus^2 = GM_\oplus/d_\oplus^3$, where $d_\oplus$ is the distance of the Earth from the origin of the CW frame, $M_\oplus \sim 5.97\times 10^{24}$ kg is the mass of the Earth and $G = 6.67\times 10^{-11}$ kg^{-1}m^{3}sec^{-2} Newton's gravitational constant. We find $d_\oplus \sim 5.2\times 10^7$ km. We define $\epsilon = \omega_\oplus^2/\Omega^2 \simeq 7.16\times 10^{-5}$ which is just the ratio of the tidal forces due to the Earth and the Sun. The CW equations including the Earth's field take the form:

$$\begin{aligned} \ddot{x} - 2\Omega\dot{y} - 3\Omega^2 x + \epsilon\Omega^2(x - x_\oplus) &= 0, \\ \ddot{y} + 2\Omega\dot{x} + \epsilon\Omega^2(y - y_\oplus) &= 0, \\ \ddot{z} + \Omega^2(1+\epsilon)z &= 0. \end{aligned} \tag{3}$$

Note that the compounded flexing due to the combined field of Earth and Sun is a nonlinear problem; it is infact a three body problem. We however solve this problem approximately. Assuming that both effects are small we may linearly add the flexing vectors due to the Sun and Earth; the nonlinearities appear at higher orders in α and ϵ. These would modify the flexing but we may neglect this effect because of the smallness.

We then seek perturbative solutions to Eq. (3) to the first order in ϵ. We write, $x = x_0 + \epsilon x_1, y = y_0 + \epsilon y_1, z = z_0 + \epsilon z_1$ where x_0, y_0, z_0 are solutions at the zeroth order given by Eq.(2). With the initial conditions: $x_1 = y_1 = z_1 = \dot{x}_1 = \dot{y}_1 = \dot{z}_1 = 0$ at $t = 0$, we solve the equations for x_1, y_1 and z_1. We then add the perturbation $\epsilon\mathbf{r}_1 = \epsilon(x_1, y_1, z_1)$ to the Keplerian orbit of each spacecraft to obtain the orbits in the combined gravitational field of the Sun and Earth [4]. The time-delay required for the TDI needs to be known to at least to 1 part in 10^8 - about few metres - for the laser frequency noise to be suppressed. In order to guarantee such level of accuracy, we numerically integrate the null geodesics followed by the laser ray. Our numerical scheme is accurate to better than 10 metres. In figure 1 we plot the rate of change of armlengths (flexing) for the six optical links. Including the Earth's field the flexing still remains $\lesssim 6$ m/sec in the first two years and increases to $\lesssim 8$ m/sec in the third year. Earlier estimates were ~ 10 m/sec.

3. Algebraic approach to TDI: polynomial equations and the module of syzygies

For stationary LISA in flat spacetime, there are only three delay operators corresponding to the three armlengths and the time-delay operators commute. However, in a more

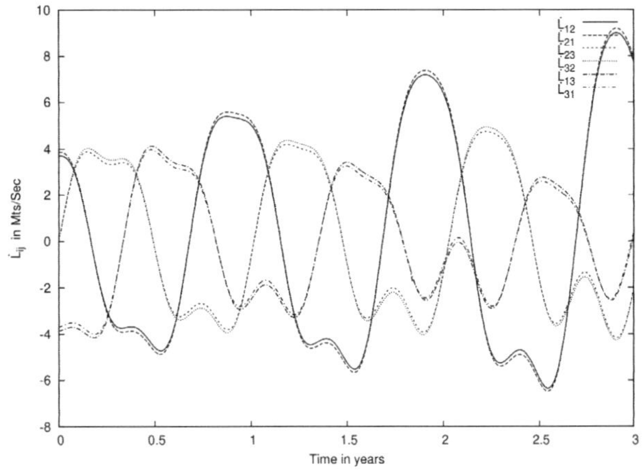

Figure 1. The rate of change of armlengths for the six links is shown in units of m/sec for $\phi_0 = 0$. This rate of change is less than 6 m/sec upto the second year and increases to a maximum of about 8 m/sec in the third year.

realistic LISA model the arms flex and the up-down links are unequal which gives six non-commuting time-delay operators - the second generation TDI.

We follow the notation and conventions of [8] and [3] which are the simplest for our purpose. The six links are denoted by $U^i, V^i, i = 1, 2, 3$. The time-delay operator for the link U^2 from S/C 1 to S/C 2 or $1 \longrightarrow 2$ is denoted by x in [8] and so on in a cyclic fashion. The delay operators in the other sense are denoted by l, m, n; the link $-V^1$ from $2 \longrightarrow 1$ by l and similarly the links V^2, V^3 are defined through cyclic permutation. Figure 2 depicts the optical links as described.

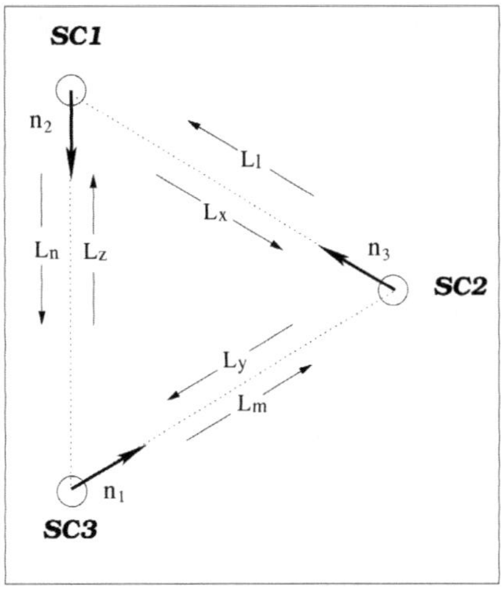

Figure 2. A schematic diagram of LISA's optical links.

Let $C_i(t) = \Delta\nu_i(t)/\nu_0$ represent the laser frequency noise in S/C i. Let j be the delay operator corresponding to the variable armlength $L_j(t)$, i.e. $jC_i(t) = C_i(t - L_j(t))$. Then we have, $U^1 = C_1 - zC_3$, $V^1 = lC_2 - C_1$. The other links in terms of $C_i(t)$ are obtained by cyclic permutations. Also in the U^i, V^i we have not included contributions from the secondary noises, gravitational wave signal etc. since here our aim is to deal with laser frequency noise only. Any observable X is written as: $X = p_i V^i + q_i U^i$, where $p_i, q_i, i = 1, 2, 3$ are polynomials in the variables x, y, z, l, m, n. Thus X is specified by giving the six tuple polynomial vector (p_i, q_i). Writing out the (V_i, U_i) in terms of the laser noises $C_i(t)$, and in order that the laser frequency noise cancel for arbitrary functions $C_i(t)$, the polynomials (p_i, q_i) must satisfy the equations:

$$\begin{aligned} p_1 - q_1 + q_2 x - p_3 n &= 0, \\ p_2 - q_2 + q_3 y - p_1 l &= 0, \\ p_3 - q_3 + q_1 z - p_2 m &= 0. \end{aligned} \tag{4}$$

The solutions to these equations as realised in earlier works are important, because they consist of polynomials with lowest possible degrees and thus are simple [2].

Since these are linear equations they define a homomorphism of modules as follows: Eliminating p_1 and p_2 from the three equations (4) while respecting the order of the variables we get:

$$\psi(x, y, z, l, m, n) \equiv p_3(1 - nlm) + q_1(z - lm) + q_2(xl - 1)m + q_3(ym - 1) = 0 \,. \tag{5}$$

Consider the polynomial ring $\mathcal{Q}(x, y, z, l, m, n) \equiv \mathcal{K}$, in general non-commutative, of polynomials in the six variables x, y, z, l, m, n and coefficients in the rational field $\mathcal{Q}$. The operators x, y, z, l, m, n play the role of indeterminates. Eq. (5) defines a homomorphism $\varphi : \mathcal{K}^4 \longrightarrow \mathcal{K}$ where any polynomial vector $(p_3, q_1, q_2, q_3) \in \mathcal{K}^4$ is mapped to the polynomial $\psi(x, y, z, l, m, n) \in \mathcal{K}$. The set of noise free TDI combinations is just the kernel of this homomorphism $\varphi^{-1}(0) \subset \mathcal{K}^4$ which is a submodule of $\mathcal{K}^4$. This is called the *first module of syzygies* over the polynomial ring $\mathcal{K}$. This homomorphism can be extended to $\mathcal{K}^6$ via the elimination equations for p_1 and p_2. Thus one obtains a module of noisefree TDI observables $\mathcal{M} \subset \mathcal{K}^6$ which is isomorphic to $\varphi^{-1}(0)$.

In case of non-commuting operators the problem is far more complex than the commutative case. If we follow on the lines of the commutative case, the first step would be to find a Gröbner basis for the ideal generated by the coefficients appearing in Eq. (5), namely, the set of polynomials $\{1 - nlm, z - lm, (xl - 1)m, ym - 1\}$. Here we do not even know whether the Gröbner basis for this ideal would be finite - the algorithm for finding the basis may not terminate. However, here we show that simplifications are possible because of the inherent symmetries in the problem and so the ring $\mathcal{K}$ can be quotiented by a certain ideal, simplifying the algebraic problem. One then needs to deal with a 'smaller' ring which is perhaps simpler. We now estimate the level of non-commutativity of these operators in the context of our LISA model.

4. The role of symmetries and the construction of the quotient ring

The level of non-commutativity can be found by computing commutators which occur in several of the well known TDI observables like the Michelson, Sagnac etc. We find that given our model of LISA, we require to go only upto the first order in $\dot{L}$; we find for our model $\ddot{L} \sim 10^{-6}$ metres/sec^2 and thus even if one considers say 6 successive optical paths, that is, about $\Delta t \sim 100$ seconds of light travel time, $\Delta t^2 \ddot{L} \sim 10^{-2}$ metres. This is well below few metres and thus can be neglected in the residual laser noise computation. Moreover, $\dot{L}^2$ terms (and higher order) can be dropped since they are of the order of $\lesssim 10^{-15}$ (they come with a factor $1/c^2$) which is much smaller than 1 part in 10^8. The calculations which follow neglect these terms.

Applying the operators twice in succession and dropping higher order terms as explained above,

$$\begin{aligned} k_2 k_1 C &= C(t - L_{k_1}(t - L_{k_2}) - L_{k_2}), \\ &\approx C(t - L_{k_1} - L_{k_2}) + L_{k_2} \dot{L}_{k_1} \dot{C}(t - L_{k_1} - L_{k_2}). \end{aligned} \tag{6}$$

The above formula can be easily generalised by induction to n operators. In several of the TDI observables, commutators play a major role. In the Sagnac observable, for example, the following commutator occurs:

$$\begin{aligned} lmnxyz - xyzlmn &= (L_l + L_m + L_n)(\dot{L}_x + \dot{L}_y + \dot{L}_z) \\ &\quad -(L_x + L_y + L_z)(\dot{L}_l + \dot{L}_m + \dot{L}_n), \end{aligned} \tag{7}$$

where it is understood that the LHS acts on C while the RHS multiplies $\dot{C}$ at an appropriately delayed time. Note that the term in C cancels out on the RHS. Further, we observe the following approximate symmetries in our model:

$$\dot{L}_x \approx \dot{L}_l, \quad \dot{L}_y \approx \dot{L}_m, \quad \dot{L}_z \approx \dot{L}_n, \tag{8}$$

In our model, $|\dot{L}_x - \dot{L}_l| \lesssim 0.8$ m/sec and the sum $|(\dot{L}_x + \dot{L}_y + \dot{L}_z) - (\dot{L}_l + \dot{L}_m + \dot{L}_n)| \lesssim 1$ m/sec upto the first three years in our model. Thus the commutators:

$$[x, l] \approx 0, [y, m] \approx 0, [z, n] \approx 0, \quad [xyz, lmn] \approx 0. \tag{9}$$

The near vanishing of the above commutators implies that a vast simplification in the algebra is possible. In particular the pairs $(x, l), (y, m), (z, n)$ approximately commute.

Apart from the above approximate symmetries there are other 'exact' symmetries (after dropping terms in $\dot{L}^2$ and $\ddot{L}$ and higher order) which lead to vanishing commutators. Consider the commutator $[x_1 x_2 ... x_n, y_1 y_2 ... y_n]$ where $n \geq 2$ and x_k or y_m represents any one of the delay operators x, y, z, l, m, n. Upto the order of approximation we are working in:

$$[x_1 x_2 ... x_n, y_1 y_2 ... y_n] = \sum_{k=1}^{n} L_{x_k} \sum_{m=1}^{n} \dot{L}_{y_m} - \sum_{m=1}^{n} L_{y_m} \sum_{k=1}^{n} \dot{L}_{x_k}. \tag{10}$$

From this equation it immediately follows that if the operators $y_1, y_2, ..., y_n$ are a permutation of the operators $x_1, x_2, ..., x_n$, and then the commutator,

$$[x_1 x_2 ... x_n, y_1 y_2 ... y_n] = 0. \tag{11}$$

We describe below how these vanishing commutators can be used to simplify the algebra.

The commutators can be used to construct an ideal $\mathcal{U} \subset \mathcal{K}$. We first construct the ideal $\mathcal{U}$ generated by the commutators such as those given in Eq. (9) and Eq. (11), that is, the hybrid and the exact ones. Then we quotient the ring $\mathcal{K}$ by $\mathcal{U}$, thereby constructing a smaller ring $\mathcal{K}/\mathcal{U} \equiv \bar{\mathcal{K}}$. Further, we can translate the equations given in Eq. (4) to ones over $\bar{\mathcal{K}}$ as follows: Each polynomial $\bar{p} \in \bar{\mathcal{K}}$ is an equivalence class of polynomials $p + \mathcal{U} \subset \mathcal{K}$. The solution set of such polynomial vectors, which we now denote by $(\bar{p}_i, \bar{q}_i) \in \bar{\mathcal{K}}^6$ still form a module over $\bar{\mathcal{K}}$. A future goal would be to construct this module or obtain the generators of this module.

The simplification in the ring $\mathcal{K}$ can be seen as follows: Consider just two of the operators x and l and consider the possible polynomials or monomials (polynomials with just one term) in them in $\mathcal{K}$ and $\bar{\mathcal{K}}$ for different degrees. We have $xl = lx + xl - lx = lx + [x, l] = lx$ in $\bar{\mathcal{K}}$ (we have dropped the overbars to avoid clutter). So we gain by the fact that the number of monomials in $\bar{\mathcal{K}}$ are reduced, we can identify lx with xl as just a single monomial. The higher the degree of the monomials, the greater is the number of identifications and more is the simplification in the algebra.

5. Residual laser frequency noise in the Sagnac observable

By the time LISA flies the expectations are for the laser frequency noise estimate to reduce to say $\widetilde{\Delta\nu} \sim 10\mathrm{Hz}/\sqrt{\mathrm{Hz}}$ [9]. If we divide this number by the laser frequency $\nu_0 \sim 3 \times 10^{14}$ Hz, we obtain the noise estimate in the fractional Doppler shift C with the power spectral density (PSD), $S_C(f) = \langle|\tilde{C}(f)|^2\rangle \sim 10^{-27}$ Hz^{-1}, where $\tilde{C}(f)$ is the Fourier transform of $C(t)$.

The modified Sagnac first generation TDI observable α is given by the six tuple polynomial vector $\alpha = (\kappa, \kappa l, \kappa lm, \eta, \eta zy, \eta z)$, where $\kappa = 1 - zyx$ and $\eta = 1 - lmn$. The flexing leads to the residual term $\Delta C = [zyx, lmn]C_1 = \Delta t \dot{C}_1$ where by Eq. (7):

$$\Delta t(t) = \frac{1}{c^2}[(L_x + L_y + L_z)(\dot{L}_l + \dot{L}_m + \dot{L}_n) - (L_l + L_m + L_n)(\dot{L}_x + \dot{L}_y + \dot{L}_z)], \tag{12}$$

Then the PSD of ΔC is $S_{\Delta C}(f;t) = 4\pi^2 \Delta t(t)^2 f^2 S_C(f)$. This noise must be compared with the secondary noise [1] which is given by $S_\alpha(f) = 4\sin^2(3\pi f L_0)\{[8\sin^2 3\pi f L_0 + 16\sin^2 \pi f L_0]S_{acc} + 6S_{opt}\}$, where $S_{acc} = 2.5 \times 10^{-48}(f/1\mathrm{Hz})^{-2}\mathrm{Hz}^{-1}$ and $S_{opt} = 1.8 \times 10^{-37}(f/1\mathrm{Hz})^2\mathrm{Hz}^{-1}$. In the Figure 3 we plot $S_\alpha(f)$ and $S_{\Delta C}(f;t)$ at three epochs an year apart. For more details see [4].

We see that, clearly the residual laser frequency noise is few orders of magnitude below the secondary noises. Analogous result was demonstrated in [10]. Since the other Sagnac variables β and γ are obtained by cyclic permutations of the spacecraft, the residual laser noise is similarly suppressed in them. Note that the only polynomial that occurs in ΔC is the commutator $[zyx, lmn]$ which is in $\mathcal{U}$ and so α satisfies the three equations (4) approximately, therefore is an element of the module we are seeking.

6. Concluding remarks

We have described in this paper the algebraic approach to the problem of TDI in the general case of the LISA model in which the up and down links are unequal and the arms flex. This gives rise to a polynomial ring in six variables which in general do not

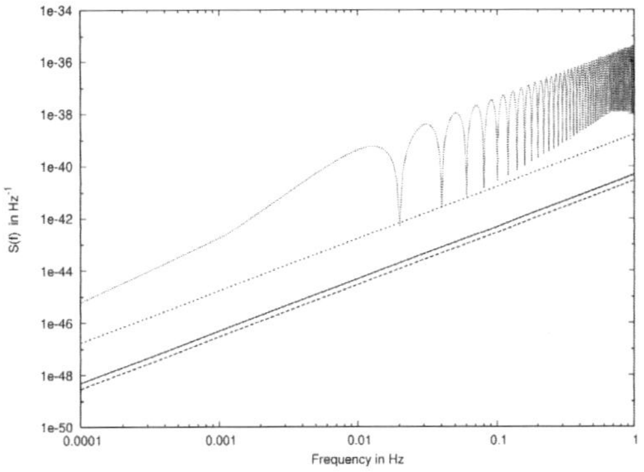

Figure 3. The 'top' curve shows the PSD $S_\alpha(f)$ of the secondary noises. The straight lines are the PSDs of the residual noise at three epochs chosen an year apart. Clearly the residual laser noise is seen to be adequately below the secondary noises.

commute. However, we emphasise that symmetries exist in the physics and show how they could be exploited to simplify the algebra -the symmetries essentially simplify the ring. It may be possible to extend the Gröbner basis methods which worked so well in the commutative case.

We have computed the residual laser frequency noise in one of the important TDI variables, namely, the Sagnac. The residual noise is satisfactorily suppressed because of the optimised orbits and the symmetries. Thus by optimising LISA's orbits it may be possible to suppress the laser frequency noise in other of TDI observables as well [4].

Acknowledgments
The author would like to thank the Indo-French Centre for the Promotion of Advanced Research (IFCPAR) project no. 3504-1 under which this work has been carried out. Much of this work is in collaboration with J-Y Vinet and R. Nayak.

[1] Bender P *et al.* 2000 *LISA: A Cornerstone Mission for the Observation of Gravitational Waves* System and Technology Study Report ESA-SCI (2000) 11
[2] Armstrong J W 2008 *lrr-2006-1 : http://relativity.livingreviews.org/Articles/lrr-2008-2*
[3] Dhurandhar S V, Nayak K R and Vinet J-Y 2002 *Phys. Rev* D **65** 102002
[4] Dhurandhar S V, Vinet J-Y and Nayak K R 2008 *Class. Quantum Grav.* **25** 245002
[5] Petiteau A, Auger G, Halloin H, Jeannin O, Pagnol E, Pireaux S, Regimbau T and Vinet J-Y 2008 *Phys. Rev.* D **77** 023002
[6] Dhurandhar S V, Nayak K R, Koshti S and Vinet J-Y 2005 *Class. Quantum Grav.* **22** 481; Nayak K R, Koshti S, Dhurandhar S V and Vinet J-Y 2006 *Class. Quantum Grav.* **22** 1763
[7] Clohessy W H and Wiltshire R S 1960 *Journal of Aerospace Sciences* 653;
Vallado D A 2001 *Foundations of Astrodynamics and Applications* (2nd edition 2001, Microcosm Press Kluwer)
[8] Nayak K R and Vinet J-Y 2004, *Phys. Rev.* D **70** 102003
[9] Brillet A (private communication)
[10] Shaddock D A, Tinto M, Eastabrook F B and Armstrong J W 2003 *Phys. Rev.* D **68** 061303

Journal of Physics: Conference Series **154** (2009) 012048 doi:10.1088/1742-6596/154/1/012048

The gravitational wave emission from white dwarf interactions in globular clusters

P Lorén–Aguilar[1,2], J A Lobo[3,2], J Isern[3,2] and E García–Berro[1,2]

[1]Departament de Física Aplicada, Escola Politècnica Superior de Castelldefels, Universitat Politècnica de Catalunya, Av. del Canal Olímpic, s/n, 08860 Castelldefels, Spain

[2]Institute for Space Studies of Catalonia, c/Gran Capità 2–4, Edif. Nexus 104, 08034 Barcelona, Spain

[3]Institut de Ciències de l'Espai, CSIC, Campus UAB, Facultat de Ciències, Torre C-5, 08193 Bellaterra, Spain

E-mail: loren@fa.upc.edu

Abstract. In the dense central regions of globular clusters close encounters of two white dwarfs are relatively frequent. The estimated frequency is one or more strong encounters per star in the lifetime of the cluster. Such encounters should be then potential sources of gravitational wave radiation. Thus, it is foreseeable that these collisions could be either individually detected by LISA or they could contribute significantly to the background noise of the detector. We compute the pattern of gravitational wave emission from these encounters for a sufficiently broad range of system parameters, namely the masses, the relative velocities and the distances of the two white dwarfs involved in the encounter.

1. Introduction

Gravitational waves are a direct consequence of the General Theory of Relativity. Many efforts have been done so far to detect these elusive waves, but the intrinsic experimental difficulties have hampered their detection. However, with the advent of the current generation of terrestrial gravitational wave detectors, like LIGO [1] — which already has completed its fifth science run — VIRGO [2], GEO600 [3], or TAMA [4], and of space-borne interferometers like LISA [5, 6], gravitational wave astronomy is already a reality. It is therefore important to properly compute with the highest possible accuracy the gravitational wave pattern of all the sources of gravitational waves with potential astrophysical interest [7].

The Galactic population of field binary white dwarf is one of the most promising sources of gravitational waves, since it was shown long ago that LISA could detect them individually [8]. It is important to realize that field binary white dwarfs are the natural outcome of the evolution of many binary systems and it can be shown that due to tidal interactions their orbits are usually circular. Thus, the gravitational wave pattern of these systems is easy to compute since the two white dwarfs can be safely assumed to be point masses. Due to the ensuing emission of gravitational waves, the orbital separation of these systems decreases and the final fate of these systems is to coalesce. Consequently, the gravitational wave emission resulting form the merger of such systems has been the subject of some recent studies [9, 10]. These studies show that the resulting pattern of gravitational wave emission is rather simple. In particular, it has been shown that these waveforms do not show a prominent peak and that the gravitational wave signal rapidly fades in a few orbital periods.

Globular clusters are old stellar systems which contain as many as 10^7 stars. The typical stellar densities are $\sim 10^3$ pc^{-3} and, due to their long lifetimes, the central regions of globular clusters contain many collapsed and degenerate objects, such as white dwarfs. In such a dense environment, where stars can pass very close one to another, every star is expected to suffer one or more strong encounters during its lifetime. In fact, it has been predicted [11] that dynamical interactions in globular clusters can form double white dwarf systems which could be eventually detectable by LISA. Moreover, in globular clusters, dynamical interactions can form binary white dwarf systems at an enhanced rate compared to the field [12]. Additionally, such systems are generally formed with non-zero eccentricities. This is an important issue because tidal interactions in eccentric systems substantially modify the gravitational wave emission pattern, thus providing us with a unique opportunity to use LISA to check our understanding of the interiors of white dwarfs [11]. Here we study the dynamical interaction of white dwarfs in globular clusters and we compute the associated gravitational wave radiation. Due to the complicated dynamics of the process, numerical simulations of such encounters are essential to obtain realistic sets of waveforms. In order to do so, we have performed a series of SPH simulations of strong encounters of white dwarfs in globular cluster, covering a broad enough range of system parameters, which include impact parameter and relative stellar velocities for two representative masses of the white dwarfs involved in the interaction.

2. Input physics and method of calculation

We follow the hydrodynamic evolution of the interacting white dwarfs using Smoothed Particle Hydrodynamics (SPH), a Lagrangian particle numerical code [13, 14]. The fact that the method is totally Lagrangian and does not require a grid makes it specially suitable for studying an intrinsically three-dimensional problem. We will not describe in detail the most basic equations of our numerical code, since this is a well-known technique. Instead, the reader is referred to Ref. [15] where the basic numerical scheme for solving the hydrodynamic equations can be found, and to Ref. [9] where all the relevant input physics for the problem at hand and of our specific implementation of this technique are explained in depth. We compute the gravitational wave emission in the slow-motion, weak-field quadrupole approximation [16]. The dimensionless wave strain, h, in the transverse-traceless gauge is given by:

$$h_{jk}^{\rm TT}(t,\mathbf{x}) = \frac{2G}{c^4 d}\frac{\partial^2 Q_{jk}^{\rm TT}(t-R)}{\partial t^2} \tag{1}$$

where $t - R = t - d/c$ is the retarded time, d is the distance to the observer and $Q_{jk}^{\rm TT}(t-R)$ is the quadrupole moment of the mass distribution. For the case of a collection of SPH particles it can be shown (Lorén–Aguilar *et al* 2005) that:

$$\ddot{Q}_{jk}^{\rm TT}(t) \approx P_{ijkl}(\mathbf{N}) \sum_{p=1}^{n} m(p)[2\mathbf{v}^k(p)\mathbf{v}^l(p) + \mathbf{x}^k(p)\mathbf{a}^l(p) + \mathbf{x}^l(p)\mathbf{a}^k(p)] \tag{2}$$

where P_{ijkl} is the transverse-traceless projection operator onto the plane orthogonal to the outgoing wave direction, $\mathbf{N}$, $m(p)$ is the mass of each SPH particle, and $\mathbf{x}(p)$, $\mathbf{v}(p)$ and $\mathbf{a}(p)$ are, respectively, its position, velocity and acceleration.

In order to compute a large enough set of gravitational wave patterns we have relaxed two initial white dwarf models with masses $0.6\,M_\odot$ and $0.8\,M_\odot$, respectively. The mass of the SPH particles were the same in both cases in order to avoid numerical artifacts. Consequently, the $0.6\,M_\odot$ white dwarf was relaxed using 2.0×10^4 particles, whereas the $0.8\,M_\odot$ white dwarf needed 2.6×10^4 particles. For the interactions in which two equal-mass white dwarfs were involved

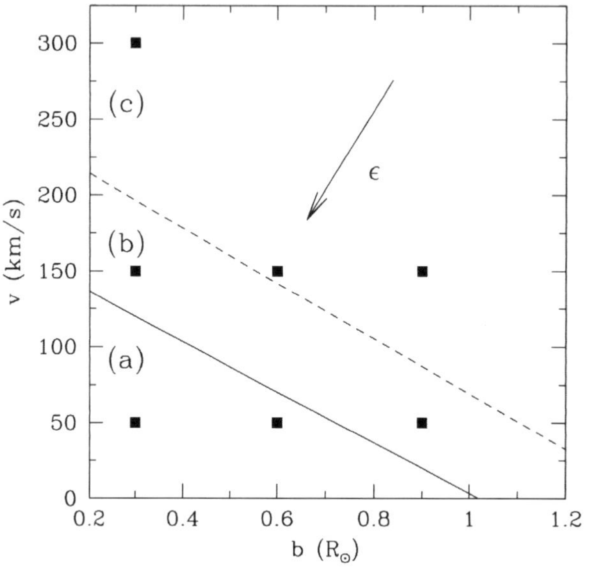

Figure 1. Outcomes of the white dwarf close encounters for the $0.6\,M_\odot + 0.8\,M_\odot$ case as a function of the initial velocity v_i and impact parameter b. In regions (a) and (b) the outcome of the interaction is a merger. In region (a) we obtain head-on impacts, whereas in region (b) a lateral collision occurs. Region (c) corresponds to the region in which an eccentric binary system is formed. The arrow points the direction of increasing eccentricity.

we have used the otherwise typical $0.6\,M_\odot$ model, whereas for the other cases we have used the $0.6\,M_\odot$ and the $0.8\,M_\odot$ models. We have fixed the initial distance x between the stars and their spin (fixed to typical values in field white dwarfs) allowing the impact parameter b (distance between stars on the y direction) and the initial velocity $\vec{v}_i = (v_i, 0, 0)$ to be our free parameters. We have performed 11 simulations with impact parameters ranging from $0.3\,R_\odot$ to $0.9\,R_\odot$ and initial velocities from 50 to 150 km/s. These values of the initial relative velocities and impact parameters are typical of globular clusters.

3. Results

We have obtained three different kinds of behavior, depending on the initial parameters adopted for the close encounter — see Fig. 1 where we show schematically the three different behaviors. When the initial velocity v_i is relatively small and the impact parameters are not very large the gravitational forces bend the initial trajectories and the result of the interaction is a head-on collision. These close encounters correspond to initial velocities and impact parameters of region (a) in Fig. 1. For increasing initial velocities there is a transition region in which the initial trajectories are less affected by the gravitational interaction and the final outcome turns out to be a lateral collision. These interactions occur in the region labelled as (b) in Fig. 1. Finally, for sufficiently large initial velocities and above a threshold impact parameter which depends on the initial relative velocity of the interacting white dwarfs, the stars do not collide but, instead, form an eccentric binary system. This kind of interactions correspond to the region labelled as (c) in Fig. 1. The solid and dashed lines in Fig. 1 separate, respectively, the region of head-on collisions and lateral collisions and the region of lateral collisions and the region in which an eccentric binary system is formed. Finally, the arrow in figure 1 points the direction in which the eccentricity of the newly formed binary system increases.

In this work we will focus in those systems in which an eccentric binary system is formed.

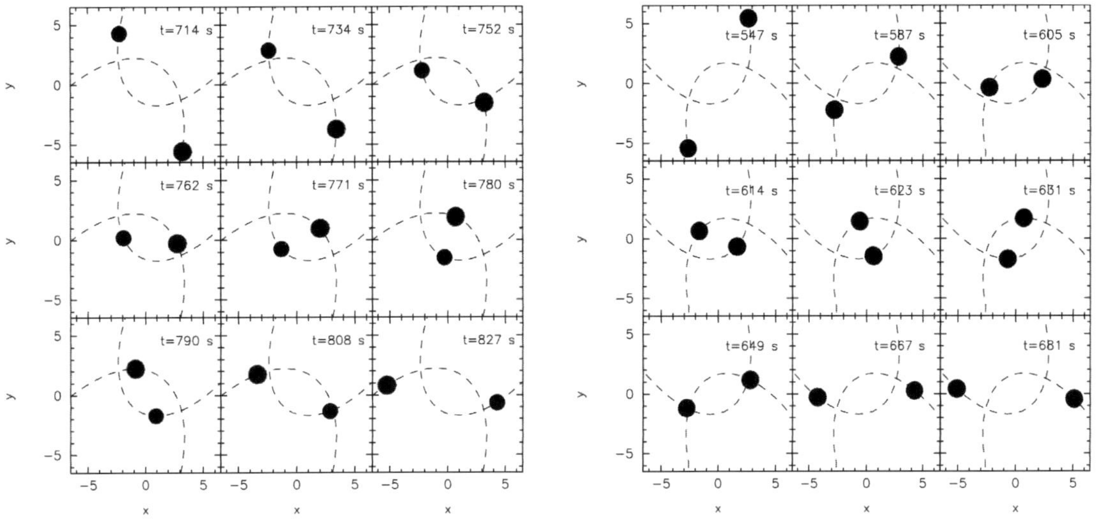

Figure 2. Temporal evolution of the close encounters of the double white dwarf systems discussed here. The left panels show the evolution of the positions of the particles for the $0.6 + 0.8\,M_\odot$ system. The right panels show the evolution of the $0.6 + 0.6\,M_\odot$ system. In both cases the initial conditions were $v_i = 150$ km/s and $b = 0.5\,R_\odot$. The positions of the particles have been projected onto the xy plane. The units of positions are 10^9 cm. Times are shown in the right upper corner of each panel. These figures have been done using the visualization tool SPLASH [17].

That is, those systems which occupy the region of sufficiently large initial velocities and impact parameters in Fig. 1. The rest of the cases shown in Fig. 1 will be discussed at length elsewhere. Fig. 2 shows two typical examples of the temporal evolution of some of the close encounters studied here. The left panels show the case in which two white dwarfs of 0.6 and $0.8\,M_\odot$ are involved and the right panels the case in which two equal-mass white dwarfs interact. The trajectories of the centers of mass of each individual white dwarf are shown using dashed lines. These lines can be used as a visual aid to follow the trajectories of the intervening white dwarfs. For the sake of definitenes we have chosen to show two cases in which the initial conditions are the same, namely $v_i = 150$ km/s and $b = 0.5\,R_\odot$. As can be seen, the systems are well separated and the white dwarfs describe elliptical orbits around the center of mass. In both panels it can be appreciated that no mass transfer occurs between both components. This is so even at the minimum separation of the system. In all the cases studied here, at early times the system does not radiate gravitational waves because the accelerations are very small. Once the stars approach sufficiently each other, the gravitational wave signal rapidly grows. In the case of a head-on collision the signal first grows and then suddenly fades away. In the case of a lateral collision the gravitational wave emission is more irregular and does present several amplitude peaks. This occurs because in a lateral collision before the final merger, the two interacting white dwarfs describe a few orbits of decreasing separation, in which sizeable mass transfer occurs between the two components of the system. For the sake of conciseness we do not show here the corresponding waveforms and we refer the interested reader to a forthcoming publication. Finally, in those cases in which an eccentric binary forms the gravitational wave patterns present a regular, periodical behavior. In Fig. 3 we show the gravitational wave signal emitted by the systems shown in Fig. 2. In particular the dimensionless strains h_+ and $h_\times$ in

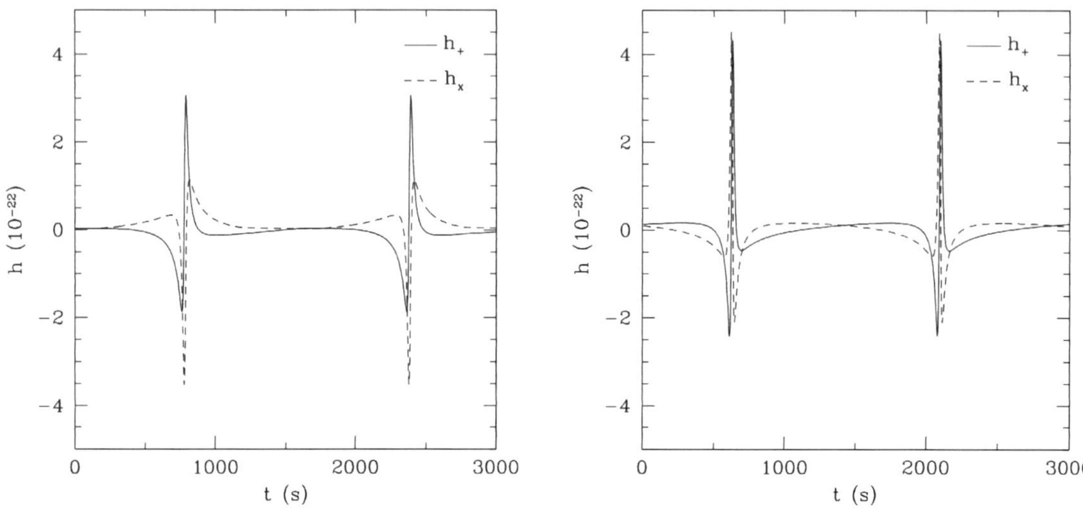

Figure 3. Gravitational wave emission for the cases shown in Fig. 2. The left panel corresponds to the case in which two white dwarfs of masses 0.6 and 0.8 $M_\odot$, respectively, interact to form an eccentric binary. The right panel corresponds to the case in which two equal-mass white dwarfs of 0.6 $M_\odot$ interact to form again an eccentric binary. The relative initial velocities and impact parameter were in both cases $v_i = 150$ km/s and $b = 0.5\, R_\odot$, respctively. The dimensionless strains h_+ and $h_\times$ are measured in units of 10^{-22}. The source is located at a distance of 10 kpc.

units of 10^{-22} are shown. As can be seen the gravitational waveforms are characterized by a sharp peak, which occurs when the white dwarfs are at the minimum orbital separation. These systems will remain bound and after several periods the emission of gravitational waves will circularize and shrink their orbits. However, our numerical method does not allow to follow the evolution of these systems for long times and, thus, we only show in Fig. 3 the first period.

4. Conclusions

We have computed the emission of gravitational waves arising from the dynamical interaction of two white dwarfs in the dense central regions of a globular cluster. These interactions are thought to be quite frequent and thus it is important to characterize the resulting gravitational waveforms. We have done so for a sufficiently broad range of initial relative velocities and impact parameters and for two different component masses. In one set of calculations we have computed the gravitational wave pattern emitted in a close encounter of two equal-mass white dwarfs of 0.6 $M_\odot$, the typical white dwarf mass. The second set of calculations corresponds to close encounters in which an otherwise typical 0.6 $M_\odot$ white dwarf and a rather massive 0.8 $M_\odot$ companion are involved. We have found that there are three possible different outcomes, depending on the initial relative velocity and impact parameter of the system, which lead to distinctive patterns of gravitational wave radiation. In particular, we have characterized the range of parameters for which head-on mergers, lateral collisions and formation of an eccentric binary system occur. For the cases in which the outcome is a collision (either a head-on one or a lateral one) we find that signal is a strong burst of gravitational waves, followed in some cases by a series of smaller peaks. Such events will not be detectable by LISA. On the contrary if the initial conditions are such that the dynamical interaction leads to the formation of an eccentric binary system, the signal is periodical and will probably be detectable with LISA.

Journal of Physics: Conference Series **154** (2009) 012048 doi:10.1088/1742-6596/154/1/012048

Acknowledgments
Part of this work was supported by MEC grants AYA05–08013–C03–01 and 02, ESP2007–61593, by the European Union FEDER funds and by the AGAUR.

References
[1] Abramovici A *et al* 1992 *Science* **256** 325
[2] Acernese F *et al* 2004 *Class. & Quantum Grav.* **21** S385
[3] Willke B *et al* 2004 *Class. & Quantum Grav.* **21** S417
[4] Takahashi R *et al* 2004 *Class. & Quantum Grav.* **21** S403
[5] Bender P L *et al* 2000 *LISA: a conerstone mission for the observation of gravitational waves* ESA-SCI(2000)11 System and Technology Study Report
[6] Danzmann K & Rüdiger A 2003 *Class. & Quantum Grav.* **20** S1
[7] Hughes S A 2006 *Laser Interferometer Space Antenna* Eds.: S.M. Merkowitz & J.C. Livas (AIP Conf. Proc., 823 Melville: New York) p 13
[8] Bender P L *et al* 1998 *"LISA: Laser Interferometer Space Antenna for the detection and observation of GW"*, Pre-fase A Report (Garching: Max Planck-Institut für Quantenoptik)
[9] Guerrero J García–Berro E & Isern J 2004 *A&A* **413** 257
[10] Lorén–Aguilar P Guerrero J Isern J Lobo J A & García-Berro E 2005 *MNRAS* **356** 627-636
[11] Willems *et al* 2007 *ApJ* **665** L59
[12] Shara M M & Hurley J R 2002 *ApJ* **571** 830
[13] Lucy L B 1977 *AJ* **82** 1013
[14] Gingold R A & Monaghan J J 1977 *M.N.R.A.S.* **181** 375
[15] Benz W 1990 in *"The numerical modelling of Nonlinear stellar pulsations"* Ed.: J Buchler (Dordrecht: Kluwer Academic Publishers) 269
[16] Misner C W Thorne K S & Wheeler J A 1973 *"Gravitation"* (New York: W.H. Freeman)
[17] Price D J 2007 *P.A.S.A.* **24** 159

Journal of Physics: Conference Series **154** (2009) 012049 doi:10.1088/1742-6596/154/1/012049

Merger of Massive Black Holes using N-Body Simulations with Post-Newtonian Corrections

Miguel Preto[1], Ingo Berentzen[1], Peter Berczik[1,2], David Merritt[3] and Rainer Spurzem[1]

...

[1] Astronomisches Rechen-Institut, Zentrum für Astronomie, Mönchhofstr. 12-14, 69120 Heidelberg, Germany
[2] Main Astronomical Observatory, National Academy of Sciences of Ukraine, 27 Akademika Zabolotnoho St., 03680 Kyiv, Ukraine
[3] Center for Computational Relativity and Gravitation, Rochester Institute of Technology, 78 Lomb Memorial Drive, Rochester, NY 14623, USA

E-mail: `miguelp@ari.uni-heidelberg.de`

Abstract.
We present preliminary results from self-consistent, high resolution direct N-body simulations of massive black hole binaries in mergers of galactic nuclei. The dynamics of the black hole binary includes the full Post-Newtonian corrections (up to 2.5PN) to its equations of motion. We show that massive black holes starting at separations of 100 pc can evolve down to gravitational-wave-induced coalescence in less than a Hubble time. The binaries, in our models, often form with very high eccentricity and, as a result, reach separations of 50 Schwarzschild radius with eccentricities which are clearly distinct from zero — even though gravitational wave emission damps the eccentricity during the inspiral. These deviations from exact circular orbits, at such small separations, may have important consequences for LISA data analysis.

1. Introduction

Massive black holes (henceforth MBHs) are thought to be ubiquitous in the centers of galaxies. During the last decade, a great deal of effort was made observationally to characterize the physical properties of galactic nuclei and their black holes [1]. A number of phenomenological relations were unveiled: the masses of the putative central MBH appear to be correlated with some physical properties of the host nucleus, namely the luminosity [2] and the velocity dispersion [3, 4]. Although there has been no direct detection yet of a MBH, there is ample observational evidence for the presence of huge mass densities inside the central inner few parsecs of galactic nuclei. In particular, (i) the orbits of single stars indicates the presence of $\sim 4\times10^6 M_\odot$ within the innermost 0.001 pc of the Milky Way [5]; (ii) stellar velocity dispersions in the inner few parsecs of nearby galaxies shows evidence for the presence of $10^{6-9} M_\odot$ in dark mass [1]; (iii) the high luminosities of distant active galactic nuclei and quasars are interpreted as resulting from accretion of gas onto a MBH [6]. Furthermore, dynamical arguments show that no stellar-like systems with such high densities could likely subsist for astrophysically long periods of time [7, 8]. Massive galaxies, however, typically do not evolve in isolation and undergo several merger events during their lifetime. If each of the merging galaxies harbors a MBH at the center, then a MBH binary is expected to form in the core of the resulting remnant. If such MBHBs form and are then able to merge within a Hubble time, gravitational waves (henceforth: GWs) will be detectable by LISA and would constitute an unambiguous demonstration of the existence of these objects.

The paradigm for MBH binary evolution consists essentially of three distinct phases [9]: (i) Right after the merger the two MBHs sink on a dynamical friction time scale, towards the center of the remnant where they form a bound pair; (ii) Once such bound pair is formed, its semimajor axis continues to shrink by dynamical friction until the binary becomes *hard*; afterwards, the binary continues to harden by the ejection of field stars; (iii) When the binary's separation becomes small enough, gravitational wave emission may drive the final stage of relativistic inspiral and coalescence. The transition from (i) to (ii) is understood to be unproblematic, as it typically occurs on a cosmologically short time scale ($\leq 10^{8-9}$ yr.) as long as the mass ratio is $q = M_1/M_2 \geq 0.1$ [10]; the subsequent transition from Phase (ii) to Phase (iii), however, could constitute a bottleneck for the evolution towards relativistic inspiral and coalescence. The binary stalling problem is especially severe in the case of both purely stellar (ie. with no gas), spherically symmetric galaxy models. In this case, the pool of stars available for close interaction with the MBH binary is very rapidly exhausted (roughly on the order of a local dynamical time scale), and leaves the binary dynamically isolated at separations of the order of a parsec, corresponding roughly to the "hard binary separation" $a_h \equiv G\mu_{red}/4\sigma^2$, where μ_{red} is the binary's reduced mass and σ is the velocity dispersion of the surrounding stellar field. This is the so-called *Final Parsec Problem* (hereafter FPP). If the time scale on which stars can diffuse in angular momentum space to enter the binary's loss cone is longer than a Hubble time, then the MBH binary evolution will stall and it will not become a GW source detectable by LISA. There are many proposed solutions to the FPP: dissipation of the binary's orbital energy in a gaseous circumbinary disk [11, 12, 13], massive perturbers [14], the infall of a third (Intermediate-)MBH, a triaxial potential may keep the loss cone full [15, 16, 17], etc.

Here we present preliminary results of *N*-body simulations of idealized models of merging spherical galactic nuclei, starting from an initial separation of tens of parsec. We follow the subsequent formation and evolution of the MBH binary until it reaches the final stages of relativistic inspiral at separations of a few Schwarschild radii ($\sim$ few $\times 10^{-7}$ pc, if the MBHs have masses $\sim 10^6 M_\odot$). The nucleus that results from the merger has a pronounced triaxial structure; triaxiality facilitates the MBH binary's transition to the GW regime, presumably by inducing a collisionless regime which is capable of repopulating the loss cone due to centrophilic orbits. In sec. 2, we describe the galaxy merger set-up and the formation of the bound MBH pair. In sec. 3, we show that the MBH binary reaches the GW-dominated inspiral phase while the merger remnant still retains some degree of triaxiality. In sec 4, we show that the MBH binary may reach the LISA band with an eccentricity different from zero. We conclude with Sec. 5.

2. Simulation of the merger of two spherical galactic nuclei

A number of studies were dedicated to the FPP for MBH binaries — *placed* in the center of a spherical model of a galaxy merger remnant — using the Fokker-Planck formalism, *N*-body simulations, or both [18, 19, 20, 21, 22, 23]. In those models, the loss cone is mostly empty and repopulation of loss cone orbits is driven by two-body relaxation in the diffusive regime [24, 19]. The characteristic time scale for this process is the *relaxation time*, which is proportional to the total number N of stars in the nucleus. For the number of stars present in a typical galactic nucleus, this generally exceeds the Hubble time by a large margin — the exception being the most compact and dense nuclei. From cosmological studies, we expect that the mass function of MBH binaries peaks in the $10^5 - 10^6 M_\odot$ range and at high redshift $z \geq 4$ [25]; in such cases, the diffusive refilling of the loss cone may be marginally efficient, even in the spherical case, to drive the MBHs to coalescence in less than a Hubble time [23].

Nevertheless, the spherical approximation is not only a *worst case scenario* from the point of view of the loss cone dynamics, but it is also highly unrealistic in the sense that no merger remnant will ever exhibit such a high degree of symmetry. It is known that some significant (transient) triaxiality results almost inevitably from a galaxy merger [26] — as well as some net rotation, details depending on stellar and gas content, orbital parameters of the merger, etc. We have recently modelled the galactic nucleus remnant with rotating King models; in some cases, the rotation was high enough that the model was unstable to the formation of a rotating, triaxial bar-like feature. In such setting, the evolution of the MBH binary did not stall at $a \approx a_h$, but did inspiral down to coalescence on timescales well below a Hubble time [16, 17].

Notwithstanding the improvement of the latter models as compared to spherical ones, we have decided to take the next step further and set up a number of galaxy mergers [27]. We let two galactic nuclei to merge on nearly parabolic orbits, since this seems to be a typical configuration for mergers as is observed in state-of-the-art cosmological simulations [28]. Each galactic nucleus is represented by a spherically symmetric model with a power law density cusp of logarithmic slope γ at the center, and a break radius which we set to unity in model units [29, 30]; incidentally, this also permits us to study the mass deficits resulting from the ejection of stars by the binary. A massive point particle represents the MBH which, at the beginning, is placed in the center of each nucleus with zero velocity with respect to each nucleus. The total mass in stars of each galactic nucleus is unity; we adopt units where $G = 1$. In our sample, we have currently some sixty simulations in which we vary essentially four parameters: the total mass of the MBH binary $M_{12} = M_1 + M_2$ in units of the total stellar mass in each merger progenitor, its mass ratio $q = M_1/M_2 < 1$, the central slope γ of the stellar density of each galaxy and the total number $N = N_1 + N_2$ of stars in the two galaxies. We have generated a sample of simulations for the following parameter range: $5 \times 10^{-3} \leq M_{12} \leq 6 \times 10^{-2}$, $0.2 \leq q \leq 1$, $0.5 < \gamma \leq 1.2$ and $64K \leq N \leq 256K$. Particle numbers in this range are still orders of magnitude lower than the number of stars present in typical galactic nuclei. We are currently extending this study up to $N \sim 10^6$ and find that the rates of hardening are essentially independent of N, thereby showing that the dynamical results obtained can be extrapolated to real galaxies with the inevitable caveat that a lot of physics is excluded from the model [27].

After the two nuclei have merged, the MBHs sink to the center of the remnant where they quickly form a bound pair. In Fig. 1, we show the statistics for the average eccentricity with which the MBHs form a bound pair in our sample. There is a broad range of possible eccentricities, but with a clear bias towards the formation of binaries with very high eccentricity $e \geq 0.8$. This tendency for the formation of highly eccentric binaries results naturally from the (cosmologically motivated) initial conditions with which we set up the merger: the nuclei fall into each other on near-parabolic orbits; therefore the MBHs, being much heavier than the stars, also approach each other on near-parabolic orbits despite the perturbations from the surrounding stellar field. In our previous work, we have modelled the galactic nucleus of a merger remnant with rotating King models; in such models, the MBH binaries also tend to become bound with very high eccentricity [16, 17]. In the latter work, it has been shown that the hardening rate due to the superelastic scattering of field stars is essentially independent of the eccentricity. It is well known that the emission of GWs has a very steep dependence on the pericenter distance and thus a highly eccentric binary inspirals on a much shorter time scale [31].

3. Orbital evolution of the massive black hole binary

We have implemented the relativistic effects to the MBH binary only, by using the Post-Newtonian (hereafter PN) equations of motion written in the center of mass frame including all

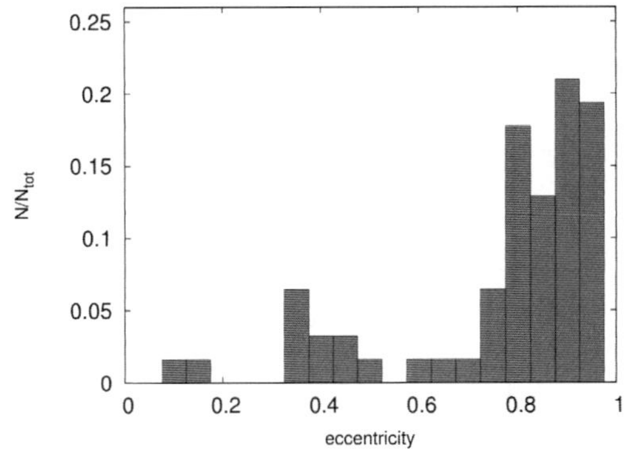

Figure 1. Histogram representing the eccentricity distribution with which the MBH binaries settle when they form a bound pair. This value is an average value taken over a few N-body time units, after the binary becomes "hard", *i.e.* after $t \approx 120(\times 0.0167$ Myr. (see Fig 3a)

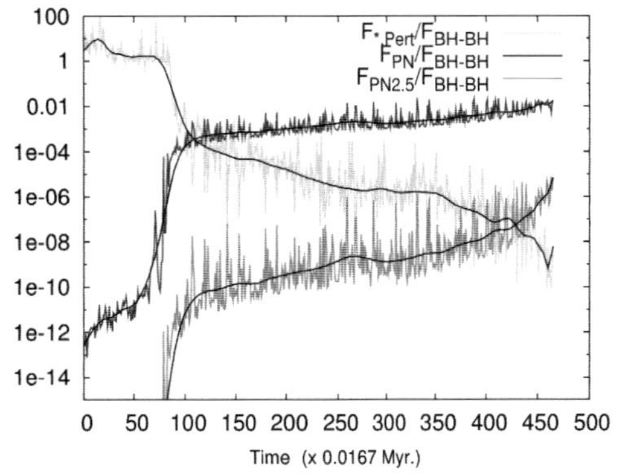

Figure 2. Fractional value of the perturbing accelerations, normalized by the Newtonian acceleration between the MBHs. From: (a) the stars (green), (b) the total 1PN+2PN+2.5PN terms (blue), (c) the 2.5PN term only (magenta). The perturbing forces are always orders of magnitude weaker than the dominant Keplerian force.

terms up to 2.5PN order:

$$\frac{d\mathbf{v}}{dt} = -\frac{GM_{12}}{r^2}\left[(1+\mathcal{A})\mathbf{n} + \mathcal{B}\mathbf{v}\right] + \mathcal{O}(1/c^6), \tag{1}$$

where $\mathbf{n} = \mathbf{r}/r$, the coefficients $\mathcal{A}$ and $\mathcal{B}$ are complicated expressions of the binary's relative separation and velocity [32, 33, 34]. The Post-Newtonian approximation is a power series expansion in $1/c$: the 0th order term corresponds to the dominant Newtonian acceleration; the even order 1PN and 2PN terms are conservative and proportional to $1/c^2$ and $1/c^4$ respectively, and they are responsible for the precession of the pericenter; finally, the lowest order dissipative 2.5PN term which is proportional to $1/c^5$ causes the loss of orbital energy and of angular momentum due to radiation reaction. We treat the MBHs as point particles and thus we neglect any spin-orbit or spin-spin coupling. The runs were carried — either with the direct-summation codes NBODY4 [35] or φ-GRAPE [36] — on the high-performance GRAPE-6A cluster at the Astronomisches Rechen-Institut (Heidelberg). The main difference is that while NBODY4 includes the regularization of close encounters and tight binaries, φ-GRAPE uses softening. The softening in φ-GRAPE is sufficiently small to reproduce the refilling of the loss cone by two-body relaxation. With the parallel version of φ-GRAPE, we can study models with very large N and the results agree with those of NBODY4 as far as single stars and distant encounters are concerned. The MBH pair motion, once it becomes a hard binary, is integrated with the 4th order Hermite scheme, after performing a transformation to KS regularized variables [35, 37]. The PN corrections are treated as perturbations to the dominant Newtonian acceleration in exactly the same manner as the perturbations from the field of stars. In Fig. 2, we show that both type of perturbations are, up to the very late stages of the relativistic inspiral, always much weaker than the dominant Newtonian acceleration. Although the PN formalism is, strictly speaking, valid only for an isolated binary, the relative weakness of both type of perturbations to the Newtonian binary allows us to safely adopt the linear approximation and add both contributions to obtain the total perturbation. The 1PN and 2PN terms are conservative and thus do not directly drive the relativistic inspiral; nevertheless, these terms encode the relativistic advance of pericenter

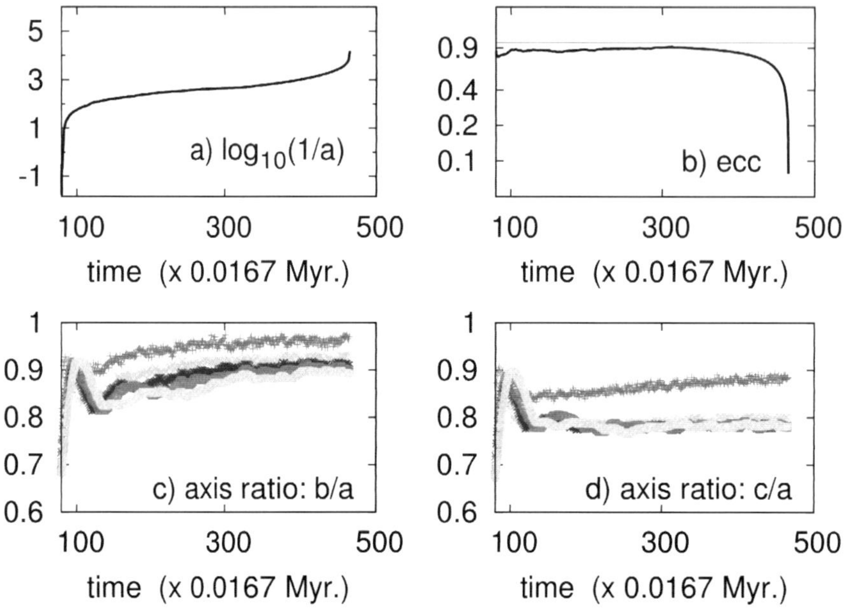

Figure 3. (a) Evolution of the inverse semimajor axis; (b) evolution of the eccentricity; (c) the evolution of the intermediate axis ratio b/a; (d) the evolution of the minor axis ratio c/a. The different colors represent five different radii: $r = 0.2$ pc (red), $r = 0.6$ pc (green), $r = 1$ pc (dark blue), $r = 1.4$ pc (magenta), and $r = 1.8$ pc (light blue).

and thus cause a slight change to the orbital frequency. This modified orbital frequency acts effectively as a secular perturbation and, as a result, it indirectly affects the dissipation of energy and angular momentum (or, equivalently, the eccentricity); for this reason, it is crucial to include them in our treatment if we want to predict the eccentricity with which the binary reaches the LISA band [17, 38].

We next describe a run that is typical of our sample. In this fiducial run, we set the black hole masses M_1 and M_2 of the two MBH particles to be 1% and 2.5% ($q = 0.4$) of the total mass in stars. We employ a total of $N = 128K$ particles to represent the total number of equal mass stars (half on each nucleus). The mass ratios between the MBHs and field stars are $M_1/m_* = 640$ and $M_2/m_* = 1600$. Both nuclei have central densities with logarithmic slope $\gamma = 1$. Since LISA will be sensitive to the inspiral of MBH binaries of total mass $M_{12} \leq 10^7 M_\odot$, we have chosen to scale our model in such a way that the lighter MBH has $M_1 = 10^6 M_\odot$ in physical units. As a result, one spatial N-body unit corresponds to 5 pc and one N-body time unit corresponds to $\approx 1.67 \times 10^4$ yr. The resulting speed of light in our model units is $c = 1022$. We place initially the centers of the two nuclei at a separation of 20 model units, i.e. 100 pc. The half-mass radius of each nucleus is, for the fiducial run with $\gamma = 1$, $r_{1/2} \approx 2.41$ in model units, or some 12 pc; this means that the nuclei are well separated in their initial configuration. The total mass in stars, initially enclosed within the inner 12 pc of each nucleus, is $5 \times 10^7 M_\odot$.

After roughly 80 N-body time units (or ~ 1.3 Myr), the MBHs have reached the center of the remnant and form a bound pair. At a first stage, they eject a large amount of mass ($M_{ej} \sim M_{12}$) in stars and its semimajor axis has shrunk by a factor of ~ 50 by time $t = 100$ (or some 1.67 Myr). Subsequently, the hardening rate slows down considerably but does never stall.

The binary forms with an eccentricity close to 0.9, which is typical of our simulation sample,

Journal of Physics: Conference Series **154** (2009) 012049 doi:10.1088/1742-6596/154/1/012049

and grows slowly over time before it eventually starts to circularize due to radiation reaction. In Fig. 3a, we can see that the MBHs coalesce after roughly 450 N-body units, or some 7.5 Myr. We should provide a word of caution, though, since the inspiral time scale is very sensitive to the eccentricity evolution as a function of the binary's semimajor axis [31]. The study of the dependence of the evolution of the eccentricity as a function of the initial conditions, and especially as a function of the total number N of stars (which, for a fixed total mass, determine the mass ratio between the MBHs and the stars) is the subject of ongoing work [27].

In panels 3c and 3d, we can observe the evolution of the axis ratio for the mass distribution of the merger remnant at several different radii. The radius of influence r_h of the binary — radius enclosing $2M_{12}$ of the stellar mass — is in N-body (physical) units 0.2 (5 pc). The overall shape of the cluster is markedly triaxial throughout the simulation; at the inner radius $r = 1$ represented in the figure, the triaxiality parameter $T \equiv (a^2 - b^2)/(a^2 - c^2)$ decreases faster but is still $\approx 0.1 - 0.25$ by the time the binary decouples from the stellar environment at time $t \approx 410 - 420$ (6.9 Myr).

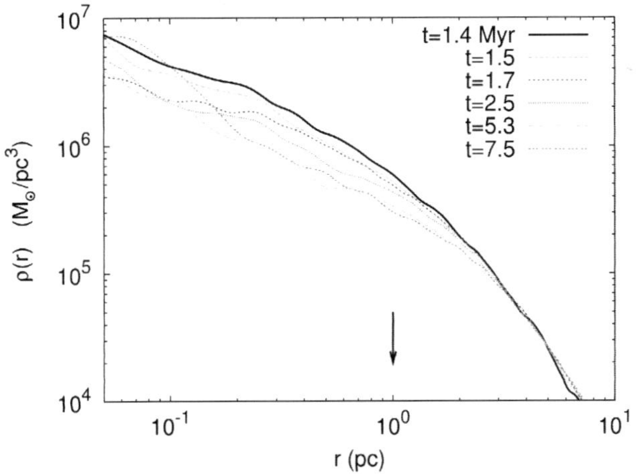

Figure 4. The scouring effect on the central density cusp of the merger remnant as the MBH binary semimajor shrinks through the ejection of stars. The mass loss extends to regions well outside of r_h, as expected if the loss cone is kept full by a centrophilic family of orbits associated with the triaxial stellar potential. The arrow signals the binary's radius of influence.

As the semimajor axis of the binary shrinks through the ejection of stars, the central density decreases progressively. If the triaxial nature of the potential drives the collisionless refilling of the loss cone, it is expected that a fair amount of stars supplied to the binary originate from a region well outside its influence radius r_h [15]. On the contrary, in the case of an empty loss cone in a spherical symmetric nucleus, the scouring effects on the central density cusp would be limited roughly to the region interior to r_h [39]. In our model, the influence radius is initially $r_h \sim 0.18$ (~ 1 pc) and evolves only very slowly to $r_h \sim 0.22$ at the end. From the inspection of Fig. 4, we can immediately see that the density cusp is being depleted out to radii of $\sim$ few$\times r_h$, and this is a generic result in our simulation sample [27]. Furthermore, we measure the amount of mass ejected by the binary, by $t = 320$, from the region interior to $3r_h$ to be $\approx 2M_{12}$; in a spherical nucleus, we expect a much smaller value $M_{ej} \approx 0.5M_{12}$ [39]. Note that this high rate of mass ejection may help to explain the existence of large mass deficits in some bright elliptical galaxies.

We can also observe that towards the end of the run, when the MBH pair has almost completely decoupled from the galactic nucleus ($t \sim 410$ or ~ 5.3 Myr) and is evolving mainly in isolation under the effect of radiation reaction, the density cusp does not decay anymore. In fact, the cusp is starting to grow again, though at a much slower rate than that with which it decayed before, as expected in the case of a single MBH.

Can we extrapolate these results to real galaxies? The answer to this question depends on the extent to which the evolution of the orbital elements (hardening rate and eccentricity evolution) converges for the range of particle particle number used in our simulations. Our preliminary results indicate that the hardening rate does indeed converge once we reach $N \approx 0.125 - 0.25 \times 10^6$. This is presumably due to the fact that the repopulation of the loss cone orbits happens in a regime that is essentially collisionless as a result of the nonlinear dynamics of a centrophilic family of orbits supported by the triaxial remnant nucleus.

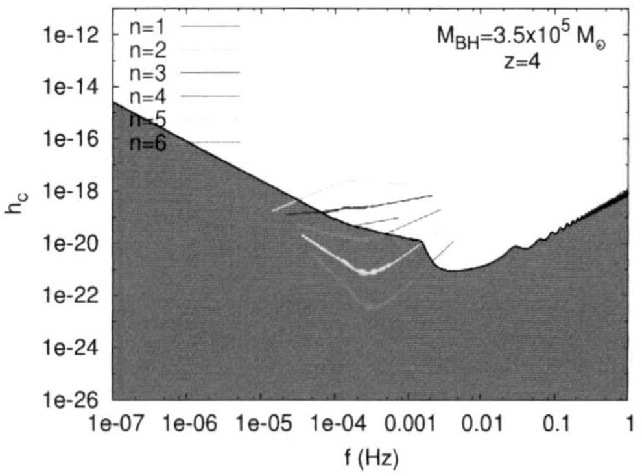

Figure 5. Locus of the MBH binary, for the first six harmonics, in the LISA sensitivity band during their final inspiral and coalescence. The dimensionless strain h_c is plotted against the observed frequency. The total mass of the binary is $M_{12} = 3.5 \times 10^5 M_\odot$.

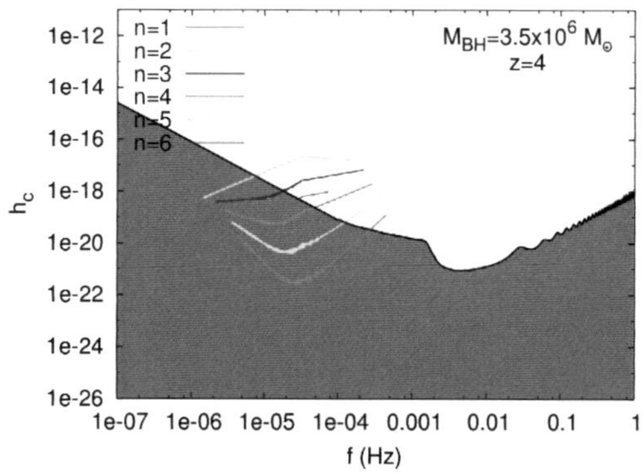

Figure 6. The same as Fig. 5, but for $M_{12} = 3.5 \times 10^6 M_\odot$.

4. Gravitational Wave Signal

The inspiral and merger of MBHs are expected to be one of the brightest sources of GWs to be detected by LISA [40]. In most of the literature, the strain amplitude of the GWs is estimated under the assumption of circular orbits. This has been motivated by the idea that MBH binaries should have been completely circularized by the time they enter the LISA band at $f \sim 10^{-4}$ Hz [41]. In our runs, the binaries form typically with very large eccentricities and usually reach small separations (e.g. 100 R_{Schw}) with non-negligible eccentricty. Depending on the redshift of the source, this implies that they may enter the LISA band with significant power distributed among the higher harmonics $f_n = nf_{orb}/(1+z)$, with $n \geq 2$ (note that $n = 2$ for a circular orbit) [42]. In Figs. 5 and 6, we plot the characteristic strain amplitudes h_c for the our fiducial run when the model is scaled in such a way that the total mass of the BH binary is $M_{12} = 3.5 \times 10^5 (3.5 \times 10^6) M_\odot$ and if the redshift of the source were $z = 4$. We conclude that several higher harmonics stand above the LISA sensitivity curve. We suggest that MBH binary orbits with non-vanishing eccentricity may be seriously considered for the LISA data analysis.

5. Conclusions

We present preliminary results from a set of N-body models of merging galactic nuclei. The generic outcome of these mergers is the formation of a triaxial remnant. The MBHs present in these models often form highly eccentric binaries and, driven by the large pool of stars in centrophilic orbits associated to the triaxial potential, do not stall at a separation $a \approx a_h$. Those binaries that get formed with the highest eccentricity, $e \geq 0.9$, can decay from initial separations of ~ 100 pc down to gravitational-wave-induced coalescence within $\sim 10^7$ yr. These binaries may reach the LISA band with non-negligible eccentricity and thus it may be important to consider finite-eccentricity effects in the data analysis. The amount of stellar mass ejected by the binary throughout the inspiral is higher than estimates made for spherical nuclei — this may help explaining the mass deficits observed in some of the brightest elliptical galaxies.

Acknowledgments
This work is supported by DLR (Deutsches Zentrum für Luft- und Raumfahrt), the Volkswagen Foundation (GRACE Ref. I/80 041-043) and SFB 439 of DFG. MP acknowledges the Symposium organizers for financial support to attend the conference.

References

[1] Ferrarese, L. and Ford, H., 2005, *Space Sci. Rev.*, **116**, 523-624
[2] Magorrian, J. *et al.*, 1998, *Astron. Journ.*, **115**, 2285-305
[3] Ferrarese, L. and Merritt, D., 2000, *ApJL*, **539**, L9-12
[4] Gebhardt, K. *et al.*, 2000, *ApJL*, **539**, L13-16
[5] Gillessen, S. *et al.*, 2008, arXiv:0810.4674
[6] Krolik, J.H., 1998, Active Galactic Nuclei (Princeton: Princeton Univ. Press)
[7] Maoz, E., 1998, *ApJL*, **494**, L181-4
[8] Miller, M.C., 2006, *MNRAS*, **367**, L32-6
[9] Begelman M.C., Blandford R.D. and Rees M.J., 1980, *Nature*, **287**, 307-9
[10] Merritt, D., 2006, *Rep. Prog. Phys.*, **69**, 2513-79
[11] Armitage, P. and Natarajan, P., 2005, *ApJ*, **634**, 921-27
[12] Mayer, L., Kazantzidis, S., Madau, P., Colpi, M., Quinn, T. and Wadsley, J. 2007, *Science*, **316**, 1874
[13] Cuadra, J., Armitage, P.J., Alexander, R.D. and Begelman, M.C., 2008, arXiv:0809.0311
[14] Perets, H.B. and Alexander. T., 2008, *ApJ*, **677**, 146-59
[15] Merritt D. and Poon M.Y., 2004, *ApJ*, **606**, 788-98
[16] Berczik, P., Merritt, D., Spurzem and R., Bischoff, H.-P., 2006, *ApJ*, **642**, L21-4
[17] Berentzen, I., Preto, M., Merritt, D., Berczik, P. and Spurzem, R., 2008 *submitted to ApJ*
[18] Yu, Q., 2002, *MNRAS*, **331**, 935-58
[19] Milosavljević M. and Merritt, D. 2003, *ApJ*, **596**, 860-78
[20] Hemsendorf M., Sigurdsson S., Spurzem R., 2002, *ApJ*, **581**, 1256-70
[21] Makino, J. & Funato, Y., 2004, *ApJ*, **602**, 93-102
[22] Berczik, P., Merritt, D. and Spurzem, R., 2005, *ApJ*, **633**, 680-7
[23] Merritt D., Mikkola S and Szell A., 2007, *ApJ*, **671**, 53-72
[24] Cohn, H. and Kulsrud, R., 1978, *ApJ*, **226**, 1087-108
[25] Sesana, A., Volonteri, M. and Haardt, F., 2007, *MNRAS*, **377**, 1711-16
[26] Moore B, Kazantzidis S., Diemand J. and Stadel J., 2004, *MNRAS*, **354**, 522-28
[27] Preto, M., Berentzen, I., Berczik, P., Merritt, D. and Spurzem, R., 2008 *to be submitted*
[28] Khochfar, S. and Burkert, A., 2006, *Astron. Astrop.*, **445**, 403-12
[29] Dehnen, W., 1993, *MNRAS*, **265**, 250-6
[30] Tremaine, S., *et al.*, *Astron. Journ.*, **107**, 634-44
[31] Peters, P.C., 1964, *Phys. Rev.* B, **136**, 1224-32
[32] Blanchet, L., 2006, *Living Rev. Relativity*, **9**, 4. [1]
[33] Blanchet, L. and Iyer, B.R., 2003, *Clas. Quant. Grav.*, **20**, 755-76
[34] Kupi, G, Amaro-Seoane, P. and Spurzem, R., 2006, *MNRAS*, **371**, L45-9
[35] Aarseth, S.J., 1999, *Publ. Astron. Soc. Pac.*, **111**, 1333-46
[36] Harfst, S., Gualandris, A., Merritt, D., Spurzem, R., Portegies Zwart, S. and Berczik, P. 2007, *New Astron.*, **12**, 357-77
[37] Aarseth, S.J., 2003, Gravitational N-body Simulations (Cambridge: Cambridge Univ. Press)
[38] Berentzen, I., Preto, M., Berczik, P., Merritt, D. and Spurzem, R., 2008, *Astronomische Nachrichten*, **329**, 904-7
[39] Merritt, D. 2006, *ApJ*, **648**, 976-86
[40] Danzmann, K., 1997, Clas. Quantum Grav., 14, 1399-404
[41] Sesana, A., Haardt, F., Madau, P. and Volonteri, M., 2005, *ApJ*, **623**, 23-30
[42] Peters, P.C. and Mathews, J., 1963, *Phys. Rev.*, **131**, 435-40

[1] URL (cited on June 2007):http://www.livingreviews.org/lrr-2006-4

Journal of Physics: Conference Series **154** (2009) 012050 doi:10.1088/1742-6596/154/1/012050

Gravitational radiation characteristics of nonspinning black-hole binaries

B J Kelly, J G Baker, W D Boggs, J M Centrella, J R van Meter and S T McWilliams

NASA Goddard Space Flight Center, Greenbelt MD 20771, U.S.A.

E-mail: bernard.j.kelly@nasa.gov, john.g.baker@nasa.gov, william.d.boggs@nasa.gov, joan.m.centrella@nasa.gov, james.r.vanmeter@nasa.gov, sean.t.mcwilliams@nasa.gov

Abstract. We present a detailed descriptive analysis of the gravitational radiation from binary mergers of non-spinning black holes, based on numerical relativity simulations of systems varying from equal-mass to a 6:1 mass ratio. Our analysis covers amplitude and phase characteristics of the radiation, suggesting a unified picture of the waveforms' dominant features in terms of an implicit rotating source, applying uniformly to the full wavetrain, from inspiral through ringdown. We construct a model of the late-stage frequency evolution that fits the $\ell = m$ modes, and identify late-time relationships between waveform frequency and amplitude. These relationships allow us to construct a predictive model for the late-time waveforms, an alternative to the common practice of modelling by a sum of quasinormal mode overtones. We demonstrate an application of this in a new effective-one-body-based analytic waveform model.

1. Outline

Since the discovery of the moving puncture recipe, numerical research into the merger of black-hole binaries (BHBs) has exploded [1, 2, 3, 4]. Moving past the canonical case of the equal-mass, nonspinning binary, assuming zero eccentricity, and taking advantage of the overall mass-scaling nature of the problem, BHBs occupy a seven-parameter space (mass ratio, spins of each hole). With LIGO being retooled as Enhanced LIGO, and LISA anticipated to fly in the next decade, it is imperative that we explore this parameter space as quickly and efficiently as possible, to supply the gravitational-wave detector community with the information they need for data analysis and detector tuning.

In this paper, we outline recent work investigating the one-dimensional parameter space of nonspinning binaries of varying mass ratio, with a focus on the relationships between harmonic modes and between mass ratios, somewhat complementary to work by [5]. The material in this brief article is a summary of numerical experiments and results more fully described in [6].

2. Numerical Methods

We evolved BHBs with mass ratios of 1:1, 2:1, 4:1 and 6:1 using the Goddard `Hahndol` code [7], which solves Einstein's equations on a Cartesian spatial grid, with adaptive mesh refinement supplied by the `Paramesh` infrastructure [8]. Initial data is of the "puncture" type [9], with momenta determined from a 2PN-accurate circular-orbit approximation [10], with constraints solved using the `AMRMG` multigrid elliptic solver [11]. Our evolution uses a modified form of the

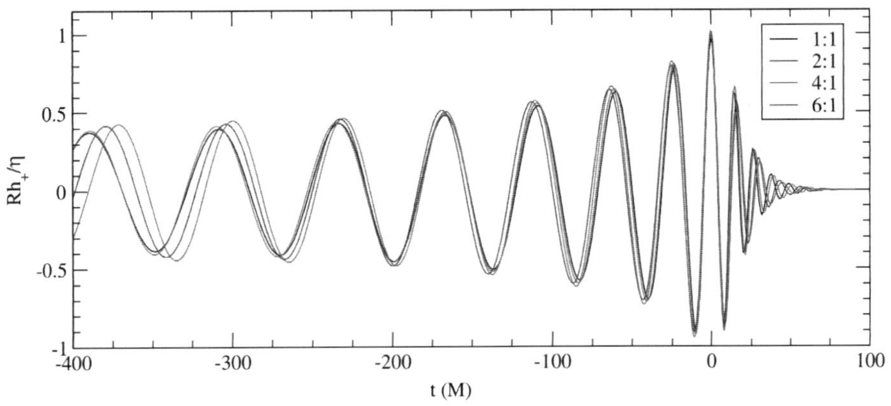

Figure 1. Gravitational-wave signals along polar axis from mass ratios 1:1, 2:1, 4:1 and 6:1. [Reprinted from [6].]

BSSN equations [12, 13], with 1+log slicing and Gamma-driver shift conditions appropriate for the "moving puncture" recipe [14].

3. Numerical Relativity Waveforms for Unequal-Mass BHBs

The first thing we note about the waveforms of the different mass ratios is that they show remarkable consistency, if viewed along the polar axis. Fig. 3 shows the "plus" polarisation of the waveform strain, h_+, as measured along the orbital axis, with amplitudes scaled by the *symmetric mass ratio* $\eta \equiv m_1 m_1/M^2$. The waveform shape is remarkably coherent across mass ratios from $200M$ before merger to the merger itself (taken here as the time of peak radiation amplitude). After the merger, the waveform is dominated by the fundamental quasinormal mode (QNM) ringdown frequency of the end-state hole, which depends strongly on the spin of the hole.

It is not surprising, perhaps, that the waveforms look so similar when they are all dominated by the quadrupole $(\ell, m) = (2, \pm 2)$ mode, which is insensitive to mass asymmetries. Viewed far off-axis, where the (2,2) mode no longer dominates, the signals can look very different. However we retain much simplicity if we decompose the signal into its harmonic modes:

$$h \equiv \sum_{\ell m} h_{\ell m} \, {}_{-2}Y^{\ell,m}(\theta, \varphi). \tag{1}$$

On performing this decomposition, we note that, after an initial pulse of high-frequency junk radiation, the strongest modes are circularly polarized. For instance, we can write the *strain-rate* – the first time-derivative of the waveform strain – in the form $\dot{h}_{\ell m} = A_{\ell m} e^{im\phi}$.

4. The Implicit Rotating Source Picture

Looking at gravitational-wave modes across harmonics reveals several general trends. For example, looking at the power emitted, mode by mode, we find that it is partitioned by modes according to leading-post-Newtonian (PN) order. The left panel of Fig. 2 shows the power for several leading modes for the 4:1 merger. The solid lines are the raw numerical data, while the dashed lines are the leading-order post-Newtonian predictions for the emitted power, as found in [15], calculated using the orbital frequency extracted from the (2,2) mode.

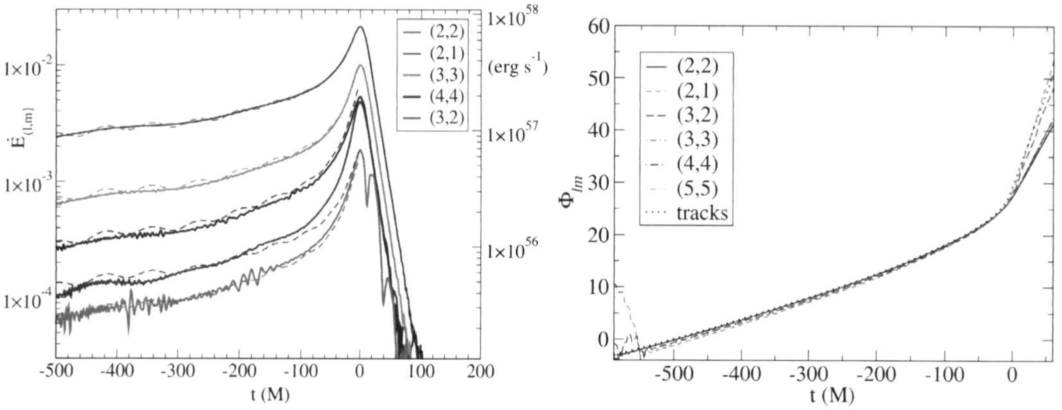

Figure 2. Left panel: Power in each mode for 4:1 mass ratio. Right panel: Rotational phase of each mode for 4:1 mass ratio. [Reprinted from [6].]

Similarly, we find that the phases of different modes are simply related through merger. While each mode accumulates phase at a different rate, we can define a "rotational phase" for the system:

$$\Phi_{\ell m} = \phi_{\ell m}/m + \text{ const.} \tag{2}$$

The right panel of Fig. 2 shows this rotational phase calculated for all the dominant modes of the 4:1 run. After initial noise (due to an initial burst of junk radiation, which is not circularly polarised), the different phase estimates agree very well, up to a common "elbow", where the binary merges. The result is a picture of BHB as a slowly changing rigid rotator – an "implicit rotating source"– generating gravitational waves all through inspiral, merger & ringdown such that each mode is tied to the corresponding multipole moment of the radiating source [16].

5. Full Late-Merger Frequency Model

One common feature of the different modes is that each (ℓ, m) mode's frequency ω increases monotonically until it plateaus after the peak amplitude, associated with the merger. The characteristic shape of the frequency-time curve is very similar for the strongest modes. It can be well modelled through merger by a smooth step-like function:

$$\Omega(t) = \Omega_i + (\Omega_f - \Omega_i)\left(\frac{1 + \tanh[\ln\sqrt{\kappa} + (t - t_0)/b]}{2}\right)^{\kappa}. \tag{3}$$

With this model, the rotational frequency Ω grows at an exponential rate from its initial value Ω_i, with an associated e-folding time $b/2\kappa$. The growth rate (chirp rate) is at its maximum $\dot{\Omega}_0$ at time t_0, and then decays exponentially with e-folding time $b/2$ to the final frequency Ω_f.

The left panel of Fig. 3 shows the leading modes in the 4:1 mass ratio case, from $60M$ before merger to $40M$ after merger, with the frequency fit (3) overlaid. This fitting works well over all mass ratios examined, and appears to be valid for more extreme mass ratios as well. The right panel of Fig. 3 shows the rotational frequency Ω_{22} as derived from the dominant (2,2) modes for the 1:1, 2:1, 4:1 & 6:1 numerical evolutions, as well as for the extreme 100:1 mass ratio[1]. The dashed horizontal line indicates the QNM ringdown frequency for an end-state Schwarzschild black hole. As might be expected, as the mass ratio increases, the final hole's spin decreases toward the Schwarzschild limit.

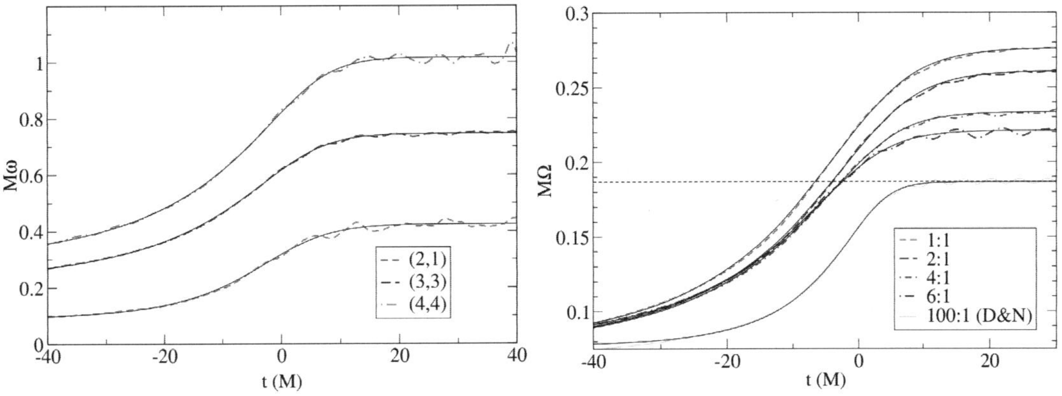

Figure 3. Left panel: fitting $\Omega_{\ell m}$ for leading modes in 4:1 mass ratio to functional form (3). Right panel: fitting Ω_{22} for all mass ratios. [Reprinted from [6].]

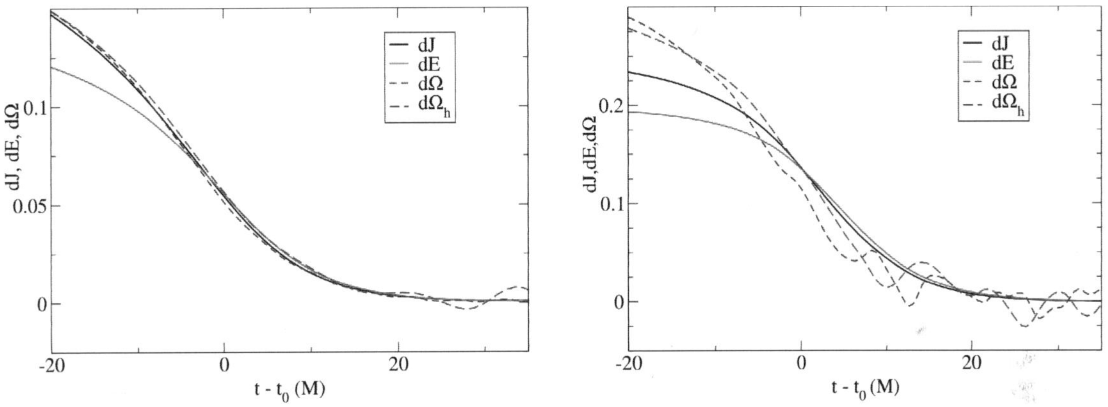

Figure 4. Decay of $E_{\ell m}$, $J_{\ell m}$, & $\Omega_{\ell m}$ for 1:1 (2,2) mode (left), 4:1 (2,1) mode (right). Both time axes are shifted relative to the peak chirp time t_0. [Reprinted from [6].]

6. Late-Merger $E(\Omega)$ & $J(\Omega)$ Relations

A consequence of the IRS picture is that both the energy and angular momentum flux assume simple forms near merger:

$$\frac{dE_{\ell m}}{dt} \propto \Omega \frac{d\Omega}{dt}, \quad \frac{dJ_{\ell m}}{dt} \propto \frac{d\Omega}{dt}. \tag{4}$$

This can be seen in Fig. 4, which shows the decay of the modal contributions to radiated E and J^z towards their final values for the equal-mass (left panel) and 4:1 (right panel) cases, as well as the decay of two measures of the rotational frequency, Ω from the strain-rate, and Ω_h from the strain (these should differ only at 2.5 PN order).

As a consistency check of this picture, if we assume that $dJ/d\Omega \sim$ constant near merger, we can use the frequency fit (3) to derive the approximate delay between peak chirp (rate of increase

[1] 100:1 data supplied by T. Damour and A. Nagar

of Ω) and peak radiation amplitude:

$$t_{peak} - t_0 \approx -\frac{\dot{\Omega}_0^2}{\Omega_0 \dddot{\Omega}_0} > 0. \tag{5}$$

Inserting values of Ω_0 and its derivatives from the frequency fit yield values of this difference of between $\sim 2-3M$, in good agreement with the fitted values for t_0 itself.

7. Relevance to LISA: New BHB Waveform Templates

New insights into gravitational waveforms from mergers can lead to better templates for data analysis. Although pure numerical waveforms are rarely long enough to cover the full visible lifetime of a LISA observation, they can be combined with post-Newtonian pre-merger waveforms for a hybrid template.

The simplest way to achieve this is to "stitch" the two segments together in such a way as to maintain an everywhere smooth waveform. The point or interval over which the waveforms can be stitched should be one where *both* post-Newtonian and numerical results can be trusted. An early example of this for the equal-mass case was supplied by Baker et al. [17].

The drawback with this direct approach is that each point in parameter space (mass ratio here) requires its own numerical waveform of limited duration to be attached to the post-Newtonian pre-merger waveform. As it is impractical to carry out a full evolution for every mass ratio of interest, we must develop instead a better analytic insight.

Several semi-analytical methods have been proposed involving the combination of effective-one-body (EOB) theory [18] and multiple quasinormal ringdown modes (QNMs) [19, 15], with flexibility parameters tuned using numerical merger data. The current work, with the empirical functional frequency (3), has allowed an alternative to QNMs, with a late-time strain-rate amplitude given in terms of the rotational frequency:

$$A_{\ell m} \approx \sqrt{16\pi \xi_{\ell m} \Omega_{\ell m} \dot{\Omega}_{\ell m}}, \tag{6}$$

where the constant ξ can be fixed by demanding continuity of amplitude at a matching frequency. Fig. 7 shows the modelled frequency and strain-rate amplitude (dotted lines) compared with the numerical results (solid lines) for the 4:1 evolution. We see that this new model encodes both frequency and amplitude well.

8. Summary

At the time of writing, fully numerical evolution of black-hole-binary spacetimes is possible to high accuracy. Nevertheless, LISA data analysis demands waveforms of much greater duration than is numerically feasible; additionally an understanding is required of waveforms from systems over a seven-parameter configuration space (more is eccentricity is considered crucial). To fill this gap in computational practicality, we are drawn to the development of analytical models of the gravitational-wave signals.

In this paper, we have introduced a set of numerical evolutions probing one dimension of parameter space – mass ratio. We have noted common characteristics of the dominant waveform modes, and suggested a useful heuristic picture of the radiation as being produced by an "implicit rotating source" through inspiral, merger, and ringdown. We have shown how this picture can be used to develop an alternative model to a sum of quasinormal modes for the transition through merger and ringdown. This model can then be combined with effective-one-body theory to produce analytic waveform templates that seem to capture well the important characteristics of the merger. A fuller discussion of these issues can be found in [6].

Work is underway to assess the utility of these templates in data analysis, and to extending our analysis to other parts of parameter space, involving spinning black holes.

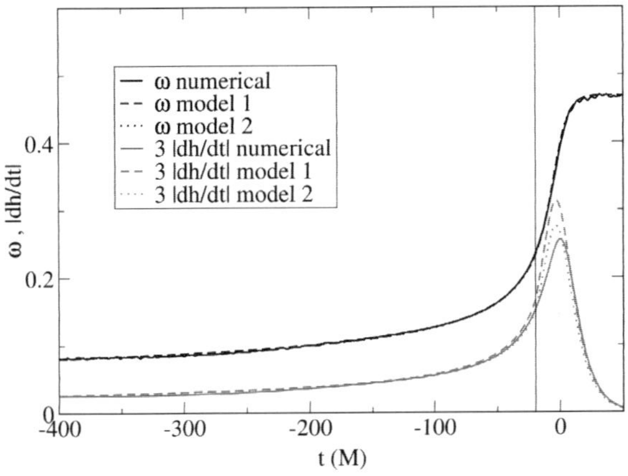

Figure 5. Frequency and strain-rate amplitude for the IRS modelled (2,2) waveform (dotted and dashed lines) compared with the numerical waveform (solid lines), for the 4:1 merger. $t = 0$ is the time of peak amplitude. [Reprinted from [6].]

Acknowledgments
We thank Emmanuele Berti and Alessandra Buonanno for interesting discussions, Luciano Rezzolla for useful comments, and Thibault Damour and Alessandro Nagar for supplying data from the 100:1 case. This work was supported in part by NASA grant 05-BEFS-05-0044 and 06-BEFS06-19. The simulations were carried out using Project Columbia at the NASA Advanced Supercomputing Division (Ames Research Center) and at the NASA Center for Computational Sciences (Goddard Space Flight Center). B.J.K. was supported by the NASA Postdoctoral Program at the Oak Ridge Associated Universities.

References
[1] Brügmann B *et al* 2004 *Phys. Rev. Lett.* **92** 211101
[2] Pretorius F 2005 *Phys. Rev. Lett.* **95** 121101
[3] Campanelli M *et al* 2006 *Phys. Rev. Lett.* **96** 111101
[4] Baker J G *et al* 2006 *Phys. Rev. D* **73** 104002
[5] Berti E *et al* 2007 *Phys. Rev. D* **76** 064034
[6] Baker J G *et al* 2008 *Phys. Rev. D* **78** 044046
[7] Imbiriba B *et al* 2004 *Phys. Rev. D* **70** 124025
[8] MacNeice P *et al* 2000 *Computer Physics Comm.* **126** 330
[9] Brandt S R and Brügmann B 1997 *Phys. Rev. Lett.* **78** 3606
[10] Kidder L E 1995 *Phys. Rev. D* **52** 821
[11] Brown J D and Lowe L L 2005 *J. Comp. Phys.* **209** 582
[12] Shibata M and Nakamura T 1995 *Phys. Rev. D* **52** 5428
[13] Baumgarte T W and Shapiro S L 1998 *Phys. Rev. D* **59** 024007
[14] van Meter J R *et al* 2006 *Phys. Rev. D* **73** 124011
[15] Buonanno A *et al* 2007 *Phys. Rev. D* **76** 104049
[16] Thorne K S 1980 *Rev. Mod. Phys.* **52** 299
[17] Baker J G *et al* 2007 *Phys. Rev. D* **75** 124024
[18] Buonanno A and Damour T 1999 *Phys. Rev. D* **59** 084006
[19] Pan Y *et al* 2008 *Phys. Rev. D* **77** 024014

Possible discovery of nonlinear tail and quasinormal modes in black hole ringdown

Satoshi Okuzumi[1], **Kunihito Ioka**[2], **and Masa-aki Sakagami**[1]

[1] Graduate School of Human and Environmental Studies, Kyoto University, Yoshida-nihonmatsu-cho, Sakyo-ku, Kyoto, 606-8501, Japan

[2] Theory Division, KEK (High Energy Accelerator Research Organization), 1-1 Oho, Tsukuba 305-0801, Japan

E-mail: satoshi.okuzumi@ax2.ecs.kyoto-u.ac.jp

Abstract. We discuss the nonlinear evolution of black hole ringdown in the framework of higher-order metric perturbation theory. By solving the initial-value problem of a simplified nonlinear field model analytically as well as numerically, we find that (i) second-order quasinormal modes (QNMs) are indeed excited at frequencies different from those of first-order QNMs, as predicted recently. We also find serendipitously that (ii) late-time evolution is dominated by a new type of power-law tail at late times. This "second-order power-law tail" decays more slowly than any late-time tails known in the first-order (i.e., linear) perturbation theory, and is generated at the wavefront of the first-order perturbation by an essentially nonlinear mechanism. These nonlinear components should be particularly significant for binary black hole coalescences, and could open a new precision science in gravitational wave studies.

1. Introduction

Direct detection of gravitational waves is one of the most exciting challenges in astrophysics today. It will not only enable to verify general relativity, but will also provide a new observational window Future gravitational wave detectors such as LISA should make it a reality in the near future.

The "ringdown" of black holes is an important target of gravitational-wave observations. It is known theoretically that black holes have characteristic damping oscillation modes, called *quasinormal modes* (QNMs) [1, 2]. The remarkable property of black hole QNMs is that their frequencies as well as damping rates are uniquely determined by the black hole parameters. This means that the parameters of black holes can be directly measured by observing their ringdown through the gravitational waves.

Usually, black hole ringdown is understood in the framework of *linear* perturbation theory. Although the linear theory is clearly useful, the results of recent numerical simulations imply that the *nonlinearity* of the ringdown may be also important. It has been revealed that quasinormal ringing radiated from merged binary black holes carry away a large fraction ($\sim 1\%$) of the initial rest mass energy of the system (see, e.g., [3]). This suggests that nonlinear components of the radiation will have amplitude as large as $\sim 10\%$ of the first-order amplitude. This strongly suggests that the second- and higher-order perturbations should contribute the ringdown waveform to that extent. These components will provide us with some additional information on the merger events and the nonlinearity of general relativity.

Despite its potential importance, the nonlinearity of black hole ringdown has been largely ignored. Only recently, one of the authors (K.I.) and his coworker have for the first time applied second-order perturbation theory [4, 5] for the ringdown of Schwarzschild black holes [6, 7]. Because of the nonlinearity of the Einstein equation, two normal modes in first-order perturbations, say $\cos(\Omega_1 t)$ and $\cos(\Omega_2 t)$, couple each other, yielding their "sum tone" $\propto \cos[(\Omega_1 + \Omega_2)t]$ and "difference tone" $\propto \cos[(\Omega_1 - \Omega_2)t]$ in second-order ones. The authors found that the sum and difference tones of the first-order QNMs also satisfy the boundary conditions of QNMs, meaning that they *are* the QNMs in the second-order perturbation theory. They concluded that if these "second-order QNMs" are excited in a binary black hole merger, they will have enough energy to be detected by future gravitational-wave detectors. Detection of the second-order QNMs has many possible applications, such as measuring the distance from the source system, testing the nonlinearity of general relativity, and rejecting fake events in ringdown searches.

Questions have remained unanswered. Are the second-order QNMs really excited in *initial value* problems? Is it only the second-order QNMs that arise from the second-order perturbations? In this study, we give answers to these questions by solving an initial-value problem for nonlinear black hole ringdown. Our analytic calculation is based on a time-domain analysis developed for linear black hole perturbations [8, 9]. As far as we know, this analysis has been never applied to nonlinear black hole perturbations. Also, we employ a simple nonlinear field model having key properties in common with black hole perturbations. This nonlinear model is appealing, since the original equations for black hole perturbations are too complicated to be treated analytically, and since current numerical relativity would not be accurate enough to resolve the nonlinear evolution of black hole ringdown [6]. Our model does not only enable to solve the problem analytically, but also allow us to confirm the results with fully nonlinear simulations in a good accuracy.

Using this simple model, we have found that the second-order QNMs are not only actually excited, but are also accompanied by a power-law tail which is *essentially different* from those known in the linear perturbation theory [10]. This tail, to which we shall refer as "second-order power-law tail," decays more slowly than any of the first-order tails, and even dominates the *fully nonlinear* evolution of the model field at late times. This means a surprising fact that the first-order theory *fails* to predict the late-time behavior of the ringdown.

2. Nonlinear scalar field model

Now we introduce our nonlinear field model. To avoid oversimplification, we require the model to satisfy that (i) its first- and second-order perturbations obey (1+1)-dimensional wave equations with a potential barrier; and that (ii) the effective source term for the second-order perturbation is quadratic in the first-order one. These are the essential properties of black hole perturbations [5]. As one of the simplest models, we adopt a nonlinear, (1+1)-dimensional scalar field obeying

$$\left[-\partial_t^2 + \partial_x^2 - V(x)\right]\Phi(t,x) = F(x)\Phi(t,x)^2, \tag{1}$$

where $V(x)$ is a potential term and $F(x)$ is a function that characterizes the nonlinearity of the system. We assume that $V(x)$ has a peak at $x = 0$ and vanishes at $|x| \gg 1$. Since x corresponds to the tortoise coordinate r_* for black holes, we refer to the limits $x \to -\infty$ and $x \to +\infty$ as "horizon" and "infinity," respectively. We also assume an observer's position near infinity, $x \gg 1$.

It is easily checked that this model meets the above requirements. We expand $\Phi(t,x)$ in terms of an expansion parameter ϵ to write

$$\Phi(t,x) = \epsilon\phi^{(1)}(t,x) + \epsilon^2\phi^{(2)}(t,x) + \cdots. \tag{2}$$

Substituting this into Eq. (1), we obtain an infinite set of perturbation equations, the first two of which read

$$\Big[-\partial_t^2 + \partial_x^2 - V(x)\Big]\phi^{(1)} = 0, \tag{3}$$

$$\Big[-\partial_t^2 + \partial_x^2 - V(x)\Big]\phi^{(2)} = S^{(2)}(t,x), \tag{4}$$

where $S^{(2)}(t,x) = F(x)(\phi^{(1)})^2$. It is evident that equations (3) and (4) satisfy the above requirements (i) and (ii).

We choose $F(x)$ so that the source term $S^{(2)}$ satisfies the same asymptotic behavior at $x \to \infty$ as that of second-order black hole perturbations at $r \to \infty$. As one of the simplest forms, we adopt $F(x) = (|x|+1)^{-2}$.

In the following, we consider an initial-value problem of Eq. (1) with the following initial conditions:

$$\Phi(0,x) = f(x), \qquad \partial_t\Phi(0,x) = g(x), \tag{5}$$

where $f(x)$ and $g(x)$ are arbitrary functions vanishing far away from the potential peak.

2.1. Analytic calculation

Since all the perturbations $\phi^{(j)}$ $(j = 1, 2, \cdots)$ obey linear equations with the same operator $-\partial_t^2 + \partial_x^2 - V(x)$, their evolution is determined by a single Green's function. The retarded Green's function $G(\tau; x, x')$ for $\phi^{(j)}$ is the solution to

$$\Big[-\partial_t^2 + \partial_x^2 - V(x)\Big]G(t-t'; x, x') = \delta(t-t')\delta(x-x'), \tag{6}$$

under the retarded boundary condition $G(\tau; x, x') = 0$, $\tau < 0$. Using this Green's function, the solutions to Eqs. (3) and (4) with the initial conditions (5) are formally written as

$$\phi^{(1)}(t,x) = \int_{-\infty}^{\infty} dt'dx' G(t-t'; x, x') I(t', x'), \tag{7}$$

$$\phi^{(2)}(t,x) = \int_{-\infty}^{\infty} dt'dx' G(t-t'; x, x') S^{(2)}(t', x'), \tag{8}$$

where $I(t,x) = -f(x)\delta'(t) - g(x)\delta(t)$.

Unfortunately, the exact expression of the Green's function $G(\tau; x, x')$ is unknown so far. We have derived an approximate form of $G(\tau; x, x')$ with the "asymptotic approximation" developed by Andersson [9]. The detailed calculation of Eqs. (7) and (8) using the approximate Green's function is shown in reference [11]. Here we show the final results of the calculation. The first-order perturbation $\phi^{(1)}(t,x)$ near infinity $(x \gg 1)$ is expressed as a superposition of the first-order QNMs truncated at the light cone $t = x$,

$$\phi^{(1)}(t,x) \approx \theta(t-x)\sum_{n=0}^{\infty}\Big[C_n e^{s_n(t-x)} + C_n^* e^{s_n^*(t-x)}\Big] \tag{9}$$

where θ is the step function, and s_n and C_n are the eigenvalues and excitation coefficient of the QNMs, respectively. s_n is related to a more familiar "quasinormal frequency" ω_n by $\omega_n = is_n$. The information of initial data $f(x)$ and $g(x)$ is implicitly contained in C_n. We have found that the asymptotic solution of the second-order perturbation $\phi^{(2)}(t,x)$ is of the form

$$\phi^{(2)}(t,x) \approx \theta(t-x)\Big[\phi^{(2)}_{1Q}(t,x) + \phi^{(2)}_{2Q}(t,x) + \phi^{(2)}_{T}(t,x)\Big] \tag{10}$$

where $\phi_{1Q}^{(2)}$ is again a superposition of first-order QNMs, and

$$\phi_{2Q}^{(2)}(t,x) \approx \sum_{n,n'} \left[C_n C_{n'} A_{nn'} e^{(s_n+s_{n'})(t-x)} + C_n C_{n'}^* B_{nn'} e^{(s_n+s_{n'}^*)(t-x)} + \text{c.c.} \right], \tag{11}$$

$$\phi_{T}^{(2)}(t,x) \approx \sum_{n,n'} \left(\frac{C_n C_{n'}}{s_n + s_{n'}} + \frac{C_n C_{n'}^*}{s_n + s_{n'}^*} + \text{c.c.} \right) \left(\frac{1}{t-x+2} - \frac{1}{t+x+2} \right), \tag{12}$$

are the contributions from the second-order QNMs and power-law tail, respectively. The excitation coefficients of the second-order QNMs, $A_{nn'}$ and $B_{nn'}$, can be expressed as a function of s_n and $s_{n'}$ [11]. Equation (12) implies that this tail behaves as t^{-1} for $x < t \lesssim 2x$, and as t^{-2} for $t \gtrsim 2x$ (note that we have assumed that $x \gg 1$, but not assumed that $t \gg x$).

We emphasize that the second-order power-law tail $\phi_T^{(2)}$ is *essentially different* from the tails known in first-order perturbation theory [10]. First, although the first-order tails are known to decay more quickly than t^{-2}, our tail decays *more slowly than* t^{-2}. Second, while the behavior of the first-order tails depends on the form of $V(x)$, that of our tail is *independent of* it. In fact, it can be shown that the behavior of $\phi_T^{(2)}$ is completely determined by the long-range behavior of the source term $S^{(2)}(x)$, or $F(x)$. This fact manifests that our tail is intrinsically nonlinear. Remember that we have carefully chose the functional form of $F(x)$ so that the asymptotic behavior of $S^{(2)}(x)$ in our model corresponds to that in original black hole perturbations. Therefore, it is concluded that black hole ringdown *should* posses the same type of second-order tail.

2.2. Numerical calculation

To check the validity of the above analysis, we have performed two types of numerical calculations. In the first type, we integrated the first- and second-order perturbation equations (3) and (4). In the second type, we integrate the full-order equation (1). In this paper, we only show the result of the second type of simulation (for the full results, see reference [11]).

As an initial condition, we chose a momentarily stationary Gaussian wave packet, $f(x) = \exp[-(2.0x)^2]$, $g(x) = 0$. As the potential term $V(x)$, we adopted the Pöschl-Teller potential $V_{\rm PT}(x) \equiv V_0/\cosh^2(Kx)$, where V_0 and K^{-1} are the height and the width of the potential barrier, respectively. We set $V_0 = 5.0$ and $K = 1.0$ in this study. The eigenvalues s_n of first-order QNMs for $V_{\rm PT}$ are known analytically [12], and the quasinormal frequency $\omega_0 = is_0$ of the least-damped mode is calculated to be $\omega_0 = 2.18 - 0.50i$ for the above set of parameters. It follows that the frequencies of the least-damped second-order QNMs are $4.36 - 1.0i$ ("sum tone" mode) and $-1.0i$ ("difference tone" mode).

Figure 1 shows the numerical waveform of the full nonlinear field Φ observed at $x = 5.0 \equiv x_{\rm obs}$. It is observed that the waveform for early times ($x_{\rm obs} \lesssim t \lesssim 12$) is well fitted by the least-damped first-order QNM ($\omega_0 = 2.18 - 0.50i$). At late times ($t \gtrsim 12$), however, the numerical waveform does not agree with the first-order QNM. Surprisingly, it is the second-order power-law tail [Eq. (12)] that agrees with the waveform (although difficult to see in the figure, the second-order QNMs are also found to be contained in the waveform). This means that the late-time behavior of the nonlinear field is dominated by the *second-order* perturbation, not by the first-order one. We have also confirmed that the second-order tail agrees with the full-order waveform even at $t \approx 50$, which strongly suggests that third and higher-order perturbations are negligible for all times.

3. Conclusion and Discussion

In this study, we have investigated the nonlinear evolution of black hole ringdown using second- and higher-order perturbation theory and a simplified nonlinear field model for black hole metric

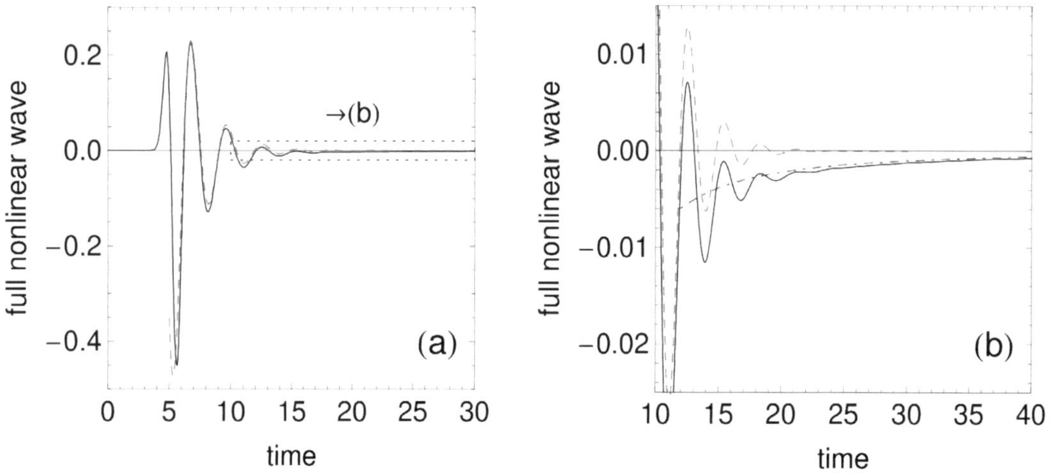

Figure 1. Temporal evolution of the full nonlinear field. The solid curve is the numerical waveform observed at $x = 5$. The dashed curve is an analytical fit by the least-damped first-order QNM. Figure (b) enlarges the region enclosed by the dotted box in figure (a). The dot-dashed curve in figure (b) is an analytical fit by the second-order power-law tail [Eq. (12)].

perturbations. We have proven that second-order QNMs, whose existence has been predicted by recent works [6, 7], do appear in the evolved second-order perturbation. As a bonus, we have discovered a new type of power-law tail appearing with the second-order QNMs . This power-law tail, to which we have referred as the second-order power-law tail, decays more slowly than any of the tails in the first-order theory, and even dominates the fully nonlinear evolution at late times. In other words, the first-order perturbation theory *fails* to predict the late-time evolution of the ringdown, and higher-order corrections *must* be taken into account. Also, we have found that the behavior of the second-order tail is determined by the asymptotic form of an effective source term, i.e., a term expressing the nonlinearity of the system. Since the asymptotic form of the source term in our model is the same as that in black hole perturbations, we conclude that the second-order power-law tail as well as the second-order QNMs will certainly appear in *real* black hole ringdown. These nonlinear components could open a new precision science in gravitational-wave studies.

Is it possible to detect the second-order power-law tail with LISA? To address this issue, we need to estimate the energy carried out by the second-order tail assuming astrophysically realistic events (e.g., binary black hole mergers), as already done for second-order QNMs [6, 7]. Also, we will need a special data analysis to extract the second-order tail from detector outputs. While the matched filtering technique has been developed for extracting QNMs [13], there seems to be only a few techniques for power-law tails [14]. If the second-order tail is detectable, it could be used to distinguish true signals and spurious ones in black hole ringdown search, where fake reduction and event identification are crucial [7, 13].

Acknowledgments
We thank H. Nakano for fruitful discussions. S.O. is supported by Grants-in-Aid for JSPS Fellows (20·4371) from the Ministry of Education, Culture, Sports, Science and Technology (MEXT) of Japan.

Journal of Physics: Conference Series **154** (2009) 012051 doi:10.1088/1742-6596/154/1/012051

References

[1] H-P. Nollert, Classical Quantum Gravity **16**, R159 (1999).
[2] K. D. Kokkotas and B. G. Schmidt, Living Rev. Relativity **2**, 2 (1999).
[3] E. Berti, V. Cardoso, J. A. Gonzalez, U. Sperhake, M. Hannam, S. Husa, and B. Brügmann, Phys. Rev. D **76**, 064034 (2007).
[4] R. J. Gleiser, C. O. Nicasio, R. H. Price, and J. Pullin, Classical Quantum Gravity **13**, L117 (1996).
[5] R. J. Gleiser, C. O. Nicasio, R. H. Price, and J. Pullin, Phys. Rep. **325**, 41 (2000).
[6] K. Ioka and H. Nakano, Phys. Rev. D **76**, 061503(R) (2007).
[7] H. Nakano and K. Ioka, Phys. Rev. D **76**, 084007 (2007).
[8] E. W. Leaver, Phys. Rev. D **34**, 384 (1986).
[9] N. Andersson, Phys. Rev. D **55**, 468 (1997).
[10] E. S. C. Ching, P. T. Leung, W. M. Suen, and K. Young, Phys. Rev. Lett. **74**, 2414 (1995); Phys. Rev. D **52**, 2118 (1995).
[11] S. Okuzumi, K. Ioka, and M-a. Sakagami, Phys. Rev. D **77**, 124018 (2008).
[12] V. Ferrari and B. Mashhoon, Phys. Rev. Lett. **52**, 1361 (1984); Phys. Rev. D **30**, 295 (1984).
[13] Y. Tsunesada, N. Kanda, H. Nakano, D. Tatsumi, and the TAMA Collaboration, Classical Quantum Gravity **22**, S1129 (2005).
[14] L. Blanchet and B. S. Sathyaprakash, Phys. Rev. Lett **74**, 1067 (1995).

An efficient numerical calculation of gravitational waves from extreme mass ratio inspirals

Ryuichi Fujita[1,2], Wataru Hikida[2] and Hideyuki Tagoshi[2]

[1]Raman Research Institute, Bangalore 560 080, India

[2]Department of Earth and Space Science, Graduate School of Science, Osaka University, Toyonaka 560-0043, Japan

E-mail: draone@rri.res.in

Abstract. Gravitational waves from extreme mass ratio inspirals are one of the important sources of LISA. We should calculate these waves so accurately that we can extract physical information of source by data analysis. Recently, we developed an efficient numerical method to compute gravitational waves from binary systems in which a point particle moves in circular orbits on the equatorial plane of the black hole. In this paper, we apply this method to compute gravitational waves from binary systems in which a point particle moves in general bound geodesic orbits of the black hole. We check the accuracy of our code using spherical symmetry of Schwarzschild black hole such that energy flux radiated by a point particle is independent of the inclination angle from the equatorial plane of black hole. We find that the accuracy of our code may be limited only by truncation of ℓ, k and n-modes, where ℓ is the degree of the spin-weighted spheroidal harmonics, and k and n are harmonics of the polar and radial motion, respectively. Then we evaluate the rate of change of three constants of motion, energy, angular momentum and the Carter constant, due to the emission of gravitational waves from a particle around Kerr black hole. This is the first time to compute the rate of change of the Carter constant using the adiabatic approximation. We also show that we can calculate gravitational waves accurately even in the case of high eccentric orbits. In this work, we truncate ℓ mode up to 20 and estimated that relative accuracy of our numerical results are better than 10^{-5} even in the high eccentric case, $e = 0.9$. Our numerical code may be useful to make templates of extreme mass ratio inspirals.

1. Introduction

Gravitational waves radiated from a particle orbiting around a Kerr black hole are one of the main targets of space-based gravitational observatory projects such as the Laser Interferometer Space Antenna(LISA)[1]. Such binary systems are called the extreme mass ratio inspirals(EMRIs). Observing gravitational waves from EMRI, we may be able to obtain information of the central black hole's spacetime such as mass, spin and the mass distribution of compact objects in the center of galaxy. In order to obtain information from EMRI, we have to achieve the phase accuracy of theoretical gravitational wave forms within one radian over the total cycle of wave, $\sim 10^5$.

Because of its extreme mass ratio, gravitational waves from EMRIs are well approximated by the black hole perturbation formalism. The conventional equation of the black hole perturbation is the Teukolsky equation[2]. There are a lot of previous works that compute gravitational waves from EMRI analytically and numerically[3, 4]. For simple orbits such as circular or equatorial orbits around black hole, numerical calculations achieved the accuracy around 10^{-5} which may be sufficient to detect gravitational waves. For more general orbits, Drasco and Hughes computed gravitational waves from EMRI[5]. However, their computational time and numerical accuracy seem to be insufficient to detect gravitational waves. It might be useful if we can compute gravitational waves more efficiently.

In this paper, we numerically compute the rates of change of constants of motion due to the emissions of gravitational waves from EMRI. The calculations are based on the method which was developed by Fujita and Tagoshi[6, 7] for computation of the homogeneous solutions of the Teukolsky equation. We will show that we can compute gravitational waves from EMRI more accurately and efficiently than past works.

2. Gravitational waves from EMRIs

The Teukolsky equation describes the gravitational perturbation of a Kerr black hole using the Newman-Penrose variables, Ψ_0 and Ψ_4, which is the gauge-invariant variables corresponding to some tetrad components of the perturbed Weyl curvature[8, 9]. The Weyl scalar Ψ_4 is related to the amplitudes of gravitational wave at infinity as

$$\Psi_4 \to \frac{1}{2}(\ddot{h}_+ - i\ddot{h}_\times), \quad \text{for } r \to \infty. \tag{1}$$

The master equation for Ψ_4 can be separated into a radial and a angular part if we expand Ψ_4 in Fourier-harmonic modes as

$$(r - ia\cos\theta)^4\, \Psi_4 = \sum_{\ell m} \int_{-\infty}^{\infty} d\omega e^{-i\omega t + im\varphi} \;{}_{-2}S^{a\omega}_{\ell m}(\theta) R_{\ell m\omega}(r), \tag{2}$$

where the angular function ${}_{-2}S^{a\omega}_{\ell m}(\theta)$ is the spin-weighted spheroidal harmonic with spin $s = -2$, and M and aM are the mass and the angular momentum of the black hole, respectively. The radial function $R_{\ell m\omega}(r)$ satisfies the radial Teukolsky equation,

$$\Delta^2 \frac{d}{dr}\left(\frac{1}{\Delta}\frac{dR_{\ell m\omega}}{dr}\right) - V(r) R_{\ell m\omega} = T_{\ell m\omega}, \tag{3}$$

where $\Delta = r^2 - 2Mr + a^2$ and the source term $T_{\ell m\omega}$ is constructed from the tetrad components of the energy momentum tensor[10]. The potential $V(r)$ is given by

$$V(r) = -\frac{K^2 - 2is(r-M)K}{\Delta} - 4is\omega r + \lambda, \tag{4}$$

where $K = (r^2 + a^2)\omega - ma$ and λ is the eigenvalue of ${}_{-2}S^{a\omega}_{\ell m}(\theta)$.

We solve the radial Teukolsky equation by using the Green function method. We introduce two kinds of solution of the homogeneous Teukolsky equation. One of them has purely outgoing property at infinity and the other has purely ingoing property at the horizon. Then the asymptotic property of the solution at the horizon is expressed as

$$R_{\ell m\omega}(r \to r_+) \equiv Z^{\mathrm{H}}_{\ell m\omega} \Delta^2 e^{-iPr^*}. \tag{5}$$

The solution at infinity is expressed as

$$R_{\ell m\omega}(r \to \infty) \equiv Z^{\infty}_{\ell m\omega} r^3 e^{i\omega r^*}. \tag{6}$$

For the bound geodesic orbits, the amplitude of partial wave $Z^{\infty/\mathrm{H}}_{\ell m\omega}$ can be expanded by using a Fourier series as[11]

$$Z^{\infty,\mathrm{H}}_{\ell m\omega} \equiv \sum_{kn} \tilde{Z}^{\infty,\mathrm{H}}_{\ell mkn} \delta(\omega - \omega_{mkn}), \tag{7}$$

where $\omega_{mkn} = m\Omega_\phi + k\Omega_\theta + n\Omega_r$, and where Ω_r is the frequencies of r-motion, Ω_θ is the frequencies of θ-motion and Ω_ϕ is the frequencies of ϕ-motion.

Then the gravitational wave form at infinity is expressed as

$$h_+ - ih_\times = -\frac{2}{r}\sum_{\ell mkn}\frac{\tilde{Z}^\infty_{\ell mkn}}{\omega^2_{mkn}}\frac{{}_{-2}S^{a\omega_{mkn}}_{\ell m}(\theta)}{\sqrt{2\pi}}e^{i\omega_{mkn}(r^*-t)+im\phi}. \tag{8}$$

Moreover, the time-averaged rates of change for the three constants of motion due to the emission of the gravitational waves are expressed as[12]

$$\begin{aligned}
\left\langle\frac{d\mathcal{E}}{dt}\right\rangle &= -\mu^{-1}\sum_{\ell mkn}\frac{1}{4\pi\omega^2_{mkn}}\left(\left|\tilde{Z}^\infty_{\ell mkn}\right|^2 + \alpha_{\ell mkn}\left|\tilde{Z}^{\rm H}_{\ell mkn}\right|^2\right),\\
\left\langle\frac{d\mathcal{L}_z}{dt}\right\rangle &= -\mu^{-1}\sum_{\ell mkn}\frac{m}{4\pi\omega^3_{mkn}}\left(\left|\tilde{Z}^\infty_{\ell mkn}\right|^2 + \alpha_{\ell mkn}\left|\tilde{Z}^\infty_{\ell mkn}\right|^2\right),\\
\left\langle\frac{d\mathcal{C}}{dt}\right\rangle &= -2\left\langle a^2\mathcal{E}\cos^2\theta\right\rangle\left\langle\frac{d\mathcal{E}}{dt}\right\rangle + 2\left\langle\mathcal{L}_z\cot^2\theta\right\rangle\left\langle\frac{d\mathcal{L}_z}{dt}\right\rangle\\
&\quad -\sum_{\ell mkn}\frac{k\Upsilon_\theta}{2\pi\omega^3_{mkn}}\left(|Z^\infty_{\ell mkn}|^2 + \alpha_{lm\omega_{mkn}}\left|Z^{\rm H}_{\ell mkn}\right|^2\right).
\end{aligned} \tag{9}$$

Here $\langle\ldots\rangle$ means the infinite time average.

3. Numerical methods and results

Practical numerical calculations of gravitational waves from EMRIs are summarized as follows.

- Calculate constants of motion $(\mathcal{E}, \mathcal{L}_z, \mathcal{C})$ and fundamental frequencies $(\Omega_\phi, \Omega_\theta, \Omega_{\rm r})$ for given orbital parameters $(p, e, \theta_{\rm inc})$, where p is semi-latus rectum, e is eccentricity and $\theta_{\rm inc}$ is inclination angle from the equatorial plane of black hole.
- Derive orbit in frequency domain.
- Calculate two kind of the homogeneous solution of the Teukolsky equation.
- Integrate the Green function over the source term.
- Evaluate gravitational waves using Eq.8 and Eq.9.

Using the methods described in Ref.[11], we evaluate constants of motion, fundamental frequencies and orbit in frequency domain. And we solve the homogeneous Teukolsky equation using Mano-Suzuki-Takasugi(MST) method[13]. In MST method, we expand the homogeneous solution in terms of hypergeometric functions as $R_{\ell m\omega}(r) \sim \sum_n a_n F(\alpha_n, \beta_n, \gamma_n; r)$. Although the application of MST method was limited to the analytical calculation using the low frequency expansion, Fujita and Tagoshi showed that MST method is also useful in numerical calculation[6, 7]. They showed that the homogeneous solutions converges very fast, and evaluated gravitational waves from a compact star moving in circular and equatorial orbits around Kerr black hole very accurately. In this paper, we apply their numerical method to compute the homogeneous solutions of the Teukolsky equation in the case of the generic bound geodesic orbits.

In this section, we show our numerical results such as the time-averaged rates of change of constants of motion, energy $\langle d\mathcal{E}/dt\rangle^\infty$, angular momentum $\langle d\mathcal{L}_z/dt\rangle^\infty$ and the Carter constant $\langle d\mathcal{C}/dt\rangle^\infty$ due to gravitational wave radiated to infinity when a compact star traces generic orbits around Kerr black hole. In numerical calculation, we have to truncate the mode summation in terms of ℓ, m, k and n-mode when we estimate gravitational waves using Eq.8 and Eq.9. We truncate the mode summation of k and n-mode when the relative accuracy with respect to total flux reaches 10^{-10}. In Figure 1, we show the behavior of the modal energy flux, $\langle d\mathcal{E}_{\ell mkn}/dt\rangle^\infty_{\rm GW}$, of gravitational waves radiated to infinity in the case of Schwarzschild black hole. From Figure 1, we find that we have to compute larger n-mode as eccentricity or ℓ becomes large. In Figure 2 and Figure 3, we show the behavior of the modal energy flux, $\langle d\mathcal{E}_{\ell mk}/dt\rangle^{\infty}_{\rm GW}$, of gravitational waves radiated to infinity in the case of Schwarzschild black hole. From these figure, we find that the peak location of k-mode is around $k = \ell - m$. In Kerr black hole case, these properties of n and k-mode are almost the same as that of Schwarzschild black hole, but the spectrum of k-mode of Kerr black hole is broader than that of Schwarzschild black hole. As for m-mode, we compute all m-mode from ℓ to $-\ell$. And we set the maximum of ℓ to be $\ell_{\rm max} = 20$. The details of the truncation rule are discussed in Ref. [14].

In order to check the accuracy of our numerical code, we compute the rate of change of the energy in the case of Schwarzschild black hole. In the Schwarzschild case, the rate of change of the energy is independent of the inclination angle from the equatorial plane of black hole. In Table 1, we check the accuracy of our code comparing the rate of change of the energy in the case of non-equatorial orbits with that of equatorial orbit. In this table, the values in the square bracket are the estimated truncation errors up to $\ell = 20$-mode. And we find that the relative errors between the equatorial plane case and the non-equatorial plane case are around $10^{-8} - 10^{-11}$. Thus, the relative errors in Table 1 are consistent with the truncation errors of summation over ℓ, k and n-mode. In Table 2, we show the rate of change of three constants of motion in the case of Kerr black hole. This is the first time to compute the rate of change of the Carter constant using the adiabatic approximation. The truncation errors of ℓ-mode in Table 2 are around $10^{-6} - 10^{-11}$. Thus numerical accuracies in the case of $e = 0.1$ are limited by the truncation of k and n-mode, $\sim 10^{-10}$, and in the case of $e = 0.7$ and $e = 0.9$ are limited by the truncation of ℓ-mode.

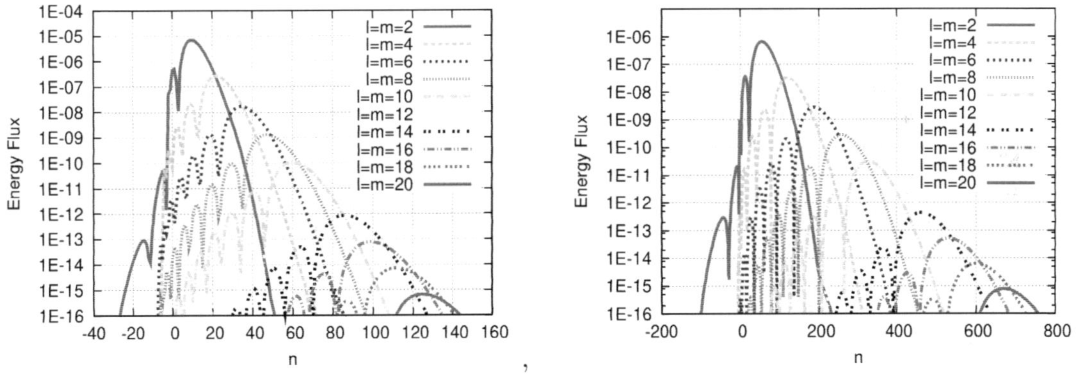

Figure 1. Modal energy flux $\langle d\mathcal{E}_{\ell m0n}/dt\rangle_{\rm GW}$ at infinity, radiated by a particle around a Schwarzschild black hole. In this figure, the orbital radius is $10M$, the inclination angle from the equatorial plane of black hole is $20°$, and the orbital eccentricity is 0.7 for the left figure and 0.9 for the right figure respectively.

4. Summary and future work

We developed an efficient numerical method to compute gravitational waves from bound geodesic orbits. In our method, we compute the homogeneous solution of the Teukolsky equation using Mano-Suzuki-Takasugi method. We checked the accuracy of our code using the fact that the rate of change of a particle's energy is independent of the inclination angle from the equatorial plane of Schwarzschild black hole. We found that the accuracy of our code may be limited only by the truncation of summation over ℓ, k and n-mode. In this paper, we set the relative errors

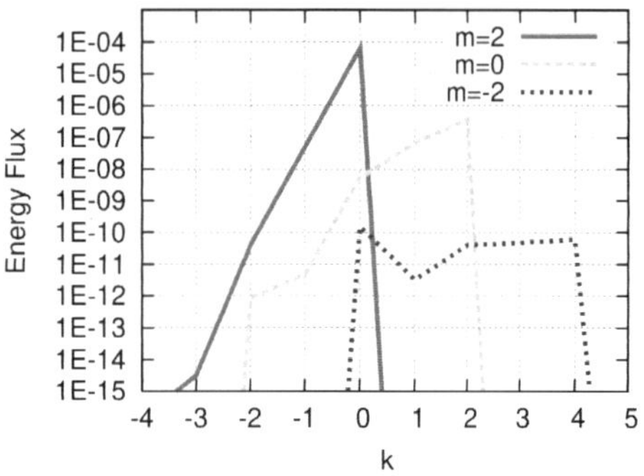

Figure 2. Modal energy flux $\langle d\mathcal{E}_{2mk}/dt\rangle_{\rm GW}$ at infinity, radiated by a particle around a Schwarzschild black hole. In this figure, the orbital radius is $10M$, the orbital eccentricity is 0.7 and the orbital inclination angle is $\theta_{\rm inc} = 20°$.

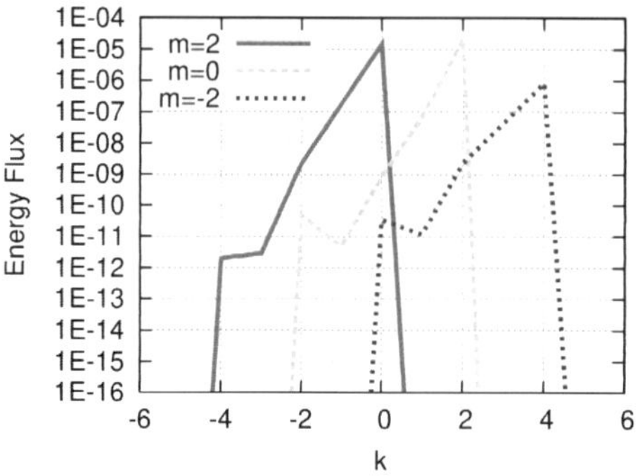

Figure 3. Modal energy flux $\langle d\mathcal{E}_{2mk}/dt\rangle_{\rm GW}$ at infinity, radiated by a particle around a Schwarzschild black hole. In this figure, the orbital radius is $10M$, the orbital eccentricity is 0.7 and the orbital inclination angle is $\theta_{\rm inc} = 70°$.

Table 1. Time-averaged rates of change of the energy of a particle due to the emission of gravitational waves to infinity from a Schwarzschild black hole. We compare the result for the equatorial plane case with the one for non-equatorial plane. Relative error in square brackets is an order of magnitude estimate for the fractional accuracy for the case of the equatorial plane, which is determined by truncating number of ℓ-mode. Here we set to $\ell_{\rm max} = 20$.

a/M	p/M	e	$\theta_{\rm inc}$	$\langle d\mathcal{E}/dt\rangle^{\infty}$	Relative error
0	10	0.1	0°	$-6.31752474714\times10^{-5}$	$[10^{-11}]$
0	10	0.1	20°	$-6.31752474712\times10^{-5}$	4.1×10^{-12}
0	10	0.1	45°	$-6.31752474688\times10^{-5}$	4.2×10^{-11}
0	10	0.1	70°	$-6.31752474700\times10^{-5}$	2.3×10^{-11}
0	10	0.5	0°	$-9.27335012129\times10^{-5}$	$[10^{-11}]$
0	10	0.5	20°	$-9.27335011990\times10^{-5}$	1.5×10^{-10}
0	10	0.5	45°	$-9.27335011942\times10^{-5}$	2.0×10^{-10}
0	10	0.5	70°	$-9.27335011362\times10^{-5}$	8.3×10^{-10}
0	10	0.9	0°	$-4.19426469281\times10^{-5}$	$[10^{-8}]$
0	10	0.9	20°	$-4.19426468367\times10^{-5}$	2.2×10^{-9}
0	10	0.9	45°	$-4.19426468292\times10^{-5}$	2.4×10^{-9}
0	10	0.9	70°	$-4.19426453712\times10^{-5}$	3.7×10^{-8}

of k and n-mode are 10^{-10} and the maximum value of ℓ-mode to be 20. Then we found that the accuracies of our code are better than 10^{-5} even in the case of $e = 0.9$. Thus our code will be useful to detect gravitational waves from EMRIs by LISA data analysis.

We also computed the rate of change of the Carter constant using adiabatic approximation. This is the first time to evaluate the rate of change of the Carter constant accurately.

Using our code, we will estimate gravitational waves from more general orbits such that eccentricity is higher than 0.9 and inclination angle is 90° in the future work. And we will compute gravitational waves including adiabatic evolution of constants of motion and construct

Table 2. Time-averaged rates of change of three constants of motion, energy $\langle d\mathcal{E}/dt\rangle^{\infty}$, angular momentum $\langle d\mathcal{L}_z/dt\rangle^{\infty}$ and the Carter constant $\langle d\mathcal{C}/dt\rangle^{\infty}$ due to gravitational wave radiated to infinity per unit mass in the case of some generic orbits around Kerr black hole. The orbital parameters used here are the same as those used by Drasco and Hughes in Ref. [5] except for the case of $e = 0.9$, which they did not show. Our results are consistent with their ones, except for the rate of change of the Carter constant, $\langle d\mathcal{C}/dt\rangle^{\infty}$. Each number in square brackets is an order of magnitude estimate for the fractional accuracy of the preceding number, which is determined by truncating number of ℓ-mode. Here we set to $\ell_{\max} = 20$. We note that the case of $q = 0.9M$, $p = 6M$, $e = 0.9$ and $\theta_{\text{inc}} = 80°$ is out of angle of stable bound orbits.

e	θ_{inc}	$\langle d\mathcal{E}/dt\rangle^{\infty}$	$\langle d\mathcal{L}_z/dt\rangle^{\infty}$	$\langle d\mathcal{C}/dt\rangle^{\infty}$
0.1	20°	$-5.87363797768 \times 10^{-4}[10^{-11}]$	$-8.53727877914 \times 10^{-3}$	$-5.24019849268 \times 10^{-3}$
0.1	40°	$-6.18322941189 \times 10^{-4}[10^{-11}]$	$-7.63099400995 \times 10^{-3}$	$-2.02271137751 \times 10^{-2}$
0.1	60°	$-6.83348194356 \times 10^{-4}[10^{-11}]$	$-6.07829110810 \times 10^{-3}$	$-4.32194649697 \times 10^{-2}$
0.1	80°	$-8.05858116215 \times 10^{-4}[10^{-10}]$	$-3.62538057535 \times 10^{-3}$	$-7.18520475478 \times 10^{-2}$
0.7	20°	$-7.73120284923 \times 10^{-4}[10^{-7}]$	$-6.69307191178 \times 10^{-3}$	$-3.88677844044 \times 10^{-3}$
0.7	40°	$-8.74643775649 \times 10^{-4}[10^{-7}]$	$-6.52867874371 \times 10^{-3}$	$-1.62235256695 \times 10^{-2}$
0.7	60°	$-1.14681348743 \times 10^{-3}[10^{-7}]$	$-6.37991880771 \times 10^{-3}$	$-4.14597930483 \times 10^{-2}$
0.7	80°	$-2.71871174658 \times 10^{-3}[10^{-6}]$	$-8.40111270562 \times 10^{-3}$	$-1.34124921753 \times 10^{-1}$
0.9	20°	$-3.22207770684 \times 10^{-4}[10^{-6}]$	$-2.36899718598 \times 10^{-3}$	$-1.35639225718 \times 10^{-3}$
0.9	40°	$-3.81207517166 \times 10^{-4}[10^{-6}]$	$-2.43210898466 \times 10^{-3}$	$-5.96021788095 \times 10^{-3}$
0.9	60°	$-5.54821575240 \times 10^{-4}[10^{-6}]$	$-2.67958515939 \times 10^{-3}$	$-1.70800085390 \times 10^{-2}$
0.9	80°	–	–	–

efficient templates for LISA data analysis.

Acknowledgments

W.H. was supported by the JSPS Research Fellowships for Young Scientists, No. 1756 and also supported by the 21st Century COE program "Towards a New Basic Science; Depth and Synthesis" at Osaka university from the Ministry of Education, Culture, Sports, Science and Technology of Japan. H.T's work was supported in part by JSPS, KAKENHI, Nos.16540251 and 20540271. R.F. thank the organizers of the 7th LISA Symposium for the financial support to attend the conference and for the conference itself that we got together.

References

[1] LISA web page : *http://lisa.jpl.nasa.gov/*, *http://www.lisa.aei-hannover.de/*
[2] Teukolsky S A 1973 *Astrophys. J.* **185** 635–647.
[3] Glampedakis K 2005 *Class. Quantum Grav.* **22** S605-S659.
[4] Sasaki M and Tagoshi H 2003 *Living Rev. Relativity* **6** 6
[5] Drasco S and Hughes S A 2006 *Phys. Rev. D* **73** 024027
[6] Fujita R and Tagoshi H 2004 *Prog. Theor. Phys.* **112** 415–450.
[7] Fujita R and Tagoshi H 2005 *Prog. Theor. Phys.* **113** 1165–1182.
[8] Chandrasekhar S 1983 *Mathematical theory of black holes* (New York: Oxford University Press)
[9] Nakamura T, Oohara K, and Kojima Y 1987 *Prog. Theor. Phys. Suppl.* **90** 1
[10] Mino Y, Sasaki M, Shibata M, Tagoshi H and Tanaka T 1997 *Prog. Theor. Phys. Suppl.* **128** 1
[11] Drasco S and Hughes S A 2004 *Phys. Rev. D* **69** 044015
[12] Sago N, Tanaka T, Hikida W, and Nakano H 2005 *Prog. Theor. Phys.* **114** 509–514.
[13] Mano S, Suzuki H, and Takasugi E 1996 *Prog. Theor. Phys.* **95** 1079–1096.
[14] Fujita R, Hikida W and Tagoshi H *in preparation*

Journal of Physics: Conference Series **154** (2009) 012053 doi:10.1088/1742-6596/154/1/012053

Simulations of Extreme-Mass-Ratio Inspirals Using Pseudospectral Methods

Priscilla Cañizares, Carlos F. Sopuerta

Institut de Ciències de l'Espai (CSIC-IEEC), Facultat de Ciències, Campus UAB, Torre C5 parells, E-08193 Bellaterra, Spain

E-mail: `pcm@ieec.uab.es`, `sopuerta@ieec.uab.es`

Abstract. Extreme-mass-ratio inspirals (EMRIs), stellar-mass compact objects (SCOs) inspiralling into a massive black hole, are one of the main sources of gravitational waves expected for the Laser Interferometer Space Antenna (LISA). To extract the EMRI signals from the expected LISA data stream, which will also contain the instrumental noise as well as other signals, we need very accurate theoretical templates of the gravitational waves that they produce. In order to construct those templates we need to account for the gravitational backreaction, that is, how the gravitational field of the SCO affects its own trajectory. In general relativity, the backreaction can be described in terms of a local *self-force*, and the foundations to compute it have been laid recently. Due to its complexity, some parts of the calculation of the self-force have to be performed numerically. Here, we report on an ongoing effort towards the computation of the self-force based on time-domain multi-grid pseudospectral methods.

1. Introduction

Extreme-Mass-Ratio Inspirals (EMRIs) are one of the main source of gravitational waves for the planned Laser Interferometer Space Antenna (LISA) [1]. EMRIs are binary systems composed by a stellar compact object (SCO), like a neutron star or a stellar black hole, with masses in the range $m = 1 - 10^2 M_\odot$, inspiralling into a massive black hole (MBH), with masses in the range $M = 10^4 - 10^7 M_\odot$, located at galactic centers. Then, the mass ratios involved are $\mu = m/M \sim 10^{-7} - 10^{-3}$. During the inspiral phase the system is driven by the emission of gravitational radiation, and hence there is loss of energy and angular momentum that makes the orbit shrink until the SCO plunges into the MBH. It is expected that LISA will be able to detect $10 - 10^3$ EMRI$/yr$ up to distances with $z \lesssim 1$ [2, 3] (see also [4]). In order to extract the signals produced by these systems, which will be buried in instrumental noise and the gravitational wave foreground (produced by compact binaries in the LISA band), and extract relevant physical information from them, we need to have an *a priori* theoretical knowledge of the gravitational waveforms with a high degree of precision. While techniques for constructing templates good enough for detection are getting ready, methods to build templates good enough for extraction of physical information are not yet fully developed. The main difficulty being that one requires a more precise treatment of the self-gravity of the SCO and its impact on the gravitational waveform.

Frequency-domain approaches provide very accurate results in the computation of quasinormal modes and frequencies and also in the computation of the self-force for EMRIs with moderate eccentricities. However, the frequency domain approach has more difficulties when dealing with EMRIs with highly eccentric orbits, which are of interest for LISA, since one has to sum over a large number of modes to obtain a good accuracy, and convergence may be an issue. On the other hand, time-domain approaches are not affected much by the eccentricity of the orbit and may be more efficient for the case of high-eccentricity EMRIs. However, in the time domain numerical approach one has to deal with different physical scales (both spatial and temporal) present in the problem due to extreme mass ratios. Specifically, we have to computationally resolve large wavelength scales (comparable to the massive black hole) and also the scales in the vicinity of the SCO, which are crucial for evaluating the self-force. This means, in practice, that we have to resolve numerically large scales and small scales inside the same computational domain. Here we introduce a new time-domain scheme towards the computation of the self-force based on the PseudoSpectral Collocation (PSC) method (see, e.g. [5]). A remarkable property of this new scheme is that it avoids having to introduce in the problem a small spatial scale associated with the presence of the SCO (see, e.g. [6]). This is done by using multiple domains and locating the particle in the interface bewteen two of them. Therefore, we just need the resolution to describe the field near the particle, but not the particle itself, which makes the computation more efficient. Here, we describe how to apply this technique to the computation of the self-force on a charged scalar particle in circular orbits around a non-rotating black hole and show some results of the numerical implementation.

2. Study of a Simplified Model

We study a simplified model of an EMRI consisting of a particle with charge q, associated with a scalar field $\Phi(x^\mu)$, orbiting a non-rotating MBH. In this model, the particle generates the scalar field, which in turn influences the particle trajectory (this effect can be described by a local force, called the *self-force*). Therefore, this model contains all the ingredients of the gravitational case (see details in [7, 8]), where the particle motion is influenced by its own gravitational field instead of a scalar field.

In the scalar field model the spacetime geometry is not dynamical. In our problem it has to describe the geometry of the MBH and therefore will be given by the Schwarzschild black hole metric:

$$ds^2 = f(-dt^2 + dr^{*2}) + r^2 d\Omega^2\,, \quad d\Omega^2 = d\theta^2 + \sin^2\theta d\varphi^2\,, \tag{1}$$

where $(x^\mu) = (t, r, \theta, \varphi)$ are the Schwarzschild coordinates, $f = 1 - 2M/r$, M is the MBH mass, and $r^* = r + 2M\ln(r/2M - 1)$ is the *tortoise* coordinate. The scalar field equation is a wave-like equation with a singular source term due to the energy density of the particle's scalar charge (see, e.g. [9])

$$\mathrm{g}^{\alpha\beta}\nabla_\alpha\nabla_\beta\Phi(x) = -4\pi\rho = -4\pi q\int_\gamma d\tau\,\delta_4(x, z(\tau))\,, \tag{2}$$

where $x^\mu = z^\mu(\tau)$ is the trajectory of the particle (without loss of generality, the orbital plane is taken to be the equatorial plane); ∇_μ denotes the associated canonical connection; τ denotes proper time along the particle timelike worldline γ, and $\delta_4(x, z(\tau))$ is the invariant Dirac functional in Schwarzschild spacetime (see, e.g. [9]) and describes the singular structure of the source term. The equations of motion for the particle, obtained from energy-momentum conservation, are

$$m\frac{du^\mu}{d\tau} = F^\mu = q\,(\mathrm{g}^{\mu\nu} + u^\mu u^\nu)\nabla_\nu\Phi\,, \quad u^\mu = \frac{dz^\mu}{d\tau}\,, \tag{3}$$

where m is the particle mass and u^μ its 4-velocity. The force F^μ diverges on the particle worldline, γ, due to the singularity that carries the particle field, and therefore it must be regularized [10]. An analysis of the solutions of (2) and (3) reveals (see [9] and references therein) that the field, Φ, can be split into two parts [11]: $\Phi = \Phi^{\mathrm{R}} + \Phi^{\mathrm{S}}$. Where Φ^{S} is the singular part, which contains the singular structure of the field and satisfies the same field equation, and $\Phi^{\rm R}$ is the regular part, which satisfies an homogeneous wave equation:

$$g^{\alpha\beta}\nabla_\alpha\nabla_\beta\Phi^{\rm S} = -4\pi\rho\,, \qquad g^{\alpha\beta}\nabla_\alpha\nabla_\beta\Phi^{\rm R} = 0\,. \tag{4}$$

The regular field is solely responsible of the deviation of the particle from geodesic motion around the MBH. Hence, the *self-force* which acts on the particle is $F^{\rm R}_\mu = q(g^{\mu\nu} + u^\mu u^\nu)\nabla_\nu\Phi^{\rm R}$.

In order to solve the field equation (2) we can make use of the spherical symmetry of the Schwartzchild metric and decompose Φ into scalar spherical harmonics. The equations for the different harmonic coefficients, $\Phi^m_\ell(t,r)$, decouple and satisfy 1+1 wave-like equations of the form

$$\left\{-\frac{\partial^2}{\partial t^2} + \frac{\partial^2}{\partial r^{*2}} - V_\ell(r)\right\}(r\Phi^m_\ell) = f\, S^m_\ell\,\delta(r - r_p(t))\,, \tag{5}$$

where $V_\ell(r)$ is the Regge-Wheeler potential for scalar fields on the Schwarzschild geometry and S^m_ℓ is the harmonic coefficient of the singular source term generated by the particle, which is given by:

$$V_\ell(r) = f\left[\frac{\ell(\ell+1)}{r^2} + \frac{2M}{r^3}\right]\,, \quad S^m_\ell = -\frac{4\pi q f(r_p)}{r_p u^t}\,\bar{Y}^m_\ell(\theta,\varphi_p)\,, \tag{6}$$

where an overbar denotes complex conjugation. The solution of (5) is finite and continuous at the particle location, but it is not be differentiable in the sense that the evaluation of the radial derivative of the scalar field from the left and from the right of the particle yields different values, that is, there is a jump [see equation (12) bellow]. Nevertheless, one can see that the contribution to the self-force of each harmonic is finite, it is the sum over ℓ that becomes infinite. Taking advantage of this fact, a method to extract the singular part of the self-force mode by mode was developed, the so-called *mode-sum* regularization scheme [12, 13, 14]. Given the multipolar decomposition of the gradient of the scalar field

$$\Phi_\alpha(x^\mu) = \sum_{\ell=0}^{\infty}\Phi^\ell_\alpha(x^\mu)\,, \quad \text{where} \quad \Phi^\ell_\alpha(x^\mu) = \nabla_\alpha\sum_{m=-\ell}^{\ell}\Phi^m_\ell(t,r)\,Y^m_\ell(\theta,\phi)\,, \tag{7}$$

each Φ^ℓ_α is finite at the particle location. Using the mode-sum scheme, the regularized part of the gradient of the scalar field can be written as follows

$$\Phi^{\rm R}_\alpha(z^\mu(\tau)) = \lim_{x^\mu\to z^\mu(\tau)}\sum_{\ell=0}^{\infty}\left(\Phi^\ell_\alpha(x^\mu) - \Phi^{{\rm S},\ell}_\alpha(x^\mu)\right)\,, \tag{8}$$

and the multipoles of the singular part, $\Phi^{{\rm S},\ell}_\alpha$, are known in a neighborhood of the particle worldline and are given by [14]:

$$\lim_{x^\mu\to z^\mu(\tau)}\Phi^{{\rm S},\ell}_\alpha = q\left[\left(\ell+\frac{1}{2}\right)A_\alpha + B_\alpha + \frac{C_\alpha}{\ell+\frac{1}{2}} - \frac{2\sqrt{2}D_\alpha}{(2\ell-1)(2\ell+3)} + \ldots\right]\,, \tag{9}$$

where A_α, B_α, C_α, D_α, ... are called the regularization parameters [13, 14, 15]. They are independent of ℓ, but depend on the particle trajectory. The singular part corresponds to the first three terms which lead to quadratic, linear, and logarithmical divergences. The remaining terms form a convergent series that does not contribute to the self-force (each of them). Since the only non-vanishing component of the regularization coefficients is the radial one, this is the only component of the self-force that is actually singular, and hence the only one to be regularized.

3. Description of the Simulations

We begin by defining the spatial domain where we want to obtain our solution: $r^* \in (-\infty, +\infty)$, where $r^* \to -\infty$ corresponds to the horizon location ($r = 2M$), while $r^* \to +\infty$ corresponds to spatial infinity. For computational reasons, we have to take a *truncated* radial domain (we could avoid this by compactifying the radial domain): $r^* \in \Omega = [r^*_{\rm H}, r^*_{\infty}]$, and hence we need to prescribe outgoing boundary conditions at the end points of this domain:

$$\partial_t(r\,\Phi^m_\ell(t,r^*_{\rm H})) - \partial_{r^*}(r\,\Phi^m_\ell(t,r^*_{\rm H})) = 0\,, \quad \partial_t(r\,\Phi^m_\ell(t,r^*_{\infty})) + \partial_{r^*}(r\,\Phi^m_\ell(t,r^*_{\infty})) = 0\,. \tag{10}$$

On the other hand, in order to numerically compute the solutions, it is necessary to discretize our equations in time and space. We use the PSC for the spatial discretization, and the time discretization is performed by using a Runge-Kutta algorithm. Then, the harmonic components of the scalar field are approximated by an expansion in Chebyshev polynomials,

$$\Phi^m_{\ell,N}(t,r^*) = \sum_{n=0}^{N} a_n(t) T_n(X(r^*))\,, \quad X \in [-1,1]\,, \tag{11}$$

and the coefficients $a_n(t)$ are determined, in terms of a system of ordinary differential equations (ODEs), by forcing the equation (5) to be satisfied at a given set of collocation points $\{X_k\}_{k=0,1,\ldots,N}$. For our calculations we took a *Lobatto-Chebyshev* grid, which includes the boundary points as nodes of the grid. The system of ODEs is solved by using a Runge-Kutta algorithm, typically a RK4 scheme.

The singular source term (6) in our equation (5) would in principle spoil the exponential convergence of the PSC method. In order to avoid this, and also in order to avoid a *small scale* associated with the presence of the particle, we use a multi-domain scheme and locate the particle in the interface between two such domains. More specifically, we divide Ω into D sub-domains $\Omega = \cup_i [r^{*\,L}_i, r^{*\,R}_i]$ $(i = 1, \ldots, D)$, with $r^{*\,R}_{i-1} = r^{*\,L}_i$, and evolve the equation (without the source term) in each sub-domain independently. In doing so we need to communicate the different subdomains, in particular the subdomains to the left and to the right of the particle, by using appropriate boundary conditions. These conditions can be derived from the equation (5), and can be written as junction conditions that describe the possible jumps in the solution ($[\lambda]_p = \lim_{r^*\to r^*_p} \lambda_+(t,r^*) - \lim_{r^*\to r^*_p} \lambda_-(t,r^*)$) across the radial particle location (see [16] for details). In the case of circular orbits they are

$$[\,\Phi^m_\ell(t,r^*)\,]_p = 0\,, \quad [\,\partial_t \Phi^m_\ell(t,r^*)\,]_p = 0\,, \quad [\,\partial_{r^*}\Phi^m_\ell(t,r^*)\,]_p = f S^m_\ell\,. \tag{12}$$

Apart from these techniques, we also use the Fast Fourier Transform in order to evaluate derivatives of the scalar field in a quick way (see, e.g. [5]). Moreover, in order to reduce possible spurious high-frequency components of the numerical solution, we apply an exponential spectral filter to the solutions after every time step of the RK algorithm.

4. Summary of Results

We have developed a numerical code that implements the previous ideas in order to compute the self-force on a charged scalar particle in circular geodesics around a non-rotating MBH. In order to validite the code we have made a series of tests. In particular, we have evolved a Gaussian pulse in flat space ($V_l = 0$) and also in the Schwarzschild geometry ($V_l \neq 0$), that is, we have evolved the equation (5) without a particle. We have done this using different numbers of collocation points, N, in order to check that we obtain the expected convergence of the PSC method for smooth solutions. In Figure 1 (top left) we show that indeed an estimation of the truncation error, $\ln|a_N|$, falls exponentially as the number of collocation points increases.

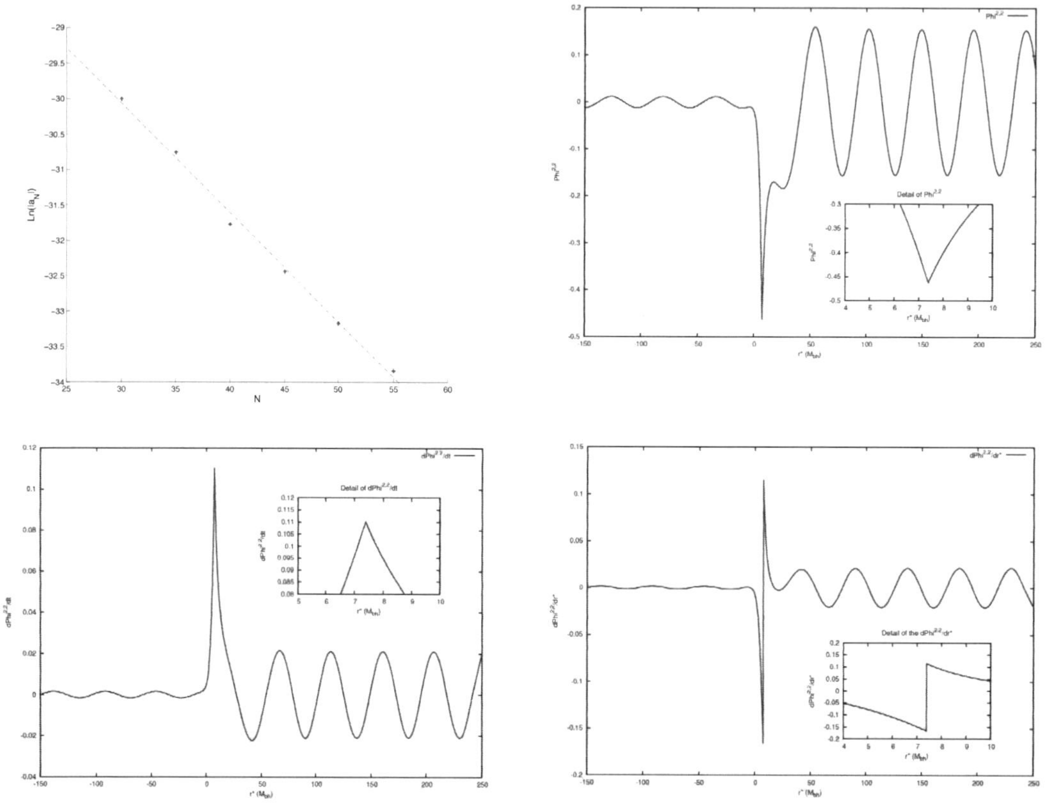

Figure 1. *Top left:* Estimation of the truncation error, $\ln |a_N|$, versus the number of collocation points, N. The points lie aroud a straight line with negative slope, showing that the expected exponential convergence is obtained. The other figures show snapshots of the evolution of the $(2,2)$ mode (with $r_p = 6M$) and the detailed structure near the particle. They show the continuity of Φ^m_ℓ (*top right*) and $\partial_t \Phi^m_\ell$ (*bottom left*), and the jump in $\partial_{r^*} \Phi^m_\ell$ (*botton right*).

On the other hand, we have performed simulations including the charged particle in order to develop and test the communication between the subdomains, in particular between those around the particle where jumps in the solution can occur. In practice, this communication is implemented by using a penalty method, in such a way that the jump conditions in equation (12) are enforced in a dynamical way, by introducing new terms into the resulting ODEs that drive the solution towards the satisfaction of these conditions. These new terms have to be callibrated against the number of collocation points and the size of the time step. We have found that the penalty method, in combination with the use of a spectral filter, works very well and can produce accurate solutions to our equations. In order to illustrate this fact we have included three plots in Figure 1 corresponding to snapshots of the evolution of the $(\ell, m) = (2,2)$ mode at the last stable circular orbit ($r = 6M$), which is probably the most demanding situation. At the top right, we show a snapshot of the scalar field, Φ^2_2, where it can be seen that the field it is continuous at the particle although it is not differentiable. At the bottom left we have a snapshot for the time derivative of the scalar field, $\partial_t \Phi^2_2$, which is also continuous (this only happens for the case of circular orbits [6]) but non-differentiable. Finally, at the bottom right, we show a snapshot of the radial derivative of the field, $\partial_{r^*} \Phi^2_2$, where we can see the ability our numerical techniques in resolving the jump accross the particle. In the small box inside this figure we zoom in near the particle location to show how well the method can deal with the sharp jump and at the same time produce smooth waves.

5. Conclusions and Future Work

We have presented a new time-domain computational framework to compute the self-force for EMRIs. We have implemented this new method numerically for the case of a charged scalar particular moving in circular geodesics around a non-rotating MBH. This method is based on the PSC method. One of the main ingredients of this technique is to consider a multi-domain grid and to locate the particle in the interface between two subdomains. In this way, we solve equations with smooth solutions inside each subdomain, which preserves the powerful convergence properties of the PSC method and, at the same time, we avoid introducing a small scale related to the size of the SCO. This in turn makes the algorithm quite efficient as compared to other treatments of the particle. The effects in the solution due to the presence of the particle source term appear in our method through the imposition of a set of jump conditions. Here, we have presented tests of the convergence of the numerical code we have built and we have also shown the ability of the code to resolve the field near the particle, including the jump in the radial derivative, and at the same time produce smooth waves. Calculations of the self-force on the particle are underway and will be presented elsewhere [17].

In this work, we have restricted ourselves to the case of circular orbits, where the particle is always at the same radial position. In order to extend our calculations to generic eccentric orbits we need to introduce a method to deal with a moving particle, either by moving our domains or by changing the coordinate system so that the particle is always located at the same node. This is a project presently under development. Given that the astrophysically interesting EMRIs involve spinning MBHs, another important task for the future is to study how to implement all these techniques for the case in which the MBH is represented by the Kerr solution.

Acknowledgments

We would like to thank Leor Barack, José Luis Jaramillo, Eric Poisson, for helpful discussions. CFS acknowledges support from the Ramón y Cajal Programme of the Ministry of Education and Science of Spain and by a Marie Curie International Reintegration Grant (MIRG-CT-2007-205005/PHY) within the 7th European Community Framework Programme.

References

[1] LISA: `http://www.esa.int`, `http://lisa.jpl.nasa.gov`
[2] Gair J R *et al.* 2004 *Class. Quant. Grav.* **21** S1595–S1606 (*Preprint* gr-qc/0405137)
[3] Hopman C and Alexander T 2006 *Astrophys. J.* **645** L133–L136 (*Preprint* astro-ph/0603324)
[4] Amaro-Seoane P *et al.* 2007 *Class. Quant. Grav.* **24** R113–R169 (*Preprint* astro-ph/0703495)
[5] Boyd J P 2001 *Chebyshev and Fourier Spectral Methods* 2nd ed (New York: Dover)
[6] Sopuerta C F, Sun P, Laguna P and Xu J 2006 *Class. Quant. Grav.* **23** 251–286 (*Preprint* gr-qc/0507112)
[7] Mino Y, Sasaki M and Tanaka T 1997 *Phys. Rev.* **D55** 3457–3476 (*Preprint* gr-qc/9606018)
[8] Quinn T C and Wald R M 1997 *Phys. Rev.* **D56** 3381–3394 (*Preprint* gr-qc/9610053)
[9] Poisson E 2004 *Living Rev. Rel.* **7** 6 (*Preprint* gr-qc/0306052)
[10] Quinn T C 2000 *Phys. Rev.* **D62** 064029 (*Preprint* gr-qc/0005030)
[11] Detweiler S and Whiting B F 2003 *Phys. Rev.* **D67** 024025 (*Preprint* gr-qc/0202086)
[12] Barack L and Ori A 2000 *Phys. Rev.* **D61** 061502 (*Preprint* gr-qc/9912010)
[13] Barack L, Mino Y, Nakano H, Ori A and Sasaki M 2002 *Phys. Rev. Lett.* **88** 091101 (*Preprint* gr-qc/0111001)
[14] Barack L and Ori A 2002 *Phys. Rev.* **D66** 084022 (*Preprint* gr-qc/0204093)
[15] Detweiler S, Messaritaki E and Whiting B F 2003 *Phys. Rev.* **D67** 104016 (*Preprint* gr-qc/0205079)
[16] Sopuerta C F and Laguna P 2006 *Phys. Rev.* **D73** 044028 (*Preprint* gr-qc/0512028)
[17] Cañizares P and Sopuerta C F 2008 *in preparation*

AUTHOR INDEX

AUTHOR INDEX